The Role of Entomopathogenic Fungi in Agriculture

Series

Progress in Mycological Research

- Fungi from Different Environments (2009)
- Systematics and Evolution of fungi (2011)
- Fungi from Different Substrates (2015)
- Fungi: Applications and Management Strategies (2016)
- Advances in Macrofungi: Diversity, Ecology and Biotechnology (2019)
- Advances in Macrofungi: Industrial Avenues and Prospects (2021)
- Advances in Macrofungi: Pharmaceuticals and Cosmeceuticals (2021)
- Fungal Biotechnology: Prospects and Avenues (2022)
- Bioprospects of Macrofungi: Recent Developments (2023)
- Ecology of Macrofungi: An Overview (2024)
- Advances in *Cordyceps* Research: Prospects and Avenues (2025)
- The Role of Entomopathogenic Fungi in Agriculture (2025)

Series: Progress in Mycological Research

The Role of Entomopathogenic Fungi in Agriculture

Editors:

Sunil K. Deshmukh

Scientific Advisor

Greenvention Biotech Pvt. Ltd., Pune, India

Kandikere R. Sridhar

Professor, Department of Biosciences

Mangalore University, Mangalore, India

CRC Press

Taylor & Francis Group

Boca Raton London New York

CRC Press is an imprint of the

Taylor & Francis Group, an **informa** business

A SCIENCE PUBLISHERS BOOK

First edition published 2025
by CRC Press
2385 NW Executive Center Drive, Suite 320, Boca Raton FL 33431

and by CRC Press
4 Park Square, Milton Park, Abingdon, Oxon, OX14 4RN

CRC Press is an imprint of Taylor & Francis Group, LLC

Library of Congress Cataloging-in-Publication Data (applied for)

ISBN: 978-1-032-81322-6 (hbk)
ISBN: 978-1-032-82576-2 (pbk)
ISBN: 978-1-003-50522-8 (ebk)

DOI: 10.1201/9781003505228

Typeset in Times New Roman
by Prime Publishing Services

Preface

Fungi contribute to the different spheres of the global economy through ecosystem services encompassing nutrition, health, biomaterials, and the circular economy (e.g., antibiotics, beverages, biopesticides, drugs, enzymes, fermented foods, metabolites, organic acids, pigments, and recycling of organic wastes) (Niego et al., 2023). Biopesticides, a broad term, control pests using biochemicals, microbes, and bioactive compounds produced by genetically engineered plant species. A variety of microorganisms are known to naturally regulate the insect vectors affecting agricultural crops. Knowledge of natural biocontrol agents paves the way to propose and fabricate desired microbes against specific insect pests. Such microbes include bacteria, fungi, protozoans, and viruses. These microbes are crucial in safeguarding agricultural produce through plant protection. Biological control of insect pests is the order of the day in comparison with chemical pesticides, which are detrimental to the environment and, in turn, to life. One group of biocontrol microbes, entomopathogenic fungi, have a crucial role in regulating insect pests in the agricultural sector (Bamisile et al., 2021; Shamim et al., 2024). Fungal biopesticides are vital to integrated pest management (IPM) programs. The common entomopathogenic fungi used in IPM include *Akanthomyces muscarius, Beauveria bassiana, Cordyceps fumosorosea, Metarhizium anisopliae, Metarhizium brunneum, Metarhizium robertsii, Nomuraea rileyi, Purpureocillium iilacinum,* and *Trichoderma harzianum*. A variety of *Cordyceps* have been implemented as entomopathogens in a wide range of humid climatic regions and tropical forests (North America, Europe, East Asia, and Southeast Asia) (Qu et al., 2022). They are highly valued biological control agents owing to their specificity against the target agricultural insect pests. To achieve the goal of managing insect pests, mass production of entomopathogenic fungal mycelial and spore biomass is necessary. Several successful attempts have been made for mass production of these fungi in a variety of solid-state and liquid-state fermentations (Silva et al., 2023; Jaronski, 2014; Velozo et al., 2023; Zhang et al., 2023).

This book, *The Role of Entomopathogenic Fungi in Agriculture,* offers different facets of investigation into the role of fungi in modern agriculture. Fifteen chapters written by 58 researchers from six countries delve into different aspects of entomopathogens. This volume highlights five subdivisions in basic and applied areas: (1) Science, History, and Exploration; (2) Integrated Pest Management; (3) Biocontrol of Insect Pests; (4) Role of Secondary Metabolites; (5) Culturing Entomopathogens. The first subsection in three chapters records the science behind entomopathogenic fungi, historical perspectives, and exploration aspects; the second subsection in two chapters deals with integrated pest management and biocontrol of insects by natural enemies; six chapters in the third subsection discuss entomopathogenic fungi as bioweapons, their specificity, roles of *Nomuraea*, roles of *Trichoderma*, control of *Aedes* mosquitoes, and control of groundnut pests; the fourth subsection in two chapters presents metabolic profiles of entomopathogenic fungi and the involvement of their secondary metabolites in pest control; and the last subsection in two chapters offers insights into possible mass production of *Beauveria* and mass spore production of entomopathogenic fungi by solid-substrate culture.

In recent years, entomopathogenic fungi have created interest in agricultural insect pest control approaches. We hope that the information provided by the authors and editors in the following chapters will be useful for readers with an interest in mycology and allied areas (applied biology, biochemistry, biotechnology, botany, ecology, entomology, environmental biology, field biology, forestry, and zoology).

The editors appreciate the excellent gestures of the contributors and reviewers for the submission, evaluation, and revision of chapters on time. We are grateful to the CRC Press, Boca Raton, USA, for its productive collaboration in fulfilling the many formalities necessary to publish this book on the stipulated date.

Pune, India
Mangalore, India
March 30, 2024

Sunil K. Deshmukh
Kandikere R. Sridhar

References

Bamisile, B.S., Akutse, K.S., Siddiqui, J.A. and Xu, Y. (2021). Model application of entomopathogenic fungi as alternatives to chemical pesticides: Prospects, challenges, and insights for next-generation sustainable agriculture. Front. Plant Sci., 12: 741804. https://doi.org/10.3389/fpls.2021.741804

Jaronski, S.T. (2014). Mass production of entomopathogenic fungi-state of the art. pp. 317–357. *In*: Morales-Ramos, J., Rojas, G., Shapiro-Ilan (eds.). Mass Production of Beneficial Organisms: Invertebrates and Entomopathogens, Academic Press. https://doi.org/10.1016/B978-0-12-391453-8.00011-X

Neigo, A.G.T., Lambert, C., Mortimer, P., Thongklang, N., Rapior, S. et al. (2023). The contribution of fungi to the global economy. Fungal Divers., 121: 95-137. https://doi.org/10.1007/s13225-023-00520-9

Qu, S.-L., Li, S.-S., Li, D. and Zhao, P.-J. (2022). Metabolites and their bioactivities from the genus *Cordyceps*. Microorganisms, 10(8): 1489. https://doi.org/10.3390/microorganisms10081489

Shamim, M., Kumar, D., Kumar, M., Srivastava, D., Kumar, S. et al. (2024). Role of fungi in biocontrol of diseases in cereal crops. pp. 79-96. *In*: Singh, S.K., Kumar, D., Shamnim, M. and Sharma, R. (eds.). Applied Mycology for Agriculture and Foods. Apple Academic Press, Canada and USA.

Silva, J.N., Mascarin, G.M., Lopes, R.B. and Freire, D.M.G. (2023). Production of dried *Beauveria bassiana* conidia in packed-column bioreactor using agro-industrial palm oil residues. Biochem. Eng. J., 198: 1–8. https://doi.org/10.1016/j.bej.2023.109022

Velozo, S.G.M., Velozo, M.R., Domingues, M.M., Becchi, L.K., de Carvalho, V.R., et al. (2023). From the dual cyclone harvest performance of single conidium powder to the effect of *Metarhizium anisopliae* on the management of *Thaumastocoris peregrinus* (Hemiptera: Thaumastocoridae). PLoS One, 18(3): 1–13. https://doi.org/10.1371/journal.pone.0283543

Zhang, Y., Li, K., Zhang, C., Liao, H. and Li, R. (2023). Research progress of *Cordyceps sinensis* and its fermented mycelium products on ameliorating renal fibrosis by reducing epithelial-to-mesenchymal transition. J. Inflamm. Res., 16: 2817–2830. https://doi.org/10.2147/JIR.S413374

Contents

List of Contributors

Aguilar-Marcelino, L.
Centro Nacional de Investigación Disciplinaria en Sanidad e Inocuidad Animal, INIFAP, Km 11 Carretera Federal Cuernavaca-Cuautla, No. 8534, C.P. 62550, Jiutepec, Estado de Morelos, México.

Andrade-Hoyos, P.
Instituto Nacional de Investigaciones Forestales Agrícolas y Pecuarias, INIFAP, Campo Experimental Zacatepec, Morelos, México.

Arvind Bharani, R.S.
Institute of Obstetrics and Gynaecology, Madras Medical College, Egmore, Chennai-600008, Tamil Nadu, India.

Bitencourt, R.O.B
Universidade Estadual do Norte Fluminense Darcy Ribeiro, Department of Entomology and Plant Pathology, Av. Alberto Lamego 2000, Parque California, Campos dos Goytacazes, RJ 28013-602, Brazil.

Buenrostro-Figueroa, J.J.
Laboratorio de Biotecnología y Bioingeniería, Centro de Investigación en Alimentación y Desarrollo, Delicias, Chihuahua, Mexico, 33088.

Burla, Sashidhar
Director, Techno-Marketing, ATGC Biotech Pvt. Ltd., Hyderabad-500014, India.

Calderón-Cortés, N.
National School of Higher Studies Morelia Unit, National Autonomous University of Mexico (UNAM), Morelia 58190, Michoacán, Mexico.

Castañeda-Ramírez, G.S.
Centro de Investigación en Biotecnología (CEIB), Universidad Autónoma del Estado de Morelos (UAEM), C.P. 62209. Cuernavaca, Morelos, México.

Chandan, B.S.
Department of Applied Zoology, Mangalore University, Mangalagangotri, Mangalore, Karnataka, India.

Chowdhury, S.
Symbiotic Sciences Pvt. Ltd., Plot No 575, Symbiotic Sciences, 22, Udyog Vihar Industrial Area Phase VI, Sector 37, Gurugram, Haryana-122004, India.

Dorantes-Jiménez, J.
Mexican Association of Swiss Cattle Breeders of Registration, Ciudad de México 03400, Mexico.

Ernesto Favela-Torres, E.
Departamento de Biotecnología, Universidad Autónoma Metropolitana, Iztapalapa, Mexico City, Mexico, 09340.

Galappaththi, M.C.A.
Harry Butler Institute, Murdoch University, Murdoch-6150 WA, Australia.

García-Pérez, F.
Universidad NovaUniversitas, Oaxaca-Puerto Ángel, Ocotlán de Morelos 71513, Oaxaca, Mexico.

Gaurav, K.
Department of English, School of Social Science and Humanities, Lovely Professional University, Phagwara-144411, Punjab, India.

Ghorui, M.
Symbiotic Sciences Pvt. Ltd., Plot no 575, Symbiotic Sciences, 22, Udyog Vihar Industrial Area Phase VI, Sector 37, Gurugram, Haryana-122004, India.

Gómez-Domínguez, N.S.
Instituto Politécnico Nacional (IPN), Biotic Product Development Center (CEPROBI), Yautepec - Jojutla 62739 San Isidro, Morelos, Mexico.

Granados-Echegoyen, C.A.
CONAHCYT-Instituto Politécnico Nacional (IPN), CIIDIR-Oaxaca, Santa Cruz Xoxocotlán 71230, Oaxaca, Mexico.

John, A.
Departmnet of Computional Biology, Saveetha School of Engineering, Saveetha Institute of Medical and Technical Sciences (SIMATS), Chennai-602105, Tamil Nadu, India.

Karunarathna, S.C.
Center for Yunnan Plateau Biological Resources Protection and Utilization, College of Biological Resource and Food Engineering, Qujing Normal University, Qujing 655011, China.

Kothari, M.
Fungal Biotechnology and Invertebrate Pathology Laboratory, Department of Biological Sciences, Rani Durgavati University, Jabalpur-482001, Madhya Pradesh, India.

Kumar, A.
Department of Horticulture, School of Agriculture, Lovely Professional University, Phagwara, Punjab, 144411, India.

Landero-Valenzuela, N.
Department of Horticulture, Universidad Autónoma Agraria Antonio Narro, Calzada Antonio Narro, 25294, Coahuila, Mexico.

Leyte-Lugo, M.
Investigadora por México CONAHCYT Comisionado al Departamento de Sistemas Biológicos, División de Ciencias Biológicas y de la Salud, Universidad Autónoma Metropolitana–Xochimilco (UAM–X), C.P. 04960. Mexico City, Mexico.

Loera-Alvarado, E.
CONAHCYT-Universidad Autónoma Chapingo, Centro Regional Universitario Centro Occidente (CRUCO), Morelia, Michoacán, 58170, Mexico.

Lumyong, S.
Department of Biology, Faculty of Science, Chiang Mai University, Chiang Mai 50200, Thailand.

Luna-Cruz, A.
CONAHCYT-Instituto de Investigaciones Químico-Biológicas, Universidad Michoacana de San Nicol´as de Hidalgo, Morelia-58030 Michoacán, México.

Mendoza de Gives, P.
Laboratory of Helminthology, National Centre for Disciplinary Research in Animal Health and Innocuity (INIFAP-Mexico), Jiutepec-62550, Morelos, Mexico.

Méndez-González, F.
Laboratorio de Biotecnología y Bioingeniería, Centro de Investigación en Alimentación y Desarrollo, Delicias, Chihuahua, Mexico, 33088.

Mishra, M.
Department of Entomology, School of Agriculture, Lovely Professional University, Phagwara-144411, Punjab, India.

Molina-Gayosso, E.
Universidad Politécnica de Puebla (UPP), Tercer Carril del Ejido, Serrano s/n, Cuanalá, 72640 Puebla, Pue.

Mustak, M.S.
Department of Applied Zoology, Mangalore University, Mangalagangotri, Mangalore, Karnataka, India.

Nagaraju, A.
Department of Biochemistry, Mangalore University, Mangalagangotri, Mangalore, Karnataka, India.

Namasivayam, S.K.R.
Centre for Applied Research, Saveetha School of Engineering, Saveetha Institute of Medical and Technical Sciences (SIMATS), Chennai-602105, Tamil Nadu, India.

Paez-Leon, S.,
Centro Nacional de Investigación Disciplinaria en Sanidad e Inocuidad Animal, INIFAP, Km 11 Carretera Federal Cuernavaca-Cuautla, No. 8534, C.P. 62550. Jiutepec, Estado de Morelos, México.
Centro de Investigación en Biotecnología (CEIB), Universidad Autónoma del Estado de Morelos (UAEM), C.P. 62209. Cuernavaca, Morelos, México.

Palacios-Espinosa, J.F.
Departamento de Sistemas Biológicos, División de Ciencias Biológicas y de la Salud, Universidad Autónoma Metropolitana–Xochimilco (UAM–X), C.P. 04960. Mexico City, Mexico.

Paula, A.R.
Universidade Estadual do Norte Fluminense Darcy Ribeiro, Department of Entomology and Plant Pathology, Av. Alberto Lamego-2000, Parque California, Campos dos Goytacazes, RJ 28013-602, Brazil.

Priyashantha, A.K.H.
Department of Biology, Faculty of Science, Chiang Mai University, Chiang Mai 50200, Thailand.

Reyes-Estébanez, M.
Research Center in Environmental Microbiology and Biotechnology (CIMAB), Autonomous University of Campeche, San Francisco de Campeche 24079, Campeche, Mexico.

Ribeiro, A.
Universidade Estadual do Norte Fluminense Darcy Ribeiro, Department of Entomology and Plant Pathology, Av. Alberto Lamego 2000, Parque California, Campos dos Goytacazes, RJ 28013-602, Brazil.

Ribeiro, J.P.
Universidade Estadual do Norte Fluminense Darcy Ribeiro, Department of Entomology and Plant Pathology, Av. Alberto Lamego 2000, Parque California, Campos dos Goytacazes, RJ 28013-602, Brazil.

Rivera-Jiménez, M.N.
Agrosis mg S.A.de C.V./Biosoluciones Orgánico Minerales, Dirección de Investigación Agrícola, Villahermosa, Tabasco, México.

Samuels, R.I.
Universidade Estadual do Norte Fluminense Darcy Ribeiro, Department of Entomology and Plant Pathology, Av. Alberto Lamego 2000, Parque California, Campos dos Goytacazes, RJ 28013-602, Brazil.

Sánchez-Hernández, C.
Universidad NovaUniversitas, Oaxaca-Puerto Ángel, Ocotlán de Morelos-71513, Oaxaca, Mexico.

Sandhu, S.S.
Bio-Design Innovation Centre, Rani Durgavati University, Jabalpur 482001, Madhya Pradesh, India

Sarmiento-López, L.G.
Departamento Académico de Ciencias Naturales y Exactas, Universidad Autónoma de Occidente Unidad Regional Los Mochis. Sinaloa, 81223, Mexico.

Seethapathy, P.
Department of Plant Pathology, Amrita School of Agricultural Science, Coimbatore, Tamil Nadu India.

Siddaraju, M.N.
Department of Botany, University College Mangalore, Karnataka, India.

Silva, G.A.
Universidade Estadual do Norte Fluminense Darcy Ribeiro, Department of Entomology and Plant Pathology, Av. Alberto Lamego-2000, Parque California, Campos dos Goytacazes, RJ 28013-602, Brazil.

Silva, L.E.I.
Universidade Estadual do Norte Fluminense Darcy Ribeiro, Department of Entomology and Plant Pathology, Av. Alberto Lamego 2000, Parque California, Campos dos Goytacazes, RJ 28013-602, Brazil.

Singh, J.
Department of Horticulture, School of Agriculture, Lovely Professional University, Phagwara, Punjab, 144411, India.

Vallejo-Pérez, M.
CONAHCYT-Universidad Autónoma de San Luis Potosí, Coordinación para la Innovación y Aplicación de la Ciencia y la Tecnología, San Luis Potosí, C.P. 78210 S.L.P. México.

Vásquez-López, A.
Instituto Politécnico Nacional (IPN), CIIDIR-Oaxaca, Santa Cruz Xoxocotlán 71230, Oaxaca, México.

Varshan, G.S.A.
Centre for Applied Research, Saveetha School of Engineering, Saveetha Institute of Medical and Technical Sciences (SIMATS), Chennai-602105, Tamil Nadu, India.

Vera-Reyes, I.
Department of Biosciences and Agrotechnology, Applied Chemistry Research Center, Saltillo-25294, Coahuila, Mexico.

Vikanksha
Department of Horticulture, School of Agriculture, Lovely Professional University, Phagwara, Punjab-144411, India.

1 The Science behind Entomopathogenic Fungi: Mechanisms and Applications

Maunata Ghorui,[1] Shouvik Chowdhury[1] and Sashidhar Burla[2*]

1. Introduction

Agriculture faces significant challenges in mitigating losses in crop yield worldwide. These challenges are attributed to various biotic and abiotic stressors, prominent among these are insect pests. Insects, constituting a diverse taxonomic group with an estimated one million recognized species, have a wide range of hosts, encompassing crops, weeds, trees, and medicinal plants. While most insect species fulfil beneficial ecological roles, a fraction exert detrimental effects on agricultural productivity, human health, and socioeconomic welfare. Notably, herbivorous insects are implicated in causing approximately 18% of overall agricultural damage worldwide (Jankielsohn, 2018).

The impact of insect pests extends beyond field crops, as they also infest stored food commodities, leading to substantial losses and compromising food quality. Every year, between 20% and 40% of global crop production is lost due to pest infestations. Invasive insects cause around $70 billion in damage each year (Researchers Helping Protect Crops from Pests, 2023). Taxonomically, insect pests inflicting damage exceeding 10% are categorized as major pests, while those causing harm in the range of 5 to 10% are designated minor pests (Sharma et al., 2017).

To confront these challenges, extensive efforts are directed toward pest management strategies. Annual global pesticide usage surpasses two billion tons, encompassing a repertoire of chemical compounds such as bactericides, fungicides, herbicides, and insecticides (Baron et al., 2018). Noteworthy among these are chlorantraniliprole, cyantraniliprole, novaluron, neonicotinoids, fipronil, farnesyl acetate, emamectin benzoate, phoxim, and pyrethroids, along with juvenile hormone analogs including methoprene, fenoxycarb, and pyriproxyfen. However, the pervasive utilization of these chemical agents incurs deleterious consequences, ranging from ecological perturbations to the development of pest resistance (Rust et al., 2016; Lawler, 2017). Synthetic pesticides, notably insecticides, pose inherent risks to the environment, manifesting in water and soil contamination and exerting selection pressures favouring resistance evolution among target pest populations. Furthermore, the indiscriminate application of broad-spectrum insecticides compromises beneficial organisms, including natural enemies of crop pests (Omkar, 2016). In response to these

[1] Symbiotic Sciences Pvt. Ltd., Plot no 575, Pace City-II, Sector 37, Gurugram, Haryana-122001, India.
[2] Director, Techno-Marketing, ATGC Biotech Pvt. Ltd., Hyderabad-500014, India.
* Corresponding author: bshashidhar12@gmail.com

challenges, concerted scientific endeavours are directed towards devising sustainable and eco-friendly pest management strategies. Biological control represents a promising avenue harnessing natural enemies such as predators, parasitoids, and pathogens for pest suppression. Prominent in focused pest management, entomopathogens - which include bacteria, fungi, viruses, protists, and nematodes — result in host-death (Silva et al., 2020). Myco-biocontrol, using fungi to reduce insect populations and minimise agricultural losses, is receiving a lot of attention. For biocontrol, more than 800 fungal species from various genera have been isolated (Shin et al., 2020; Sinha et al., 2016). EPFs are found in terrestrial ecosystems worldwide, with tropical forests showing the highest diversity. Certain species have even been shown to adapt to harsh environments, like the Arctic tundra (Hughes et al., 2004).

EPF constitute a distinct category of microorganisms found in soil that invade and eliminate insect pests and other arthropods by penetrating their cuticles (Mantzoukas et al., 2022). Arthropod elimination is achieved through a parasitizing process (Litwin et al., 2020; Sharma et al., 2023). Though EPF is commonly found in the remnants of arthropods, it predominantly dwells in the soil (Behie and Bidochka, 2014). According to Araújo and Hughes (2016), these fungi are divided into six groups as they do not belong to a single monophyletic group: Oomycetes (12 species), Chytridiomycota (65 species), Microsporidia (339 species), Basidiomycota (238 species), Ascomycota (476 species), and Entomophtoromycota (474 species).

The most common species from genera within Ascomycota, such as *Beauveria* (including *B. bassiana* and *B. brongniartii*), *Metarhizium* (including *M. anisopliae, M. robertsii, M. brunneum, M. lepidiotae, M. globosum, M. acridum, M. majus, M. flavoviride, M. rileyi, M. pingshaense, M. lepidiotae,* and *M. guizhouense*), *Isaria* (including *I. fumosorosea, I. farinosa,* and *I. tennuipes*), *Ophiocordyceps* (including *O. sinensis*, and *O. unilateralis*), *Cordyceps* (such as *C. militaris*), *Torubiella* (such as *T. ratticaudata*), *Pochonia* (such as *P. chlamydosporia*), *Lecanicillium* (such as *L. lecani*, and *L. longisporum*), *Hirsutella* (including *H. thompsonii, H. nodulosa,* and *H. aphidis*), and *Paecilomyces variotii* and *Purpureocillium lilacinum*, are recorded in the scientific literature that constitutes the EPF (Khan et al., 2012; Tkaczuk et al., 2015; Jaihan et al., 2016). EPF presents a viable economic and ecologically sustainable approach to Integrated Pest Management (IPM). This chapter addresses EPF mechanisms of action, production of EPF, crop protection, IPM, and mechanisms of action and production of toxic substances.

2. Mechanisms of Action of Entomopathogenic Fungi

A variety of methods, including starvation and toxin synthesis, are employed by EPF to kill insects. EPF generates a variety of toxins as well as extracellular enzymes like chitinases and proteases that help breach the host's physical barriers. The cuticle is the primary barrier against infection in insects, serving as the main pathway for fungal penetration. Therefore, breaching the cuticle requires a physical or enzymatic action.

As heterotrophic animals, fungi primarily acquire energy by absorbing the organic molecules made by other organisms. Exogenous carbon sources, like chitin in the case of *B. bassiana* conidia, are necessary for spore or conidia germination. Additionally, insect epicuticular lipids may help the fungus attach to the host cuticle

(Ferron, 1978). According to Lecuona et al. (1997), insect epicuticular lipids may have two functions: first, they may give rare conidia readily available energy sources for germination; second, they may have antifungal properties that prevent hyphal development. Proteases, lipases, and chitinases are examples of hydrolytic enzymes that are synthesised and break down the cuticle to release nutrients for the fungus. Lipid degradation in the outer layer of the cuticle also occurs prior to fungal penetration. The mechanism of EPF infection in insects or other arthropods is shown in Figure 1.

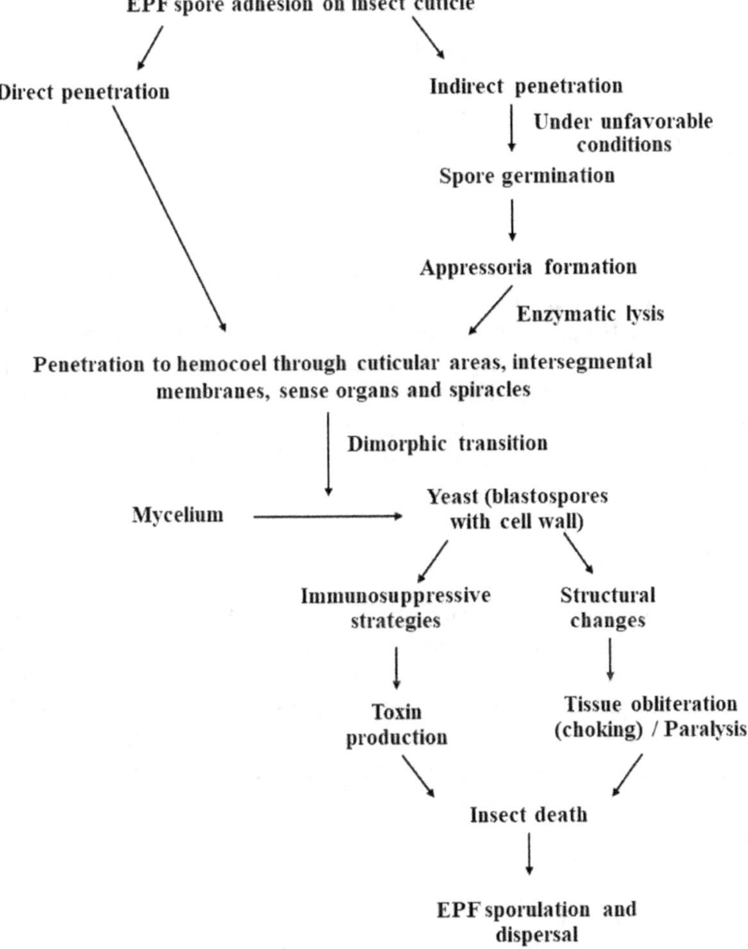

Fig. 1 Mechanism of action by entomopathogenic fungi on insects.

2.1 *Spore Adhesion and Spore Germination*

In most fungi, the infective unit is typically a spore, namely a conidium. Pre-germination conidia swell and stick to the cuticle or exude sticky mucus. There are multiple steps involved in the spore's attachment to the host's cuticle and eventual germination. Infection may occur directly by encountering infectious cadavers and vulnerable hosts, or indirectly by encountering spores in the air or spores that are deposited on plants or soil particles (Hesketh et al., 2009).

2.1.1 Adhesion Mechanisms

Fungal propagules adhere to the host cuticle at the first stage of infection, which is made possible by adhesion processes that are influenced by the characteristics of the conidial cell wall (Boucias et al., 1988). This mechanism involves the interaction of proteins found in the conidia with the hydrophobic surface of the exoskeleton of the susceptible insect (Fang et al., 2005). Adhesins are chemicals produced by the fungus that facilitate the adherence between spores and the insect cuticle. For instance, adhesion and virulence in *Metarhizium anisopliae* are significantly influenced by a kind of adhesin termed MAD1, which is found on the surface of conidia (Wang and St. Leger, 2007a).

2.1.2 Spore Germination

The penetration phase begins when the conidium germinates into a short germ tube and forms tiny swellings called appressoria under favourable conditions, which include humidity, temperature, and appropriate nutrients on the cuticle (Téllez-Jurado et al., 2009). Certain fungi, such as some Entomophthorales, are able to break through the cuticle right out of the germ tube without developing appressoria (Hajek and Delalibera, 2009). To create a strong attachment and physical penetration into the host, the appressorium clings to the cuticle and extends an infection peg (Shah and Pell, 2003; Hajek and Delalibera, 2009).

2.1.3 Recognition of Receptors

Pre-germinated spores adhere to the epicuticle on the surface of some Hypocreales taxa, including *Beauveria*, *Metarhizium*, and *Isaria*, by recognising glycoprotein receptors in the insect. Studies have shown that different types of spores exhibit varied adhesion preferences to hydrophobic and hydrophilic surfaces (Holder and Keyhani, 2005). For example, *Beauveria* exhibits a hydrophobic nature in its conidia, attributed to the presence of cysteine-rich proteins called hydrophobins in the cell wall. However, hydrophilic conidia are displayed by *Verticillium lecanii* (Inglis et al., 2001).

2.1.4 Penetration into the Hemocoel

Hemocoel penetration refers to the process by which the EPF invades the insect's hemocoel. Subsequently, the hyphae penetrate the integument layers through the enzymatic dissolution of chitin and protein, initially ramifying in the cuticle before reaching the haemocoel and internal organs. Insect cuticles are comprised of a network of polysaccharide polymers embedded in a protein matrix (up to 70%) (Vega et al., 2009). The cuticle is divided into three layers: the procuticle, which is a matrix of proteins and chitin; the epicuticle, and the outermost envelope, also known as the cuticulin layer (Locke, 2001; Klowden, 2013). The procuticle consists of an outside layer called the exocuticle and an inner layer called the endocuticle. Different insect anatomical locations and developmental stages have different cuticular protein compositions (Gilbert et al., 2005). Epidermal cells are situated at the base of the procuticle, and the hemocoel is located beneath them.

2.1.5 *Enzymatic Cuticle Degradation*

The cuticle needs to be broken through mechanical pressure and the release of enzymes for fungi to feed and colonise it (Vega et al., 2009). Cuticle characteristics such as thickness, sclerotization, and the presence of nutritional and antifungal chemicals influence the fungal penetration mechanism (Charnley, 2003). Cuticle disintegration is dependent on enzymes such as lipases, proteases, and chitinases (Monzón, 2002). This process frequently proceeds in a lipase-protease-chitinase sequence (Prior et al., 1988), which may be aided by the production of organic acids like oxalic acid. PR1 protease in *M. anisopliae* and over-expression of the gene encoding *B. bassiana* chitinase are significant virulence factors, hastening insect death (St Leger et al., 1996; Fan et al., 2007). The secretion of these hydrolytic enzymes underscores their role in fungal virulence, offering potential for selecting strains for biological insecticides.

2.1.6 *Alternate Modes*

In addition to penetration through cuticular areas and intersegmental membranes, EPF can also infect insects via sense organs and spiracles (St Leger, 1996). Despite higher humidity in the digestive tract facilitating rapid spore germination, digestive fluids may degrade spores or hyphae, causing toxicity rather than mycosis in some cases (Charnley, 1992).

2.2 *Replication of EPF in Hemocoel*

EPF employs a remarkable strategy to multiply within the insect's body, ensuring its persistence and effectiveness as a bioinsecticide

2.2.1 *Fungal Dimorphic Transition in Hemocoel*

Many fungi, especially Entomophthorales, undergo a dimorphic transition from mycelium to yeast after infiltrating the hemocoel. In this phase, they develop as protoplasts (blastospores) devoid of cell walls, which helps them evade detection by circulating hemocytes (Vinson, 1993). Many EPF prefer spores to grow within the hemocoel of infected insects because they are hydrophilic and vegetative fungal propagules (Boucias and Pendland, 1991; Humber, 2008). Some species, such as Nomuraea rileyi, have cryptic surface residues that prevent insect hemocytes from phagocytozing their spores, making them invisible to humoral lectins (Boucias et al., 1988). Benefits of this cellular form include increased rates of nutrient uptake and immune system evasion (the insect's immune system normally identifies fungal cells via cell wall epitopes).

2.2.2 Fungal Strategies against Insect Immune Response

Toxin synthesis and cell wall structural alterations are examples of defensive and immunosuppressive tactics used by fungal infections to subdue insect immune system response mechanisms. These include employing cyclic depsipeptides like destruxins, which cause paralysis in insects by activating calcium channels, and synthesising extra proteases to weaken the humoral immune system. Along with some protein-based macromolecules, such as melanizing proteins from *B. bassiana*, a glycoprotein from *B. sulphurescens*, and hirsutellin A from *H. thompsonii*, several low-molecular-weight secondary metabolites isolated from insect pathogens have shown insecticidal activity (Gillespie and Claydon, 1989; Mollier et al., 1994; Fuguet and Vey, 2004). These insecticidal compounds work in a variety of ways, but they frequently kill insects directly by concentrating on immune system cells specifically designed to stop invasive fungal structures from being attacked (Téllez-Jurado et al., 2009). Among EPF, the development of toxins is a common trait. The mechanism of action of destruxins, which block the synthesis of DNA, RNA, and proteins in insect cells, has been thoroughly explored. Toxins may also harm the insect's Malpighian tubules and muscular system, which would impair excretion and make it difficult for it to eat and move (Pal et al., 2007).

2.2.3 EPF Colonization within Insect Bodies

The invasion by fungal mycelium persists until the insect is extensively colonized, resulting in a firm texture upon touch. Finally, under favourable temperature and humidity conditions, hyphae can breach the insect's integument, leading to fungal emergence. Emergence typically occurs in less sclerotic regions, like intersegmental membranes or spiracles, depending on the host and developmental stage. Since *B. bassiana* and *B. brongniartii* synthesise antibiotics such as oosporein to inhibit the growth of opportunistic organisms, hyphae crossing the integument may stay in the vegetative phase and begin sporulation within 24 to 48 hours (Srivastava et al., 2009). Eventually, conidiophores develop into asexual spores that act as infectious agents for spreading. The development of epizootics is ultimately influenced by environmental conditions that are critical to conidia generation, survival, and germination (Fuxa and Tañada, 1987). Although sporulation mostly happens in cadavers, it can also happen in live insects. Depending on the properties of the spore and sporangium, spore dispersal can be either active or passive (Vega et al., 2009).

2.2.4 Insect Immune Response to Fungal Infection

Subsequently, septicemia occurs once the fungus evades the insect's immune defences (Eilenberg and Michelsen, 1999). The insect can react by utilising cellular processes like phagocytosis and encapsulation, humoral mechanisms including phenol oxidases, lectins, proteins, and defence peptides, or a mix of the two.

2.2.5 Reversion to Mycelial Growth upon Nutrient Depletion

Upon depletion of nutrients, particularly nitrogen sources, yeast phases revert to mycelial growth, as observed in *Entomophthora thripidum* (Freimoser et al., 2003).

2.2.6 Physiological Effects on Insect

The host's death is caused by tissue obliteration, often due to choking, and the toxins produced by the fungus. Mycosis induced by fungal infection manifests in physiological symptoms such as seizures, lack of coordination, altered behaviour, and paralysis in insects. Death ensues from a combination of factors, including physical tissue damage, toxicity, cell dehydration due to fluid loss, and nutrient consumption.

3. Production of Entomopathogenic Fungi

Nutritional components like carbohydrates, proteins, lipids, and nucleic acids are essential for microbial growth, with elements such as carbon, hydrogen, nitrogen, sulphur, and phosphorus playing vital roles in host-pathogen interactions and self-defence mechanisms (Raimbault, 1998). Growth characteristics, along with growth substances, are essential in tolerance selection studies. The media used for growth, storage, and transport of microorganisms can be in solid or liquid form. Nutritional studies have been conducted on the production and sporulation of filamentous fungi such as *B. bassiana, M. anisopliae*, and *I. fumosorosea* (Kumar and Mukherjee, 1996). *B. bassiana* spores for coffee berry borer biocontrol are mainly produced using a simple sterilization technique with cooked rice in bottles used for field spray applications. Rice media is washed with a 1% oil-water suspension to harvest the spores (Antía et al., 1992). However, the aqueous spore suspension must be used immediately to prevent germination, as spore viability diminishes rapidly due to the high moisture content in the bottles. Similar studies have been conducted on EPF for controlling date palm pests, particularly with *B. bassiana* and *M. anisoplia* (Latifian et al., 2013).

Jaronski and Mascarin (2017) outlined broad strategies for the production and formulation of hypocrealean fungal propagules specifically aimed at insect control. Firstly, maintaining genetically uniform cultures is essential to establishing a primary "mother culture" from which subsequent production is derived. Single-spore or single-colony isolation methods are employed to ensure genetic uniformity, with caution taken to limit successive passages on artificial media to prevent morphological and virulence alterations (Jackson et al., 1997; Butt et al., 2007; Shah et al., 2007; Wang et al., 2002, 2005; Ansari and Butt, 2011). Preservation methods, such as low-temperature storage or desiccation, are employed to maintain viability and inhibit genetic variation (Humber, 2008). Process sterility is crucial to prevent contamination, achieved through sterilization of fermentation medium, air, and equipment. Nutrient optimization is key for maximizing propagule yield and quality, with dissolved oxygen being a critical factor in aerobic fermentation. Response surface methodology aids in efficiently determining optimal parameters (Jackson et al., 1997; Garza-López et al., 2011; Tlécuitl-Beristain et al., 2009; Prakash et al., 2008). Strain selection is vital, as different strains may respond differently to fermentation conditions, impacting propagule yield (Jaronski and Mascarin, 2017).

The selection process for fungal isolates must consider not only their virulence but also their ability to form stable propagules that can be economically mass-produced. These propagules should have long-term stability, be compatible with existing application technologies, possess acceptable environmental and toxicology profiles, and consistently perform well under typical environmental conditions considering the target insect(s) (Jackson et al., 2009). However, commercial success is complicated when considering a single product for targeting multiple insect species or crops. The high cost of registration in North America or the European Union may require using a single strain against multiple targets, leading to compromises between efficacy and other criteria, particularly mass production (Jaronski and Mascarin, 2017). Selecting the right propagule involves assessing its intended application, effectiveness, tolerance to desiccation and heat, germination and infection speed, environmental stability, reproductive capacity, and resistance to UV radiation. This evaluation considers the fungus's natural ability to produce the chosen propagule (Jackson et al., 1997; Jackson and Jaronski, 2009; Fernandes et al., 2015).

Grain, vegetable waste, seeds, rice husk, sawdust, and liquid media (coconut water, rice wash water, and rice cooked water) are among the products and byproducts of agriculture evaluated for their potential to produce large quantities of three EPF: *B. bassiana*, *P. fumosoroseus*, and *V. lecanii*. *P. fumosoroseus* and *V. lecanii* produced more spores in sorghum than in wheat, while *B. bassiana* produced the most spores in wheat. Furthermore, ladies' fingers, carrots, and jack seeds encouraged the three fungi's vigorous development and sporulation. It was found that coconut water encouraged the fungi to proliferate and sporulate maximally (Sahayaraj and Namasivayam, 2008).

Cowpea, maize, sorghum, and oat were the four-grain media tested for growth and multiplication of *B. brongniartii*. With the highest vertical growth of fungus (6.60–6.97 cm) in test tubes, cowpea was shown to be the ideal substrate. Among cowpeas, the KH I isolate had the highest spore count (2.30×10^7 conidia/ml). Field trials demonstrated the effectiveness of double application of *B. brongniartii* (KH I) at a rate of 10^{14} spores/ha. This application resulted in a 44.85% reduction in tuber damage based on weight at the time of harvesting, compared to the control. Additionally, a 33.99% reduction in grub population was observed in the treated plots compared to the untreated check plots (Soni et al., 2017).

Another study mass-produced *M. anisopliae* using various grains as substrates. Testing was done on a variety of grains and liquid media, including Sabouraud's Dextrose Broth (SDB) and Potato Dextrose Broth (PDB). The findings indicated that, out of 10^3 dilutions, green gram and sorghum had the greatest conidial count (67.6×10^3 spores/ml). Furthermore, it was shown that SDB produced considerably more spores than PDB. On SDB media, the greatest conidial count (63.7×10^3 spores/ml) was recorded, with PDB in 10^3 dilutions following (Agale et al., 2018).

A study provided a protocol employing appropriate media to achieve optimal production of *B. bassiana* biomass, conidial count, and germination. Results indicated that broken rice emerged as the most effective substrate, yielding the highest biomass, conidial count, and germination rates (0.62 g, 10.92×10^7 conidia/ml, and 86.94%, respectively), followed by sorghum (0.54 g, 7.35×10^7 conidia/ml, and 77.43%) and maize (0.37 g, 6.05×10^7 conidia/ml, and 72.44%) (Rai et al., 2021).

A recent study commercially produced large quantities of resilient infective propagules from South African strains of *M. robertsii* and *M. pinghaense*. As solid fermentation substrates, three grain products, flaked oats, flaked barley, and rice, were examined. Conidial suspensions and liquid fungal cultures of blastospores were the two inoculation techniques employed. Conidial suspension inoculation showed higher contamination levels compared to blastospore inoculation. Flaked oats were not suitable for the growth of either fungus, while flaked barley prefered *M. robertsii* over *M. pinghaense*. Rice grains were effective for conidial production of both strains, with *M. pinghaense* yielding an average of 8.2 g ± 4.38 g and *M. robertsii* yielding 6 g ± 2 g of dry conidia harvested from the substrate (Mathulwe et al., 2022).

Chandwani et al. (2022) provided a summary of the large-scale cultivation of *B. bassiana* using readily available agricultural and industrial waste to reduce production costs and enhance the cost-effectiveness of producing potent spores, exploring both solid and liquid media options to expedite the commercialization of *B. bassiana*.

Another study developed a cost-effective solid-state fermentation (SSF) method for large-scale production of *Purpureocillium lilacinum* PL1 conidia to control *Aphis devastans* infestations in okra cultivation. Rice and maize were identified as highly suitable substrates, yielding conidia densities exceeding 2×10^{10} conidia/g. The impact of agricultural phytosanitary agents on *P. lilacinum* PL1 growth rates was assessed, with certain pesticides showing no effect and fungicides causing complete inhibition. Laboratory tests demonstrated that 1×10^7 conidia/ml of *P. lilacinum* PL1 reduced *A. devastans* nymph populations by 88.66%. Field trials in okra plantations revealed a significant 72.87% reduction in pest nymph populations after two applications of *P. lilacinum* PL1 at a concentration of 1×10^7 conidia/ml (Thi et al., 2023).

When *B. bassiana* was inoculated onto samples of white rice and allowed to incubate at 100% moisture content, the highest conidial output was produced by several isolates, such as ARSEF 3462. Significantly, isolates of *B. bassiana*, *M. anisopliae*, and *A. album* showed a high level of resistance to heat and UV-B radiation, but conidia of *S. lanosoniveum* and *L. aphanocladii* did not germinate following heat treatment (Rangel et al., 2023).

Another study tested eight isolates of six EPFs on white or brown rice under varying moisture conditions. Conidial production was generally higher on white rice compared to brown rice, except for one fungal species. The 100% moisture condition favoured higher conidial production for certain isolates, while the addition of peanut oil enhanced yield for another isolate. With 100% water added to white rice, one isolate of *B. bassiana* produced the highest amount of conidia (1.3×10^{10} conidia/g substrate) (Rangel et al., 2023).

4. Applications and Formulations for Entomopathogenic Fungi

More than 170 strains have been developed into mycopesticides and are currently accessible for commercial purposes (Bamisile et al., 2021). Various mycopesticides have been developed from fungal species such as *B. bassiana*, *B. brongniartii*, *M. anisopliae*, and *I. fumosorosea*, which are commonly used for pest control. *M. anisopliae* strains have been commercialized for combating various pests and disease vectors (Akutse et al., 2020), and products based on *M. anisopliae* and *B. bassiana* have been registered in several countries (Zimmermann, 2007a, b; Wraight et al., 2000; Faria and Wraight, 2007). *B. bassiana* and *B. brongniartii* spores have been effectively formulated into mycopesticides in numerous countries and are widely used for controlling pest insects (Wraight et al., 2000; Faria and Wraight, 2007). *L. lecanii* has been studied and formulated as a mycoinsecticide for controlling aphids and scale insects, with products like Vertalec and Mycotal registered in various

European countries and beyond (Shah and Pell, 2003; Yeo et al., 2003). Mycotrol, derived from *B. bassiana*, was registered in 1999 for controlling aphids, grasshoppers, thrips, and whiteflies, among other pests (Bradley et al., 1992). Green Muscle, another mycoinsecticide, was developed to combat desert locust outbreaks and consists of dried conidia of *M. anisopliae* var. *acridum* mixed with kerosene or diesel oil (Lomer et al., 2001). In Russia, Boverin, a *B. bassiana*-based mycopesticide, was extensively utilized to control the Colorado potato beetle and the codling moth. Other similar products, like BotaniGard and Mycotrol-O, are available for use in glasshouses and by organic farmers in the United States and elsewhere, providing viable alternatives to synthetic insecticides.

Currently available biopesticide formulations on the market come in wettable powders, granules, tablets, oil-based suspensions, and biopolymers for different uses, with stability during storage being a critical consideration (Mascarin et al., 2013). Granular formulations are preferred for soil insects, with inert or nutritive granular materials coated with conidia. Microsclerotia, a stable storage form of some *Metarhizium* species, show promise for soil applications and have been used in controlling arboreal pests like the Asian long-horned beetle (Jackson et al., 2009). The development of effective formulations is crucial for the successful utilization of commercial biopesticides. Factors influencing the development of commercial formulations encompass various considerations, including:

(a) Product shelf life is crucial. Long shelf life is highly desirable, enabling formulations to maintain efficacy over multiple cropping seasons benefitting both manufacturers and end users.
(b) Fungal propagules should be stored in a dormant yet viable state, preferably at room temperature rather than under refrigeration. Optimal storage conditions are achieved by incorporating desiccants and oxygen scavengers into packaging. Recent advancements include the development of wax-based carriers for aerial conidia and oil-based formulations for blastospores (Meikle et al., 2008).
(c) It is imperative for formulations to be easily suspended in carriers, typically aqueous, for spray applications while also remaining viable on treated substrates withstanding exposure to UV radiation, rainfall, and other environmental stressors. Incorporating fungal propagules with various additives like wetting agents, suspension agents, and dispersants is necessary, with non-ionic wetting agents often required for proper suspension (Bateman et al., 1996). While aerial conidia are hydrophobic and challenging to suspend in aqueous carriers, spray-dried blastospores are more hydrophilic but may need wetting agents for dispersion (Inglis et al., 1996). Oil formulations are suitable for suspending hydrophobic aerial conidia, enhancing their application in low-volume scenarios.

Challenges to commercialization:

(a) High contamination by saprophytic fungi.
(b) Low viability of microbes, leading to reduced efficiency.
(c) Limited tolerance to adverse field conditions, such as UV exposure, high temperature, drought, and surface application (Immediato et al., 2015; Mascarin and Jaronski, 2016).
(d) Powder formulations are prone to moisture absorption, which contaminates and diminishes the viability of the active ingredient.

(e) Oil formulations suffer from poor spraying characteristics, and powder formulations leave residual talc on plant surfaces after spraying (Jaronski and Mascarin, 2017).

These conditions reduce the effectiveness of EPF and necessitate frequent applications, leading to increased costs. To overcome these challenges, researchers have turned to biopolymer-based formulations (Kulinets, 2015; Nicodemus and Bryant, 2008). Biopolymers, derived from renewable sources like plants, animals, and microorganisms, offer an eco-friendly and biodegradable solution. They enhance the stability, adherence, and persistence of EPF in field conditions by shielding them from environmental stressors and creating a favourable microenvironment for growth and insecticidal action (Friuli et al., 2022). Biopolymer-based formulations present a sustainable alternative to synthetic insecticides by reducing harm to non-target organisms and decreasing reliance on non-renewable resources (Kumar et al., 2022). By reducing the initial amount of bioinsecticide required and the frequency of application, these formulations make EPF more cost-effective. Furthermore, the protective effect of biopolymers may extend the shelf life of EPF-based products, addressing another limitation in their use as biopesticides (Friuli et al., 2021).

Various methods are used to introduce EPF into plants, such as leaf spraying, stem injection, seed treatment, foliar application, flower treatment, seed soaking, and soil irrigation (Quesada-Moraga et al., 2006; Lopez and Sword, 2015; Muvea et al., 2014; Greenfield et al., 2016; Bamisile et al., 2018a, b; Rondot and Reineke, 2018; González et al., 2020). The specific plant portion targeted for endophytic colonisation, or the kind of insect to be controlled, root eaters, stem borers, or leaf-chewing insects, for example, may influence the choice of injection technique. (Bamisile et al., 2018 b).

Foliar spraying of EPF like *M. brunneum* and *B. bassiana* can transiently establish an endophytic presence in plants like alfalfa, tomatoes, sweet peppers, and melons (Jaber and Araj, 2018). Insect pests on the leaf surface (phylloplane) can be effectively controlled by foliar spraying of spore solutions (Vega et al., 2009). Insects that burrow into and eat inside leaves, roots, stems, seeds, and rhizomes are the target of endophytic application of EPF (Resquín-Romero et al., 2016). Artificial inoculation of EPF into tomato plants has demonstrated effective control of *Tuta absoluta* (leaf miner) (Klieber and Reineke, 2015; Resquín-Romero et al., 2016). The application frequencies and timings of these mycoinsecticides are comparable to those of traditional insecticides (Wraight et al., 2000; Shah and Pell, 2003).

Most EPF formulations available on the market have a shelf life of three to six months and contain spore concentrations ranging from 10^9 to 10^{10} spores per gram. The formulation, extent of infestation, insect species, and ambient conditions impact the application dosage. The label for mycoinsecticide contains comprehensive usage instructions.

5. Major Applications of Entomopathogenic Fungi

5.1 General Benefits

The colonization of plants by EPF offers various benefits, including promoting plant growth, defending against insect pests, inducing systemic resistance, antagonizing plant pathogens, and mitigating the effects of abiotic stress. Alongside these advantages, fungal endophytes have gained attention for their ability to produce secondary metabolites, potentially surpassing chemical pesticides. Additionally, they exhibit potential in pharmaceutical and medical settings as immune-suppressing, anticancer, antidiabetic, and antibacterial medicines (Kabaluk and Ericsson, 2007; Kim et al., 2008; Vega et al., 2009; Elena et al., 2011; Sasan and Bidochka, 2012; Liao et al., 2014; Gouda et al., 2016; Vega, 2018; Fadiji and Babalola, 2020 a,b; Prajapati et al., 2024).

5.2 Integrated Pest Management

EPF serve as crucial elements in IPM strategies, functioning as biological control agents against arthropods and insect pests. They also constitute essential components of myco-insecticides used in horticulture, forestry, and agriculture. EPF application in pest management provides a practical and environmentally friendly substitute for chemical control techniques. Over 800 species of EPF have been described and identified as promising biocontrol agents. The studies on insect pathogenic fungi in the 1980s aimed to find methods for managing silkworm diseases. Bassi's discovery in 1835 of the white muscardine fungus's germ theory, later named *Beauveria bassiana* in his honour, laid the groundwork for using insect-infecting fungi in pest management (Gilbert and Gill, 2010). EPF has been actively employed to manage numerous economically significant crop pests for approximately two centuries. *B. bassiana* was initially isolated and identified around 170 years ago, while *B. brongniartii and M. anisopliae* have been in use for over 110 and 130 years, respectively (Zimmermann, 2007a, b). These fungal species, alongside other hypocrealean fungi like *I. fumosorosea, M. brunneum, M. robertsii,* and *H. thompsonii,* are commonly deployed against a wide range of arthropod pests (Dara, 2019). EPF primarily acts through inundative approaches and has demonstrated efficacy against various insects, including Homoptera (particularly aphids, whiteflies, cicadas, scale insects, locusts, thrips, grubs, moths, and mites), Lepidoptera (particularly larvae), Hymenoptera (bees), Coleoptera (beetles), Diptera (flies and mosquitoes), and tephritid fruit flies (Gulzar et al., 2021). Additionally, these fungi exhibit pathogenicity against phytopathogenic nematodes and other soil-borne pests (Pocasangre et al., 2000). Their mode of action varies, from causing starvation to toxin production, with mechanisms including enzymatic dissolution of the insect cuticle and production of toxins that induce host paralysis and eventual death (Table 1).

Table 1 Some commonly utilized entomopathogenic fungi

Entomopathogenic fungus	*Host*
Beauveria bassiana	Sucking pests and caterpillars infest various crop plants, including the yellow stem borer and leaf folder of rice, white grub of groundnut, sugarcane pyrilla, coconut rhinoceros beetle, caterpillars of pulses, tomato, and cotton, diamondback moth, and leaf-eating caterpillars of tobacco and sunflower.

Contd.

Table 1 *Contd.*

Entomopathogenic fungus	*Host*
Metarrhizium anisopliae	Coconut rhinoceros beetle, groundnut cutworm, rice brown plant hopper, diamondback moth, early shoot borer, top shoot borer, and internode borer of sugarcane.
Lecanicelium lecanii	Whiteflies, aphids, thrips, brown plant hoppers, scale insects, mealybugs, and other sucking insect pests
Hirsutella thompsonii	Hoppers, bug pests, whiteflies, red mites, and similar insects.
Paecilomyces fumosoroseus	Yellow and red mites, whiteflies, and related pests.
Nomuraea rileyi	Pod borers, cutworms, cabbage borers, and similar pests.

Nanopesticides for plant protection and pest control introduce novel strategies for developing active compounds at the nanoscale. Research indicates that metal nanoparticles show efficacy against various insects and pests, with silver nanoparticles emerging as particularly potent insecticidal agents. Biologically synthesized silver nanoparticles are becoming increasingly popular due to their ecological safety. These nanoparticles offer a sustainable approach to managing insect pest populations. EPF-based nanosilver exhibits high efficiency against diverse pest species even at very low concentrations (Bihal et al., 2023).

5.2.1 Beekeeping

Parasites, particularly Varroa destructors, and predators pose a significant threat to the health, welfare, and productivity of *Apis mellifera* worldwide, leading to substantial hive losses in beekeeping globally (Lozano et al., 2019; Hernández-Rodríguez et al., 2021). The predominant method of parasite control currently relies on chemical drugs. However, the emergence of parasiticide drug resistance has become a prevalent issue in bee farms across various countries (Semalulu et al., 1992; Maggi et al., 2008). In contrast, EPF have not exhibited such resistance issues, and studies assessing their environmental impact have found no adverse effects on ecosystems and mammals (Ebani and Mancianti, 2021; Pietropaoli and Formato, 2021). Therefore, the utilization of EPF holds promise as an eco-friendly approach for controlling parasites in beekeeping, offering prospects for the future (Bava et al., 2021; Bava et al., 2022).

5.2.2 Control of Malaria Mosquito Larvae

Spores of *B. bassiana* and *M. anisopliae,* when formulated in ShellSol T, showed increased efficacy against *Anopheles gambiae* s.s. larvae in comparison to spores that were not formulated. Moreover, these formulated spores displayed enhanced longevity when deployed in field conditions in Kenya (Bukhari et al., 2011).

5.2.3 Elimination of Toxic Pollutants

Nonylphenols, including para-nonylphenols (4-NP) commonly used in household products, pose significant environmental risks due to their toxicity to aquatic life and disruption of the endocrine system in humans and animals (Acir and Guenther, 2018). *M. robertsii* and *M. brunneum* have demonstrated the ability to degrade 4-NP with over 90% substrate loss within 24 hours (Różalska et al., 1970). During their degradation process, these fungi generate derivatives, which finally turn nonylphenol into CO_2 and H_2O (Różalska et al., 2015; Nowak et al., 2019). Triazines are herbicides that are carcinogenic and have endocrine-disrupting effects. *M. brunneum* can partially degrade ametryn, a triazine herbicide (Szewczyk et al., 2018). Dibutyltin (DBT), which is frequently released into the environment, through PVC materials or as a result of TBT degradation, is used as a catalyst in esterification reactions and as a curing agent in silicone rubbers. Aerated cultures of *M. robertsii* may totally breakdown DBT despite its immunotoxic, neurotoxic, and prooxidative characteristics, even at a starting concentration of 20 mg/l (Siewiera et al., 2017a, b; Stolarek et al., 2019). EE2, a synthetic estrogen used in hormonal contraception, found extensively in the environment is resistant to degradation. *M. robertsii* can efficiently remove up to 85% of EE2 from the growth medium (Różalska et al., 2015). *B. bassiana* MTCC 4580 efficiently eliminated four industrial dyes (Reactive remazol red, Indanthrene blue, Yellow 3RS, and Vat Novatic Grey) dyes at concentration of 500 mg/L, achieving a removal rate of up to 97% by the process of bioaccumulation (Gola et al., 2017b).

5.2.4 Removal of Heavy Metals

EPF assists in phytoremediation, removing heavy metals from the environment (Farias et al., 2019). *B. bassiana* efficiently removes heavy metals like Cu, Ni, Cd, Zn, Cr, and Pb from contaminated water, with removal rates between 58% to 75% (Gola et al., 2017a). Though lead induces morphological changes in *B. bassiana*, it exhibits low toxicity (Gola et al., 2017b; Wang et al., 2019a, 2019b). *B. bassiana* and *M. anisopliae* biomass effectively adsorb lead and cadmium from water, with *B. bassiana* showing higher adsorption due to cell wall differences (Hussein et al., 2011). Heavy metals inhibit the growth of EPFs like *I. farinosa*, *I. fumosorosea*, and *I. tennuipes* (Tkaczuk et al., 2019). Despite this, EPF, when part of fungal consortia with biochar, enhances plant metal absorption and translocation, as seen with *Jacaranda mimosifolia* (Farias et al., 2019).

5.2.5 Medicinal and Therapeutic Properties

EPF offers a diverse range of bioactive secondary metabolites with antimicrobial, anti-insect, and anticancer properties (Gouda et al., 2016; Fadiji and Babalola, 2020 a,b). *Isaria* spp. has yielded over 70 novel secondary metabolites, with compounds like beauvericin showing insecticidal, antibacterial, antiviral, and cytotoxic properties (Weng et al., 2019; Lu et al., 2016). Fumosorinone, isolated from *I. fumosorosea*, exhibits potential as a non-competitive inhibitor of protein tyrosine phosphatase 1B, suggesting medicinal applications for treating diabetes and related metabolic disorders (Liu et al., 2015). Peroxyergosterol, another compound from *I. fumosorosea*, has shown cytotoxicity against cancer cells and potential for vitamin A production (Sheu et al., 1999; Takei et al., 2005). EPF are crucial in mass-producing drugs, enzymes, antibodies, supplements, and industrial products, contributing significantly to fields such as medicine, agriculture, and industry (Latz et al., 2018; Sahay et al., 2017).

6. Impact of Toxic Substances on Entomopathogenic Fungi

Pesticides, including herbicides, fungicides, and insecticides, are major contributors to soil and water contamination due to the accumulation of their residues, impacting various living organisms. These chemicals have detrimental effects on EPF, such as *B. bassiana* and *M. anisopliae*, which are commonly used for pest control.

6.1 Fungicides

These chemicals exhibit varying levels of compatibility depending on their concentration and the EPF being used.

Chlorothalonil, a widely used broad-spectrum fungicide, severely inhibits conidial germination of *I. farinose, M. anisopliae,* and *B. bassiana* even at low concentrations (Demirci et al., 2011; Celar and Kos, 2016; Fiedler and Sosnowska, 2017) by inhibiting growth and sporulation.

Chlorothalonil and iprodione have a minimal effect on the growth and viability of the microsclerotia of *M. brunneum*, unlike their impact on conidial germination and mycelial growth (Wu et al., 2019).

Captan significantly suppresses conidial germination of *B. bassiana, I. fumosorosea*, and *L. longisporum* but only marginally impairs mycelial development (Shah et al., 2009; Dara, 2017). Additionally, Captan applied in field doses significantly affects the mortality of *Tenebrio molitor* larvae when combined with *B. bassiana* (Dara, 2017).

Copper-based fungicides containing over 40% metallic copper significantly affect the growth of *I. fumosorosea*, while those with lower copper doses impact the infection capacity of the EPF (D'Alessandro et al., 2011; Demirci et al., 2011; Avery et al., 2013). Additionally, copper oxide, containing high metallic copper percentages, has a toxic effect on *B. bassiana* and *L. muscarium* (Kouassi et al., 2003; Ali et al., 2013).

Strong growth inhibition of EPF is observed in dithiocarbamate fungicides containing metals such as manganese and zinc. For example, Mancozeb suppresses the germination of *I. farinosa* conidia at low concentrations and inhibits the growth of *B. bassiana* and *I. farinosa* at suggested field doses (Todorova et al., 1998; Kouassi et al., 2003). In the same way, low dosages of Mancozeb and Propineb prevent *B. bassiana* from sporulating. (Celar and Kos, 2016).

6.2 Pesticides

Quizalofop-p-ethyl and glyphosate, when applied at recommended field doses, inhibited the growth of *H. nodulosa* by 63.4% and 52.1%, respectively, and *Beauveria bassiana* by 43.9% and 6.5%, respectively, over a 20-day cultivation period (Tkaczuk et al., 2015). Among herbicides, Ametryn, an s-triazine herbicide, significantly inhibited the growth of *M. brunneum*, interfered with its glucose degradation and carbon and nitrogen metabolism, and induced oxidative stress in its mycelia (Szewczyk et al., 2018).

6.3 Insecticides

Organophosphates and chloroorganic compounds, significantly inhibit the growth of *B. bassiana* and *M. anisopliae* (Amutha et al., 2010; Asi et al., 2010). Carbamates and pyrrole insecticides also adversely affect the mycelial growth and conidial germination of these fungi (Asi et al., 2010). Furthermore, exposure to these insecticides induces detoxification mechanisms in EPF, including the upregulation of cytochrome P450 genes and antioxidant systems. Neem oil, a natural insecticide derived from plants, was found to adversely affect EPF, reducing the germination ratio, vegetative growth, and conidia production in *M. anisopliae* and *B. bassiana* (Hirose et al., 2001). Interactions between EPF and insecticides can have complex effects on pest control. For example, the botanical insecticides Tondexir and Palizin, when applied at field doses, didn't significantly inhibit fungal germination but reduced the susceptibility of *Galleria mellonella* larvae to *B. bassiana* infection. Palizin also reduced *L. lecanii's* effectiveness against *G. mellonella* larvae (Sohrabi et al., 2019). Additionally, proper application timing is crucial; synergistic effects between α-cypermethrin and *B. bassiana* were observed only when the EPF was applied first with an interval of at least 48 hours between applications (Meyling et al., 2018).

6.4 Industrial Dyes

Reactive remazol red, Indanthrene blue, Yellow 3RS, and Vat Novatic Grey reduced mycelium biomass, with Vat Novatic Grey showing the highest toxicity to *B. bassiana* (Gola et al., 2017b).

6.5 Biofertilizers

Multibion TM, despite being considered environmentally harmless, can impact soil, inhabiting EPF by reducing germination, vegetative growth, and conidia production in *M. anisopliae* and exhibiting moderate toxicity to *B. bassiana* (Hirose et al., 2001).

6.6 Anthropogenic Compounds

Endocrine disrupting compounds (EDCs) such as nonylphenols were found to be toxic to *Metarhizium*, causing significant changes in pellet morphology, metabolic activity, hyphae structure, and cell wall composition (Różalska et al., 2014). Another EDC, dibutyltin, and its by-product monobutyltin induced nitrosative and oxidative stress, along with lipid profile alterations in *M. robertsii* (Siewiera et al., 2017a,b; Stolarek et al., 2019).

7. Challenges in Utilizing Entomopathogenic Fungi

Most EPF exhibit limited growth outside the bodies of their insect hosts, thereby minimizing disturbance to soil carbon and nitrogen content (Ownley et al., 2008). Their utilization in IPM programs is significant due to their host-specific nature, effectiveness in small quantities, and rapid degradation. Several drawbacks associated with the utilization of EPF include high contamination by saprophytic fungi, leading to reduced efficiency due to competition for resources. Additionally, the low viability of EPF microbes contributes to decreased effectiveness. EPF also exhibits limited tolerance to adverse field conditions such as UV exposure, high temperatures, drought, and surface application. Powder formulations, commonly used for EPF, are susceptible to moisture absorption, which contaminates and diminishes the viability of the active ingredient. Furthermore, oil formulations suffer from poor spraying characteristics, while powder formulations may leave residual talc on plant surfaces after application (Immediato et al., 2015; Mascarin and Jaronski, 2016; Jaronski and Mascarin, 2017). Their narrow host range can sometimes lead to outbreaks of secondary pests. EPF inoculum typically has a short lifespan and requires a longer period (2–3 weeks) to control insect pests, particularly under suboptimal environmental conditions such as low relative humidity and non-optimal temperatures (Islam et al., 2021). Moreover, the cost of commercial formulations is often high, and there is a risk of environmental contamination with mycotoxins produced by EPF, including aflatoxins, fumonisins, and citrinins (Sinha et al., 2016). In some instances, EPF may inadvertently affect non-targeted species such as predators, parasitoids, earthworms, or honeybees. Farmers' expectations regarding the efficacy of EPF may be high, leading to comparisons with synthetic insecticides. However, their efficacy can be compromised by unsystematic handling and a lack of expertise, resulting in unfulfilled expectations among farmers (Jaber and Ownley, 2018). These challenges highlight the need for careful consideration and management when utilizing EPF in pest control strategies.

8. Future Directions

While chemical pesticides are still used to control insects, other methods, such the usage of transgenic plants, are becoming more popular due to the harmful effects of synthetic chemicals. However, reliance on transgenic plants or chemical pesticides is not sustainable due to constant insect evolution (Wheelis, 2002). In contrast, the global biopesticide market, valued at approximately US$ 7.7 billion with a growth rate of 14.1% annually, represents a more sustainable approach, with biopesticides posing minimal threats to human and animal health, aiding in pesticide resistance management, and being environmentally friendly (Sharma et al., 2020).

EPF offers a sustainable solution to pest control, with advantages including effectiveness in organic farming, safety for humans and non-target organisms, and environmental sustainability (Litwin et al., 2020). With their insecticidal efficiency and ability to reduce reliance on chemical treatments, EPF products can be useful substitutes for chemical insecticides in IPM programmes (Sandhu et al., 2012). Additionally, certain EPFs have been shown to have synergistic effects when mixed with synthetic pesticides (Islam et al., 2021), and endophytic EPFs have been shown to promote plant development without having any adverse consequences (Vega, 2018).

To promote the adoption of EPF for pest control, farmers should be educated about their benefits and provided with financial subsidies and technological knowledge for handling and application. Protocols ensuring food safety and environmental protection must be followed diligently before EPF application. Additionally, ongoing research efforts should focus on improving EPF formulations for longer shelf life, reduced human allergenicity, and enhanced safety for non-target species (Sinha et al., 2016).

Further studies are needed to explore the ecological roles of fungi as endophytes, rhizosphere colonizers, and plant growth promoters, with the aim of enhancing pest control efficacy and crop yield (Vega et al., 2009). Molecular research should prioritize enhancing fungal strains resistance, broadening host range, and developing effective metabolites and combinations with insecticides for IPM programs (Sinha et al., 2016). It is essential to conduct pathogenicity and toxicity tests on non-target organisms, including vertebrates, to ensure the safety of EPF applications (Fadiji and Babalola, 2020 a,b; Adeleke and Babalola, 2021). Although the presence of risks is understood, and that they cannot be completely eradicated, efforts should be focused on putting in place appropriate precautions for application and manufacture to avoid adverse outcomes. The many benefits of using bioproducts derived from these microorganisms justify their promotion, even though there are no current set standards for the approval or use of fungal biocontrol agents (Fadiji and Babalola, 2020b). It will take more than just lab or field testing to address the underlying issues; policy and regulatory changes will also be necessary. To effectively utilise the potential applications of these microbes, issues related to the economy, society, and politics must also be resolved.

Research on fungal endophytes has predominantly focused on co-culturing methods in-vitro to assess their antagonistic effects against targeted pathogens. While many studies have demonstrated these effects, few have thoroughly examined the physiological changes in colonized plants. Another common approach involves comparing treated and untreated seedlings after artificial inoculation with pathogens to assess various parameters. However, understanding the mechanisms underlying the effects of endophytic fungi on host physiology and volatility levels remains limited and inconsistent across different environmental conditions (Fontana et al., 2009).

Recent studies have delved into the ecology of fungal endophyte-plant host specificity and its impacts on multiple trophic levels (Hartley and Gange, 2009). Molecular mechanisms underlying fungal endophyte-induced host plant defence have received increased attention (Zheng and Dicke, 2008). Genetic engineering of fungal endophytes has aimed to enhance plant yields and defence systems, leading to the development of recombinant fungal strains with improved virulence (Wang and St Leger, 2007b; Chen et al., 2015). However, the use of recombinant endophytic fungi may pose risks to pollinators and beneficial insects (Fadiji and Babalola, 2020b).

Efforts have been made to integrate endophytic fungi like *B. bassiana* with chemical pesticides to enhance resistance management strategies and reduce ecosystem pollution (Al-Ani et al., 2021). Autodissemination strategies involving EPF combined with semiochemicals and other natural enemies have also shown promise (Vega et al., 2018).

Advancements in microscopy techniques, metagenomic analysis, and molecular techniques have enabled a more comprehensive understanding of endophytic colonization, plant-microbe interactions, and fungal microbiomes. Omics technologies have revolutionized research on plant-microbe interactions, offering insights into genomics, proteomics, and transcriptomics. With continued research and technological advancements, endophytic fungi and their bioactive compounds hold potential as alternatives to inorganic fertilizers and chemical pesticides (Fadiji and Babalola, 2020b).

Conclusion

Agriculture confronts formidable challenges in managing crop yield losses worldwide, primarily attributed to diverse stressors like insect pests and synthetic pesticides. Synthetic pesticides, particularly insecticides, pose environmental risks through water and soil contamination and drive resistance evolution in target pest populations. Additionally, their indiscriminate action compromises beneficial organisms, including natural enemies of crop pests. In response, scientific efforts are focused on developing sustainable and eco-friendly pest management strategies, with biological control emerging as a promising avenue. Myco-biocontrol, utilizing fungi to reduce insect populations and mitigate crop losses, has gained significant attention. With over 800 fungal species isolated from insects for biocontrol purposes, EPF represents a distinct category found primarily in soil, eliminating insects by invading their cuticles.

Efforts in mass-producing EPF involve utilizing various agricultural and liquid media, ensuring genetically uniform cultures to prevent alterations in morphology and virulence. Strain selection is critical, considering their response to fermentation conditions and ability to form stable propagules economically. Preservation methods like low-temperature storage or desiccation maintain viability and prevent genetic variation. Over 170 strains of EPF have been developed into mycopesticides, offering commercial solutions in various formulations. Challenges in formulations include contamination, low microbial viability, and limited tolerance to adverse field conditions. Various methods introduce EPF into plants, including leaf spraying, stem injection, and seed treatment, depending on targeted pests and plant types. However, challenges persist in ensuring efficacy and environmental stability.

Despite these challenges, EPF-based pest management holds significant promise for sustainable agriculture. EPF finds diverse applications across agriculture, medicine, and environmental remediation. In agriculture, they play a vital role in integrated pest management, offering a sustainable alternative to chemical control methods. EPF colonize plants, promoting growth, defending against pests, and mitigating abiotic stress. EPF also shows promise in beekeeping for controlling parasites like *Varroa destructor*, offering eco-friendly solutions without the risk of resistance. In medicine, EPF-derived compounds exhibit antimicrobial, antidiabetic, antitumor, and immunosuppressive properties, hinting at their potential in pharmaceutical applications. Furthermore, EPF contributes to environmental remediation by degrading toxic pollutants like nonylphenols, herbicides, and heavy metals, offering a sustainable approach to pollution management.

However, the presence of pesticides, including fungicides, herbicides, and insecticides, poses significant challenges to EPFs such as *B. bassiana* and *M. anisopliae*, widely used for pest control. Fungicides like chlorothalonil and Captan inhibit conidial germination and mycelial growth of EPF, while copper-based fungicides and dithiocarbamates affect their growth and infection capacity. Herbicides such as quizalofop-p-ethyl and glyphosate inhibit EPF growth, while insecticides like organophosphates and neem oil adversely affect their mycelial growth and conidial germination. Additionally, industrial dyes and biofertilizers like Multibion TM can reduce EPF biomass and exhibit toxicity. Moreover, endocrine-disrupting compounds like nonylphenols and dibutyltin induce stress and alter metabolic activity in EPF.

Other challenges include that EPF are typically confined to their insect hosts, compete with saprophytic fungi, reduce efficacy, and have limited tolerance to environmental conditions like UV exposure and drought. Powder formulations, commonly used for EPF, are prone to moisture absorption and poor spraying, while oil formulations have application issues. Additionally, EPF may inadvertently affect non-target species and have a short lifespan, requiring longer control periods. High costs, the risk of mycotoxin contamination, and farmer expectations add complexity to their utilization emphasizing the need for careful management in pest control strategies.

To encourage EPF adoption, educating farmers about their benefits and providing support in handling and application is crucial. Research should focus on enhancing EPF formulations, understanding their ecological roles, improving host range, and ensuring safety for non-target species. Molecular research and regulatory interventions are needed to maximize their potential while addressing economic and political constraints. Recent studies explore endophytic fungi's impacts on plant physiology and host defence mechanisms, with genetic engineering aiming to enhance plant yields and resistance. Integration with chemical pesticides and autodissemination strategies show promise, while advancements in microscopy and omics technologies offer deeper insights into plant-microbe interactions, paving the way for EPF as a sustainable alternative to conventional pesticides and fertilizers.

References

Acir, I. and Guenther, K. (2018). Endocrine-disrupting metabolites of alkylphenol ethoxylates – A critical review of analytical methods, environmental occurrences, toxicity, and regulation. Sci. Total Environ., 635: 1530–1546. https://doi.org/10.1016/j.scitotenv.2018.04.079

Adeleke, B.S. and Babalola, O. O. (2021). Biotechnological overview of agriculturally important endophytic fungi. Hortic. Environ. Biotechnol., 62(4): 507–520. https://doi.org/10.1007/s13580-021-00334-1

Agale, S., Gopalakrishnan, S., Ambhure, K., Chandravanshi, H., Gupta, R. and Wani, S.P. (2018). Mass Production of Entomopathogenic Fungi (*Metarhizium anisopliae*) using Different Grains as a Substrate. Int. J. Curr. Microbiol. Appl. Sci., 7(1): 2227–2232. https://doi.org/10.20546/ijcmas.2018.701.268

Akutse, K.S., Subramanian, S., Maniania, N.K., Dubois, T. and Ekesi, S. (2020). Biopesticide research and product development in africa for sustainable agriculture and food security – Experiences from the international centre of insect physiology and ecology (icipe). Front. Sustain. Food Syst. 4:563016. doi: 10.3389/fsufs.2020.563016

Al-Ani, L.K.T., Surono, S., Aguilar-Marcelino, L., Salazar-Vidal, V., Becerra, A.G., and Raza, W. (2021). Role of useful fungi in agriculture sustainability. In Fungal biology Al-Ani, L.K.T., Surono, Aguilar-Marcelino, L., Salazar-Vidal, V.E., Becerra, A.G. and Raza, W. (Eds) (pp. 1–44). Springer, Cham. https://doi.org/10.1007/978-3-030-60659-6_1

Ali, S., Huang, Z. and Ren, S. (2013). Effect of fungicides on growth, germination and cuticle-degrading enzyme production by *Lecanicillium muscarium*. Biocontrol Sci. Technol., 23(6): 711–723. https://doi.org/10.1080/09583157.2013.794258

Amutha, M., Banu, J.G., Surulivelu, T. and Gopalakrishnan, N. (2010). Effect of commonly used insecticides on the growth of white muscardine fungus, *Beauveria bassiana* under laboratory conditions. J Biopest., 3(1 Special Issue): 143–146.

Ansari, M.A. and Butt, T.M. (2011). Effects of successive subculturing on stability, virulence, conidial yield, germination and shelf-life of entomopathogenic fungi. J. Appl. Microbiol,, 110(6), 1460–1469. https://doi.org/10.1111/j.1365-2672.2011.04994.x

Antía, O.P., Posada, F.J., Bustillo, A.E., and Gonzáles, M.T. (1992). Producción en finca del hongo *Beauveria bassiana* para el control de la broca del café. Centro Nacional de Investigaciones de Café (Cenicafé). Pp. 1–12.

Araújo, J.P.M. and Hughes, D. (2016). Diversity of entomopathogenic fungi. In Advances in Genetics 94: 1–39. https://doi.org/10.1016/bs.adgen.2016.01.001

Asi, M.R., Bashir, M.K., Afzal, M., Ashfaq, M. and Sahi, S.T. (2010). Compatibility of entomopathogenic fungi, *Metarhizium anisopliae* and *Paecilomyces fumosoroseus* with selective insecticides. Pak. J. Bot., 42(6): 4207–4214. https://www.cabdirect.org/abstracts/20113034436.html

Avery, P.B., Pick, D.A., Aristizábal, L.F., Kerrigan, J.R., Powell, C.A., Rogers, M.E. and Arthurs, S. (2013). Compatibility of *Isaria fumosorosea* (Hypocreales: Cordycipitaceae) Blastospores with Agricultural Chemicals Used for Management of the Asian Citrus Psyllid, *Diaphorina citri* (Hemiptera: Liviidae). Insects, 4(4): 694–711. https://doi.org/10.3390/insects4040694

Bamisile, B.S., Akutse, K.S., Siddiqui, J.A. and Xu, Y. (2021). Model application of entomopathogenic fungi as alternatives to chemical pesticides: Prospects, challenges, and insights for next-generation sustainable agriculture. Front. Plant Sci. 12:741804. doi: 10.3389/fpls.2021.741804

Bamisile, B.S., Dash, C.K., Akutse, K.S., Keppanan, R., Afolabi, O.G., Hussain, M., Qasim, M. and Wang, L. (2018a). Prospects of endophytic fungal entomopathogens as biocontrol and plant growth promoting agents: An insight on how artificial inoculation methods affect endophytic colonization of host plants. Microbiol. Res., 217: 34–50. https://doi.org/10.1016/j.micres.2018.08.016

Bamisile, B.S., Dash, C.K., Akutse, K.S., Keppanan, R. and Wang, L. (2018b). Fungal endophytes: Beyond herbivore management. Front. Microbiol. 9:544. doi: 10.3389/fmicb.2018.00544

Baron, N.C., Costa, N.T.A., Mochi, D.A. and Rigobelo, E.C. (2018). First report of *Aspergillus sydowii* and *Aspergillus brasiliensis* as phosphorus solubilizers in maize. Ann. Microbiol., 68(12): 863–870. https://doi.org/10.1007/s13213-018-1392-5

Bateman, R., Carey, M., Batt, D., Prior, C., Abraham, Y.J., Moore, D., Jenkins, N.E., and Fenlon, J.S. (1996). Screening for Virulent Isolates of Entomopathogenic Fungi Against the Desert Locust, *Schistocerca gregaria Forskal*. Biocontrol Sci. Technol., 6(4): 549–560. https://doi.org/10.1080/09583159631181

Bava, R., Castagna, F., Piras, C., Musolino, V., Lupia, C., Palma, E., Britti, D. and Musella, V. (2022). Entomopathogenic fungi for pests and predators control in beekeeping. Vet. Sci., 9(2), 95. https://doi.org/10.3390/vetsci9020095

Bava, R., Castagna, F., Piras, C., Palma, E., Cringoli, G., Musolino, V., Lupia, C., Perri, M., Statti, G., Britti, D. and Musella, V. (2021). In Vitro Evaluation of Acute Toxicity of Five Citrus spp. Essential Oils towards the Parasitic Mite *Varroa destructor*. Pathogens, *10*(9): 1182. https://doi.org/10.3390/pathogens10091182

Behie, S.W., and Bidochka, M. (2014). Ubiquity of Insect-Derived Nitrogen Transfer to Plants by Endophytic Insect-Pathogenic Fungi: an Additional Branch of the Soil Nitrogen Cycle. Appl. Environ. Microbiol., 80(5): 1553–1560. https://doi.org/10.1128/aem.03338-13

Bihal, R., Al-Khayri, J.M., Banu, A.N., Kudesia, N., Ahmed, F.K., Sarkar, R., Arora, A. and Abd-Elsalam, K.A. (2023). Entomopathogenic Fungi: an Eco-Friendly synthesis of sustainable nanoparticles and their nanopesticide properties. Microorganisms, 11(6): 1617. https://doi.org/10.3390/microorganisms11061617

Boucias, D.G., Pendland, J.C. (1991). Attachment of Mycopathogens to Cuticle. In: The Fungal Spore and Disease Initiation in Plants and Animals. Cole, G.T., Hoch, H.C. (eds) Springer, Boston, M.A. (pp. 101–127). https://doi.org/10.1007/978-1-4899-2635-7_5

Boucias, D.G., Pendland, J.C. and Latgé, J.P. (1988). Nonspecific factors involved in attachment of entomopathogenic deuteromycetes to host insect cuticle. Appl. Environ. Microbiol., 54(7): 1795–1805. https://doi.org/10.1128/aem.54.7.1795-1805.

Bradley, C., Black, W., Kearns, R. and Wood, P. (1992). Role of production technology in mycoinsecticide development. In: Frontiers of Industrial Mycology Leatham G (ed). (pp. 160–173). Springer, Boston, https://doi.org/10.1007/978-1-4684-7112-0_11

Bukhari, T., Takken, W. and Koenraadt, C.J.M. (2011). Development of *Metarhizium anisopliae* and *Beauveria bassiana* formulations for control of malaria mosquito larvae. Parasit Vectors, 4(1): https://doi.org/10.1186/1756-3305-4-23

Butt, T.M., Wang, C., Shah, F.A. and Hall, R. (2007). Degeneration of entomogenous fungi. In An Ecological and Societal Approach to Biological Control. J. Eilenberg, J. and Hokkanen, H.M.T. (Eds.) (pp. 213–226). Springer Dordrecht. https://doi.org/10.1007/978-1-4020-4401-4_10

Celar, F. and Kos, K. (2016). Effects of selected herbicides and fungicides on growth, sporulation and conidial germination of entomopathogenic fungus *Beauveria bassiana*. Pest Manag. Sci., 72(11): 2110–2117. https://doi.org/10.1002/ps.4240

Chandwani, S., Naik, H.Y., Gamit, H.A., Chandarana, K.A. and Amaresan, N. (2022). Mass multiplication, production cost analysis, and marketing of Beauveria. In: Agricultural Microbiology Based Entrepreneurship. Microorganisms for Sustainability, vol 39, Amaresan, N., Dharumadurai, D., Babalola, O.O. (eds). (pp. 251–266). Springer, Singapore. https://doi.org/10.1007/978-981-19-5747-5_16

Charnley, A.K. (1992). Mechanism of fungal pathogenesis in insects with particular reference to locusts. In Biological control of locusts and grasshoppers (pp. 191–199). Melkshan, UK: CAB International.

Charnley, A. (2003). Fungal pathogens of insects: Cuticle degrading enzymes and toxins. In Advances in Botanical Research 40: 241–321. https://doi.org/10.1016/s0065-2296(05)40006-3

Chen, X., Lin, L., Hu, Q., Zhang, B., Wu, W., Jin, F. and Jiang, J. (2015). Expression of dsRNA in recombinant *Isaria fumosorosea* strain targets the TLR7 gene in *Bemisia tabaci*. BMC Biotechnol., 15(1). https://doi.org/10.1186/s12896-015-0170-8

D'Alessandro, C.P., Padín, S.B., Urrutia, M.I., and Lastra, C.C.L. (2011). Interaction of fungicides with the entomopathogenic fungus *Isaria fumosorosea*. Biocontrol Sci. Technol., 21(2), 189–197. https://doi.org/10.1080/09583157.2010.536200

Dara, S.K. (2017). Compatibility of the Entomopathogenic Fungus *Beauveria bassiana* with Some Fungicides Used in California Strawberry. Open Plant Sci. J., 10(1): 29–34. https://doi.org/10.2174/1874294701710010029

Dara, S.K. (2019). Non-Entomopathogenic roles of entomopathogenic fungi in promoting plant health and growth. Insects, 10(9), 277. https://doi.org/10.3390/insects10090277

Demirci, F.Y., Muştu, M., Kaydan, M.B. and Ülgentürk, S. (2011). Effects of some fungicides on *Isaria farinosa*, and in vitro growth and infection rate on *Planococcus citri*. Phytoparasitica, 39(4): 353–360. https://doi.org/10.1007/s12600-011-0168-2

Ebani, V.V. and Mancianti, F. (2021). Entomopathogenic fungi and bacteria in a veterinary perspective. Biology, 10(6): 479. https://doi.org/10.3390/biology10060479

Eilenberg, J. and Michelsen, V. (1999). Natural host range and prevalence of the Genus *Strongwellsea* (Zygomycota: entomophthorales) in Denmark. J. Invertebr. Pathol., *73*(2): 189–198. https://doi.org/10.1006/jipa.1998.4795

Elena, G.J., Beatriz, P.J., Perticari, A. and Roberto, E.L. (2011). *Metarhizium anisopliae* (Metschnikoff) *Sorokin* promotes growth and has endophytic activity in tomato plants. Adv. Biol. Res.. 5 (1): 22–27, http://www.idosi.org/abr/5(1)/3.pdf

Fadiji, A.E. and Babalola, O.O. (2020a). Elucidating mechanisms of endophytes used in plant protection and other bioactivities with multifunctional prospects. Front. Bioeng. Biotechnol. 8:467. doi: 10.3389/fbioe.2020.00467

Fadiji, A.E., and Babalola, O.O. (2020b). Exploring the potentialities of beneficial endophytes for improved plant growth. Saudi J. Biol. Sci. 27(12): 3622–3633. https://doi.org/10.1016/j.sjbs.2020.08.002

Fan, Y., Fang, W., Guo, S., Pei, X., Zhang, Y., Xiao, Y., Li, D., Jin, K., Bidochka, M. and Pei, Y. (2007). Increased insect virulence in *Beauveria bassiana* strains overexpressing an engineered Chitinase. Appl. Environ. Microbiol., 73(1): 295–302. https://doi.org/10.1128/aem.01974-06

Fang, W., Leng, B., Xiao, Y., Jin, K., Ma, J., Fan, Y., Feng, J., Yang, X., Zhang, Y. and Pei, Y. (2005). Cloning of *Beauveria bassiana* Chitinase Gene Bbchit1 and its application to improve fungal strain virulence. Appl. Environ. Microbiol., 71(1): 363–370. https://doi.org/10.1128/aem.71.1.363-370.2005

Faria, M. and Wraight, S. P. (2007). Mycoinsecticides and Mycoacaricides: A comprehensive list with worldwide coverage and international classification of formulation types. Biol Con., 43(3): 237–256. https://doi.org/10.1016/j.biocontrol.2007.08.001

Farias, C.P., Alves, G.S., De Oliveira, D.C., De Melo, E.I. and Azevedo, L.C.B. (2019). A consortium of fungal isolates and biochar improved the phytoremediation potential of *Jacaranda mimosifolia D. Don* and reduced copper, manganese, and zinc leaching. J Soil Sediment., *20*(1): 260–271. https://doi.org/10.1007/s11368-019-02414-3

Fernandes, É.K.K., Rangel, D.E., Braga, G.Ú.L., and Roberts, D.W. (2015). Tolerance of entomopathogenic fungi to ultraviolet radiation: a review on screening of strains and their formulation. Curr. Genet, 61(3): 427–440. https://doi.org/10.1007/s00294-015-0492-z

Ferron, P. (1978). Biological control of insect pests by entomogenous fungi. Annu. Rev. Entomol., 23(1): 409–442. https://doi.org/10.1146/annurev.en.23.010178.002205

Fiedler, Z., and Sosnowska, D. (2017). Side effects of fungicides and insecticides on entomopathogenic fungi in vitro. J. Plant Prot. Res., 57(4):355-360. https://doi.org/10.1515/jppr-2017-0048

Fontana, A., Reichelt, M., Hempel, S., Gershenzon, J. and Unsicker, S.B. (2009). The Effects of Arbuscular Mycorrhizal Fungi on Direct and Indirect Defence Metabolites of Plantago lanceolata L. J. Chem. Ecol,, 35(7): 833–843. https://doi.org/10.1007/s10886-009-9654-0

Freimoser, F.M., Grundschober, A., Tuor, U. and Aebi, M. (2003). Regulation of hyphal growth and sporulation of the insect pathogenic fungus *Entomophthora thripidum in vitro*. FEMS Microbiol. Lett., 222(2): 281–287. https://doi.org/10.1016/s0378-1097(03)00315-x

Friuli, M., Cafarchia, C., Lia, R.P., Otranto, D., Pombi, M. and Demitri, C. (2022). From tissue engineering to mosquitoes: biopolymers as tools for developing a novel biomimetic approach to pest management/vector control. Parasit Vectors, 15(1): https://doi.org/10.1186/s13071-022-05193-y

Friuli, M., Nitti, P., Aneke, C.I., Demitri, C., Cafarchia, C. and Otranto, D. (2021). Freeze-drying of *Beauveria bassiana* suspended in Hydroxyethyl cellulose based hydrogel as possible method for storage: Evaluation of survival, growth and stability of conidial concentration before and after processing. Results Eng., 12: 100283. https://doi.org/10.1016/j.rineng.2021.100283

Fuguet, R. and Vey, A. (2004). Comparative analysis of the production of insecticidal and melanizing macromolecules by strains of *Beauveria* spp.: in vivo studies. J. Invertebr. Pathol., 85(3): 152–167. https://doi.org/10.1016/j.jip.2004.03.001

Fuxa, J.R. and Tañada, Y. (1987). Epizootiology of insect diseases. John Wiley & Sons; https://ci.nii.ac.jp/ncid/BA03580396

Garza-López, P.M., Königsberg, M., Gómez-Quiroz, L.E. and Loera, O. (2011). Physiological and antioxidant response by *Beauveria bassiana* Bals (Vuill.) to different oxygen concentrations. World J. Microbiol. Biotechnol., 28(1): 353–359. https://doi.org/10.1007/s11274-011-0827-y

Gilbert, L.I., Iatrou, K. and Gill, S.S. (2005). Comprehensive molecular insect science. http://ndl.ethernet.edu.et/bitstream/123456789/6876/1/molecular2005.pdf.pdf

Gilbert, L.I. and Gill, S.S. (2010). Insect control: biological and synthetic agents. pp.490. http://ci.nii.ac.jp/ncid/BB02762497

Gillespie, A.T. and Claydon, N. (1989). The use of entomogenous fungi for pest control and the role of toxins in pathogenesis. Pestic. Sci., 27(2): 203–215. https://doi.org/10.1002/ps.2780270210

Gola, D., Malik, A., Namburath, M. and Ahammad, S.Z. (2017a). Removal of industrial dyes and heavy metals by *Beauveria bassiana*: FTIR, SEM, TEM and AFM investigations with Pb(II). Environ. Sci. Pollut. Res., *25*(21): 20486–20496. https://doi.org/10.1007/s11356-017-0246-1

Gola, D., Malik, A., Namburath, M. and Ahammad, S.Z. (2017b). Removal of industrial dyes and heavy metals by *Beauveria bassiana*: FTIR, SEM, TEM and AFM investigations with Pb(II). Environ. Sci. Pollut. Res., 25(21): 20486–20496. https://doi.org/10.1007/s11356-017-0246-1

González, Y.R., Taibo, A.D., Jiménez, J.A., and Portal, O. (2020). Endophytic establishment of *Beauveria bassiana* and *Metarhizium anisopliae* in maize plants and its effect against *Spodoptera frugiperda* (J. E. Smith) (Lepidoptera: Noctuidae) larvae. Egypt. J. Biol. Pest Control, 30(1): https://doi.org/10.1186/s41938-020-00223-2

Gouda, S., Das, G., Sen, S.K., Shin, H. and Patra, J.K. (2016). Endophytes: a treasure house of bioactive compounds of medicinal importance. Front. Microbiol. 7:1538. doi: 10.3389/fmicb.2016.015387.

Greenfield, M., Gómez-Jiménez, M.I., Ortiz, V., Vega, F.E., Kramer, M. and Parsa, S. (2016). *Beauveria bassiana* and *Metarhizium anisopliae* endophytically colonize cassava roots following soil drench inoculation. Biol. Control, 95: 40–48. https://doi.org/10.1016/j.biocontrol.2016.01.002

Gulzar, S., Wakil, W. and Shapiro-Ilan, D. I. (2021). Combined effect of entomopathogens against *Thrips tabaci* Lindeman (Thysanoptera: Thripidae): Laboratory, greenhouse and field trials. insects, 12(5): 456. https://doi.org/10.3390/insects12050456

Hajek, A.E. and Delalibera, Í. (2009). Fungal pathogens as classical biological control agents against arthropods. BioControl, 55(1): 147–158. https://doi.org/10.1007/s10526-009-9253-6

Hartley, S. and Gange, A.C. (2009). Impacts of plant symbiotic fungi on insect herbivores: Mutualism in a multitrophic context. Annu. Rev. Entomol., 54(1): 323–342. https://doi.org/10.1146/annurev.ento.54.110807.090614

Hernández-Rodríguez, C.S., Marín, Ó., Calatayud, F., Mahiques, M.J., Mompó, A., Segura, I., Simó, E. and González-Cabrera, J. (2021). Large-Scale Monitoring of Resistance to Coumaphos, Amitraz, and Pyrethroids in Varroa destructor. Insects, 12(1): 27. https://doi.org/10.3390/insects12010027

Hesketh, H., Roy, H.E., Eilenberg, J., Pell, J.K. and Hails, R.S. (2009). Challenges in modelling complexity of fungal entomopathogens in semi-natural populations of insects. BioControl, 55(1): 55–73. https://doi.org/10.1007/s10526-009-9249-2

Hirose, E., Neves, P.M.O.J., Zequi, J.a.C., Martins, L.H., Peralta, C.H. and Moino, A. (2001). Effect of Biofertilizers and neem oil on the entomopathogenic fungi *Beauveria bassiana* (Bals.) *Vuill.* and *Metarhizium anisopliae* (Metsch.) Sorok. Braz. Arch. Biol. Technol., 44(4): 419–423. https://doi.org/10.1590/s1516-89132001000400013

Holder, D.J. and Keyhani, N.O. (2005). Adhesion of the Entomopathogenic Fungus *Beauveria (Cordyceps) bassiana* to Substrata. Appl. Environ. Microbiol., 71(9): 5260–5266. https://doi.org/10.1128/aem.71.9.5260-5266.2005

Hughes, W.O.H., Thomsen, L., Eilenberg, J. and Boomsma, J.J. (2004). Diversity of entomopathogenic fungi near leaf-cutting ant nests in a neotropical forest, with particular reference to *Metarhizium anisopliae* var. *anisopliae*. J. Invertebr. Pathol., 85(1): 46–53. https://doi.org/10.1016/j.jip.2003.12.005

Humber, R.A. (2008). Evolution of entomopathogenicity in fungi. J. Invertebr. Pathol., 98(3): 262–266. https://doi.org/10.1016/j.jip.2008.02.017

Hussein, K.A., Hassan, S.H. and Joo, J.H. (2011). Potential capacity of *Beauveria bassiana* and *Metarhizium anisopliae* in the biosorption of Cd2+ and Pb2+. J. Gen. Appl. Microbiol., 57(6): 347–355. https://doi.org/10.2323/jgam.57.347

Immediato, D., Camarda, A., Iatta, R., Puttilli, M.R., Ramos, R.A.N., Di Paola, G., Giangaspero, A., Otranto, D. and Cafarchia, C. (2015). Laboratory evaluation of a native strain of *Beauveria bassiana* for controlling *Dermanyssus gallinae* (De Geer, 1778) (Acari: Dermanyssidae). Vet. Parasitol., 212(3–4): 478–482. https://doi.org/10.1016/j.vetpar.2015.07.004

Inglis, G.D., Goettel, M.S., Butt, T.M. and Strasser, H. (2001). Use of hyphomycetous fungi for managing insect pests. In Fungi as biocontrol agents: progress, problems and potential (pp. 23-69). Wallingford UK: CABI publishing. https://doi.org/10.1079/9780851993560.0023

Inglis, G.D., Johnson, D.L. and Goettel, M.S. (1996). Effect of Bait Substrate and Formulation on Infection of Grasshopper Nymphs by *Beauveria bassiana*. Biocontrol Sci. Technol., 6(1): 35–50. https://doi.org/10.1080/09583159650039511

Islam, W., Adnan, M., Shabbir, A., Naveed, H., Abubakar, Y. S., Qasim, M., Tayyab, M., Noman, A., Nisar, M.S., Khan, K.A. and Ali, H. (2021). Insect-fungal-interactions: A detailed review on entomopathogenic fungi pathogenicity to combat insect pests. Microb. Pathog., 159: 105122. https://doi.org/10.1016/j.micpath.2021.105122

Jaber, L.R. and Araj, S. (2018). Interactions among endophytic fungal entomopathogens (Ascomycota: Hypocreales), the green peach aphid *Myzus persicae* Sulzer (Homoptera: Aphididae), and the aphid endoparasitoid *Aphidius colemani Viereck* (Hymenoptera: Braconidae). Biol. Control, 116: 53–61. https://doi.org/10.1016/j.biocontrol.2017.04.005

Jaber, L.R. and Ownley, B.H. (2018). Can we use entomopathogenic fungi as endophytes for dual biological control of insect pests and plant pathogens? Biol. Control, 116: 36–45. https://doi.org/10.1016/j.biocontrol.2017.01.018

Jackson, M.A., Dunlap, C.A. and Jaronski, S.T. (2009). Ecological considerations in producing and formulating fungal entomopathogens for use in insect biocontrol. BioControl, 55(1): 129–145. https://doi.org/10.1007/s10526-009-9240-y

Jackson, M. A. and Jaronski, S. T. (2009). Production of microsclerotia of the fungal entomopathogen *Metarhizium anisopliae* and their potential for use as a biocontrol agent for soil-inhabiting insects. Mycol. Res., 113(8): 842–850. https://doi.org/10.1016/j.mycres.2009.03.004

Jackson, M.A., McGuire, M., Lacey, L.A., and Wraight, S.P. (1997). Liquid culture production of desiccation tolerant blastospores of the bioinsecticidal fungus *Paecilomyces fumosoroseus*. Mycol. Res., 101(1): 35–41. https://doi.org/10.1017/s0953756296002067

Jaihan, P., Sangdee, K. and Sangdee, A. (2016). Selection of entomopathogenic fungus for biological control of chili anthracnose disease caused by *Colletotrichum* spp. Eur. J. Plant Pathol., 146(3): 551–564. https://doi.org/10.1007/s10658-016-0941-7

Jankielsohn, A. (2018). The importance of insects in agricultural ecosystems. Adv. Entomol., 06(2): 62–73. https://doi.org/10.4236/ae.2018.62006

Jaronski, S.T. and Mascarin, G.M. (2017). Mass production of fungal entomopathogens. In Microbial Control of Insect and Mite Pests, From Theory to Practice, Lacey, L.A., (pp. 141–155). https://doi.org/10.1016/b978-0-12-803527-6.00009-3

Kabaluk, J.T. and Ericsson, J.D. (2007). *Metarhizium anisopliae* seed treatment increases yield of field corn when applied for wireworm control. Agron. J., 99(5): 1377–1381. https://doi.org/10.2134/agronj2007.0017n

Khan, S., Guo, L., Maimaiti, Y., Mijit, M. and De-Wen, Q. (2012). Entomopathogenic fungi as microbial biocontrol agent. Mol. Plant Breed. 3(7): 63–79. https://doi.org/10.5376/mpb.2012.03.0007

Kim, J.J., Goettel, M.S. and Gillespie, D.R. (2008). Evaluation of *Lecanicillium longisporum, Vertalec®* for simultaneous suppression of cotton aphid, Aphis gossypii, and cucumber powdery mildew, *Sphaerotheca fuliginea*, on potted cucumbers. Biol. Control, 45(3): 404–409. https://doi.org/10.1016/j.biocontrol.2008.02.003

Klieber, J. and Reineke, A. (2015). The entomopathogen *Beauveria bassiana* has epiphytic and endophytic activity against the tomato leaf miner Tuta absoluta. J. Appl. Entomol., 140(8): 580–589. https://doi.org/10.1111/jen.12287

Klowden, M.J. (2013). Physiological systems in insects. Academic press. https://doi.org/10.1016/c2011-0-04120-0

Kouassi, M., Coderre, D. and Todorova, S. (2003). Effects of the timing of applications on the incompatibility of three fungicides and one isolate of the entomopathogenic fungus *Beauveria bassiana (Balsamo) Vuillemin* (Deuteromycotina). J. Appl. Entomol., 127(7): 421–426. https://doi.org/10.1046/j.1439-0418.2003.00769.x

Kulinets, I. (2015). Biomaterials and their applications in medicine. In Regulatory Affairs for Biomaterials and Medical Devices, Amato, S.F. and Ezzell, R.M. Jr (Eds.) (pp. 1–10). Woodhead Publishingm. https://doi.org/10.1533/9780857099204.1

Kumar, R., Kumar, N., Rajput, V.D., Mandzhieva, S., Minkina, T., Saharan, B.S., Kumar, D., Sadh, P.K. and Duhan, J.S. (2022). Advances in biopolymeric nanopesticides: A new eco-friendly/eco-protective perspective in precision agriculture. Nanomaterials, 12(22): 3964. https://doi.org/10.3390/nano12223964

Kumar, R.N. and Mukerji, K.G. (1996). Integrated Disease Management Future Perspectives, In: Advances in Botany, Mukerji, K.G., Mathur, B., Chamala, B.P. and Chitralekha, C. (Eds.), (pp 335-347), APH Publishing Corporation, New Delhi. https://www.scirp.org/reference/referencespapers?referenceid=2694124

Latifian, M., Rad, B., Amani, M. and Rahkhodaei, E. (2013). Mass production of entomopathogenic fungi *Beauveria bassiana* (Balsamo) by using agricultural products based on liquid-solid diphasic method for date palm pest control. Int. J. Agri. Crop Sci., 5(19), 2337–2341.

Latz, M., Jensen, B., Collinge, D.B. and Jørgensen, H.J.L. (2018). Endophytic fungi as biocontrol agents: elucidating mechanisms in disease suppression. Plant Ecol. and Divers., 11(5–6): 555–567. https://doi.org/10.1080/17550874.2018.1534146

Lawler, S.P. (2017). Environmental safety review of methoprene and bacterially-derived pesticides commonly used for sustained mosquito control. Ecotox. Environ. Safe., 139: 335–343. https://doi.org/10.1016/j.ecoenv.2016.12.038

Lecuona, R.E., Clément, J., Riba, G., Joulie, C., and Juárez, P. (1997). Spore germination and hyphal growth of *Beauveria* sp. on insect lipids. J. Econ. Entomol., 90(1): 119–123. https://doi.org/10.1093/jee/90.1.119

Liao, X., O'Brien, T.R., Fang, W. and St Leger, R.J. (2014). The plant beneficial effects of *Metarhizium* species correlate with their association with roots. Appl. Microbiol. Biotechnol., 98(16): 7089–7096. https://doi.org/10.1007/s00253-014-5788-2

Litwin, A., Nowak, M. and Różalska, S. (2020). Entomopathogenic fungi: unconventional applications. Rev. Environ. Sci. Biotechnol., 19(1): 23–42. https://doi.org/10.1007/s11157-020-09525-1

Liu, L., zhang, J., Chen, C., Teng, J., Wang, C. and Luo, D. (2015). Structure and biosynthesis of fumosorinone, a new protein tyrosine phosphatase 1B inhibitor firstly isolated from the entomogenous fungus *Isaria fumosorosea*. Fungal Genet. Biol., 81: 191–200. https://doi.org/10.1016/j.fgb.2015.03.009

Locke, M. (2001). The Wigglesworth Lecture: Insects for studying fundamental problems in biology. J. Insect Physiol., 47(4–5): 495–507. https://doi.org/10.1016/s0022-1910(00)00123-2

Lomer, C.J., Bateman, R., Johnson, D.L., Langewald, J. and Thomas, M. B. (2001). Biological control of locusts and grasshoppers. Annu. Rev. Entomol., 46(1): 667–702. https://doi.org/10.1146/annurev.ento.46.1.667

Lopez, D.C., and Sword, G.A. (2015). The endophytic fungal entomopathogens Beauveria bassiana and *Purpureocillium lilacinum* enhance the growth of cultivated cotton (*Gossypium hirsutum*) and negatively affect survival of the cotton bollworm (*Helicoverpa zea*). Biol. Control, 89: 53–60. https://doi.org/10.1016/j.biocontrol.2015.03.010

Lozano, A., Hernando, M., Uclés, S., Hakme, E. and Fernández-Alba, A.R. (2019). Identification and measurement of veterinary drug residues in beehive products. Food Chem., 274: 61–70. https://doi.org/10.1016/j.foodchem.2018.08.055

Lu, C.W., Lin, H., Chen, B. and Jow, G. (2016). Beauvericin-induced cell apoptosis through the mitogen-activated protein kinase pathway in human nonsmall cell lung cancer A549 cells. J. Toxicol. Sci., 41(3): 429–437. https://doi.org/10.2131/jts.41.429

Maggi, M., Ruffinengo, S., Damiáni, N., Sardella, N.H. and Eguaras, M.J. (2008). First detection of *Varroa destructor* resistance to coumaphos in Argentina. Exp. Appl. Acarol., 47(4): 317–320. https://doi.org/10.1007/s10493-008-9216-0

Mantzoukas, S., Kitsiou, F., Natsiopoulos, D.A. and Eliopoulos, P.A. (2022). Entomopathogenic fungi: interactions and applications. Encyclopedia, 2(2): 646–656. https://doi.org/10.3390/encyclopedia2020044

Mascarin, G.M. and Jaronski, S.T. (2016). The production and uses of *Beauveria bassiana* as a microbial insecticide. World J. Microbiol. Biotechnol., 32(11): https://doi.org/10.1007/s11274-016-2131-3

Mascarin, G.M., Kobori, N.N., De Jesus Vital, R.C., Jackson, M.A., and Quintela, E.D. (2013). Production of microsclerotia by Brazilian strains of *Metarhizium* spp. using submerged liquid culture fermentation. World J. Microbiol. Biotechnol., 30(5): 1583–1590. https://doi.org/10.1007/s11274-013-1581-0

Mathulwe, L.L., Malan, A.P. and Stokwe, N.F. (2022). Mass Production of Entomopathogenic Fungi, *Metarhizium robertsii* and *Metarhizium pinghaense*, for Commercial Application Against Insect Pests. JoVE, 181. https://doi.org/10.3791/63246

Meikle, W.G., Mercadier, G., Holst, N. and Girod, V. (2008). Impact of two treatments of a formulation of *Beauveria bassiana* (Deuteromycota: Hyphomycetes) conidia on *Varroa* mites (Acari: Varroidae) and on honeybee (Hymenoptera: Apidae) colony health. Exp. Appl. Acarol., 46(1–4):105–117. https://doi.org/10.1007/s10493-008-9160-z

Meyling, N.V., Arthur, S.D., Pedersen, K.E., Cedergreen, N. and Fredensborg, B.L. (2018). Implications of sequence and timing of exposure for synergy between the pyrethroid insecticide alpha-cypermethrin and the entomopathogenic fungus *Beauveria bassiana*. Pest Manag. Sci., 74(11): 2488–2495. https://doi.org/10.1002/ps.4926

Mollier, P., Lagnel, J., Fournet, B., Aioun, A. and Riba, G. (1994). A glycoprotein highly toxic for *Galleria mellonella* larvae secreted by the entomopathogenic fungus *Beauveria sulfurescens*. J. Invertebr. Pathol., 64(3): 200–207. https://doi.org/10.1016/s0022-2011(94)90175-9

Monzón, A. (2002). Production, use and quality control of entomopathogenic fungi in Nicaragua. Manejo Integrado de Plagas (Costa Rica) No. 63: 95–103 https://repositorio.catie.ac.cr/handle/11554/6723

Muvea, A., Meyhöfer, R., Subramanian, S., Poehling, H., Ekesi, S. and Maniania, N.K. (2014). Colonization of onions by endophytic fungi and their impacts on the biology of *Thrips tabaci*. PLoS One, 9(9):e108242. https://doi.org/10.1371/journal.pone.0108242

Nicodemus, G.D. and Bryant, S.J. (2008). Cell encapsulation in biodegradable hydrogels for tissue engineering applications. Tissue Eng. Part B Rev. 14(2): 149–165. https://doi.org/10.1089/ten.teb.2007.0332

Nowak, M., Soboń, A., Litwin, A. and Różalska, S. (2019). 4-n-nonylphenol degradation by the genus *Metarhizium* with cytochrome P450 involvement. Chemosphere, 220: 324–334. https://doi.org/10.1016/j.chemosphere.2018.12.114

Omkar (2016). Ecofriendly pest management for food security. Pp 750. Elsevier Inc. https://doi.org/10.1016/c2014-0-04228-1

Ownley, B.H., Griffin, M.R., Klingeman, W.E., Gwinn, K.D., Moulton, J.K. and Pereira, R.M. (2008). *Beauveria bassiana*: Endophytic colonization and plant disease control. J. Invertebr. Pathol., 98(3): 267–270. https://doi.org/10.1016/j.jip.2008.01.010

Pal, S., St Leger, R.J. and Wu, L.P. (2007). Fungal Peptide Destruxin A Plays a Specific Role in Suppressing the Innate Immune Response in *Drosophila melanogaster*. J. Biol. Chem., 282(12): 8969–8977. https://doi.org/10.1074/jbc.m605927200

Pietropaoli, M. and Formato, G. (2021). Formic acid combined with oxalic acid to boost the acaricide efficacy against *Varroa destructor* in *Apis mellifera*. J. Apic. Res., 61(3): 320–328. https://doi.org/10.1080/00218839.2021.1972634

Pocasangre, L.E., Sikora, R.A., Vilich, V. and Schuster, R.P. (2000). Survey of banana endophytic fungi from central America and screening for biological control of the burrowing nematode (*Radopholus similis*). Infomusa, 9(1): 3–5. https://www.cabdirect.org/abstracts/20003014169.html

Prajapati, J., Sheth, R., Bhatt, R.K., Chavda, K., Solanki, Z.S., Rawal, R. and Goswami, D. (2024). Biodiversity and biotechnological applications of host-specific endophytic fungi for sustainable agriculture and allied sector. In Sustainable Agricultural Practices, Kumar, A., White, J.F. and Singh, J. (Eds.) (pp. 101–124). Academic Press.

Prakash, G.V. S. B., Padmaja, V. and Kiran, R. (2008). Statistical optimization of process variables for the large-scale production of Metarhizium anisopliae conidiospores in solid-state fermentation. Bioresour. Technol;, 99(6): 1530–1537. https://doi.org/10.1016/j.biortech.2007.04.031

Prior, C., Jollands, P. and Patourel, G.L. (1988). Infectivity of oil and water formulations of *Beauveria bassiana* (Deuteromycotina: Hyphomycetes) to the cocoa weevil pest *Pantorhytes plutus* (Coleoptera: Curculionidae). J. Invertebr. Pathol., 52(1): 66–72. https://doi.org/10.1016/0022-2011(88)90103-6

Quesada-Moraga, E., Ruiz-García, A. and Santiago-Álvarez, C. (2006). Laboratory evaluation of entomopathogenic fungi *Beauveria bassiana* and *Metarhizium anisopliae* against puparia and adults of *Ceratitis capitata* (Diptera: Tephritidae). J. Econ. Entomol., 99(6): 1955–1966. https://doi.org/10.1093/jee/99.6.1955

Rai, R., Pandey, R. and Tamta, A.K. (2021). Mass Multiplication of Entomopathogenic Fungi *Beauveria Bassiana* with agroindustrial wastes. Indian J. Entomol. Indian, 83(4), 644–645. https://doi.org/10.5958/0974-8172.2020.00229.1

Raimbault, M. (1998). General and microbiological aspects of solid substrate fermentation. Electron. J. Biotechnol., 1(2): 174–188. https://doi.org/10.2225/vol1-issue3-fulltext-9

Rangel, D.E., Acheampong, M.A., Bignayan, H.G., Golez, H.G. and Roberts, D.W. (2023). Conidial mass production of entomopathogenic fungi and tolerance of their mass-produced conidia to UV-B radiation and heat. Fungal Biol., 127(12): 1524–1533. https://doi.org/10.1016/j.funbio.2023.07.001

Researchers helping protect crops from pests. (2023). National Institute of Food and Agriculture. https://www.nifa.usda.gov/about-nifa/blogs/researchers-helping-protect-crops-pests

Resquín-Romero, G., Garrido-Jurado, I., Delso, C., Ríos-Moreno, A. and Quesada-Moraga, E. (2016). Transient endophytic colonizations of plants improve the outcome of foliar applications of mycoinsecticides against chewing insects. J. Invertebr. Pathol., 136: 23–31. https://doi.org/10.1016/j.jip.2016.03.003

Rondot, Y. and Reineke, A. (2018). Endophytic *Beauveria bassiana* in grapevine *Vitis vinifera* (L.) reduces infestation with piercing-sucking insects. Biol. Control, 116: 82–89. https://doi.org/10.1016/j.biocontrol.2016.10.006

Różalska, S., Bernat, P., Michnicki, P. and Długoński, J. (2015). Fungal transformation of 17α-ethinylestradiol in the presence of various concentrations of sodium chloride. Int Biodeterior Biodegradation, 103: 77–84. https://doi.org/10.1016/j.ibiod.2015.04.016

Różalska, S., Glińska, S. and Długoński, J. (2014). *Metarhizium robertsii* morphological flexibility during nonylphenol removal. Int Biodeterior Biodegradation, 95: 285–293. https://doi.org/10.1016/j.ibiod.2014.08.002

Różalska, S., Pawłowska, J., Wrzosek, M., Tkaczuk, C. and Długoński, J. (1970). Utilization of 4-n-nonylphenol by *Metarhizium* sp. isolates. Acta Biochim. Pol., 60(4): https://doi.org/10.18388/abp.2013_2040

Rust, M.K., Lance, W.R. and Hemsarth, H. (2016). Synergism of the IGRs methoprene and pyriproxyfen against larval cat fleas (Siphonaptera: pulicidae). J Med Entomol., 53(3): 629–633. https://doi.org/10.1093/jme/tjw010

Sahay, H., Yadav, A.N., Singh, A.K., Singh, S., Kaushik, R. and Saxena, A.K. (2017). Hot springs of Indian Himalayas: potential sources of microbial diversity and thermostable hydrolytic enzymes. 3 Biotech, 7(2): https://doi.org/10.1007/s13205-017-0762-1

Sahayaraj, K. and Namasivayam, S.K.R. (2008). Mass production of entomopathogenic fungi using agricultural products and by products. Afr. J. Biotechnol., 7(12): 1907–1910. https://doi.org/10.5897/ajb07.778

Sandhu, S.S., Sharma, A.K., Beniwal, V., Goel, G., Batra, P., Kumar, A., Jaglan, S., Sharma, A. and Malhotra, S. (2012). Myco-Biocontrol of Insect pests: factors involved, mechanism, and regulation. J. Pathog., 2012: 1–10. https://doi.org/10.1155/2012/126819

Sasan, R.K. and Bidochka, M. (2012). The insect-pathogenic fungus *Metarhizium robertsii* (Clavicipitaceae) is also an endophyte that stimulates plant root development. Am. J. Bot., 99(1): 101–107. https://doi.org/10.3732/ajb.1100136

Semalulu, S.S., MacPherson, J.M., Schiefer, H.B. and Khachatourians, G.G. (1992). Pathogenicity of *Beauveria bassiana* in Mice. J Vet Med B Infect Dis Vet Public Health, *39*(1–10): 81–90. https://doi.org/10.1111/j.1439-0450.1992.tb01141.x

Shah, F.A., Allen, N., Wright, C.J. and Butt, T.M. (2007). Repeated *in vitro* subculturing alters spore surface properties and virulence of *Metarhizium anisopliae*. FEMS Microbiol. Lett., 276(1): 60–66. https://doi.org/10.1111/j.1574-6968.2007.00927.x

Shah, F.A., Ansari, M.A., Watkins, J.E., Phelps, Z., Cross, J.V. and Butt, T.M. (2009). Influence of commercial fungicides on the germination, growth and virulence of four species of entomopathogenic fungi. Biocontrol Sci. Technol., 19(7): 743–753. https://doi.org/10.1080/09583150903100807

Shah, P.A. and Pell, J.K. (2003). Entomopathogenic fungi as biological control agents. Appl. Microbiol. Biotechnol., 61(5–6): 413–423. https://doi.org/10.1007/s00253-003-1240-8

Sharma, A., Sharma, S. and Yadav, P.K. (2023). Entomopathogenic fungi and their relevance in sustainable agriculture: A review. Cogent Food and Agric., 9(1): https://doi.org/10.1080/23311932.2023.2180857

Sharma, L., Bohra, N., Rajput, V.D., Quiroz-Figueroa, F.R., Singh, R.K. and Marques, G. (2020). Advances in entomopathogen isolation: a case of bacteria and fungi. Microorganisms, *9*(1): 16. https://doi.org/10.3390/microorganisms9010016

Sharma, S., Kooner, R. and Arora, R. (2017). Insect pests and crop losses. In Breeding Insect Resistant Crops for Sustainable Agriculture, Arora, S. and Sandhu, S. (pp. 45–66). Springer Singapore. https://doi.org/10.1007/978-981-10-6056-4_2

Sheu, J., Chang, K. and Duh, C. (1999). A Cytotoxic 5α,8α-Epidioxysterol from a Soft Coral Sinularia Species. J. Nat. Prod., 63(1): 149–151. https://doi.org/10.1021/np9903954

Shin, T.Y., Lee, M.R., Park, S.E., Lee, S.J., Kim, W.J. and Kim, J.S. (2020). Pathogenesis-related genes of entomopathogenic fungi. Arch. Insect Biochem. Physiol., 105(4): https://doi.org/10.1002/arch.21747

Siewiera, P., Różalska, S. and Bernat, P. (2017a). Efficient dibutyltin (DBT) elimination by the microscopic fungus *Metarhizium robertsii* under conditions of intensive aeration and ascorbic acid supplementation. Environ Sci Pollut Res Int, 24(13): 12118–12127. https://doi.org/10.1007/s11356-017-8764-4

Siewiera, P., Różalska, S. and Bernat, P. (2017b). Estrogen-mediated protection of the organotin-degrading strain *Metarhizium robertsii* against oxidative stress promoted by monobutyltin. Chemosphere, 185: 96–104. https://doi.org/10.1016/j.chemosphere.2017.06.130

Silva, A.C.L., Silva, G.A., Abib, P.H.N., Carolino, A.T. and Samuels, R.I. (2020). Endophytic colonization of tomato plants by the entomopathogenic fungus *Beauveria bassiana* for controlling the South American tomato pinworm, Tuta absoluta. CABI Agric. and Biosci., 1(1): https://doi.org/10.1186/s43170-020-00002-x

Sinha, K.K., Choudhary, A.K. and Kumari, P. (2016). Entomopathogenic fungi. In Ecofriendly pest management for food, Omkar (Ed.) (pp. 475–505). https://doi.org/10.1016/b978-0-12-803265-7.00015-4

Sohrabi, F., Jamali, F., Morammazi, S., Saber, M. and Kamita, S.G. (2019). Evaluation of the compatibility of entomopathogenic fungi and two botanical insecticides tondexir and palizin for controlling *Galleria mellonella* L. (Lepidoptera: Pyralidae). Crop Prot., 117: 20–25. https://doi.org/10.1016/j.cropro.2018.11.012

Soni, S., Mehta, P.K., Chandel, R.S. and Kalia, M. (2017). Mass production and field evaluation of *Beauveria brongniartii* (Saccardo) against *Brahmina coriacea* (Hope) whitegrubs in potato. Potato J., 44(1): 45–51. https://www.cabdirect.org/cabdirect/abstract/20173300966

Srivastava, C., Maurya, P., Sharma, P. and Mohan, L. (2009). Prospective role of insecticides of fungal origin: Review. Entomol Res., 39(6): 341–355. https://doi.org/10.1111/j.1748-5967.2009.00244.x

St Leger, R., Joshi, L., Bidochka, M. and Roberts, D.W. (1996). Construction of an improved mycoinsecticide overexpressing a toxic protease. Proceedings of the National Academy of Sciences of the United States of America, 93(13), 6349–6354. https://doi.org/10.1073/pnas.93.13.6349

St Leger, R.J. (1996). Integument as a barrier to microbial infections. In Physiology of the Insect Epidermis (pp. 284–306). Australia: CSIRO.

Stolarek, P., Różalska, S. and Bernat, P. (2019). Lipidomic adaptations of the *Metarhizium robertsii* strain in response to the presence of butyltin compounds. Biochim. Biophys. Acta, 1861(1), 316–326. https://doi.org/10.1016/j.bbamem.2018.06.007

Szewczyk, R., Kuśmierska, A. and Bernat, P. (2018). Ametryn removal by *Metarhizium brunneum*: Biodegradation pathway proposal and metabolic background revealed. Chemosphere, 190: 174–183. https://doi.org/10.1016/j.chemosphere.2017.10.011

Takei, T., Yoshida, M., Ohnishi-Kameyama, M. and Kobori, M. (2005). Ergosterol Peroxide, an Apoptosis-Inducing Component Isolated from *Sarcodon aspratus*(Berk.) S. Ito. Biosci. Biotechnol. Biochem., 69(1): 212–215. https://doi.org/10.1271/bbb.69.212

Téllez-Jurado, A., Ramírez, M.G.C., Flores, Y.M., Torres, A.A. and Arana-Cuenca, A. (2009). Mecanismos de acción y respuesta en la relación de hongos entomopatógenos e insectos. Rev. Mex. Micol., 30(30): 73–80. http://repositoriodigital.academica.mx/jspui/handle/987654321/82800

Thi, H.N., Nguyễn, Q.N., Thi, N.Q.D., Nguyễn, N.T. and Duy, A. (2023). Mass production of entomopathogenic fungi *Purpureocillium lilacinum* PL1 as a biopesticide for the management of *Amrasca devastans* (Hemiptera: Cicadellidae) in okra plantation. Egypt. J. Bìol. Pest Control., 33(1): https://doi.org/10.1186/s41938-023-00730-y

Tkaczuk, C., Harasimiuk, M., Król, A. and Bereś, P. (2015). The effect of selected pesticides on the growth of entomopathogenic fungi *Hirsutella nodulosa* and *Beauveria bassiana*. J. Ecol. Eng., 16:177–183. https://doi.org/10.12911/22998993/2952

Tkaczuk, C., Majchrowska-Safaryan, A., Panasiuk, T. and Tipping, C. (2019). Effect of selected heavy metal ions on the growth of entomopathogenic fungi from the genus *Isaria*. Appl. Ecol. Environ. Res., 17(2): 2571–2582. https://doi.org/10.15666/aeer/1702_25712582

Tlécuitl-Beristain, S., Viniegra-González, G., Díaz-Godínez, G. and Loera, O. (2009). Medium Selection and Effect of Higher Oxygen Concentration Pulses on *Metarhizium anisopliae* var. *lepidiotum* Conidial Production and Quality. Mycopathologia, 169(5): 387–394. https://doi.org/10.1007/s11046-009-9268-7

Todorova, S., Coderre, D., Duchesne, R. and Côté, J. (1998). Compatibility of *Beauveria bassiana* with selected fungicides and herbicides. Environ. Entomol., 27(2): 427–433. https://doi.org/10.1093/ee/27.2.427

Vega, F.E. (2018). The use of fungal entomopathogens as endophytes in biological control: a review. Mycologia, 110(1): 4–30. https://doi.org/10.1080/00275514.2017.1418578

Vega, F.E., Goettel, M.S., Blackwell, M., Chandler, D., Jackson, M.A., Keller, S., Koike, M., Maniania, N.K., Monzón, A., Ownley, B.H., Pell, J.K., Rangel, D.E. and Roy, H.E. (2009). Fungal entomopathogens: new insights on their ecology. Fungal Ecol., 2(4): 149–159. https://doi.org/10.1016/j.funeco.2009.05.001

Vinson, S.B. (1993). Suppression of the insect immune system by parasitic hymenoptera. In: Insect Immunity. Series Entomologica, vol 48., Pathak, J.P.N. (eds), (pp. 171–187). Springer, Dordrecht., https://doi.org/10.1007/978-94-011-1618-3_12

Wang, C., Typas, M.A. and Butt, T.M. (2002). Detection and characterization ofpr1virulent gene deficiencies in the insect pathogenic fungus *Metarhizium anisopliae*. FEMS Microbiol. Lett., 213(2): 251–255. https://doi.org/10.1111/j.1574-6968.2002.tb11314.x

Wang, C. and St Leger, R. J. (2005). Developmental and Transcriptional Responses to Host and Nonhost Cuticles by the Specific Locust Pathogen *Metarhizium anisopliae* var. *acridum*. Eukaryot. Cell, 4(5): 937–947. https://doi.org/10.1128/ec.4.5.937-947.2005

Wang, C. and St Leger, R.J. (2007a). The MAD1 Adhesin of *Metarhizium anisopliae* Links Adhesion with Blastospore Production and Virulence to Insects, and the MAD2 Adhesin Enables Attachment to Plants. Eukaryot. Cell, 6(5): 808–816. https://doi.org/10.1128/ec.00409-06

Wang, C. and St Leger, R.J. (2007b). A scorpion neurotoxin increases the potency of a fungal insecticide. Nat. Biotechnol. 25(12): 1455–1456. https://doi.org/10.1038/nbt1357

Wang, X., Xu, J., Wang, X., Qiu, B., Cuthbertson, A.G.S., Du, C., Wu, J. and Ali, S. (2019a). *Isaria fumosorosea*-based zero-valent iron nanoparticles affect the growth and survival of sweet potato whitefly, *Bemisia tabaci* (Gennadius). Pest Manag. Sci., 75(8): 2174–2181. https://doi.org/10.1002/ps.5340

Wang, Z., Chen, Z., Jiang, Z., Luo, P., Liü, L. et al., (2019b). Cordycepin prevents radiation ulcer by inhibiting cell senescence via NRF2 and AMPK in rodents. Nat. Commun., 10(1): https://doi.org/10.1038/s41467-019-10386-8

Weng, Q., Zhang, X., Chen, W. and Hu, Q. (2019). Secondary Metabolites and the Risks of *Isaria fumosorosea* and *Isaria farinosa*. Molecules, 24(4): 664. https://doi.org/10.3390/molecules24040664

Wheelis, M.L. (2002). Biological warfare at the 1346 siege of Caffa. Emerg. Infect. Dis., 8(9): 971–975. https://doi.org/10.3201/eid0809.010536

Wraight, S.P., Carruthers, R.I., Jaronski, S.T., Bradley, C., Garza, C. and Galaini-Wraight, S. (2000). Evaluation of the Entomopathogenic Fungi *Beauveria bassiana* and *Paecilomyces fumosoroseus* for Microbial Control of the Silverleaf Whitefly, *Bemisia argentifolii*. Biol. Control, 17(3): 203–217. https://doi.org/10.1006/bcon.1999.0799

Wu, S., Kostromytska, O.S., Goble, T.A., Hajek, A.E. and Koppenhöfer, A.M. (2019). Compatibility of a microsclerotial granular formulation of the entomopathogenic fungus *Metarhizium brunneum* with fungicides. BioControl, 65(1): 113–123. https://doi.org/10.1007/s10526-019-09983-9

Yeo, H., Pell, J.K., Alderson, P.G., Clark, S.J. and Pye, B.J. (2003). Laboratory evaluation of temperature effects on the germination and growth of entomopathogenic fungi and on their pathogenicity to two aphid species. Pest Manag. Sci., 59(2): 156–165. https://doi.org/10.1002/ps.622

Zheng, S. and Dicke, M. (2008). Ecological Genomics of Plant-Insect Interactions: From Gene to Community. Plant Physiol., 146(3): 812–817. https://doi.org/10.1104/pp.107.111542

Zimmermann, G. (2007a). Review on safety of the entomopathogenic fungi *Beauveria bassiana* and *Beauveria brongniartii*. Biocontrol Sci. Technol., 17(6): 553–596. https://doi.org/10.1080/09583150701309006

Zimmermann, G. (2007b). Review on safety of the entomopathogenic fungus *Metarhizium anisopliae*. Biocontrol Sci. Technol., 17(9): 879–920. https://doi.org/10.1080/09583150701593963

2 Historical Perspectives on Biopesticides

Matangi Mishra[1*] and Kumar Gaurav[2]

1. Introduction

Pest control involves the regulation and management of pests, encompassing animals, plants, and microorganisms, that negatively impact human activities and plant productivity (Dent et al., 1995). Human actions are dictated by the severity of the harm, which ranges from tolerance, deterrence, and management to elimination or containment of the pests. Pest control strategies are often integrated into Integrated Pest Management (IPM) approaches. Common pests in residential, rural, and urban areas include rodents, insects, mollusks, and other organisms that damage structures and plant products. Various methods, such as quarantine, repulsion, and physical or chemical removal, are employed to control these pests (Staunton et al., 2008). Additionally, biocontrol methods, including sanitation and sterilization plans, may also be utilized.

The history of insect control is a broad and continuously evolving field, marked by various methods employed by humans over the centuries. In countries like India, agriculture is a pivotal activity, not only meeting the food needs of the growing population but also contributing significantly to the national economy. The adoption of Green Revolution (GR) technology between 1960 and 2000 substantially increased crop yields per hectare and enhanced food supply in developing nations by 12–13%. India and Southeast Asia were among the first emerging economies to demonstrate the impact of GR on rice yield varieties.

The utilization of chemical fertilizers and biopesticides has had adverse effects on the environment, altering factors such as water hardness, soil fertility, insect resistance, genetic variation in plants, and the accumulation of toxic residues in animal feed and the food chain, leading to increased health concerns. Addressing these issues requires the implementation of appropriate policies. Biopesticides and biofertilizers offer environmentally friendly solutions to these challenges. With the exception of those affecting the nervous systems of pests, biopesticides are considered "mild pesticides," consisting of natural compounds that suppress the growth and proliferation of pest populations through various mechanisms. They are categorized into plant-incorporated protectants (PIP), biochemical biopesticides, and microbial biopesticides.

[1] Department of Entomology, School of Agriculture, Lovely Professional University, Phagwara-144411, Punjab, India.

[2] Department of English, School of Social Science and Humanities, Lovely Professional University, Phagwara-144411, Punjab, India.

* Corresponding author: matangi.28192@lpu.co.in

Microbial biopesticides, particularly *Bacillus thuringiensis* (Bt), dominate 90% of the market, although other microorganisms like Baculovirus, *Beauveria bassiana, Chlorella, Steinernema,* and *Nosema* have also shown significant roles. Biochemical biopesticides, as described by Kumar (2012) and Reddy and Chowdary (2021), are natural or synthetically produced substances with active components that control pests without harming humans, the environment, or non-target organisms.

Utilizing natural elements such as plants, animals, microorganisms, or minerals as biopesticides is a valuable approach to pest management. They are often used as alternatives to conventional chemical pesticides and are considered environmentally sustainable. Biopesticides, derived from microorganisms, plants, and biochemicals, offer numerous benefits, including reducing chemical exposure risk, minimizing fertilizer runoff, requiring fewer applications, having fewer adverse effects on beneficial insects, and being biodegradable. This chapter discusses the history of insect pest control and the significance of microbial and plant biopesticides.

2. History of Pest Control

2.1 Early Civilization

Pest control methods have been intertwined with agriculture since ancient times, as the necessity of pest-free crops has always been paramount. Early civilizations such as the ancient Egyptians, Greeks, and Romans employed simple techniques like manual pest removal and the use of natural substances such as sulfur and arsenic (Koul and Dhaliwal, 2003). Cats were employed to manage rodent pests in grain storage as far back as 3000 BC in Egypt, while ferrets were domesticated in Europe around 1500 BC for their mousing abilities. Mongooses were likely introduced into households by ancient Egyptians to control rats and snakes. Ducks were utilized to consume pests in Chinese rice fields around 4000 BC, as depicted in ancient cave art. In 1762, an Indian mynah was transported to Mauritius to manage locusts, and citrus trees in Burma were cultivated alongside bamboo to allow ants to control caterpillars. Ladybirds were deployed to control scale insects in California orange orchards in the 1880s, followed by further biocontrol trials.

Traditional techniques were predominantly utilized initially, as they offered relatively simple solutions such as burning or plowing under weeds and managing larger competing herbivores. Practices such as crop rotation, companion planting (mixed-cropping, intercropping, or multi-cropping), and selective breeding of pest-tolerant cultivars have long histories.

Chemical pest management gained prominence as agriculture became increasingly industrialized and mechanized during the 18th and 19th centuries, particularly with the introduction of insecticides like pyrethrum. The 20th century saw the discovery of many chemical pesticides, including DDT and herbicides, which further accelerated this trend. However, the widespread adoption of DDT, hailed as a cheap and effective pesticide, led to a decline in biocontrol experimentation. By the 1960s, issues such as chemical resistance and environmental degradation had emerged, sparking a renewed interest in biological control (van Emden et al., 2004). This resurgence in biological control was accompanied by a renewed appreciation for traditional and biocontrol pest management methods during the 20th century and beyond. Table 1 provides the chronological development of the use of insecticides in plant protection.

Table 1 History of Insecticide development

Period	*Development/Event*	*Insecticide*	*Reference*
Ancient	Manual pest removal, anduse of natural substances	Sulfur and arsenic	Koul and Dhaliwal, 2003
Middle Ages and Renaissance	Cultural and botanical insecticides	Nicotine and pyrethrum	Georghiou, 1990
19th century	Use of chemical insecticides	Paris green, and lead arsenate	Nauen, 2007
Mid 20th century	Discovery and widespread use of synthetic pesticides	DDT, organophosphates, and carbamates	Carson, 1962
1960s-1970s	Environmental concerns and regulations	DDT was banned	Pimentel (1971)
1980s-1990s	Resurgence of biological control	Increased emphasis on natural enemies	DeBach and Rosen (1991)
Late 20th century to present	Biotechnological advances	Genetic modification, and RNA interference	Romeis et al. (2011)
21st century	Sustainable pest management	Emphasis on eco-friendly alternatives and biopesticides	Kogan and Jepson (2007)

The evolution of insecticide development reflects a broader narrative of human ingenuity, environmental stewardship, and the ongoing quest for balance between agricultural productivity and ecological health. By learning from the past and embracing innovative approaches, modern pest management continues to evolve towards more sustainable and effective practices. This historical perspective underscores the importance of adaptability and foresight in addressing the complex challenges of pest control in an ever-changing world.

Farmers frequently face challenges in harvest management due to losses caused by insect pests. The exact origins of insecticide use are unknown, but it is believed to date back a considerable time. In their efforts to protect crops from insects, farmers likely turned to natural substances known to deter pests. Botanicals appear to have been among the earliest insecticides used by ancient civilizations. For example, the flowers of Dalmatian pyrethrum contain up to 1.5% pyrethrin, an active bioinsecticide. This chemical was historically used as an insecticide in China and Persia during the Middle Ages (Davies et al., 2007). Additionally, various other plants have been commonly employed as pesticides (Isman, 2006). Elderberry flowers, tobacco aqueous extract, and wormwood aqueous extract have been shown to effectively repel pests (Ignatowicz and Wesołowska, 1994; Dagaev, 1997; Sohail et al., 2012).

2.2 Chemical Pesticides

In 1000 BCE, humans began using natural compounds to combat insect pests. One such compound was inorganic sulfur, applied through fumigation. Homer referred to this "divine cleansing" process in both "The Iliad" and "The Odyssey," where sulfur was used to eliminate lice. Arsenic was first used as insect bait in the 900s AD, followed by lead arsenate ($PbHAsO_4$), borax (Na_2B4O_7), and cryolite (Na_3AlF_6) as dehydrators and cellular poisons (Popov et al., 2003).

DDT became widely employed in plant protection until the second half of the 20th century, when it was replaced by organophosphates (cyanophos, dichlorvos, and fonofos) and carbamates (aldicarb, carbaryl, and carbofuran). Despite their negative environmental impact, carbamates and organophosphates are still widely used today (comprising 19% of the global market) and play a crucial role in pest control (Casida and Durkin, 2013). Notably, the manufacture of carbaryl is linked to one of history's greatest tragedies. In 1984, an explosion at the Union Carbide Corporation (UCC) plant in Bhopal, India, released deadly methyl isocyanate (MIC), resulting in the deaths of up to 3,800 workers on the first day (Yamamoto, 1999; Broughton, 2005).

Pyrethrins were reintroduced as insecticides following the synthesis of allethrin in 1949, forming the basis for the first generation of pyrethroids. These were popular among health-conscious individuals due to their low toxicity to warm-blooded organisms. However, pyrethroids such as cypermethrin, deltamethrin, and permethrin faced significant challenges in the 1970s due to their reduced efficacy when exposed to ultraviolet light. Despite this, their unique characteristics prevented environmental accumulation, and they remain widely used today for plant protection.

Plant protection chemicals gained widespread use in the mid-19th century. Mixed copper acetoarsenite, known as Paris green, was successfully used against the Colorado potato beetle in 1871 (Alyokhin, 2009). Until the mid-20th century, Paris green was also commonly used to control malaria vectors, specifically Anopheles mosquitoes (Majori, 2012).

Neonicotinoids are currently the most widely used insecticides (Yamamoto, 1999). They move freely through plant tissues, protecting all parts of the plant. According to Goulson (2013), neonicotinoids effectively manage insect pests by acting as neurotoxins to most arthropods. They bind to nicotinic acetylcholine receptors, overstimulating nerve cells and paralyzing insects. The market for neonicotinoids has grown significantly over the past 20 years, with numerous variants available, including acetamiprid, thiamethoxam, dinotefuran, and thiacloprid.

2.3 Biopesticides

Efforts to develop and improve existing insecticides are ongoing, as insects frequently develop resistance to pesticides. This necessitates the continuous development of new pesticides. Currently, new chemical insecticides are being both developed and marketed. These include phenylpyrazoles, fourth-generation pyrethroids, avermectins, botanical insecticides such as azadirachtin, diamides, formamidine insecticides like amitraz, insect growth regulators, macrocyclic lactone insecticides, pyrazole insecticides, and spinosyns. The post-genomic era of plant protection has created opportunities for innovative formulations, including genetically engineered baculoviruses that more effectively target insect pests (Inceoglu et al., 2006; Rosell et al., 2008).

Today, biopesticides under development could become highly effective alternatives to chemical pesticides (Senthil-Nathan, 2015; Pathan et al., 2021; Hanif et al., 2022). India, in particular, has a wealth of readily available local plants that can be processed to enhance the use of biopesticides, such as beshram, neem, garlic, Triphala, and *Pinus kesia*. Currently, several biopesticides, including Bacillus thuringiensis, neem-based insecticides, NPV, and *Trichoderma*, are registered and used as biocontrol agents in India (Table 2). Examples of natural minerals, semiochemicals, and plant and insect growth regulators are also in use (Ghongade and Sangha, 2021; Singh et al., 2021).

Transgenic plants, known as plant-incorporated protectants (PIP), make plants resistant to insect attacks. The insecticidal molecules used in PIP technology include double-stranded ribonucleic acid (dsRNA), protease from Baculovirus, α-amylase inhibitors, toxic complex (Tc) proteins from Bt, Mir1-CP from maize, and compounds from *Photorhabdus* and *Xenorhabdus* (Shingote et al., 2013).

Table 2 Potential plant extracts for use as biopesticides.

Plant extract	*Application*
Adathoda Kashayam and *Pudhina* Kashayam	Leaf folder, bacterial leaf blight, and *Helminthosporium* leaf spot
Andrographis Kashayam and *Sida* Kashayam	Aphids and borers in brinjal, and ladies' finger
Barley-*Sesamum*-Horsegram Kashayam	Fruit yield-enhancer
Cow's urine arkam and sweet flag arkam	Bacterial leaf blight, *Helminthosporium* leaf spot, vein clearing disease, and *Fusarium* wilt
Garlic arkam	Leaf folder, bacterial leaf blight, and *Helminthosporium* leaf spot
Melia azedarach (Chinaberry or Persian lilac tree)	Antifeedant, growth-regulation, insect-deforming, and growth-modifying properties
Neem seed extract (for all crops)	Leaf folder, aphids, jassids, fruit borer, and stem borer
Thriphalakashayam	Bacterial leaf blight and *Helminthosporium* leaf spot

2.4 Methods of Pest Control

A number of pest control measures have been applied to control insect pests. Simple techniques like the use of natural enemies to chemical, crop rotation, mechanical, genetic, and molecular techniques have been developed (Table 3).

Table 3 Pest control methods (with specific examples).

Pest control method	*Description*	*Example*	*Reference*
Mechanical control	Physical methods to remove or exclude pests.	Insect traps: Using traps to capture and reduce pest populations.	Beroza and Green (1966)

Contd.

Table 3 *Contd.*

Pest control method	*Description*	*Example*	*Reference*
Cultural control	Practices that modify the environment to make it less favorable for pests.	Crop rotation: Alternating the type of crops planted in a field to disrupt the life cycle of pests.	Cook and Baker (1983)
Genetic control	Using genetic methods to control pests.	Sterile insect technique (SIT): Releasing sterilized male insects to reduce reproduction in pest populations.	Dyck et al. (2005)
Chemical control	Use of chemical substances to kill or repel pests.	Neonicotinoid insecticides: Neonicotinoids like imidacloprid are commonly used to control various insect pests in crops.	Jeschke et al. (2011)
RNA interference (RNAi)	Using RNA molecules to interfere with the expression of specific genes in pests.	RNAi for aphid control: Developing transgenic plants that produce RNA molecules to disrupt aphid feeding.	Baum and Roberts (2014)
Biological control	Introducing natural enemies to control pest populations		Gurr et al. (2016)

3. Impact of Chemical Pesticides

The history of pest control has not been without negative impacts, particularly in the context of widespread use of synthetic pesticides. Some notable negative consequences are as follows.

3.1 Environmental Pollution

The use of chemical pesticides, such as DDT and organophosphates, has led to environmental pollution. Runoff from fields and improper disposal of pesticides have contaminated water sources, soil, and affected non-target organisms. Learning from the negative impacts of past practices is crucial to shaping future approaches. Sustainable and integrated pest management strategies and the adopting environmentally friendly alternatives are essential for minimizing adverse effects.

3.2 Biodiversity Loss

Indiscriminate pesticide usage has contributed to the decline of non-target species, especially aquatic organisms, beneficial insects, and birds. Such ecosystem disruptions can lead to imbalances, like the loss of biodiversity and ecosystem processes.

3.3 Pesticide Resistance

Overreliance on certain pesticides has been responsible for the development of resistance in pest populations. Resistant pests are able to survive and reproduce, over time reducing the effectiveness of pesticides.

3.4 Secondary Pest Outbreaks

Eliminating specific pests through chemical means can create conditions for other pests to thrive. This can lead to secondary pest outbreaks, requiring additional control measures.

3.5 Impact on Human Health

Pesticide exposure poses health risks to farmworkers, nearby communities, and consumers. Long-term contact with certain pesticides has been linked to several health concerns, including cancer, respiratory problems, and neurological disorders.

3.6 Impact on Environmental Health

Pesticides can persist in the environment, affecting the quality of soil and water. This contamination can have long-term impacts on agricultural ecosystems and surrounding habitats. Beneficial insects, such as pollinators and natural predators, can be adversely affected by pesticide applications, disrupting ecological balances and natural pest control mechanisms.

4. Environmental Safety of Plant Biopesticides

Biopesticides are naturally occurring microorganisms that are usually nontoxic and possess biochemical properties that enable them to combat pests. These biopesticides, which can be natural enemies or their byproducts (semiochemicals), phytochemicals, or microbial products, are employed to control pests harmful to plants. They pose less risk to both the environment and public health. Several plant species, or their derived products, serve as significant sources of biopesticides. Natural compounds created through genetic modification of plants, known as plant-incorporated protectants, include examples such as the introduction of the Bt gene, protease inhibitors, lectins, chitinase, and other substances into plant genomes. This enables the resulting transgenic plants to produce chemical compounds that target and kill specific pests.

The use of biopesticides offers substantial benefits for public health and agricultural practices. Generally, they are less harmful to the environment and are designed to target a specific pest species, though occasionally, a small number of non-target organisms may also be affected. Biopesticides are often effective in small quantities and break down rapidly, leading to reduced exposure and fewer pollution issues. When used appropriately, biopesticides can make a significant contribution to Integrated Pest Management (IPM) systems.

4.1 Plant Products

Plants and insects have coexisted for over half a million years and have both developed defensive and offensive mechanisms (Senthil-Nathan, 2015). There are several plant-based biopesticides produced using neem and *Melia* spp. Many biochemicals serve to control or deter insect pests (Table 4). For example, pheromones secreted by insects are highly species-specific and can serve as substances that disrupt processes of attraction, congregation, and mating.

The use of botanicals to safeguard agricultural produce and the environment is becoming increasingly common in response to the global issue of pesticidal pollution. Neem-based products are effective against over 350 spp. of arthropods, 12 spp. of nematodes, 15 spp. of fungi, 3 varieties of viruses, 2 spp. of snails, and 1 sp. of crustacean. Neem is considered the most environmentally friendly and effective biopesticides.

Table 4 Plant-based products registered as biopesticides (modified from Salma et al., 2011).

Plant product	*Target pest*
Limonene and linalool hydroperoxides (from citrus fruits)	Fleas, aphids and mites, fire ants, several types of flies, paper wasps, and house crickets
Azadirachtin (from neem, *Azadirachta indica*)	A variety of sucking and chewing insects
Pyrethrum and pyrethrins from Chrysanthemum (*Tanacetum cinerariaefolium*)	Ants, aphids, roaches, fleas, flies, and ticks
Rotenone (from *Derris*, *Lonchocarpus Mundulea*, and *Tephrosia* spp.)	Leaf-feeding insects (aphids, asparagus beetles, bean leaf beetles, Colorado potato beetles, cucumber beetles, flea beetles, and strawberry leaf beetles), caterpillars, fleas, and lice
Ryania (from *Ryna speciosa*)	Caterpillars (European corn borer, corn earworm), and thrips
Sabadilla (from *Sabadilla officinalis*)	Squash bugs, harlequin bugs, thrips, caterpillars, leaf hoppers, and stink bugs

4.2 Vrakshayurveda

The traditional Indian body of knowledge concerning plants is known as Vrakshayurveda. It includes methods for planting seeds, growing plants, controlling pests and diseases, and preventive and proactive practices to nurture healthy plants and increase disease resistance. The Centre for Indian Knowledge Systems (CIKS), Chennai (India), has evaluated plant extracts of various species capanle of curing plant diseases (*Aloe*, *Calotropis*, custard apple, garlic, ginger, leucas, neem, onion, papaya, persian lilac, poison nut, pongam, sweetflag, tobacco, basil, turmeric, and *Vitex*).

4.3 Advantages of Biopesticides

Many plant-based biopesticides are known to disrupt growth regulation, cause gut disruptions, act as metabolic toxins, neuromuscular poisons, and function as non-specific multisite inhibitors. These mechanisms are some of the ways biopesticides affect pests (Sparks and Nauen, 2015; Dar et al., 2021). These diverse modes of action prevent the development of resistance, which is common with chemical pesticides. Consequently, there are increasing restrictions on synthetic pesticides to reduce their usage. For instance, the number of active chemicals in conventional pesticides dropped to between 250 and 1,000 from 2001 to 2009. The introduction of new conventional pesticides also declined from 70 in 2000 to 28 in 2012 (McDougall, 2013).

There is a clear correlation between the decrease in traditional pesticide usage and the rise in demand for biopesticides. Addressing pest resistance with biopesticides has become a complementary practice to using synthetic pesticides. Biopesticides offer various advantages: they are eco-friendly, biodegradable, target-specific, have minimal or no post-harvest contamination, show stability to abiotic stresses, and are compatible with other pest management strategies.

4.4 Limitations of Biopesticides

Farmers are accustomed to readily available and conveniently packaged insecticides. Although they recognize the benefits of using plant-based products instead of chemical pesticides, it will take time for these alternatives to be widely adopted and appreciated. One essential task is to process these plant-based products and provide them to farmers in a user-friendly form. In the near future, biopesticides are expected to replace chemical pesticides without negatively impacting productivity or yield, thereby maximizing their potential (Mishra et al., 2020).

However, several limitations currently hinder the widespread adoption of biopesticides: their higher cost compared to commercial pesticides, production challenges, difficulty in meeting demand, lack of standard production techniques, susceptibility to environmental factors, and slower action. Despite these challenges, crude plant extracts can offer farmers, particularly in rural areas, a practical means of protecting plants and enhancing production. Scientific advancements are expected to address these limitations in the coming years (Stevenson et al., 2017). Integrating pesticides, especially biopesticides, into Integrated Pest Management (IPM) systems maximizes their efficacy (Quarles, 2013).

5. Biopesticides in India

As of 2005, biopesticides accounted for just 2.89 percent of India's total pesticide market. However, currently, this percentage is projected to rise significantly. According to the Indian Insecticide Act 1968, only 12 biopesticides have been registered (Table 5). The main biopesticides manufactured and utilized in India are *Bacillus thuringiensis*, neem derived, NPV, and *Trichoderma*. In contrast, over 190 synthetic products are registered to be used as chemical pesticides in different parts of the world. Under the Integrated pest management, in the Indian subcontinent, 15 entomopathogenic fungi are or insect control (Pathan et al., 2021). Prominent fungi and byproducts in use as biopesticides include *Beauveria, Hirsutella, Lecanicillium*, and *Metarhizium*, with about 90, 3, 70, and 40 formulations, respectively.

Table 5 Biopesticides registered as insecticides in India {(Source: Indian Government. (1968). The Insecticides Act, 1968. Retrieved from https://www.indiacode.nic.in/bitstream/123456789/1551/1/A1968-46.pdf)}

Bacteria
Bacillus thuringiensis var. *israelensis*
Bacillus thuringiensis var. *kurstaki*
Bacillus thuringiensis var. *galleriae*
Bacillus sphaericus
Pseudomonas fluoresens
Fungi
Beauveria bassiana
Trichoderma harzianum
Trichoderma viride
Viruses
NPV of *Helicoverpa armigera*
NPV of *Spodoptera litura*
Plant-based
Cymbopogan
Neem-based pesticides

6. Conclusion and Future Concerns

The history of pest control represents a dynamic evolution, shaped by ancient practices, chemical innovations, and environmental considerations. From manual removal and botanical insecticides to the era of synthetic pesticides and the subsequent environmental awakening, the journey has culminated in Integrated Pest Management (IPM) and contemporary sustainable approaches. The resilience and adaptability of agricultural practices throughout history underscores humanity's ongoing commitment to effective, environmentally conscious pest control. As we advance into the future, the integration of cutting-edge technologies and a holistic approach heralds a promising era of balanced coexistence between agriculture and nature in the realm of pest management.

The history of pest control suggests a trajectory marked by continual innovation and sustainable practices. Emerging technologies such as artificial intelligence, precision agriculture, and nanotechnology promise to revolutionize pest management, offering more targeted and eco-friendly solutions. The integration of genetic tools, including CRISPR technology, may provide precise and species-specific control measures. As environmental awareness grows, emphasis on biological control methods and organic farming practices is likely to increase. The history of pest control serves as a roadmap, guiding us towards a future where intelligent, sustainable, and nature-aligned strategies foster a harmonious balance between agriculture and the ecosystems we depend upon.

India's abundant biodiversity is a key asset, consistently offering a broad range of biopesticides that are suitable for widespread application in agriculture. The advancing health consciousness among Indians has also led to a desire for organic food. This suggests that the biopesticide industry has enormous room to grow. The extremely diversified indigenous populations in India possess a wealth of traditional knowledge that could offer important hints for creating more advanced and potent biopesticides. The importance of residue-free produce and organic farming will undoubtedly encourage farmers to employ biopesticides more frequently in agricultural practice.

References

Abu-Dieyeh, M.H. and Watson, A.K. (2007). Efficacy of *Sclerotinia minor* for dandelion control: effect of dandelion accession, age and grass competition. Weed. Res., 47: 63–72. 10.1111/j.1365-3180.2007.00542

Alyokhin, A. (2009). Colorado potato beetle management on potatoes: current challenges and future prospects. Fruit, Veg. Cereal Sci. Biotech., 3 (1): 10–19.

Baum, J.A. and Roberts, J.K. (2014). Progress towards RNAi-mediated insect pest management. Adv. Insect Physiol., 47: 249–295.

Beroza, M. and Green, M.B. (1966). Pest control: biological, physical, and selected chemical methods. Ann. Rev. Entomol., 11(1): 183–212.

Broughton, E. (2005). The Bhopal disaster and its aftermath: a review. Environ. Health 4: 1–6.

Carson, R. (1962). Silent Spring. Houghton Mifflin, USA, p. 336.

Casida, J.E., Durkin, K.A. (2013). Anticholinesterase insecticide retrospective. Chem. Biol. Interact., 203 (1): 221–225.

Cook, R.J. and Baker, K.F. (1983). The nature and practice of biological control of plant pathogens. The American Phytopathological Society, USA, p. 539.

Dagaev, M.V. (1997). Baleen aggressors or everything about the war on cockroaches. Iayza, Lan', Moskva, pp. 112.

Dar, S.A., Wani, S.H., Mir, S.H., Showkat, A., Dolkar, T. and Dawa, T. (2021). Biopesticides: mode of action, efficacy and scope in pest management. J. Adv. Res. Biochem. Pharmacol., 4: 1–8.

Davies, T.G., Field, L.M., Usherwood, P.N. and Williamson, M.S. (2007). DDT, pyrethrins, pyrethroids and insect sodium channels. IUBMB Life, 59 (3): 151–162.

DeBach, P. and Rosen, D. (1991). Biological control by natural enemies. Cambridge University Press, UK, p. 456.

Dent, N.C., Elliott, J.A., Farrell, A.P., Gutierrez, J., van Lenteren, C., Walton, M.P. and Wratten, S.D. (1995). Integrated pest management. Chjapman& Hall, p. 356.

Dyck, V.A., Hendrichs, J. and Robinson, A.S. (2005). Sterile insect technique: principles and practice in area-wide integrated pest management. Springer Dordrecht, Ther Netherlands, p. 787.

Georghiou, G.P. (1990). Overview of insecticide resistance. American Chemical Society Symposium Series, Volume 421. 10.1021/bk-1990-0421.ch002

Ghongade, D.S. and Sangha, K.S. (2021). Efficacy of biopesticides against the whitefly, *Bemisia tabaci* (Gennadius) (Hemiptera: Aleyrodidae), on parthenocarpic cucumber grown under protected environment in India. Egypt. J. Biol. Pest Control., 31: 1–11. 10.1186/s41938-021-00365-x

Gonçalves, A.L. (2021). The use of microalgae and cyanobacteria in the improvement of agricultural practices: a review on their biofertilising, biostimulating and biopesticide roles. Appl. Sci., 11:871. 10.3390/app11020871

Goulson, D. (2013). Review: An overview of the environmental risks posed by neonicotinoid insecticides. J. Appl. Ecol., 50 (4): 977-987.

Gurr, G.M., Lu, Z., Zheng, X., Xu, H., Zhu, P. et al. (2016). Multi-country evidence that crops diversification promotes ecological intensification of agriculture. Nat. Pl., 2:16014. https://doi.org/10.1038/nplants.2016.14

Hanif, K., Zubair, M., Hussain, D., Ali, S., Saleem, M. et al. (2022). Biopesticides and insect pest management. Int. J. Trop. Insect Sci., 42: 3631-3637. https://doi.org/10.1007/s42690-022-00898-0

Ignatowicz, S. and Wesołowska, B. (1994). Potential of common herbs as grain protectants: repellent effect of herb extracts on the granary weevil, *Sitophilus granarius* (L). pp. 790–794. In: Proceedings of the 6th International Working Conference on Stored-Product Protection, 17–23 April 1994, Volume 2, Canaberra, Australia.

Inceoglu, A.B., Kamita, S.G. and Hammock, B.D. (2006). Genetically modified baculoviruses: a historical overview and future outlook. Adv. Virus Res., 68: 323–360.

Isman, M.B. (2006). The role of botanical insecticides, deterrents and repellents in modern agriculture and an increasingly regulated world. Annu. Rev. Entomol., 51 (1): 45–66.

Jeschke, P., Nauen, R., Schindler, M. and Elbert, A. (2011). Overview of the status and global strategy for neonicotinoids. J. Agric. Food Chem., 59(7), 2897–2908.

Kogan, M. and Jepson, P.C. (2007). Perspectives on the need for knowledge-intensive approaches to pest management in the developing world. Crop Protec., 26(3), 313–322.

Koul, O. and Dhaliwal, G.S. (2003). Advances in Biopesticide Research, Volume III- Predators and Parasitoids. Taylor and Francis, UK, p. 191.

Kumar, S. (2012). Biopesticides: a need for food and environmental safety. J. Biofertil. Biopestic., 3: 1–3. 10.4172/2155-6202.1000e107

Leahy, J., Mendelsohn, M., Kough, J., Jones, R. and Berckes, N. (2014). Biopesticide oversight and registration at the US Environmental Protection Agency, in Biopesticides: State of the Art and Future Opportunities, Gross, A.D., Coats, J.R., Duke, S.O. and Seiber, J.N, (Eds) (Washington, DC: ACS), pp. 3–18.

Liu, X., Cao, A., Yan, D., Ouyang, C., Wang, Q. and Li, Y. (2019). Overview of mechanisms and uses of biopesticides. Int. J. Pest. Manag., 67: 65–72. 10.1080/09670874.2019.1664789

Majori, G. (2012). Short history of malaria and its eradication in Italy. Mediterr. J. Hematol. Infect. Dis., 4(1): e2012016.

Marrone, P.G. (2019). Pesticidal natural products–status and future potential. Pest. Manag. Sci., 75: 2325–2340. 10.1002/ps.5433

McDougall, P. (2013). R&D Trends for Chemical Crop Protection Products and the Position of the European Market. A Consultancy Study Undertaken for ECPA. (https://issuu.com/cropprotection/docs/r_and_d_study_2013_v1.8_webversion_).

Mishra, J., Dutta, V. and Arora, N.K. (2020). Biopesticides in India: technology and sustainability linkages. 3 Biotech, 10: 1–12. 10.1007/s13205-020-02192-7

Nauen, R. (2007). Insecticide resistance in disease vectors of public health importance. Pest Manag Sci. 63(7):628-633.

Nuruzzaman, M., Liu, Y., Rahman, M.M., Dharmarajan, R., Duan, L., Uddin, A.F.M.J. and Naidu, R. (2019). Nanobiopesticides: composition and preparation methods, In: Nano-Biopesticides Today and Future Perspectives, Koul, O. (Ed.) (London: Academic Press), pp. 69–131.

Pathan, E.K., Ghormade, V., Tupe, S.G. and Deshpande, M.V. (2021). Insect pathogenic fungi and their applications: An Indian perspective. pp. 311-327. In: Satyanarayana, T., Deshmukh, S.K. and Deshpande, M.V. (eds.), Progress in Mycology – An Indian Perspective. Springer Singapore.

Pimentel, D. (1971). Ecological effects of pesticides on non-target species. Office of Science and Technology; 1971.

Popov, S.Ya., Dorozhkina L.A. and Kalinin V.A. (2003). Fundamentals of Chemical Plant Protection. Art-Lion, Moskva, 208 pp.

Quarles, W. (2013). New biopesticides for IPM and organic production. IPM Pract., 33: 1–20.

Radwan, E.M., El-Malla, M.A., Fouda, M.A. and Mesbah, R.A.S. (2018). Appraisal of Positive Pesticides Influence on pink bollworm larvae, *Pectinophora gossypiella* (Saunders). Egyp. Acad. J. Biol. Sci. F. Toxicol. Pest. Control., 10: 37–47. 10.21608/eajbsf.2018.17018

Ram, K. and Singh, R. (2021). Efficacy of different fungicides and biopesticides for the management of lentil wilt (*Fusarium oxysporum* f. sp. Lentis). J. Agri. Search, 8: 55–58.

Reddy, D.S. and Chowdary, N.M. (2021). Botanical biopesticide combination concept—a viable option for pest management in organic farming. Egypt. J. Biol. Pest. Control., 31: 1–10. 10.1186/s41938-021-00366-w

Romeis, J., Hellmich, R.L., Candolfi, M.P., Carstens, K., De Schrijver, A., Gatehouse, A.M., Herman, R.A., Huesing, J.E., McLean, M.A., Raybould, A. and Shelton, A.M., (2011). Recommendations for the design of laboratory studies on non-target arthropods for risk assessment of genetically engineered plants. Transgenic Res., 20(1), 1-22.

Salma, M., Ratul, C.R. and Jogen, C.K. (2011). A review on the use of biopesticides in insect pest management. Int J Sci Adv Tech., 1: 169–178.

Senthil-Nathan, S. (2015). A review of biopesticides and their mode of action against insect pests. pp. 49–63. In: Thangavel, P. and Sridevi, G. (eds.). Environmental Sustainability, Springer India.

Shingote, P.R., Moharil, M.P. and Dhumale, D.R. (2013). Distribution of vip genes, protein profiling and determination of entomopathogenic potential of local isolates of *Bacillus thuringiensis*. Bt Res., 4: 45–54.

Singh, K.D., Mobolade, A.J., Bharali, R., Sahoo, D. and Rajashekar, Y. (2021). Main plant volatiles as stored grain pest management approach: a review. J. Agric. Food Res., 4:100127. 10.1016/j.jafr.2021.100127

Sohail, M.N., Karimi, S.M., Asad, S., Mansoor, S., Zafar, Y. and Mukhtar, Z. (2012). Development of broad-spectrum insect-resistant tobacco by expression of synthetic cry1Ac and cry2Ab genes. Biotechnol. Lett., 34: 1553–1560.

Sparks, T.C. and Nauen, R. (2015). Mode of action classification and insecticide resistance management. Pest. Biochem. Physiol., 121: 122–128. 10.1016/j.pestbp.2014.1 1.014

Staunton, I., Hadlington, P. and Gerozisis, J. (2008). Urban Pest Management in Australia. UNSW Press, p. 432.

Steinkraus, D.C. and Tugwell, N.P. (1997).*Beauveria bassiana* (Deuteromycotina: Moniliales) effects on *Lygus lineolaris* (Hemiptera: Miridae). J. Entomol. Sci., 32: 79–90. 10.18474/0749-8004-32.1.79

Stevenson, P.C., Isman, M.B. and Belmain, S. (2017). Pesticidal plants in Africa: a global vision of new biological control products from local uses. Indus. Crops Products, 110: 2–9. doi: 10.1016/j.indcrop.2017.08.034

Tijjani, A., Bashir, K.A., Mohammed, I., Muhammad, A., Gambo, A., and Habu, M. (2016). Biopesticides for pests control: a review. J. Biopest. Agric., 3: 6–13.

Van Emden, H.F.; Service, M.W. (2004). Pest and Vector Control. Cambridge University Press. p. 147.

Wattimena, M.A.C. and Latumahina, F. S. (2021). Effectiveness of botanical biopesticides with different concentrations of termite mortality. J. Belantara, 4(1): 66–74. https://doi.org/10.29303/jbl.v4i1.630

Yamamoto, I. (1999). Nicotine to Nicotinoids: 1962 to 1997. p. 3–27. In: Nicotinoid Insecticides and the Nicotinic Acetylcholine Receptor. Tokyo: Springer Japan.

3 Exploration of Entomopathogenic Fungi in the Western Ghats

M.N. Siddaraju,[1*] Anitha Nagaraju,[2] B.S. Chandan[3] and Mohammed S. Mustak[3*]

1. Introduction

The Western Ghats, a UNESCO World Heritage Site, stretch 1,600 km along India's western coast and across six states (Gujarat, Maharashtra, Goa, Karnataka, Kerala, and Tamil Nadu). The area is well-known for its rich biodiversity, with many endemic species, and is regarded as one of the world's biodiversity hotspots. The Western Ghats are home to a wide variety of flora and wildlife, including plants, animals, reptiles, and fungi. Many endemic and rare species are supported by the region's rich and diverse habitats.

A varied fungal population, comprising different species that contribute to ecological processes like decomposition and mycorrhizal relationships, can be seen in the Western Ghats. Numerous fungi, including many species that have not yet been identified, find an ideal home in the region due to the dense and varied vegetation. Lichens, lichen-forming fungi, and mycorrhizal fungi, which develop symbiotic associations with plants, are a few of the noteworthy species found in the Western Ghats. In addition to adding to the biological richness of the area, these fungal species are of direct interest to the fields of ethnomycology, traditional medicine, and entomopathogenic and ecological restoration. Research is still being conducted to precisely identify and fully comprehend the fungal diversity in the Western Ghats.

Among the multiple uses of fungi, the entomopathogenic role is majorly beneficial in agriculture and forestry (Abdel-Raheem, 2020). Entomopathogenic fungi (EPF) from diverse classes play a crucial role as biological control agents for insect populations. They are among the first microorganisms that have been recognized to cause diseases in insects. These fungi exhibit a wide range of variations and infection capabilities, including both obligate and facultative pathogens. The spread of fungal diseases is prevalent among many insect species. There are about 1,000 known species of fungal phyla that can infect and kill insects, including Ascomycota, Basidiomycota, Chytridiomycota, Entomophthoromycota, and Microsporidia (Vega et al., 2012). Among these, *Beauveria bassiana* and *Metarhizium robertsii* are well-known and well-studied models of species for the control of insects (de Faria et al., 2007).

[1] Department of Botany, University College Mangalore, Karnataka, India.
[2] Department of Biochemistry, Mangalore University, Mangalagangotri, Mangalore, Karnataka, India.
[3] Department of Applied Zoology, Mangalore University, Mangalagangotri, Mangalore, Karnataka, India.
* Corresponding author: siddumn@gmail.com | msmustak@gmail.com

Agostino Bassi, in 1835, discovered that silkworm moth larvae were afflicted with a sickness, leading to the first demonstration of EPF. He called this sickness "white muscardine" based on the colour of the conidial layer that protrudes from the insect's cuticle (Zimmermann, 2007). Since then, *B. bassiana*, has been widely used against harmful insects attacking important crops such as wheat, maize, cotton, banana, and jute (Abdel-Raheem, 2020; Lopez et al., 2014, Biswas et al., 2013, Vega et al., 2012). This chapter looks at occurrences and diversity of entomopathogenic fungi in the Western Ghats of India, and their significance in insect control and metabolites.

2. Fungi of the Western Ghats

The total number of fungi found in India exceeds 27,000 species, and many more are still being discovered (Manoharachary et al., 2005). The Western Ghats are considered a rich repository of fungal diversity. While approximately 7,000 plant species may exist in this habitat, it is estimated that the fungal diversity of the area may be even wider and include at least 42,000 fungal species. The Western Ghats are thus an important and diverse ecosystem for both plants and fungi. Some of the major groups of fungi found in this region are listed in Table 1. Scientists are continuously exploring and identifying new fungal varieties. Some of the recently discovered fungi in the Western Ghats include *Asterediella pittosporacearum, Meliola arkevermae, M. colubrinicola, M. kakkachiana* var. *poochiparensis, Palawaniella jasmine* and *Sarcinella embeliae.*

Table 1 Major fungi reported from the Western Ghats.

Fungus (common name)	*Specific features*
Amanita sp. (Fly agaric or death cap)	Recognized by their characteristic features; a cap adorned with white spots and a ring encircling the stem. Some types of *Amanita* are extremely poisonous, but some species are edible.
Coprinus sp. (Inky caps)	Coprinus are notable for their swift transformation into an inky liquid after spore release. Typically found in clusters, these mushrooms have a distinct and unique appearance.
Cordyceps sp. (Caterpillar fungus)	*Cordyceps* are parasitic fungi that target insects and other arthropods. They dominate their hosts and then leave the host body to disperse the spores. In addition to their fascinating behaviour, these mushrooms have medicinal properties and are used in traditional medicine.
Cyathus sp. (Bird's nest fungi)	These little cup-shaped mushrooms look like tiny bird nests. Spore-filled "eggs" are found inside the cup that get released into the air when raindrops hit the nest.
Gibellula sp.	These fungi are parasitic and primarily infect grasshoppers and crickets.
Isaria tenuipes	These fungi are parasitic, infecting and eventually killing the host insect.
Omphalotus sp. (Lantern mushrooms)	*Omphalotus* species exhibit bioluminescence, emitting a fascinating glow in the dark. They can be seen growing on decaying wood and fallen vegetation in high humidity conditions. Some species are highly poisonous.

Contd.

Table 1 *Contd.*

Fungus (common name)	*Specific features*
Schizophyllum sp. (Split gill mushroom)	This group of fungi is characterized by a remarkable feature: their gills split lengthwise, creating the characteristic "split gill" appearance. *Schizophyllum* are widespread and play an essential role in the process of decomposition.
Termitomyces spp.	Growing in termite mounds with lengthy stipe and highly edible.

3. Some Entomopathogenic Fungi in the Western Ghats

3.1 Cordyceps

The largest genus within the *Cordyceps* family encompasses up to 200 species. This makes it the most extensive among the entomopathogenic genera, classified under Hypocreales. *Ophiocordyceps nutans* is a formidable parasitic fungus, targeting approximately 22 species of stinkbugs that pose a threat to agriculture and silviculture. Its role as a potent biocontrol agent has been well-documented by Sasaki et al. (2008). Many *Cordyceps* spp. have also been reported from lateritic scrub jungles as well as the coconut plantations of the southwest coast of India (Juliya et al., 2012; Kumar and Aparna, 2014; Prathibha, 2015; Dattaraj et al., 2018; Laya et al., 2019).

Sridhar and Karun (2012) were the first to discover *Ophiocordyceps nutans,* beneath a *Cassine glauca* tree in the Western Ghats of India. They found that the fruit bodies of *O. nutans* consistently fell in sparse concentrations (1 per 5 m^2), beneath the groves of *C. glauca* in the mixed jungle of the Makutta region in the Western Ghats. The host tree, *C. glauca,* also known as Jamrasi, is a naturally occurring tree species of the Western Ghats (Table 2). This tree serves as a shade tree in the coffee plantations and agroforests of the region. Interestingly, despite harbouring the anamorphic antagonistic endophyte *Hymenostilbe nutans*, ingestion of *H. nutans* through bark tissue proves catastrophic for stinkbugs, ultimately leading to death and the subsequent emergence of the teleomorphic *O. nutans*. Caro and O'Doherty (1999) defined 'flagship' species as those that have high-profile, noteworthy ecological roles in a specific ecosystem. *Ophiocordyceps nutans* could be considered a flagship species of the Western Ghats, similar to *Ophiocordyceps sinensis* in the Himalayas (Karun and Sridhar, 2013; Sridhar and Karun, 2017).

Table 2 *Ophiocordyceps* nutans in the Western Ghats.

Part of fungus	*Description*
Stroma	Solitary on small hosts, multiple on larger specimens (body size 20-30 mm). Length: 50-90 mm, width: 400-800 μm, colour: marasmioid-black or blackish brown, transitioning to red 8-12 mm below the fertile head.
Perithecia	Immersed, hyaline-walled. Oblique with a curved neck. Dimensions: 550-800 × 130-200 μm. Ostioles visible on the surface.
Asci	Cylindrical. Size: Up to 780 μm × 7-8 μm diameter. 8-spored
Ascospores	Easily break into 64 part-spores. Part-spores are cylindrical or slightly barrel-shaped. Ends are blunt. Dimensions: 9.5-15 × 1.5-2 μm.

3.2 *Gibellula*

The genus *Gibellula* belongs to the family Cordycipitaceae within the order Hypocreales (Ascomycota). Species within this genus give rise to one or multiple synnemata, which are compact and erect conidiomata. These synnemata harbour conidiophores with terminal vesicles resembling those of *Aspergillus* or *Penicillium*. These vesicles contain hyaline metulae, phialides, and conidia, and typically do not produce sexual structures. *Gibellula* fungal parasites are known for their interactions with insects, and recent studies reveal their ability to parasitize spiders. The infected spiders may find themselves displaced to different habitats due to the influence of *Gibellula*. Different families of spiders are found dead on the abaxial face of leaves following infection by *Gibellula*. One of the species, *Gibellula pulchra*, germinates when its fungal spores encounter a spider. It starts to grow inside the spider's body, breaking it down from the inside out. After the spider dies, stiff lavender-fruiting bodies sprout from its carcass.

3.3 *Isaria*

The genus *Isaria* also belongs to the family Cordycipitaceae within the order Hypocreales (Ascomycota). It is a multi-infectious fungus that parasitizes insects of the order Lepidoptera. *Isaria tenuipes* is known for its entomopathogenic properties, and it specifically targets insects. The life cycle of the fungus begins with infecting the insect host through cuticle penetration and subsequently proliferating within the host's body. Later, the fungus kills the insect and produces spores on or within the remains. These spores can then be dispersed into the environment for further infection and growth. This fungus is used as a biological control agent against various insect pests.

In a survey conducted by Juliya et al. (2012) in the Kerala region of the Western Ghats, a notable diversity of entomopathogenic fungi was discovered. A total of 401 fungi from 341 insect cadavers were identified during the study period. The fungi were isolated from insects representing 10 different orders, with the highest number of isolations observed in the order Lepidoptera. The majority of identified species belonged to the class Deuteromycotina.

3.4 *Bioluminescent Fungi*

In addition to their pathogenic roles, some fungi in the Western Ghats exhibit fascinating traits, such as glowing at night (e.g., *Armillaria mellea* and *Mycena chlorophos*) (Arya et al., 2021; Patil and Yadav, 2022). These glowing or bioluminescent fungi are found on decaying wood in the thick forests along the borders of Karnataka, Goa, and Maharashtra. These glowing fungi can attract both insects and their natural prey and bringing about insect death.

4. Mechanisms of Action of Entomopathogenic Fungi

Entomopathogenic fungi typically initiate infection when their disseminated spores encounter insects. Fungal spores germinate under favourable conditions such as moderate temperatures and high relative humidity. The germinating spores enter the insect's cuticle by forming appressorium and through a combination of enzymatic degradation and mechanical pressure. They secrete enzymes such as chitinases to degrade the chitin wall and proteases and lipases to degrade tissue.

After entering the hypodermis, the hyphae develop and spread throughout the insect's body, infiltrating blood cells. As the hyphae grow, they disrupt the insect's internal organs and tissues and cause damage. Simultaneously, the fungus suppresses immune response, preventing effective combating of the infection. Many times, a fungal infection weakens the insect, rendering it unable to feed or move properly. Further, the fungal growth may damage the host's vital tissues, leading to death.

Another peculiar mechanism occurs as the insect nears death. The fungus redirects its resources towards spore production, forming specialized structures called conidiophores. The conidiophores bear conidia, which are released from the insect's body. The released spores are then ready to infect other insects.

5. Significance and Applications

Applications of entomopathogenic fungi from the Western Ghats have not been extensively studied. Currently, they have been explored and identified in their natural habitats. However, many other entomopathogenic fungi have been studied and utilized for various applications, including pest control in agriculture (Table 3).

Table 3 Commercial products of entomopathogenic fungi.

Entomopathogenic fungus	*Commercial name*	*Host*
Beauveria bassiana	Mycotrol	Whiteflies and aphids
Beauveria bassiana	Botector	Seed treatments
Beauveria bassiana	BotaniGard	Whiteflies and aphids
Metarhizium anisopliae	BioBlast	Termites
Lecanicillium lecanii	Vertalec	Aphids
Beauveria bassiana	Mycotrol ESO	Nematodes
Paecilomyces fumosoroseus	PFR-97	Whiteflies
Beauveria brongniartii	Engerlingpilz	Manas larvae
Metarhizium anisopliae	Green Muscle	Grasshoppers
Beauveria bassiana	Mycotrol O	Whiteflies and aphids

5.1 Biopesticides

Entomopathogenic fungi are commonly used as biocontrol agents for many crops and in organic farming. They infect and gradually kill insects and act as an eco-friendly alternative to chemical pesticides. The fungi are effective against a wide range of pests and are precise and safe (Litwin et al., 2020). Some common fungal species used as biopesticides are presented in Table 4.

5.2 Removal of Toxins

Entomopathogenic fungi exhibit many ecologically beneficial properties, such as the removal of toxic contaminants from the environment. They can detoxify such

Table 4 Entomopathogenic Fungi and their applications.

Fungal species	*Applications*
Beauveria bassiana	Commonly used to control a wide range of insect pests, including beetles, mites, and caterpillars. Products developed from these fungi are available in various formulations, such as granules, dusts, and sprays.
Metarhizium anisopliae	Commonly used against soil-dwelling pests like white grubs and larvae of beetles.
Isaria fumosorosea	Commonly used against various pests, including whiteflies, aphids, and thrips, especially in greenhouses.

compounds as alkylphenols, organotin compounds, synthetic estrogens, pesticides, and hydrocarbons.

5.3 Secondary Metabolites

Entomopathogenic fungi produce a variety of secondary metabolites; bioactive compounds that play diverse roles in insect interactions and the environment (Fig. 1).

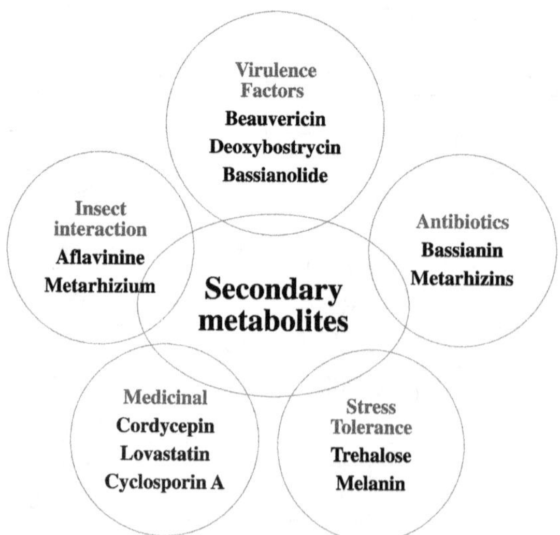

Fig. 1 Secondary metabolites of entomopathogenic fungi and their applications.

5.3.1 *Antibiotics and Antimicrobials*

***Beauvericins*:** It exhibits potent antibacterial effects against pathogenic bacteria affecting humans, animals, and plants (Meca et al., 2005).

***Cyclosporins*:** have been used medically to prevent organ transplant rejection and as immunosuppressive agents. They may also be produced by certain strains of *Beauveria bassiana.*

***Myriocin*:** exhibit antimicrobial properties and are used in pharmaceutical applications. They are derived from *Isaria sinclairii.*

***Hirsutellin A*:** has shown antimicrobial capability, particularly against Gram-positive bacteria. It is derived from the fungus *Hirsutella thompsonii.*

***Bassianin*:** is an antibiotic produced by *Beauveria bassiana.*

***Metarhizins*:** are antibiotics from *Metarhizium* spp.

5.3.2 Virulence Factors

***Beauvericin*:** It exhibits antifungal and insecticidal properties. The insecticidal properties can be seen in Beauveria strains as *Beauveria bassiana.*

***Deoxybostrycin*:** exhibits insect pathogenicity and is produced by *Metarhizium* spp.

***Bassianolide*:** enhances fungal virulence and insect infection. It is produced by *Beauveria bassiana.*

5.3.3 Insect Communication and Behaviour Modulation

***Aflavinine and Metarhizium*:** compounds influence insect host behaviour.

5.3.4 Stress Tolerance and Defense

***Trehalose*:** compounds enhance fungal tolerance to abiotic stress.

***Melanin*:** compounds provide protectio against UV radiation and oxidative stress.

5.3.5 Nutraceuticals and Medicinal Compounds for Humans

***Cordycepin*:** has antiviral and anticancer properties and is derived from *Cordyceps* spp.

***Lovastatin*:** is used to reduce cholesterol levels.

***Cyclosporin* A:** is used as an immunosuppressant. It is isolated from *Tolypocladium inflatum.*

Conclusion

The entomopathogenic fungi found in the Western Ghats hold great potential for ecological and agricultural applications. Scientists have long been fascinated by the diversity of fungi found in the Western Ghats, which may be useful for long-term pest control. Even though there is still a lot to learn about the unique characteristics of these fungi, their potential applications highlight the necessity of ongoing conservation initiatives and in-depth scientific research. We can uncover important tools for a healthier and more environmentally friendly future as we learn more about fungi biology and ecology. The unique biodiversity of the Western Ghats provides a vital resource for the identification of new entomopathogenic fungi that may aid in environmental remediation, sustainable pest control, and the development of bioactive compounds with a range of potential applications. It is essential to continue research and conservationto fully unlock the benefits offered by the fungi found in the Western Ghats ecosystem.

References

Abdel-Raheem, M. (2020). Isolation, Mass Production and Application of Entomopathogenic Fungi for Insect Pests Control. pp. 231-251. *In*: El-Wakeil, N., Saleh, M. and Abu-hashim, M. (eds.). Cottage Industry of Biocontrol Agents and Their Applications. Springer. https://doi.org/10.1007/978-3-030-33161-0_7

Arya, C.P., Ratheesh, S. and Pradeep, C.K. (2021). New record of luminescent *Mycena chlorophos* from Western Ghats of India. Studies in Fungi 6(1): 507–513. Doi 10.5943/sif/6/1/40

Biswas C., Dey, P., Satpathy, S., Satya, P. and Mahapatra, B.S. (2013). Endophytic colonization of white jute (*Corchorus capsularis*) plants by different Beauveria bassiana strains for managing stem weevil (*Apion corchori*). Phytoparasitica, 41: 17–21.

Caro, T.M. and O'Doherty, G. (1999). On the use of surrogate species in conservation biology. Conser. Biol., 13: 805–814.

Dattaraj, H.R., Jagadish, B.R., Sridhar, K.R. and Ghate, S.D. (2018). Are the scrub jungles of southwest India potential habitats of *Cordyceps*? Kavaka, -Trans. Mycol. Soc. India, 51: 20–22.

de Faria M.R. and Wraight, S.P. (2007). Mycoinsecticides and Mycoacaricides: A comprehensive list with worldwide coverage and international classification of formulation types. Biol. Cont., 43: 237–256.

Juliya, R.F., Sankaran, K.V. and Varma, R.V. (2012). Diversity of entomopathogenic fungi in the Kerala part of the Western Ghats. Ind. For., 138(2): 182–188.

Karun, N.C. and Sridhar, K.R. (2013). The stink bug fungus *Ophiocordyceps nutans* - a proposal for conservation and flagship status in the Western Ghats of India. Fungal Conserv., 3: 43–49.

Kumar, T.S. and Aparna, N.S. (2014). *Cordyceps* species as a bio-control agent against coconut root grub, *Leuopholis coneophora* Burm. J. Environ. Res. Dev., 8: 614–618.

Laya, P.K., Verma, C.K.Y., Cherian, A.K., Anees, M.M. and Rashmi, R. (2019). Observations on entomopathogenic fungus *Ophiocordyceps neovolkiana* on coconut root grub *Leucopholis coneophora*. KAVAKA – Trans. Mycol. Soc. India, 53: 55–60.

Litwin, A., Nowak, M. and Różalska, S. (2020). Entomopathogenic fungi: unconventional applications. Rev. Environ. Sci. Biotechnol., 19: 23–42. https://doi.org/10.1007/s11157-020-09525-1

Lopez D.C., Zhu-Salzman, K., Ek-Ramos, M.J. and Sword, G.A. (2014). The entomopathogenic fungal endophytes *Purpureocillium lilacinum* (formerly Paecilomyces lilacinus) and *Beauveria bassiana* negatively affect cotton aphid reproduction under both greenhouse and field conditions. PLoS One, 9: e103891.

Manoharachary, C., Sridhar, K.R., Singh, R. Adholeya, A., Suryanarayanan, T.S. et al. (2005). Fungal biodiversity: Distribution, conservation and prospecting of fungi from India. Curr. Sci., 89: 58-71.

Meca G., Sospedra I., Soriano J.M., Ritieni A., Moretti A., Manes J. (2005). Antibacterial effect of the bioactive compound beauvericin produced by *Fusarium proliferatum* on solid medium of wheat. Toxicon. 56: 349–354. doi: 10.1016/j.toxicon.2010.03.022

Patil, S.R. and Yadav, S.V. (2022). Photographic record of *Armillaria mellea* a bioluminescent fungi from Lonavala in Western Ghats, India. J. Threat. Taxa, 14(2): 20692–20694.

Prathibha, P.S. (2015). Behavioural Studies of Palm White Grubs *Leufopholis* spp. (Coleoptera: Scarabaeidae) and Evaluation of New Insecticides for their Management. Ph.D. Thesis, Agricultural Entomology, University of Agricultural Sciences, Bangalore, India, p. 119.

Sasaki F., Miyamoto, T., Yamamoto, A., Tamai, Y. and Yajima, T. (2008). Morphological and genetic characteristics of the entomopathogenic fungus *Ophiocordyceps nutans* and its host insects. Mycol. Res., 112: 1241–1244.

Sridhar, K.R. and Karun, N.C. (2017). Observations on *Ophiocordyceps nutans* in the Western Ghats. J. New Biol. Rep., 6(2): 104–111.

Vega F.E., Meyling, N.V., Luangsa-Ard, J.J. and Blackwell, M. (2012). Fungal entomopathogens. pp. 171–220. *In*: Vega, F. and Kaya, H.K. (eds.). Insect Pathology, 2nd Edition, Academic Press.

Zimmermann, G. (2007). Review on safety of the entomopathogenic fungi *Beauveria bassiana* and *Beauveria brongniartii*. Biocon. Sci. Technol., 17(5-6): 553–596.

4 Integrated Pest Management through Biological Control

Arun Kumar and Vikanksha[1] and Jatinder Singh[1*]

1. Introduction

Over the course of thousands of years, agriculture became the foundation of mankind's society, serving as a source of nourishment and subsistence. Over time, the pursuit of greater crop yields and monetary benefit led to the heavy application of synthetic pesticides, unintentionally causing deterioration of the natural environment, damage to organisms, and the evolution of pesticide resistant species. Considering such problems, Integrated Pest Management (IPM) has emerged as a disruptive method that aspires to revolutionize the way in which we preserve harvests while also protecting the tenuous equilibrium of landscapes (Brownhill et al., 2021). To understand IPM, one must study the history of pest control. Considering the broad introduction of cultivars with excellent yields and synthetic fertilizers, the Green Revolution in the mid-twentieth century greatly increased crop production (Wise et al., 2021). Pesticides were the preferred method of pest management due to their speed and efficacy. Eventually, the unforeseen implications of this method became clear. Overuse of chemical pesticides created resistance among pest varieties, caused ecological damage, and caused the extinction of ecologically important creatures (Van den Bosch, 2023). These shortcomings have led to a transition towards environmentally friendly and comprehensive approaches, embodied in integrated pest control solutions. As societies struggled with the complicated connections between plants and the many insects that endangered harvests, pest control methods evolved across centuries (Wagner et al., 2021). The history of insect pest control shows how inventive people have been in protecting their crops. It spans across periods of human history and illustrates the evolving coexistence of people and insects (Duffus et al., 2021). This chapter deals with comprehensive strategies for mitigating the impact of insect pests on agricultural productivity and ecosystem health.

1.1 *Prehistoric Agriculture*

Prior to contemporary scientific discoveries, early agricultural societies had to contend with insect infestations. The Sumerians, Egyptians, and Greeks documented their conflicts with insects, larvae, and other agricultural pests that caused damage to their

[1] Department of Horticulture, School of Agriculture, Lovely Professional University, Phagwara, Punjab-144411, India.
* Corresponding author: jatinder.19305@lpu.co.in

crops. Early agriculture included systematic investigation and the trial-and-error process, wherein specific societies developed techniques for managing pests.

1.2 Culture and Myths

In many cases, communities employed myths and cultural traditions in their efforts to control pests. To placate agricultural spirits or elicit spiritual pest protection, ceremonies, requests, and figurative offerings were widespread practices (Imai et al., 2021). These practices combined faith with pest management, showing how religion and farming were intertwined in prehistoric cultures (Olivadese and Dindo, 2023).

1.3 Initial Pharmaceutical Treatments

The use of chemical pest management may be traced back to ancient times. Arsenic-based chemicals were used by Chinese farmers for pest control as early as 1000 BCE. Sulphur and pitch were used by the ancient Greeks and Romans as insect repellents. Numerous past chemical treatments were basic and did not fully understand insect ecology and the interconnectedness of the environment.

1.4 Industrialization and Chemical Pesticides

The Industrial Revolution in the 19th century brought about advancements in agriculture and pest control. Chemical pesticides like pyrethrum and nicotine changed the game and revolutionized the era. The development of DDT in the mid-twentieth century started the contemporary chemical pesticide modern era (Zaller, 2020). DDT, along with additional chemical compounds, were initially touted as pesticide miracles, but eventually exposed as having ecological and biological risks.

1.5 Green Revolution and Extensive Agriculture

Crop varieties with higher yields, along with chemical fertilizers and pesticides, were extensively used during the Green Revolution beginning in the mid-twentieth century (Hamdan et al., 2022). This technological advance enhanced crop production and agricultural productivity but also increased substantial chemical pesticide usage, causing ecological problems, insect revival, and environmental degradation. IPM garnered recognition as a more environmentally friendly method during this period (Barrett et al., 2020).

1.6 Current Problems and Environmental Responses

Climate change, globalization, and resistant species complicate insect pest control in the 21st century. Climate change affects pest dispersal and behaviour, necessitating adaptation (Secretariat et al., 2021). Globalization allows pests to traverse boundaries, requiring cooperation among nations. In the field of pest management, the resistance that pests create brings to light the need for diversity and sustainability (Angon et al., 2023).

2. The Components of Integrated Pest Management

2.1 Pest Identification and Monitoring

Effective pest identification and monitoring are the cornerstones of every effective IPM programme and strategy. This requires identifying specific pests and knowing their daily life phases, behaviours, and linkages among agricultural system relationships (Fig. 1) (Deguine et al., 2023). Advancements in remote sensing and data analysis have improved pest tracking, enabling growers to take immediate action.

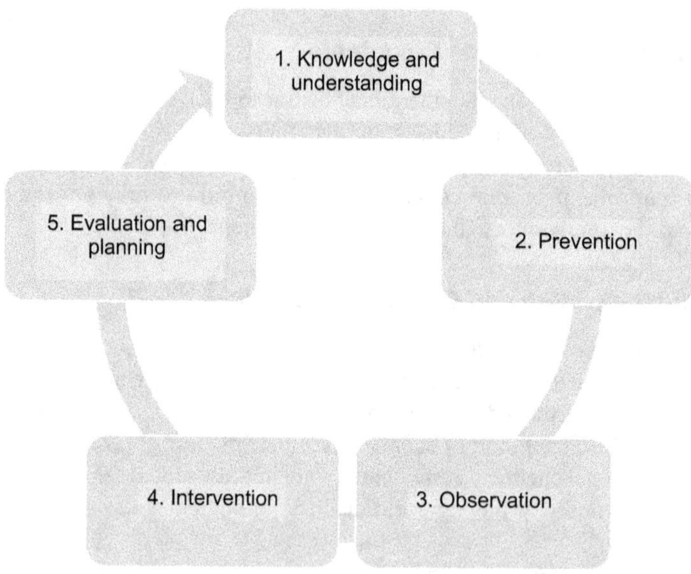

Fig. 1 Integrated Pest Management model for crop improvement.

2.1.1 Methods for Pest Identification

Species Identification: Determining the precise pest species that are causing harm. It can involve a visual inspection of pests and affected plants or using field guides and identification keys.

Life Stage Identification: Identifying pest life stage: egg, larva, pupa, or adult. Different life phases can require specific preventative actions at various level example, the nymph, larva, and adult stages.

Damage Symptoms: Tracking and recording the signs of pest damage at a specific site. Includes indications such as withering, deformation or distortion, of plant components, discolouration, and feeding damage (Christian et al., 2020).

DNA bar-coding: A newer method for species identification using gene sequencing and genome editing.

Biological Characteristics: Understanding the biology, behaviour, and life cycle of pests in their unique form. With the proper application of this knowledge, it is possible to anticipate the beginning of pest outbreaks and pinpoint vulnerable phases with which to take control measures.

Molecular Techniques: For precise identification, molecular methods like DNA barcoding by employing molecular markers and marker-assisted selection may occasionally be used, particularly when working with cryptic species or those that share comparable physical traits.

2.1.2 *Pest Monitoring*

Frequent Tracking: Agricultural sites, structures, and other pest-prone areas are to be inspected regularly.

Baits: Pheromone, sticky, and additional traps may catch pests for diagnosis and population assessment.

Over time, it is possible to monitor the distribution of pests, resources, and removal procedures in order to identify trends and evaluate the efficiency of control measures. There is a possibility that climate and other atmospheric elements might influence the behaviour and quantity of insects.

2.2 *Biological Control*

In order to effectively control pests, IPM depends on biological approaches. It is necessary to introduce or reinforce diseases, parasites, and predators in order to do this. Through the use of biological management, a balance may be created between pests and their natural enemies, therefore lowering the need for chemical control and increasing the lifespan of the control (Costa et al., 2023).

2.2.1 *Types of Biological Control Agents*

Pest-eating predators: When it comes to insects, organisms that devour insects normally ingest a broad range of insects, and they seldom focus on pests unless they are prevalent in the environment. Due to the fact that arthropod predators have shorter life cycles than other predators, their population density may vary in reaction to fluctuations in the density of their prey. They feed on a particular diversity of prey species. Numerous arthropod predators, including insects, are widely used in the process of biological control. Lady beetles, lacewings, and carnivorous mites are all used in agricultural production.

Parasitoids: Insects at an immature life stage, known as parasitoids, grow on or within an insect host, eventually killing it. Usually residing unrestricted by the host adults are potential predators. They may also consume pollen, plant nectar, honeydew, or other food sources. Parasites like wasps and sure flies control insects that are pests.

Pathogens: Bacteria, fungi, and viruses destroy insect pests. These illnesses have the potential to either kill the insect outright, slow or stop reproduction, or decrease the rate at which they feed, depending on the specific location. Furthermore, several nematode species that cause sickness or even death through their bacterial symbionts also attack insects. Diseases can spread organically among insect populations under specific environmental conditions, especially when insect density is high. The famous pesticide *Bacillus thuringiensis* (Bt) is a microbial disease.

Competitors: Specific species may compete with pests for assets resources that are necessary for growth, i.e., nutrients and host interaction, thereby lowering their reproduction and survival. Competing plant species have been used in several weed control applications.

2.2.2 Techniques in Biological Control

It is necessary to obtain an inherent adversary from the pest's native habitat in order to effectively control an imported or invading pest. Increasing the number of indigenous competitors by reproducing them in large numbers and then releasing them into the wild is one technique, i.e., increasing the effectiveness of biological enemies such as ladybeetle bugs in providing protection.

2.3 Cultural Practices

Cultural practices are utilized in IPM in an attempt to reduce and limit pest habitats (Fig. 2). Rotating crops, cross-cropping, and changing times for planting disturb pest habitats and life cycles, thereby preventing outbreaks. Such methods reduce pests and improve soil health and resilience. IPM employs cultural control measures to establish an inhospitable and non-friendly habitat for pests, reducing the need for chemical treatments. These simple, economical, and ecologically sound pest control methods are vital first lines of defence (Klassen and Vreysen, 2021).

Crop Rotation: Pests can be separated from food sources (host), or environmental factors can be altered by rotating crops to other locations. An example could be planting melons away from squash or cucumbers.

Sanitation: Pest populations can be decreased by removing weeds, agricultural residues, and other plant waste that creates habitat. Proper disposal of infested materials is crucial to preventing the growth of insect populations over time.

Trap Cropping: Trap farming is the practice of planting a specific crop to draw pests away from the main crop. By concentrating pests in a single location for easier monitoring or treatment, we can protect the main crop.

Polyculture and Diversification: By increasing the diversity of crops in a field or planting numerous crops together (polyculture) may make the environment more unsuitable and less inviting to certain pests. Additionally, this may strengthen the populations of natural predators and rivals, causing a diversity of environmental conditions that will eventually be unbearable for certain pests.

Companion Planting: Planting specific crops next to each other to benefit growth or toward offpests. For instance, growing marigolds alongside vegetables can promote nematode resistance.

Selection of Resistant Varieties: A cultural control strategy that utilizes crop varieties that are resistant or less vulnerable to particular pests. This is one method of combating pests with genetic resistance.

Mechanical Barriers: Physical barriers that keep pests away from crops include netting and row coverings. This technique works especially well for shielding plants from specific flying insects.

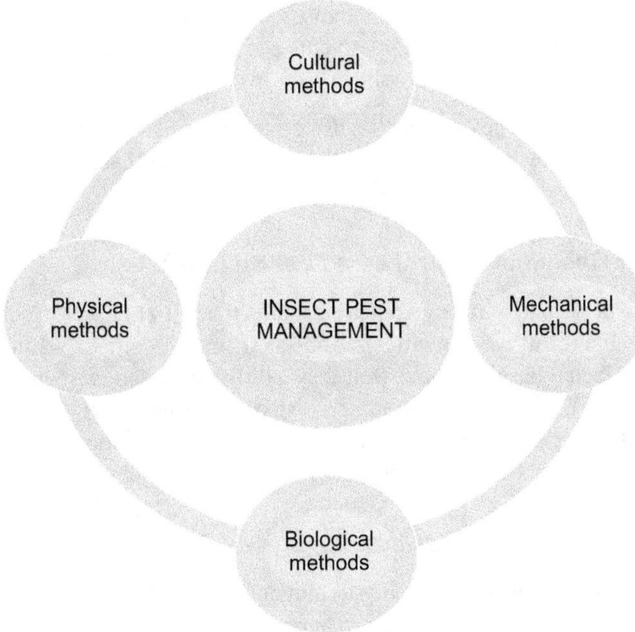

Fig. 2 Different components of insect pest management.

2.4 Mechanical Control

Mechanical pest control encompasses physical pest management (Fig. 2). Traps, obstacles, and machinery can all be used to address destructive pests and animals. Although laborious, such approaches are effective and ecologically benign alternatives to chemical pesticides, particularly in organic and small-scale farming (Maher, 2021).

2.4.1 Mechanical Control Systems

Handpicking: This entails the removal of pests by hand. For larger pests, like snails, beetles, or caterpillars, this is an effective method. Routine crop scouting enables early detection and control.

Traps: Insect pests can be captured, monitored, or tracked using a variety of traps. Examples include light traps, pitfall traps, pheromone traps, and sticky traps. Traps are very useful in the mass-capture and monitoring of particular pests.

Barriers: Plants can be protected from pests with physical barriers. Screens, netting, and row covers are a few examples. Plants can be shielded from flying insects and crawling pests with such measures.

Mulching: Plants can be protected from certain pests by creating a physical barrier around them with mulching. Furthermore, by changing soil moisture and temperature, mulches can alter the habitat of pests.

Heat Treatment: Some pests can be managed with heat, particularly those present in stored goods. For instance, heat treatments can be used to eradicate pests found in stored grain or stop them from growing.

Vacuuming: Vacuum devices can be used to remove pests from plants in specific circumstances. This technique is frequently applied to pests that are easily eliminated by suction or in greenhouse environments.

2.5 Use of Chemical Control as a Last Solution

In traditional pest control, chemicals are used as the initial phase of defence. IPM, on the other hand, utilizes chemicals only as a final recourse. While necessary, poisons are selected based on how well they kill the intended bug while having the least detrimental effect on other creatures. Also, using herbicides strategically and carefully, based on tracking information, prevents the development of resistance among plants (Desneux et al., 2022).

2.5.1 Types of Chemical Pesticides

Insecticides are chemicals formulated to terminate insect pests. Individual formulations are classified according to their chemical composition (for example, organophosphates and carbamates), their working mechanism (nerve poisons and growth regulators), or application method (foliage mists and systemic pesticides).

Herbicides: These are formulated to eliminate weeds and other target plants, some of which may harbour harmful insects.

Baiting: Use of insecticides or pesticide-containing baits to attract pests. This is frequently used to target specific pests, including rodents, cockroaches, and ants.

Fumigation: Fumigants, also known as gaseous pesticides, are applied to prevent pests from entering enclosed areas such as grain silos and storage facilities. Although it is effective against a variety of pests, careful application is necessary to protect people.

Repellents: These are pest-repelling chemicals that deter pests from treated surfaces. Personal insect repellent products frequently contain these chemicals.

Systemic Acquired Resistance (SAR) Inducers: Such substances imbue plants with systemic acquired resistance, thereby increasing their defences against pest infestations. These substances strengthen the plant's defence mechanisms.

3. Environmental Impact

Using chemical, biological, cultural, or mechanical methods and practices to get rid of insect pests can have serious effects on the surroundings. It is important to understand these consequences, to ensure sustainable practices that cause the least amount of damage to environments, non-target species, and the well-being of individuals (Selvaraj et al., 2024).

3.1 *Chemical Pesticides*

Residual Accumulation: Pesticide residue may accumulate in soil, water, and agricultural products, polluting the environment for many years.

Unintended Consequences on Ecosystems: Chemical poisons can harm bees, marine life, and native bugs.

Resistance Development: Overuse of pesticides can result in the development of immunity in insect populations, making chemical measures less successful.

3.2 *Biological Control Agents*

Non-Target Affects: Introducing carnivorous or parasitical organisms to manage pests can result in unintended ecological disruptions, particularly if these kinds of organisms consume other non-pest wildlife (Kehoe et al., 2021).

Disturbance of Native Ecosystem: Non-native biological pesticides may surpass or consume native species, disturbing ecosystem equilibrium.

3.3 *Cultural and Mechanical Controls*

Problematic soils: The use of mechanical techniques to control weeds, like heavy digging, can cause soil instability, loss of cohesiveness, and detriment to useful inhabitant creatures.

Power Intake: Methods that include machinery that demands a great deal of power, such as the operation of large machines or the movement of water, contribute to the harm that is caused to the environment by the utilisation of energy.

3.4 *Integrated Pest Management*

Lower Chemical Dependence: IPM employs fewer chemicals, which means that pesticides have a lower impact on natural surroundings.

Ecological Robustness: By treating pests in a variety of different ways integrated pest management (IPM) helps to ensure the long-term viability of ecosystems and reduces the magnitude of environmental changes.

3.5 *Implications for Biodiversity*

Disturbances in the functioning of ecosystems can arise from intensive pest management procedures, which, in turn, can impact the interrelationships between plants and pollinators, predators and prey, and biodiversity as a whole.

The detrimental consequences of careless pesticide usage include the extinction or diminishment of non-target species, thereby causing significant repercussions for the ecological balance.

3.6 Water and Air Quality

Pesticides may infiltrate water bodies via discharge, thereby polluting aquatic ecosystems and creating hazards for marine life. Airborne drift can occur as a consequence of pesticide sprinkling, potentially causing inadvertent pollution of the air and adjacent environments.

3.7 Human Welfare Issues

Inhalation of pesticides, either directly or via residue present on foodstuffs, can result in detrimental consequences for the well-being of humans.

Professional Hazards: Personnel engaged in pest management tasks may be exposed to pesticides, potentially leading to adverse health effects.

3.8 Noteworthy Climate Change Factors

The utilisation and manufacturing processes of specific commodities for pest management, including synthetic pesticides, result in the discharge of greenhouse gases, thereby exacerbating the issue of climate change.

Difficulties in Mitigation: The potential consequences of global warming on pest dispersal and behaviour could present obstacles for conventional pest control approaches, mainly chemical control.

3.9 Strategy and Administrative Elements

The implementation of rigorous rules and regulations is critical for the oversight and management of pesticide usage, thereby guaranteeing adherence to protocols that protect both human health and the environment. Governmental regulations and incentives have the potential to foster environmentally responsible insect management practices, furthering the effort of environmental preservation.

The management of insect pests, although essential to agriculture, necessitates maintaining a delicate equilibrium between effective control measures and the reduction of adverse environmental effects. It is imperative to implement environmentally friendly procedures, including IPM, and to consistently evaluate and enhance methods of pest control in order to alleviate detrimental impacts on human health, biodiversity, and ecological systems (Table 1) (Baker et al., 2020). Professionals can foster resilient and environmentally friendly farming practices by adopting and promoting comprehensive strategies and remaining well-informed regarding the ecological effects of pest management.

Table 1 Various methods of managing and preventing crop infestations by insects and pests.

Insect Pest	*Crop Infested*	*Preventive Approach*	*Specific Treatment*	*Reference*
Tomato fruit worm	Tomatoes	Biological	Release of *Trichogramma chilonsis*	Nisar et al., 2020
Diamondback Moth	Cabbage	Chemical	Application of *Bacillus thuringiensis*	Legwaila et al., 2020
Leafhopper	Grapes	Physical	Installation of sticky traps	Jambagi and Kambrekar, 2023
Cabbage aphid	Cabbage	Cultural	Companion planting with marigolds	Lopez and Liburd, 2022
Citrus leaf miner	Citrus	Mechanical	Pruning affected leaves	Matias et al., 2023
Codling moth	Apples	Integrated pest management	Release of *Trichogramma* wasps	Maggi and Chreil, 2023
Wireworm	Potatoes	Biological	Inoculation with entomopathogenic nematodes	Askar et al., 2023
Flea beetle	Brassicas	Chemical	Foliar application of pyrethroids	Hoarau et al., 2022
Green peach aphid	Peaches	Physical	Reflective mulching	Khan et al., 2020
Armyworm	Maize	Cultural	Intercropping with legumes beans	Bamboriya et al., 2022
Carrot rust fly	Carrots	Mechanical	Floating row covers	Kumar et al., 2023
Onion thrips	Onions	Integrated pest management	Biological control with predatory mites	Woldemelak, 2020
Rose aphid	Roses	Chemical	Systemic application of imidacloprid	Sridhar et al., 2022
Squash bug	Squash	Physical	Handpicking and trapping	Mani ,2022
European corn borer	Corn	Cultural	Crop rotation and trap crops	Njeru et al., 2020
Apple maggot fly	Apples	Mechanical	Mass trapping with pheromone traps	Mikhael, 2021
Colorado potato beetle	Potatoes	Biological	Release of parasitoid wasps	Weber et al., 2022
Grape berry moth	Grapes	Integrated pest management	Pheromone traps and mating disruption	Alsoliman, 2020
Mealybug	Various crops	Chemical	Neem oil spray	Gopal et al., 2021

4. Risk Assessment of Pest Management Procedures

The management of insect pests is a critical component of contemporary agriculture, serving to protect the availability of food and the production of crops. Nevertheless, pest management approaches present intrinsic hazards to human health, non-target organisms, and the environment. Conducting in-depth risk evaluation is critical to achieving a balance between efficient pest management and ecological protection.

4.1 Chemical Pesticides

Chemical pesticides were once an integral component of insect pest management due to their efficacy in providing rapid and accurate results when their consequences were not well studied. Nevertheless, there are numerous dangers associated with their use. The persistence of pesticide residue in water, soil, and crops adversely affects ecosystems and human health. The detrimental impacts of these chemicals extend beyond target vermin, impacting pollinating organisms, marine creatures, and useful insects (Dhiman et al., 2024). Furthermore, the increase in parasite populations resistant to pesticides emphasises the need for prudent and calculated application.

4.2 Biological Control Agents

The implementation of biological control agents, including parasitoids and predators, holds great potential as a substitute for chemical pesticides. Nevertheless, a comprehensive assessment of risks is essential to avoid inadvertent natural repercussions. Thorough deliberation is necessary regarding the non-target impacts of introduced plants and animals on nearby ecosystems, including disturbances to indigenous species and environmental imbalances (Carter et al., 2021). To avoid unintentional environmental changes, it is important to evaluate the host population and its growth capacity in a habitat.

4.3 Cultural and Mechanical Controls

Although mechanical and cultural limits provide ecologically beneficial choices, they nonetheless come with their own set of hazards. Soil conditions can be adversely affected by practices like excessive cultivation, resulting in runoff and disturbance to crucial microbial populations. Costly mechanical processes contribute to environmental consequences linked to energy consumption. Thorough evaluation of non-target impacts, including detrimental impacts on beneficial wildlife, is imperative to ascertain the long-term viability and effectiveness of such pest management approaches (Smith et al., 2020).

4.4 Integrated Pest Management

IPM, by adopting a holistic approach, aims to reduce hazards associated with specific techniques of pest management. Combining mechanical, chemical, biological, and cultural approaches, IPM seeks to minimize detrimental ecological effects while maximizing efficacy. Regular surveillance and leadership over the management of control, which are integral components of the IPM structure and framework, ensures that procedures for managing pests continue to be adaptable to evolving circumstances and new obstacles (Barrera, 2020).

4.5 Biodiversity

The Biodiversity repercussions of pest control methods are an essential component of risk evaluation. Unintended consequences for non-target species may include impacting beneficial invertebrates, raptors, and other fauna. Meticulous assessment of environmental disruptions and biological connections is crucial to avoid inadvertent damage to biodiversity and the long-term stability of ecosystems (Sharifi, 2023).

4.6 Human Health and Safety

Evaluation of hazards to human health associated with pest management methods is of utmost importance. Chemical traces on foodstuffs and direct contact with pesticides through administration present risks to farm labourers, citizens, and consumers. A detailed evaluation of health risks is imperative to formulate appropriate precautions and ensure safe practices (Jyotsna et al., 2023).

4.7 Water and Air Quality

The possibility of pesticide discharge and airborne dispersal contaminating lakes and rivers raises serious apprehensions regarding the purity of air and water. The primary objective of risk evaluation should be to understand the potential entry routes of pesticides into these settings, as well as the resulting consequences for air quality and the functioning of aquatic ecosystems (Rathi et al., 2021).

4.8 Climate Change Considerations

In the current epoch of climate change, it is imperative to assess the environmental impact. It is necessary to comprehend how these practices either exacerbate or alleviate the processes of climate change in order to formulate adaptable and viable approaches (Newell et al., 2022).

4.9 Regulatory Compliance

A foundational aspect of risk evaluation is enforcing adherence to rules that regulate pest management methods. By creating and maintaining awareness of legal obligations and sticking to proven, effective procedures, a structure can be established that places the preservation of the environment and safety for people at the forefront.

Risk assessment for insect eradication is an evolving and multidisciplinary endeavour that requires constant evaluation and adaptation. The adoption of sustainable and integrated pest control methods becomes crucial as agriculture progresses. By recognising and addressing the potential hazards of various insect management techniques, it is possible to promote a symbiotic relationship between agricultural output and safeguarding the environment, thereby guaranteeing an environmentally friendly and robust tomorrow for both supply chains and environments (Verma et al., 2022).

5. Limiting Factors in Insect Pest Management

There are numerous constraints that undermine the efficacy and long-term viability of insect pest control. The emergence of synthetic pesticide tolerance in the pest population is a significant obstacle that requires ongoing research and the development of new techniques and formulations. Furthermore, environmental variables, including climate change, are affecting the distribution and behaviour of pests, thereby diminishing the effectiveness of traditional chemical methods (Skendžić et al., 2021). Biological control techniques, despite their ecological friendliness, frequently face challenges in attaining the necessary precision and effectiveness to be widely implemented. Inadequate comprehension and knowledge regarding natural phenomena and pest behaviour among agricultural professionals and producers can result in poor choices. Consumer worries and legislative elements like a ban on specific chemicals regarding the harmful effects of pesticides motivate the pursuit of safer, greener techniques (Wahab et al., 2022). Poorer countries, due to financial constraints, face limitations in the availability of sophisticated, improved insect control methods. Such challenges emphasize the intricate and ever-changing characteristics of insect pest management, which requires a flexible and comprehensive approach to succeed.

Despite being an essential component of agricultural and ecological wellness, insect pest management faces a multitude of constraints that undermine its effectiveness and long-term viability.

One of the foremost obstacles encountered in pest management pertains to the potential of pests to evolve defences against chemical pesticides. The perpetual emergence of resistance requires ongoing research and the creation of novel pesticides and techniques, thereby escalating expenses and complicating application Resistance may lead to an increase in drug use, which in turn reduces the effectiveness of preventative methods and worsens health and safety concerns. (Goel et al., 2020).

Environmental variables, such as warming temperatures, have a substantial impact on the numbers and behaviour of pests. Conditions are influenced by climate change processes, resulting in unforeseeable insect interactions. Pest habited areas may expand, and reproduction rates can increase as a result of changing precipitation patterns and elevated temperatures, thereby further challenging control measures. Conventional methods for controlling pests may also be disrupted by extreme environmental phenomena (Rehman et al., 2022).

Difficulties with Biological Control: Biological control, notwithstanding its eco-friendliness, is not devoid of constraints. Biological substances occasionally exhibit a deficiency in the necessary precision and effectiveness, to be widely utilized. An additional concern is the potential for imported biological pesticides to have adverse effects on native species or the natural environment, known as "non-target consequences" (Shaw et al., 2021).

Limited comprehension and access to resources: For efficient administration of an IPM program, an understanding of parasite nature and behaviour is necessary. Producers and professionals often lack the necessary information, resulting in ineffective insect management choices. Additionally, scarcity of resources, particularly in developing nations, limits access to instructions, and sophisticated tools and technologies are a major hinderance (Ali et al., 2023).

The increasing need for more effective pest control is driven by the necessity to address concerns about the negative ecological and health effects of pesticides, as well as the growing regulatory restrictions on their use. However, the challenge of creating and implementing these strategies while maintaining their effectiveness might be difficult.

Financial limitations play a substantial role in the choice of techniques employed in insect management. Particularly for smaller businesses, the expenses associated with devising, signing up for, and implementing novel insect management techniques may prove to be unaffordable. Due to these constraints, archaic and occasionally smaller-scale ecologically conscious techniques are frequently utilized (Dyck et al., 2021).

Difficulties in collaboration and integration, or the successful execution of an effective pest management strategy that amalgamates mechanical, chemical, cultural, and biological techniques, necessitate substantial expertise and cooperation. Developing and implementing coordinated approaches may prove a formidable challenge due to their complexity, especially in areas where capital and experience are scarce.

The ramifications of pest control methods on non-target species and the surrounding ecosystem continue to be a substantial concern. Efforts to regulate parasite populations may, unintentionally, adversely affect beneficial microbes, upset the equilibrium of the environment, and diminish diversity (Van den Bosch, 2023).

6. Field Trials on Insect Pest Management

Insect pest management field trials are crucial for assessing the efficacy, feasibility, and safety of pest control techniques in real-world agricultural environments. The trials yield useful data regarding the efficacy of different insect management approaches, thereby assisting scholars and professionals to make well-informed choices regarding the incorporation and execution of said approaches (Alahmad et al., 2023). The following is a synopsis of essential factors to be considered and procedures that must be followed when undertaking insect control field trials.

6.1 Experiment Design

Randomised control trials (RCTs) are a valuable method for ensuring the integrity and dependability of research findings. By randomising the designation of both treatment and control plots, the impact of outside factors is reduced.

Repetition and Randomising: To account for variations in the environment, numerous samples of the same treatment must be used. Treatments should be assigned at random across all replicates in order to accommodate for possible spatial differences (Karp et al., 2020).

6.2 Selection of Treatments

The effectiveness of various chemical pesticides in controlling the intended vermin must be assessed. Differences in composition, application stages, and schedules should be considered.

Biological Control Agents: Evaluating the effectiveness of newly introduced or enhanced native foes. Assessment of their capacity to manage the number of insects in a manner that prevents damage to non-target species.
Cultural and Mechanical Control: Evaluating the efficacy of mechanical approaches (traps, barriers) and cultural practices (combining, crop succession) in preventing harm from insects (Gabryś and Kordan, 2022).

6.3 *Plot Size and Location*

Field Size: Selection of a suitable field area that corresponds to the magnitude of customary farming operations.

Randomized procedure: To account for possible variation in insect population by geography, remedies are assigned at random inside the experiment area (Bernauer et al., 2022).

6.4 *Monitoring and Data Collection*

Prior to the application of remedies, gathering baseline data on insect populations is necessary to establish the initial infestation level (Fig. 2).

Consistent Surveillance: Establishment of a routine monitoring timetable to observe and document the evolution of insect populations.

Collection Procedures: Utilizing suitable sampling techniques, including sweep nets, pitfall devices, or optical instruments, to precisely evaluate the quantity of pests (Montgomery et al., 2021).

6.5 *Environmental Factors*

Meteorological Conditions: As they impact the behaviour of pests and the effectiveness of treatments, document and evaluate environmental factors that occur during the trial.

Soil Well-being: Evaluation of substrate condition and quality, given its influence on the abundance of pests and the efficacy of specific pest management techniques (Hossain et al., 2023).

6.6 *Data Analysis*

Utilization of statistical methods to ascertain the significance of variations resulting from intervention (Fig. 3).

Production Evaluations: To evaluate the financial cost of insect control methods, assess crop yields, if relevant.

6.7 *Protection of the Environment*

Non-Target Impacts: Evaluating the ramifications of pest control methods on non-target organisms, such as insect flora and fauna, and the environment as a whole.

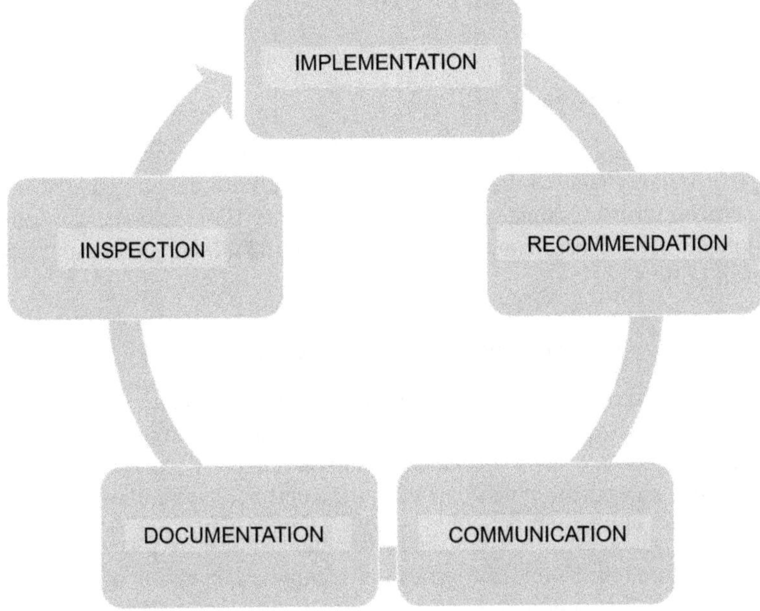

Fig. 3 Different steps of integrated pest management.

Residual examination is a crucial step in ensuring that chemical residues do not unintentionally contaminate crops (Atta et al., 2022).

6.8 *Record-Keeping and Monitoring*

Thorough Documentation: It is imperative to maintain comprehensive documentation of all outdoor trial tasks, encompassing procedures, application stages, outcome tracking, and unforeseen findings.

Concluding Summary: Consolidate a conclusive synopsis of trial results, including recommendations and suggestions for pragmatic application.

6.9 *Community Participation*

Engaging Partners: To augment the pertinence and practicality of findings, incorporate local producers, agricultural services, and other interested parties into the trial procedure.

Information Transmission: Disseminate trial results via outreach programmes, seminars, or field days encouraging agricultural organizations to implement successful pest-management techniques.

Field trials are of utmost importance in the realm of insect pest management as they serve as a critical link between experimental studies and practical farming implementation (Deguine et al., 2023). The knowledge acquired through these experiments aids in the formulation of environmentally friendly and viable solutions to insect control, thereby advancing the cause of crop safeguarding and safeguarding the environment.

7. Laboratory Assays of IPM

Laboratory assays are critical in IPM because they provide essential insights into pest biology, behaviour, and regulated management. The information gathered by these tests is crucial for the formulation and improvement of IPM methods. The following is a summary of the numerous tests performed in laboratories that are frequently employed in IPM.

7.1 Detection and the Biological Sciences of Pests

Morphology Recognition: Correctly recognising nuisance species through taxonomic references and binoculars. The molecular method involves the utilization of genetic methods such as polymerase chain reaction (PCR) to accurately identify species, with a particular emphasis on obscure species or their earliest phases.

Biological Cycle Case Studies: Identification of fragile phases for focused management through comprehension of the complete life cycles of parasites with respect to diverse surroundings (Liu and Wang, 2020).

7.2 Behaviour Evaluation

Host Selection Research: Utilizing the above-mentioned information to create tailored prevention strategies by identifying plants or plant portions that parasites favour. The investigation of oviposition and sexual behaviour is conducted in order to devise methods of interruption, including pheromone nets (Thiery et al., 2023).

7.3 Resistance Tracking

Assessing the susceptibility of parasite populations to different pesticides through bioassays. This process entails subjecting vermin to various amounts of toxins and determining the probability of survival.

Tests at the cellular level: Identification of genetic markers linked to resistance in order to comprehend causes of resistance and track dissemination (Bengtsson-Palme et al., 2023).

7.4 Measuring the Effectiveness of Control Agents

Chemical Pesticide Evaluation: Assessing the efficacy of novel or pre-existing chemical pest management agents.

Biological Control Agents: Evaluating the efficacy of naturally occurring adversaries (predators, parasitoids, and pathogens) in the management of nuisance communities.

Biopesticides: Assessing the pest control capabilities of compounds that come from natural sources (Agboola et al., 2022).

Evaluation of Toxicological Sublethal Impacts: Pesticides not only impact mortality rates, but also influence the growth and behaviour of pests. Non-target impact assessments examine the consequences of restrictions on creatures that are not their primary focus, such as beneficial organisms (Schmidt-Jeffris, 2023).

8. Greenhouse Trials of IPM

Insect pest control greenhouse experiments are crucial to sustained and productive agriculture. In a greenhouse, scientists and agriculturalists investigate insect-plant interaction and try several pest-control methods to maintain ecological equilibrium. The entire approach uses biological, chemical, cultural, physical, and integrated pest control methods to safeguard plants while not harming nature (Baker et al., 2020).

Greenhouse experiments rely on biological control. The addition of native predators or parasitoids provides a live pest defence. Scientists test pest control by introducing predatory insects or parasitic wasps into a greenhouse. Additionally, the study of microbiological insecticides, including bacteria, fungi, and viruses, strengthens the biological armoury. Unlike conventional pesticides, these tiny molecules address particular pests and are ecologically beneficial (Ahirwar et al., 2020).

Chemical control is essential to greenhouse experiments, but with an ecological focus. Scientists carefully test pesticides, from synthetic compounds to organic substances with insecticidal capabilities. IGRs' effects on pest lifespans and growth are being examined. Pest control should be done with careful chemical usage, including monitoring immediate efficacy and long-term environmental impact (Bueno-Marí et al., 2022).

Greenhouse experiments use cultural management techniques to modify the agricultural environment in order to eliminate pests. The tried-and-true rotation of crops disturbs insect life cycles, preventing the formation of colonies. Keeping the greenhouse clean and weed-free eliminates insect populations and hinders their growth. Pest control should be comprehensive and active, according to these cultural customs (Luna and House, 2020). Physical controls in greenhouse studies include obstacles, traps, and mechanical elimination. Screening or nets restrict pests from accessing agricultural products, providing simple security. Carefully positioned greenhouse nets lure and catch pests, reducing crop damage. Pest populations are immediately addressed by mechanical elimination, such as vacuuming. These hands-on methods demonstrate the adaptability of greenhouse mechanical control.

IPM relies on integrating pest control strategies. Investigators employ greenhouse experiments to find ways to combine strategies for an environmentally friendly system. Pest monitoring systems identify insect populations promptly and set action levels. IPM manages pests and reduces resistance by integrating biological, chemical, cultural, and physical control approaches (Kaur and Kaur, 2020). In greenhouse studies, scientists cultivate pest-resistant plants. Researchers use precise breeding programmes to increase plant pest resistance and reduce the need for external treatments. Sustainable agriculture emphasizes adaptable ecological systems; an appropriate preventative approach.

Greenhouse experiments demonstrate advanced management of the environment, including temperature, humidity, and illumination. Scientists study how changing these elements affects insect pest growth, behaviour, and survival. Greenhouse management can control pests and optimize plant development by studying pest biology in reaction to environmental variations (Messelink et al., 2021).

Gathering and analyzing data is crucial to greenhouse studies. Scientists closely track insect numbers, the condition of plants, and other factors. Statisticians analyze the information to find associations, trends, and connections. The methods used here turn raw data into useful knowledge, helping producers and greenhouse managers choose the best pest control options for their crops and environments (Lavik et al., 2020). Greenhouse experiments are evaluated based on cost and environmental impact. A detailed cost-effectiveness analysis considers the first investment, recurrent expenses, and crop production to determine any pest control strategy's financial viability. These approaches are also evaluated for their environmental effects to ensure longevity and the wellness of ecosystems.

Academic research, technological advancement, and sustainability in agriculture coincide in insect pest control greenhouse studies. The regulated greenhouse setting allows scientists to test several insect control methods. Natural predators, chemical treatments, cultural practices, physical controls, and integrated control of pests all contribute to a comprehensive knowledge of agriculture that is environmentally friendly. Greenhouse experiments provide promise for a sustainable and environmentally friendly agricultural future as the globe struggles to produce food while minimizing environmental damage (Adegbeye et al., 2020).

Conclusion

Ultimately, Integrated Pest Management (IPM) represents a promising solution in the pursuit of sustainable agriculture. IPM provides a practical solution to the drawbacks of traditional pest control by adopting a comprehensive and environmentally conscious strategy. As agricultural environments undergo changes globally, the concepts of IPM provide a clear path towards an era where nutritious food is attained without jeopardizing the well-being of the natural world. Although the process of achieving broad acceptance of IPM may present difficulties, the benefits, such as the development of robust ecosystems, flourishing biodiversity, and environmentally friendly agriculture, make the endeavour very worthwhile.

References

Adegbeye, M.J., Reddy, P.R.K., Obaisi, A.I., Elghandour, M.M.M.Y., Oyebamiji, K.J., Salem, A.Z.M. and Camacho-Díaz, L.M. (2020). Sustainable agriculture options for production, greenhouse gasses and pollution alleviation, and nutrient recycling in emerging and transitional nations-An overview. J. Clean. Prod., 242: 118319.

Agboola, A.R., Okonkwo, C.O., Agwupuye, E.I. and Mbeh, G. (2022). Biopesticides and conventional pesticides: Comparative review of mechanism of action and future perspectives. AROC Agric, 1: 14–32.

Ahirwar, N.K., Singh, R., Chaurasia, S., Chandra, R. and Ramana, S. (2020). Effective role of beneficial microbes in achieving the sustainable agriculture and eco-friendly environment development goals: a review. Front. Microbiol, 5: 111–123.

Alahmad, T., Neményi M. and Nyéki, A. (2023). Applying IoT Sensors and Big Data to Improve Precision Crop Production: A Review. Agronomy, 13(10): 2603.

Ali, M.A., Abdellah, I.M., and Eletmany, M.R. (2023). Towards Sustainable Management of Insect Pests: Protecting Food Security through Ecological Intensification. Int. J. Chem. Biochem. Sci. 24(4), 386–394.

Alsoliman, M. (2020). Effectiveness and synergy of some integrated management in controlling grape moths. Al Baath University - Faculty of Agriculture Engineering

Angon, P.B., Mondal, S., Jahan, I., Datto, M., Antu, U.B., Ayshi, F.J. and Islam, M.S. (2023). Integrated Pest Management (IPM) in Agriculture and Its Role in Maintaining Ecological Balance and Biodiversity. Adv Agric, 2023: https://doi.org/10.1155/2023/5546373

Askar, A.G., Yüksel, E., Bozbuğa, R., Öcal, A., Kütük, H., Dinçer, D. and İmren, M. (2023). Evaluation of entomopathogenic nematodes against common wireworm species in potato cultivation. Pathog., 12(2): 288.

Atta, A.H., Atta, S.A., Nasr, S.M. and Mouneir, S.M. (2022). Current perspective on veterinary drug and chemical residues in food of animal origin. Environ. Sci. Pollut. Res., 1–21.

Baker, B.P., Green, T.A. and Loker, A.J. (2020). Biological control and integrated pest management in organic and conventional systems. Biol. Control, 140: 104095.

Bamboriya, S.D., Bana, R.S., Kuri, B.R., Kumar, V., Bamboriya, S.D., and Meena, R.P. (2022). Achieving higher production from low inputs using synergistic crop interactions under maize-based polyculture systems. Environ. Sustain., 5(2): 145–159.

Barrera, J.F. (2020). Beyond IPM: Introduction to the theory of holistic pest management. Springer International Publishing.

Barrett, C.B., Benton, T.G., Fanzo, J., Herrero, M.T., Nelson, R., Bageant, E. and Wood, S.A. (2020). Socio-technical innovation bundles for agri-food systems transformation.

Bengtsson-Palme, J., Abramova, A., Berendonk, T.U., Coelho, L.P., Forslund, S.K., Gschwind, R. and Zahra, R. (2023). Towards monitoring of antimicrobial resistance in the environment: For what reasons, how to implement it, and what are the data needs? Environ. Int., 108089.

Bernauer, O.M., Tierney, S.M., and Cook, J.M. (2022). Efficiency and effectiveness of native bees and honey bees as pollinators of apples in New South Wales orchards. Agric. Ecosyst. Environ., 337: 108063.

Brownhill, L., Engel-Di Mauro, S., Giacomini, T., Isla, A., Löwy, M. and Turner, T.E. (Eds.). (2021). The Routledge handbook on ecosocialism. Oxon and New York: Routledge.

Bueno-Marí, R., Drago, A., Montalvo, T., Dutto, M. and Becker, N. (2022). Classic and novel tools for mosquito control worldwide. In Ecology and Control of Vector-borne Diseases (pp. 234–238). Wageningen Academic Publishers.

Carter, L., Mankad, A., Zhang, A., Curnock, M.I. and Pollard, C.R. (2021). A multidimensional framework to inform stakeholder engagement in the science and management of invasive and pest animal species. Biol. Invasions, 23: 625–640.

Christian, N., Lawrence, R. and Andrew, N.R. (2020). Integrated pest management in Northern NSW grains cropping: Lessons learnt from an industry-focussed project. Gen. App. Entomol. 48: 43–59.

Costa, C.A., Guiné, R.P., Costa, D.V., Correia, H.E. and Nave, A. (2023). Pest control in organic farming. In Advances in Resting-state Functional MRI (pp. 111–179). Woodhead Publishing.

Deguine, J.P., Aubertot, J.N., Bellon, S., Côte, F., Lauri, P.E., Lescourret, F. and Lamichhane, J.R. (2023). Agroecological crop protection for sustainable agriculture. Adv. Agron, 178: 1–59.

Desneux, N., Han, P., Mansour, R., Arnó, J., Brévault, T., Campos, M.R. and Biondi, A. (2022). Integrated pest management of *Tuta absoluta*: practical implementations across different world regions. J. Pest Sci., 95: 17–39.

Dhiman, S., Kour, J., Singh, A.D., Devi, K., Tikoria, R., Ali, M. and Bhardwaj, R. (2024). Impact of pesticide application on the food chain and food web. In Pesticides in a Changing Environment (pp. 87–118). Elsevier.

Duffus, N.E., Christie, C.R. and Morimoto, J. (2021). Insect cultural services: How insects have changed our lives and how can we do better for them. Insects, 12(5): 377.

Dyck, V.A., Hendrichs, J. and Robinson, A.S. (2021). Sterile insect technique: principles and practice in area-wide integrated pest management (p. 1216). Taylor and Francis.

Gabryś, B. and Kordan, B. (2022). Cultural control and other non-chemical methods. In Insect Pests of Potato (pp. 297–314). Academic Press.

Goel, S., Hawi, S., Goel, G., Thakur, V.K., Agrawal, A., Hoskins, C. and Barber, A.H. (2020). Resilient and agile engineering solutions to address societal challenges such as coronavirus pandemic. Mater. Today Chem., 17: 100300.

Gopal, G.S., Venkateshalu, B., Nadaf, A.M., Guru, P.N. and Pattepur, S. (2021). Management of the grape mealybug, Maconellicoccus hirsutus (Green), using entomopathogenic fungi and botanical oils: a laboratory study. Egypt. J. Biol. Pest. Co., 31: 1–8.

Hamdan, M.F., Mohd Noor, S.N., Abd-Aziz, N., Pua, T.L. and Tan, B.C. (2022). Green revolution to gene revolution: Technological advances in agriculture to feed the world. Plants, 11(10): 1297.

Hoarau, C., Campbell, H., Prince, G., Chandler, D. and Pope, T. (2022). Biological control agents against the cabbage stem flea beetle in oilseed rape crops. Biol. Control, 167: 104844.

Hossain, M.S., Small, B.C. and Hardy, R. (2023). Insect lipid in fish nutrition: Recent knowledge and future application in aquaculture. Rev. Aquac. 15(4): 1664–1685.

Imai, I., Inaba, N. and Yamamoto, K. (2021). Harmful algal blooms and environmentally friendly control strategies in Japan. Fish. Sci., 87(4): 437–464.

Jambagi, S. R., and Kambrekar, D. N. (2023). Management of Major insect pests in Grape (*Vitis vinifera* l.) Ecosystem: Novel tools and technologies. In Pests and Disease Management of Horticultural Crops. In Pests and Disease Management of Horticultural Crops (pp. 255–274) Biotech New Delhi, India-110 002.

Jyotsna, F.N.U., Ahmed, A., Kumar, K., Kaur, P., Chaudhary, M.H., Kumar, S. and Kumar, F.K. (2023). Exploring the complex connection between diabetes and cardiovascular disease: analyzing approaches to mitigate cardiovascular risk in patients with diabetes. Cureus, 15(8). doi: 10.7759/cureus.43882.

Karp, N.A., Wilson, Z., Stalker, E., Mooney, L., Lazic, S.E., Zhang, B. and Hardaker, E. (2020). A multi-batch design to deliver robust estimates of efficacy and reduce animal use–a syngeneic tumour case study. Sci. Rep. 10(1): 6178.

Kaur, T. and Kaur, M. (2020). Integrated pest management: A paradigm for modern age. Pests, Weeds and Diseases in Agricultural Crop and Animal Husbandry Production. IntechOpen. http://dx.doi.org/10.5772/intechopen.92283.

Kehoe, R., Frago, E., and Sanders, D. (2021). Cascading extinctions as a hidden driver of insect decline. Environ. Entomol., 46(4): 743–756.

Khan, A.A., Kundoo, A.A., Nissar, M. and Mushtaq, M. (2020). Sucking pests of temperate fruits. In: Sucking Pests of Crops, Omkar (eds). (pp.369–409), Springer, Singapore. https://doi.org/10.1007/978-981-15-6149-8_12

Klassen, W. and Vreysen, M.J.B. (2021). Area-wide integrated pest management and the sterile insect technique. In Sterile insect technique (pp. 75–112). CRC Press.

Kumar, M., Kumar, A., Saba, M., Prasad, S. and Mandal, S. K. (2023). Insect-Pests of Carrot and their Integrated Pests Management. Rashtriya Krishi,10 (2): 1–2

Lavik, M.S., Hardaker, J.B., Lien, G. and Berge, T.W. (2020). A multi-attribute decision analysis of pest management strategies for Norwegian crop farmers. Agric. Syst., 178: 102741.

Legwaila, M.M., Munthali, D.C., Kwerepe, B.C. and Obopile, M. (2020). Efficacy of *Bacillus Thuringiensis* (var. *Kurstaki*) against diamondback moth (*Plutella Xylostella* L.) eggs and larvae on cabbage under semi-controlled greenhouse conditions. Int. J. Insect Sci., 7(1). https://doi.org/10.4137/IJIS.S23637

Liu, J. and Wang, X. (2020). Tomato diseases and pests detection based on improved Yolo V3 convolutional neural network. Front. Plant Sci. 11:898. doi: 10.3389/fpls.2020.00898

Lopez, L. and Liburd, O.E. (2022). Can the introduction of companion plants increase biological control services of key pests in organic squash? Entomol. Exp. Appl., 170(5): 402–418.

Luna, J.M. and House, G.J. (2020). Pest management in sustainable agricultural systems. In Sustainable agricultural systems (pp. 157–173). CRC Press.

Maggi, C. and Chreil, R. (2023). Codling Moth (*Cydia pomonella*) Biology, and Integrated Pest Management. Tree Fruit Insects, (1): 1–12.

Maher, R. (2021). Does one acre fund promote sustainable and agroecological intensification? eCommons, Cornell University, Library.

Mani, M. (2022). Organic Pest Management in Horticultural Crops. Trends in Horticultural Entomology, (pp. 211–241). Springer Singapore.

Matias, P., Barrote, I., Azinheira, G., Continella, A. and Duarte, A. (2023). Citrus pruning in the mediterranean climate: A review. Plants, 12(19): 3360.

Messelink, G.J., Lambion, J., Janssen, A. and van Rijn, P.C. (2021). Biodiversity in and around greenhouses: Benefits and potential risks for pest management. Insects, 12(10): 933.

Mikhael, H. (2021). Towards sustainable management of the Mediterranean fruit fly (*Ceratitis capitata*) in apple orchard in Bane North of Lebanon (Doctoral dissertation).

Montgomery, G.A., Belitz, M.W., Guralnick, R.P. and Tingley, M.W. (2021). Standards and best practices for monitoring and benchmarking insects. Front. Ecol. Evol. 8:579193. doi: 10.3389/fevo.2020.579193

Newell, P., Daley, F. and Twena, M. (2022). Changing our ways: Behaviour change and the climate crisis. Cambridge University Press. https://doi.org/10.1017/9781009104401

Nisar, N., Shah, S.F., Sarwar, J., Amin, F. and Rasheed, I. (2020). 43. Impact of *Trichogramma chilonis* on Tomato fruit worm (*Helicoverpa Armigera* Hub.) in Tomato crop. Pure Appl. Biol., 9(1): 443-447.

Njeru, N.K., Midega, C.A., Muthomi, J.W., Wagacha, J.M. and Khan, Z.R. (2020). Impact of push–pull cropping system on pest management and occurrence of ear rots and mycotoxin contamination of maize in western Kenya. Plant Pathol., 69(9): 1644–1654.

Olivadese, M. and Dindo, M.L. (2023). Edible insects: A historical and cultural perspective on entomophagy with a focus on western societies. Insects, 14(8): 690.

Rathi, B.S., Kumar, P.S. and Vo, D.V.N. (2021). Critical review on hazardous pollutants in water environment: Occurrence, monitoring, fate, removal technologies and risk assessment. Sci. Total Environ., 797: 149134.

Rehman, A., Saba, T., Kashif, M., Fati, S.M., Bahaj, S.A. and Chaudhry, H. (2022). A revisit of internet of things technologies for monitoring and control strategies in smart agriculture. Agronomy, 12(1), 127.

Schmidt-Jeffris, R.A. (2023). Non-target pesticide impacts on pest natural enemies: Progress and gaps in current knowledge. Curr. Opin. Insect Sci., 58: 101056.

Secretariat, I.P.P.C., Gullino, M.L., Albajes, R., Al-Jboory, I., Angelotti, F., Chakraborty, S. and Stephenson, T. (2021). Scientific review of the impact of climate change on plant pests. FAO on behalf of the IPPC Secretariat.

Selvaraj, B., Nadana, S., Sabariswaran, K. and Jintae, L. (2024). Present status of insecticide impacts and eco-friendly approaches for remediation-a review.

Sharifi, A. (2023). The resilience of urban social-ecological-technological systems (SETS): a review. Sustain. Cities Soc., 99: 104910.

Shaw, B., Nagy, C. and Fountain, M.T. (2021). Organic control strategies for use in IPM of invertebrate pests in apple and pear orchards. Insects, 12(12): 1106.

Skendžić, S., Zovko, M., Živković, I.P., Lešić, V. and Lemić, D. (2021). The impact of climate change on agricultural insect pests. Insects, 12(5): 440.

Smith, D., King, R. and Allen, B.L. (2020). Impacts of exclusion fencing on target and non-target fauna: a global review. Biol. Rev., 95(6): 1590–1606.

Sridhar, V., Naik, S.O., Swathi, P. and Mani, M. (2022). Pests and Their Management in Ornamental Plants: (Rose, Jasmine, Chrysanthemum, Crossandra, Marigold, Tuberose, Carnation, China aster, Gerbera, Gladiolus, Hibiscus, etc.). Trends in Horticultural Entomology, (pp.1189–1237). Springer, Singapore. https://doi.org/10.1007/978-981-19-0343-4_52

Thiery, D., Mazzoni, V. and Nieri, R. (2023). Disrupting pest reproduction techniques can replace pesticides in vineyards. A review. Agron. Sustain. Dev., 43(5): 69.

Van den Bosch, R. (2023). The pesticide conspiracy. Univ of California Press.

Verma, A., Shameem, N., Jatav, H.S., Sathyanarayana, E., Parray, J.A., Poczai, P. and Sayyed, R.Z. (2022). Fungal endophytes to combat biotic and abiotic stresses for climate-smart and sustainable agriculture. Front. Plant Sci. 13:953836. doi: 10.3389/fpls.2022.953836.

Wagner, D.L., Grames, E.M., Forister, M.L., Berenbaum, M.R., and Stopak, D. (2021). Insect decline in the Anthropocene: Death by a thousand cuts. Proc. Natl. Acad. Sci. U.S.A. 118(2): e2023989118.

Wahab S., Muzammil K., Nasir N., Khan M.S., Ahmad M.F., Khalid M., Ahmad W., Dawria A., Reddy L.K.V., Busayli A.M. (2022). Advancement and new trends in analysis of pesticide residues in food: A comprehensive review. Plants, 11(9): 1106.

Weber, D.C., Blackburn, M.B. and Jaronski, S.T. (2022). Biological and behavioral control of potato insect pests. pp. 231–276. In: Alyokhin, A., Rondon, S.I. and Gao, Y. (eds.). Insect Pests of Potato, Global Perspectives on Biology and Management, Elsevier Science.

Wise, T.A. (2021). Old fertilizer in new bottles: Selling the past as innovation in Africa's Green Revolution. Global Development and Environment Institute, (pp 1–34). Tufts University Medford MA 02155, USA https://sites.tufts.edu/gdae/

Woldemelak, W.A. (2020). The major biological approaches in the integrated pest management of onion thrips, *Thrips tabaci* (Thysanoptera: Thripidae). J. Hortic. Res., 28(1).

Zaller, J.G. (2020). What is the problem? pesticides in our everyday life. Daily Poison: Pesticides-an Underestimated Danger, pp. 1–125. Springer Nature Switzerland AG.

5 Biocontrol of Insect Pests by Natural Enemies

Matangi Mishra[1*] and Kumar Gaurav[2]

1. Introduction

Ecosystems consist of organisms that coexist with their surroundings and compete with one another to reproduce and survive at the highest possible trophic level (Polis, 1999; Ohgushi, 2005). An ecosystem, which is a network formed by the connections between different organisms in their surroundings, hosts many interactions. Depending on their morphological, physiological, and behavioral traits, organisms can be producers, consumers, or decomposers ina given environment (Moraes et al., 2000). Biocontrol is a natural phenomenon of the regulation of biota by natural enemies (Sampaio et al., 2008). Larvae of Chrysopidae, for instance, usually cover themselves with small plant fragments, consumed prey, their carriers. These coverings serve as camouflage, defending them from natural predators. Besides this protective behavior, larvae from other genera within the same family do not carry trash on their bodies. Instead, they use different defense mechanisms, such as agility and the production of deterrent emissions, to protect themselves. Biological control of insect pests using natural enemies, known as beneficial insects, is a crucial aspect of Integrated Pest Management (IPM). Organisms such as bacteria, fungi, nematodes, protozoa, and viruses serve as natural enemies of insect pests. The use of these natural enemies has a long history in the global biocontrol of crop pests (Senthil-Nathan, 2015; Pathan et al., 2021; Riddick, 2022). Several predators possess characteristics that enable them to act as biocontrol agents for arthropods in agricultural ecosystems (Akter et al., 2019). Current knowledge indicates that the key to successfully implementing biological control strategies is focusing on predator-prey interactions (Costanza et al., 1997; Daily, 1997; Groot et al., 2002; Andrade and Romeiro, 2009).

2. Search for Prey

In addition to exhibiting various behavioral strategies for locating, selecting, and consuming prey, generalist predators can switch to alternative prey when their preferred prey becomes scarce. Most predatory insects actively hunt and capture prey using their speed and agility. Ants (Formicidae), carabid beetles (Carabidae), and true bugs (Heteroptera), such as Anthocoridae, have specialized mouthparts for

1 Department of Entomology, School of Agriculture, Lovely Professional University, Phagwara-144411, Punjab, India.
2 Department of English, School of Social Science and Humanities, Lovely Professional University, Phagwara-144411, Punjab, India.
* Corresponding author: matangi.28192@lpu.co.in

capturing prey by intercepting it. They may also use their modified front legs, which resemble raptor appendages, for this purpose (e.g., Belostomatidae). The thoracic anatomy of predatory insects like those in the Odonata and Asilidae groups is adapted for capturing prey in flight. Alternatively, predatory insects may employ a passive hunting technique, waiting for prey to come close, which conserves energy. Examples of this strategy include the raptorial prothoracic legs of Mantodea insects and the underground chambers built by pit-building antlion larvae (Myrmeleontidae), both of which wait for prey to pass by.

3. Encountering Prey

3.1 Habitat Characteristics

The occurrence, diversity, and activity of predatory insects in agroclimatic conditions are influenced by several environmental factors, including climate; habitat resources such as food sources, refuges, landscapes, and reproduction sites; intra- and inter-specific competition; and the presence of other organisms (Akter et al., 2019). In agroecosystems, temperature, relative humidity, and photoperiod are climatic parameters that affect the population dynamics of green lacewings. For instance, in Brazilian citrus farms, species like *Ceraeochrysa* sp. and *C. externa* (Chrysopidae) experienced population booms during periods of low temperature and rainfall (Neto et al., 2001; Souza and Carvalho, 2002).

A study on carabids in Brazil over two crop seasons of maize and soybeans demonstrated that these beetles were more abundant during the season with higher precipitation, with up to 2.2 times greater levels, indicating that increased relative humidity promotes higher population density (Cividanes, 2002). Carabids differ from other insect groups in their ability to quickly spread over the earth, a behavior that helps them endure unfavorable weather conditions and the effects of agricultural practices like pesticide use and ploughing (Wallin, 2002).

The physical structure of the habitat is a critical factor in managing local population density and may influence the preference of microhabitats for predators and prey (Lima, 1998). For example, the morphology and architecture of plant organs can affect predator-prey relationships by providing protection to both predators and prey. These features can also alter the microenvironment created by the plant, impacting the survival of associated microbes. According to Honek (2012), research on lady beetle habitats is essential for gaining fundamental knowledge in ecology, ecophysiology, and biogeography. This includes understanding the factors that limit the beetles' range, their food and microclimate preferences, and the identification of ecological niches.

3.2 Natural Enemies

Insect pests should attract natural enemies to the colonized environment so they can locate and establish themselves in the cultivated area. Natural enemies are typically drawn to prey that fulfills their nutritional needs, either directly or indirectly (Table 1). They can utilize chemical signals released by plants indicating herbivory, the presence of floral resources in the fields, colors, forms, and movements of the flowers, or any other visually appealing traits (Warren and James, 2008; Maffei, 2010; Hogg et al., 2011; Salamanca et al., 2015; Silva et al., 2017; Sousa et al., 2018).

The impact of kairomones is noted as a potential trapping factor for natural enemies. For instance, when *Podisus nigrispinus* was released to control leaf worms (*Thyrinteina arnobia*) on *Eucalyptus*, they preferred to stay with *Eucalyptus pellita* infested by the pest compared to other *Eucalyptus* species. In greenhouse trials, rose plant species infested with *Macrosiphum euphorbiae* attracted more female *Chrysoperla externa* insects to deposit their eggs compared to aphid-free plants (Salamanca et al., 2015).

Volatile metabolites released by tomato plants help predatory mirids, such as *Campyloneuropsis infumatus, Dicyphus cerastii, Engytatus varians, Macrolophus basicornis,* and *Nesidiocoris tenuis*, in their search for *Tuta absoluta* prey (Silva et al., 2017; Abraços-Duarte et al., 2021; Sarmah et al., 2021). Additional cues for searching and hunting prey pertain to female *Orius insidiosus* insects, which are drawn to rose plants infested with *Frankliniella insularis* or *Tetranychus urticae* under mono-herbivory conditions, or with both *F. insularis* and *T. urticae* under multiple herbivory conditions (Sousa et al., 2020).

Table 1 Overview of different natural enemies with examples and target insect pests.

Natural enemy	*Example*	*Target insect pest*
Birds	Swallos	Flying insects
Entomopathogenic fungi	*Bauveria bassiana*	Various insect pests
	Metarhizium robertsii	Parasitoids
	Metarhizium sp. BCC 4849	Spider mites
Entomopathogenic nematodes	Heterorhabditis	Soil dwelling larva and fall army worm
Parasitic wasps	*Trichogramma*	Aphids, tomato leaf minor and eggs
Parasitoids	*Trichogramma*	Moth, larva, European pepper moth, Asian corn borer, and aphids
Predatory beetles	Ground beetles	Soil dwelling pests
Predatory insects	Ladybugs and lacewings	Aphids and mites
Predatory mites	*Anystis baccarum*	Foxglove aphids
	Phytoseiusulus persimilis	Spider mites
Spiders	Jumping spiders	Various insect pests

4. Interactions of Predator and Prey

Interactions between predators and prey influence behavior and can reduce the rate of predation. For example, the presence of a predator in an environment often leads to significant behavioral changes in the prey, decreasing its vulnerability to attack. The success of predatory actions can be affected by the adaptive plasticity in prey behavior, including agility to avoid predators, camouflage, and thanatosis, which are anatomical and physiological changes (Lima, 1998).

4.1 Search-Ability and Handling Time

These two factors significantly influence population dynamics: a predator's search efficiency and the time it takes to locate prey. Understanding these factors is crucial for determining the threshold density at which a predator can effectively control pest populations and estimating its role in species population dynamics. Various variables, including predator and prey size, as well as the predator's nutritional status, can affect the handling time and consumption rate of food.

An essential determinant of predators' effectiveness in controlling pest arthropod populations is their functional response, which describes the relationship between the rate of prey consumption and prey population density. Changes in prey population dynamics may lead to corresponding changes in predator behavior. For instance, when feeding on first instar nymphs of *Ferrisia virgata*, the coccinellid *Tenuisvalvae notata* exhibited a Type III functional response, while it displayed a Type II response when preying on third instar nymphs and adult females of hemipterous species (Barbosa et al., 2014).

Similarly, Costa et al. (2017) observed a similar pattern when assessing the functional response of *Stethorust ridens* to the density of *Tetranychus bastosi*, a mite pest. The predator showed a Type III functional response when consuming the prey's eggs and larvae but shifted to a Type II response when preying on nymphs and adults of the mite.

4.2 Food Preference

While some predators have a broad prey spectrum, others are limited to specific prey types. Many ecosystems feature generalist predators capable of shifting from preferred prey to alternative species when the former becomes scarce. Consequently, the population dynamics of these predators and their prey may not always be strongly correlated. For example, larvae of the Chrysopidae family typically exhibit dietary preferences based on their ecological niche but can also consume other available prey. These larvae often target smaller and slower-moving prey, considering factors such as prey size and locomotion speed.

Coccinellids, commonly known as ladybugs, predominantly prey on Sternorrhyncha insects such as aphids, scales, whiteflies, and psyllids. However, certain species may also feed on eggs, mites, thrips, and newly hatched larvae of Lepidoptera and Coleoptera. *Cryptolaemus montrouzieri*, an Australian ladybird beetle, shows a preference for scale insects belonging to various families. Similarly, the green lacewing *Chrysoperla carnea* exhibits biocontrol capabilities against mealybugs in greenhouse crops. While aphids remain the primary food source for aphidophagous coccinellids, they may supplement their diet with other foods during larval growth and oviposition.

Research suggests that providing a diverse diet that meets specific nutritional requirements can enhance the reproductive output of generalist predators. Predator food selection directly impacts prey search and apprehension, particularly in determining predatory efficiency, especially when targeting multiple species within a given area. Thus, understanding the preferred prey species is crucial for effective implementation of biocontrol programs.

4.3 Suitability and Palatability

A generalist predator's dietary choices are influenced by physical, physiological, and behavioral factors, but it can adapt to alternative prey when its preferred food source becomes scarce. For instance, larvae of the Chrysopidae family target prey with less sclerotized integuments, facilitating easier penetration during feeding. Additionally, consuming certain prey species may not fulfill the predator's nutritional requirements for energy, development, survival, and reproduction. The nutritional quality of consumed herbivores affects the predator's predation rates and population growth, measured by metrics such as adult fecundity, development time, and survival.

Laboratory studies have shown that the diet of the coccinellid species *Eriopis connexa*, primarily consisting of aphids (*Macrosiphum euphorbiae*), promotes better development rates compared to diets consisting mainly of mite species like *Tetranychus evansi*. Furthermore, Saito and Brownbridge (2021) reported that the predatory mite *Anystis baccarum* can prey on foxglove aphids (*Aulacorthum solani*) in Canadian greenhouse crops.

In addition to prey dynamics, the host plant of herbivores can directly impact natural enemies. For instance, comparing nymphs raised on cucumber (*Cucumis sativus*) versus cabbage (*Brassica oleracea*), *Bemisia tabaci* nymphs biotype B sourced from the milkweed Euphorbia heterophylla as a nutritional source for *Chrysoperla externa* larvae led to decreased reproductive periods, egg production (up to 50% decline), and egg viability (up to 60% reduction) during predation (Silva et al., 2004).

4.4 Density of Prey

Research on the feeding behavior of predatory insects indicates that, besides the type and stage of prey development, predators adjust their food intake based on prey density in their surroundings (Acharya et al., 2020). In the dynamics of predator-prey populations, the correlation between population density and prey consumption is vital. Increased prey availability often leads to higher consumption due to heightened interaction opportunities.

In instances where aphid populations, such as *Myzus persicae* and *Schizaphis graminum*, surged past a certain threshold, larvae of *Chrysoperla externa* exhibited a progressive decrease in the number of prey individuals consumed until reaching a plateau (Fonseca et al., 2000; Barbosa et al., 2006, 2008). These findings are supported by the positive correlation observed between the population density of *M. euphorbiae* nymphs on rose plants and the consumption rate of *C. externa* larvae. Gamboa et al. (2016) noted an average consumption of 21 units by *C. externa* nymphs when aphid densities reached 40 per plant, increasing to 100 nymphs within 24 hours at a density of 160 aphids per plant.

Similarly, when preying on *S. graminum*, larvae of *Scymnus argentinicus* in their second, third, and fourth instars displayed comparable feeding behavior (Vieira et al., 1997). Santa-Cecília et al. (2001) also reported similar results for *Cycloneda sanguinea* larvae fed on *S. graminum*.

5. Interaction in Intraguild

Utilizing multiple species of natural enemies to suppress one or more herbivore species can enhance the effectiveness of biological pest control. However, the presence of such predators may impact prey density either negatively or positively (Cakmak et al., 2009). Competition between predators targeting the same prey can lead to adverse effects, potentially resulting in the elimination of one rival. Research by Souza et al. (2008) demonstrated that regardless of individual animal density or instar percentage, confining larvae of *Chrysoperla externa* and *Ceraeochrysa cubana* ensured greater survival of the former species.

In the management of herbivores, a combination of specialized or generalist enemies may be employed. For instance, species of the generalist predator genus Orius have been utilized alone or in conjunction with other predators to control thrips in various crops (Cloutier and Johnson, 1993; Chow et al., 2008, 2010; Van Houten et al., 2016). However, some correlations between *Orius* spp. and predatory mites are advised due to the preference of anthocorids for thrips-based diets over mite-based ones (Wasuwan et al., 2022).

Despite the potential benefits, further research is needed in certain predator pairings, as some predators may favor different prey over the intended target species. For instance, although *Orius laevigatus* and *Phytoseiulus persimilis* were partnered to control *Tetranychus urticae* in Brazil, *O. laevigatus* continued to engage in intraguild predation on *P. persimilis* (Venzon et al., 2001). Additionally, the variability in food availability and quality, predator population density, and size of conspecific individuals can affect cannibalism rates, which may be driven by competitive or nutritional factors influencing species coexistence and intraguild predation (Costa et al., 2003; Crumrine, 2010). Moreover, intraguild predation by Heteroptera and escape behavior of *Chrysoperla carnea* larvae in cotton plants infested by *Aphis gossypii* negatively influenced larval density (Rosenheim, 1998; Rosenheim et al., 1999). Volatile and non-volatile compounds like E-β-farnesene and 3-methyl-2-butenal are known to stimulate aphid-eating predators (Riddick, 2020).

6. Predator Ecology and Habitat Manipulation

The presence of diverse nutritional resources such as nectar, pollen, and prey, both within the crop and its surrounding areas, is crucial for ensuring the survival, development, and perpetuation of natural enemies, promoting their growth and establishing their populations (Janssen et al., 2007). For example, coccinellids may also favor honeydew, pollen, and nectar when their preferred prey declines, reducing mortality during the diapause phase and increasing energy reserves for migration and reproduction (Figueira et al., 2003; Oliveira et al., 2004; Michaud and Grant, 2005; Lundgren, 2009; Weber and Lundgren, 2009). The increased abundance of coccinellids (e.g., *C. sanguinea, E. connexa* and *H. convergens*) in Seropédica, Brazil, has been linked to the presence of coriander (*Anetum graveolens*) and fennel (*Foeniculum vulgare*), which serve as persistent sites for predators, providing nutritional resources such as pollen and prey, as well as shelter for larvae, pupae, and adults, and serving as mating and oviposition sites (Medeiros et al., 2009; Lixa et al., 2010; Resende et al.,

2010). Larvae of *E. connexa* fed on coriander nectar or pollen in Brazil successfully reached adulthood (Resende et al., 2015).

Species of the *Orius* genus, known as minute pirate bugs, benefit from plant diversity. The increased population diversity of *Orius insidiosus*, an omnivorous species, has been linked to greater access to refuge sites within coriander flowers and availability of nutritional resources such as pollen and nectar, resulting in a reduction of populations of *Frankliniella* spp. and *Thrips tabaci* associated with Apiaceae (Resende et al., 2012). Chrysopidae larvae exhibit great ecological flexibility and are typically associated with a wide range of plant species across different vegetation types, although some degree of specialization may occur (Freitas, 2002). Resende et al. (2014) observed a higher abundance of Chrysopidae members in different cropping systems involving Poaceae species. *C. externa* larvae were capable of sustaining themselves on pollen from elephant grass (*Pennisetum purpureum*) and reaching adulthood, although Oliveira et al. (2010) noted prolonged larval stages and reduced pupal viability.

Refuge sites are essential for the survival of carabid beetles in agroecosystems. For instance, soybean and maize crops established using direct planting systems and orange orchards with soil surfaces occupied by natural vegetation may exhibit higher richness and diversity of carabid populations due to practices such as direct planting and mulching, which alter soil properties favorably for carabid sheltering (Cividanes et al., 2009). The conservative biocontrol method involves introducing and maintaining regions favorable for the existence and survival of natural enemies to enhance their pest control capabilities (Barbosa, 1998; Lee and Landis, 2002; Amorós-Jiménez et al., 2020). Establishing refuge areas, such as grass or flowering plant species fencelines or beetle banks, encourages the richness and diversity of carabid populations within agroecosystems, thereby regulating pest populations (Thomas et al., 1991; Collins et al., 2002). In temperate regions, carabids may overwinter in stable habitats and colonize crops when conditions become favorable (Lövei and Sunderland, 1996; Wissinger, 1997), with forests, live fences, windbreaks, pastures, and prairies serving as protective zones for predators to colonize and act as biocontrol agents (Pfiffner and Luka, 2000; Lazzerini et al., 2007; Picault, 2011).

7. Factors Affecting the Predators

Exotic natural enemy populations often exhibit greater richness and exert a significant impact on the environment compared to native prey species, albeit they may compensate for the loss of native species diversity in the ecosystem. These relocated natural enemies can displace many native predators and negatively affect the roles of indigenous species (Evans and Toler, 2007; Finke and Snyder, 2010). Several native coccinellids have been displaced in South America by exotic species such as the Asian ladybird beetle *Harmonia axyridis*, due to intraguild predation and competition for resources (Koch, 2003; Mirande et al., 2015). *H. axyridis* accounted for 38% of coccinellids captured from fruit trees in Brazil during 2004–2006, and later comprised 91% of the total recorded species (Milléo et al., 2008). Studies conducted in Brazil and Argentina have reported intraguild predation by *H. axyridis*, which competes with *E. connexa* as a major species (Santos et al., 2009; Mirande et al., 2015).

7.1 Zoophagy

Despite the potential for antagonistic interactions between predators, including predation, cannibalism, and escape behaviors, the impact on prey species can vary under different conditions, as documented in various studies. For instance, the first instar of Miridae is often considered phytophagous, transitioning to predatory behavior as size and search capacity increase. Heteropteran species like *Macrolophus pygmaeus* and *Nesidiocoris tenuis*, commonly found together in Mediterranean protected tomato production systems, are highly effective natural enemies against *Bemisia tabaci, Trialeurodes vaporariorum*, and lepidopteran larvae. However, when both predators coexist, adult females of zoophytophagous species like *N. tenuis* can cause significant damage to tomato plants (Moreno-Ripoll et al., 2012; Sarmah et al., 2021). Additionally, variations in salivary gland enzymes in these bugs have been suggested as an adaptation to omnivorous conditions (Torres and Boyd, 2009).

Interactions with Other Organisms

Symbiotic interactions between ant species and plant lice may reduce the efficiency of natural enemies.

Phytosanitary Products

Predators are generally less susceptible to the impacts of phytosanitary products in terms of toxicity, selectivity, and ecological effects compared to parasitoids (Schäfer and Herz, 2020; Wang et al., 2022). Long-term studies in Sweden evaluating the influence of phytosanitary products on populations of Carabidae showed changes in various characteristics of the predators, although they remained significant for the control of *Rhopalosiphum* spp. Extensive insecticide use resulted in reduced activity and altered composition of carabid populations in the field (Rusch et al., 2013).

Conclusion

The biocontrol method relies on mutual density mechanisms, where natural enemies act as population density-dependent agents, their effectiveness varying with the thickness of their prey or hosts. Such biocontrol techniques play a crucial role in sustainable agricultural environments, as natural agents contribute to maintaining equilibrium in pest arthropod populations. Predatory insects, classified into different groups, are key in regulating agricultural pests, each group possessing unique bioecological characteristics used to distinguish their roles in ecosystem functions. Understanding these specific features is essential for promoting and strengthening biocontrol methods, as predators are influenced by various biotic and abiotic factors in the environment.

References

Abraços-Duarte. G., Ramos, S., Valente, F., da Silva, E.B. and Figueiredo, E. (2021). Functional response and predation rate of *Dicyphus cerastii* Wagner (Hemiptera: Miridae). Insects, 12: 530. https://doi.org/10.3390/insects12060530

Acharya, R., Hwang, H.S., Mostafiz, M.M., Yu, Y.S. and Lee, K.-Y. (2020). Susceptibility of various developmental stages of the fall armyworm, *Spodoptera frugiperda*, to Entomopathogenic Nematodes. Insects, 11: 868. doi:10.3390/insects11120868

Akter, M.S., Siddique, S.S., Momotaz, R., Arifunnahar, M., Alam, K.M. et al. (2019). Biological control of insect pests of agricultural crops through habitat management was discussed. J. Agric. Chem. Environl, 8(1): 1–13. 10.4236/jacen.2019.81001

Amorós-Jiménez, R., Plaza, M., Montserrat, M., Marcos-García, M.A. and Fereres, A. (2020). Effect of UV-absorbing nets on the performance of the aphid predator *Sphaerophoria rueppellii* (Diptera: Syrphidae). Insects, 11: 166. 10.3390/insects11030166

Andrade, D.C. and Romeiro, A.R. (2009). Serviços ecossistêmicos e sua importância para o sistema econômico e o bem-estar humano. Texto para discussão. IE/UNICAMP, Campinas, # 155, pp. 1–44.

Barbosa, L.R., Carvalho, C.F., Souza, B. and Auad, M. (2006). Influência da densidade de *Myzus persicae* (Sulzer) sobre alguns aspectos biológicos e capacidade predatória de *Chrysoperla externa* (Hagen). Acta Sci. Agron., 28(2): 227–231.

Barbosa, L.R., Carvalho, C.F., Souza, B. and Auad, M. (2008). Eficiência de Chrysoperla externa (Hagen, 1861) (Neuroptera: Chrysopidae) no controle de *Myzus persicae* (Sulzer, 1776) (Hemiptera: Aphididae) em pimentão (*Capsicum annum* L.). Ciências Agrárias • Ciênc. agrotec. 32 (4).https://doi.org/10.1590/S1413-70542008000400012

Barbosa, P. (1998). Conservation Biological Control. Elsevier Inc., p. 420.

Barbosa, P.R., Oliveira, M.D., Giorgi, J.A., Silva-Torres, C.S. and Torres, J.B. (2014). Predatory behavior and life history of *Tenuisvalvae notata* (Coleoptera: Coccinellidae) under variable prey availability conditions. Fla. Entomol., 97(3), pp.1026-1034.

Biddinger, D.J., Weber, D.C. and Hull, L.A., (2009). Coccinellidae as predators of mites: stethorini in biological control. Biol. Control 51(2): 268–283.

Cakmak, I., Janssen, A., Sabelis, M.W. and Baspinar, H. (2009). Biological control of an acarine pest by single and multiple natural enemies. Biol. Control 50(1): 60–65.

Carvalho, C.F. and Souza, B. (2009). Métodos de criação e produção de crisopídeos. In: Bueno VHP (ed) Controle biológico: produção massal e controle de qualidade, 2a ed. Ed. UFLA, Lavras, pp 77–115.

Chow, A., Chau, A. and Heinz, K.M. (2008). Compatibility of *Orius insidiosus* (Hemiptera: Anthocoridae) with *Amblyseius degenerans* (Acari: Phytoseiidae) for control of *Frankliniella occidentalis* (Thysanoptera: Thripidae) on cut roses. Biol. Control 44(2): 259–270.

Chow, A., Chau, A. and Heinz, K.M. (2010). Compatibility of *Amblyseius* (Typhlodromips) *swirskii* (Athias-Henriot) (Acari: Phytoseiidae) and *Orius insidiosus* (Hemiptera: Anthocoridae) for biological control of *Frankliniella occidentalis* (Thysanoptera: Thripidae) on roses. Biol Control 53(2): 188–196.

Cividanes, F.J. (2002). Efeitos do sistema de plantio e da consorciação soja-milho sobre artrópodes capturados no solo. Pesq Agropec Bras 37(1): 15–23.

Cividanes, F.J., Barbosa, J.C., Ide, S., Perioto, N.W. and Lara, R.I.R. (2009). Faunistic analysis of Carabidae and Staphylinidae (Coleoptera) in five agroecosystems in northeastern São Paulo state, Brazil. Pesq Agropec Bras 44(8): 954–958.

Cloutier, C. and Johnson, S.G. (1993). Predation by *Orius tristicolor* (Hemiptera: Anthocoridae) on *Phytoseiulus persimilis* (Acarina: Phytoseiidae): testing for compatibility between biocontrol agents. Environ. Entomol. 22(2): 477–482.

Collins, K.L., Boatman, N.D., Wilcox, A., Holland, J.M. and Chaney, K., (2002). Influence of beetle banks on cereal aphid predation in winter wheat. Agric Ecosyst Environ 93(1/3): 337–350.

Costa, J.F., Matos, C.H.C., de Oliveira, C.R.F., da Silva, T.G.F. and Neto, I.F.L. (2017). Functional and numerical responses of *Stethorus tridens* Gordon (Coleoptera: Coccinellidae) preying on *Tetranychus bastosi* Tuttle, Baker & Sales (Acari: Tetranychidae) on physic nut (Jatrophacurcas). Biol. Control, 111: 1–5.

Costa, R.I.F., Carvalho, C.F., Souza, B. and Loreti, J. (2003). Influência da densidade de indivíduos na criação de *Chrysoperla externa* (Hagen, 1861) (Neuroptera: Chrysopidae).

Costanza, R., d'Arge, R., De Groot, R., Farber, S., Grasso, M., Hannon, B., Limburg, K., Naeem, S., O'neill, R.V., Paruelo, J. and Raskin, R.G. (1997). The value of the world's ecosystem services and natural capital. Nature 387: 253–260.

Crumrine, P.W. (2010). Size-structured cannibalism between top predators promotes the survival of intermediate predators in an intraguild predation system. J. North Am. Benthol Soc. 29(2): 636–646.

Daily, G.C. (1997). Nature's services. Societal dependence on natural ecosystems. Island Press, Washington, DC.

Enkegaard, A., Brodsgaard, H.F. and Hansen, D.L. (2001). *Macrolophus caliginosus*: functional response to whiteflies and preference and switching capacity between whiteflies and spider mites. Entomol. Exp. Appl. 101(1): 81–88.

Evans, E.W. (2000). Egg production in response to combined alternative foods by the predator *Coccinella transversalis*. Entomol. Exp. Appl. 94(2): 141–147.

Evans, E.W. (2009). Lady beetles as predators of insects other than Hemiptera. Biol. Control. 51(2): 255–267.

Evans, E.W. and Toler, T.R. (2007). Aggregation of polyphagous predators in response to multiple prey: ladybirds (Coleoptera: Coccinellidae) foraging in alfalfa. Popul. Ecol. 49(1): 29–36.

Figueira, L.K., Toscano, L.C., Lara, F.M. and Boica Jr, A.L., (2003). Aspectos biológicos de *Hippodamia convergens e Cycloneda sanguinea* (Coleoptera: Coccinellidae) sobre *Bemisia tabaci* biótipo B (Hemiptera: Aleyrodidae). Bol. San.Veg. Plagas, 29: 3–7.

Finke, D. and Snyder, W.E. (2010). Conserving the benefits of predator biodiversity. Biol. Conserv., 143(10): 2260–2269.

Fonseca, A.R., Carvalho, C.F. and Souza, B. (2000). Resposta funcional de Chrysoperla externa (Hagen) (Neuroptera: Chrysopidae) alimentada com *Schizaphis graminum* (Rondani) (Hemiptera: Aphididae). An. Soc. Entomol. Bras. 292: 309–317.

Freitas S (2002). O uso de crisopídeos no controle biológico de pragas. In: Controle biológico no Brasil: parasitóides e predadores, Parra, J.R.P., Botelho, P.S.M., Corrêa-Ferreira, B.S. and Bento, J.M.S. (eds) Manole, São Paulo, pp. 209–224.

Gamboa, S., Souza, B. and Morales, R. (2016). Actividad depredadora de *Chrysoperla externa* (Neuroptera: Chrysopidae) sobre *Macrosiphum euphorbiae* (Hemiptera: Aphididae) en cultivo de Rosa sp. Rev Colomb Entomol 42(1): 54–58.

Golsteyn, L., Mertens, H., Audenaert, J., Verhoeven, R., Gobin, B. and Clercq, P.D. (2021). Intraguild interactions between the mealybug predators *Cryptolaemus montrouzieri* and *Chrysoperla carnea*. Insects, 12: 655. https://doi.org/10.3390/insects12070655

Groot, R., Wilson, M. and Boumans, R. (2002). A typology for the classification description and valuation of ecosystem functions, goods and services. Ecol. Econ., 41(3): 393–408.

Hodek, I. and Honek, A (1996). Ecology of coccinellidae. Kluwer Academic, London.

Hodek, I. and Honek, A. (2009). Scale insects, mealybugs, whiteflies and psyllids (Hemiptera, Sternorrhyncha) as prey of ladybirds. Biol. Control 51(2): 232–243.

Hodek, I., Van Emden, H.F. and Honek, A. (2012). Ecology and behaviour of the ladybird beetles (Coccinellidae). Wiley-Blackwell, Chichester, p. 561.10.1002/9781118223208

Hogg, B.N., Bugg, R.L. and Daane, K.M. (2011). Attractiveness of common insectary and harvestable floral resources to beneficial insects. Biol. Control 56(1): 76–84.

Honek, A. (2012) Distribution and habitats. In: Ecology and behaviour of the ladybird beetles (Coccinellidae), Hodek, I., Van Emden, H.F. and Honek, A. (eds). Wiley-Blackwell, Chichester, pp 110–140.

Janssen, A., Sabelis, M.W., Magalhães, S., Montserrat, M. and Van Der Hammen, T. (2007). Habitat structure affects intraguild predation. Ecology 88(11): 2713–2719.

Koch, R.L. (2003). The multicolored Asian lady beetle, Harmonia axyridis: a review of its biology, uses in biological control, and non-target impacts. J. Insect. Sci. 3: 1–16.

Lazzerini, G., Camera, A., Benedettelli, S. and Vazzana, C. (2007). The role of field margins in agrobiodiversity management at the farm level. Ital. J. Agron. 2(2): 127–134.

Lee, J.C. and Landis, D.A. (2002). Non-crop habitat management for carabid beetle. In: The agroecology of carabid beetles, Holland, J.M. (ed), pp. 279–303.

Lima, S.L. (1998) Nonlethal effects in the ecology of predator-prey interactions. BioScience 48(1): 25–34.

Lixa, A.T., Campos, J.M., Resende, A.L.S., Silva, J.C., Almeida, M.MT.B., and Aguiar-Menezes, E.L. (2010). Diversidade de Coccinellidae (Coleoptera) em plantas aromáticas (Apiaceae) como sítios de sobrevivência e reprodução em sistema agroecológico. Neotrop. Entomol., 39: 354–359.

Lövei, G.L. and Sunderland, K.D. (1996). Ecology and behavior of ground beetles (Coleoptera: Carabidae). Annu. Rev. Entomol. 41(1): 231–256.

Lundgren, J.G. (2009). Nutritional aspects of non-prey foods in the life histories of predaceous Coccinellidae. Biol. Control, 51(2): 294–305.

Maffei, M.E. (2010). Sites of synthesis, biochemistry and functional role of plant volatiles. S. Afr. J. Bot. 76(4): 612–631.

Martins, C.B.C., Almeida, L.M., Zonta-de-Carvalho, R.C., Castro, C.F. and Pereira, R.A. (2009). *Harmonia axyridis*: a threat to Brazilian Coccinellidae? Rev. Bras. Entomol. 53(4): 663–671.

Medeiros, M.A., Sujii, E.R. and Morais, H.C. (2009). Efeito da diversificação de plantas na abundância da traça-do-tomateiro e predadores em dois sistemas de cultivo. Hortic. Bras. 27(3): 300–306.

Michaud, J.P. and Grant, A.K. (2005). Suitability of pollen resources for the development and reproduction of *Coleomegilla maculata* (Coleoptera: Coccinellidae) under simulated drought conditions. Biol. Control 32: 363–370.

Milléo, J., Souza, J.M.T., Barbola, I.F. and Husch, P.E. (2008). *Harmonia axyridis* em árvores frutíferas e impacto sobre outros coccinelídeos predadores. Pesq. Agropec. Bras., 43(4): 537–540.

Mirande, L., Desneux, N., Haramboure, M. and Schneider, M.I. (2015). Intraguild predation between an exotic and a native coccinellid in Argentina: the role of prey density. J. Pest. Sci. 88(1): 155–162.

Moraes, C.M., Lewis, W.J. and Tumlinson, J.H. (2000). Examining plant-parasitoid interactions in tritrophic systems. An. Soc. Entomol. Bras. 29(2): 189–203.

Moreno-Ripoll, R., Agustí, N., Berruezo, R. and Gabarra, R. (2012). Conspecific and heterospecific interactions between two omnivorous predators on tomato. Biol Control 62(3): 189–196.

Neto, J.G., Carvalho, C.F., Souza, B. and Santa-Cecilia, L.V.C. (2001). Flutuação populacional de espécies de Ceraeochrysa Adams, 1982 (Neuroptera: Chrysopidae) em citros, na região de Lavras MG. Ciênc. Agrotec, 25(3): 550–559.

Obrycki, J.J., Harwood, J.D., Kring, T.J. and O'Neil, R.J. (2009). Aphidophagy by Coccinellidae: application of biological control in agroecosystems. Biol. Control 51(2): 244–254.

Ohgushi, T. (2005). Indirect interaction webs: herbivore-induced effects through trait change in plants. Ann. Rev. Ecol. Evol. Syst. 36: 81–105.

Oliveira NC, Wilcken CF, Matos CAO (2004) Ciclo biológico e predação de três espécies de coccinelídeos (Coleoptera, Coccinellidae) sobre o pulgão-gigante-do-pinus *Cinara atlantica* (Wilson) (Hemiptera, Aphididae). Rev. Bras. Entomol. 48(4): 529–533.

Oliveira, S.A., Souza, B., Auad, A.M. and Carvalho, C.A. (2010). Can larval lacewings *Chrysoperla externa* (Hagen) (Neuroptera, Chrysopidae) be reared on pollen? Rev. Bras. Entomol. 54(4): 697–700.

Pathan, E.K., Ghormade, V., Tupe, S.G. and Deshpande, M.V. (2021). Insect pathogenic fungi and their applications: An Indian perspective. pp. 311–327. In: Satyanarayana, T., Deshmukh, S.K. and Deshpande, M.V. (eds.), Progress in Mycology – An Indian Perspective. Springer Singapore.

Pfiffner, L. and Luka, H. (2000). Overwintering of arthropods in soils of arable fields and adjacent semi-natural habitats. Agric. Ecosyst. Environ. 78(3): 215–222.

Picault, S. (2011). Functional biodiversity: natural regulation of aphid populations in lettuce crops. Infos Ctifl 275: 27–35.

Polis, G.A. (1999). Why are parts of the world green? Multiple factors control productivity and the distribution of biomass. Oikos 86(1): 3–15.

Resende, A.L.S., de Haro, M.M., da Silva, V.F., Souza, B. and Silveira, L.C.P. (2012). Diversidade de predadores em coentro, endro e funcho sob manejo orgânico. Arq. Inst. Biol. 79(2): 193–199.

Resende, A.L.S., Ferreira, R.B., Silveira, L.C.P., Pereira, L.P.S., Landim, D.V. and Carvalho, C.F. (2015). Desenvolvimento e reprodução de Eriopis connexa (Germar 1824) (Coleoptera: Coccinellidae) alimentada com recursos florais de coentro (*Coriandrum sativum* L.). Entomotropica 30(2): 12–19.

Resende, A.L.S., Souza, B., Aguiar-Menezes, E.L., Oliveira, R.J. and Campos, M.E.S. (2014) Influência de diferentes cultivos e fatores climáticos na ocorrência de crisopídeos em sistema agroecológico. Arq. Inst. Biol. 81(3): 257–263.

Resende, A.L.S., Viana, A.J.S., Oliveira, R.J., Aguiar-Menezes, E.L., Ribeiro, R.L., Ricci, M.S. and Guerra, J.G.M. (2010) Consórcio couve-coentro em cultivo orgânico e sua influência nas populações de joaninhas. Hortic. Bras. 28(1): 41–46.

Riddick, E.W. (2020). Volatile and non-volatile organic compounds stimulate oviposition by aphidophagous predators. Insects, 11: 683; 10.3390/insects11100683

Riddick, E.W. (2022). Topical collection: natural enemies and biological control of plant pests. Insects, 13: 421. https://doi.org/10.3390/insects13050421

Rosenheim, J.A., Limburg, D.D. and Colfer, R.G. (1999). Impact of generalist predators on a biological control agent, *Chrysoperla carnea*: direct observations. Ecol. Appl. 9(2):409- 417.

Rosenhein, J.A. (1998) Higher-order predators and the regulation of insect herbivore populations. Annu. Rev. Entomol. 43: 421–447.

Rusch, A., Bommarco, R., Chiverton, P., Öberg, S., Wallin, H., Wiktelius, S. and Ekbom, B. (2013). Response of ground beetle (Coleoptera: Carabidae) communities to changes in agricultural policies in Sweden over two decades. Agric. Ecosyst. Environ. 176: 63–69.

Saito, T. and Brownbridge, M. (2021). Efficacy of *Anystis baccarum* against foxglove aphids, *Aulacorthum solani*, in laboratory and small-scale greenhouse trials. Insects, 12: 709. https://doi.org/10.3390/insects12080709

Salamanca, J., Pareja, M., Rodriguez-Saona, C., Resende, A.L.S. and Souza, B. (2015). Behavioral responses of adult lacewings, Chrysoperla externa, to a rose-aphid-coriander complex. Biol. Control. 80: 103–112.

Sampaio, M.V. Bueno, V.H.P., Silveira, L.C.P. and Auad, A.M. (2008). Biological control of insect pests in the Tropics. In: Del Claro K. (ed.), Encyclopedia of Life Support Systems. EOLSS Publishers, Oxford, UK.

Santa-Cecília, L.V.C., Gonçalves-Gervásio, R.D.C., Tôrres, R.M.S. and Nascimento, F.D. (2001). Aspectos bilógicos e consumo alimentar de larvas de *Cycloneda sanguinea* (Linnaeus, 1763) (Coleoptera: Coccinelidae) alimentadas com *Schizaphis graminum* (Rondani, 1852) (Hemiptera: Aphididae). Ciênc. Agrotec., 25(6): 1273–1278.

Santos, N.R.P., Santos-Cividanes, T.M., Cividanes, F.J., Anjos, A.C.R. and Oliveira, L.V.L. (2009) Aspectos biológicos de Harmonia axyridis alimentada com duas espécies de presas e predação intraguilda com Eriopis connexa. Pesq. Agropec. Bras., 44(6): 554–560.

Sarmah, N., Kaldis, A., Taning, C.N.T., Perdikis, D., Smagghe, G. and Voloudakis, A. (2021). dsRNA-Mediated Pest Management of *Tuta absoluta* Is Compatible with Its Biological Control Agent *Nesidiocoris tenuis*. Insects, 12: 274. https://doi.org/10.3390/insects12040274

Schäfer, L. and Herz, A. (2020). Suitability of European *Trichogramma* species as biocontrol agents against the tomato leaf miner *Tuta absoluta*. Insects, 11: 357. doi:10.3390/insects11060357

Senthil-Nathan, S. (2015). A review of biopesticides and their mode of action against insect pests. pp. 49–63. In: Thangavel, P. and Sridevi, G. (eds.). Environmental Sustainability, Springer India.

Silva CG, Auad AM, Souza, B., Carvalho, C.F. and Bonani, J.P. (2004). Aspectos biológicos de *Chrysoperla externa* (Hagen, 1861) (Neuroptera: Chrysopidae) alimentada com *Bemisia tabaci* (Gennadius, 1889) Biótipo B (Hemiptera: Aleyrodidae) criada em três hospedeiros. Ciênc. Agrotec., 28(2): 243–250.

Silva, D.B., Weldegergis, B.T., Van Loon, J.J. and Bueno, V.H. (2017). Qualitative and quantitative differences in herbivore-induced plant volatile blends from tomato plants infested by either *Tuta absoluta* or *Bemisia tabaci*. J. Chem. Ecol. 43(1): 55–65.

Solomon, M.E. (1949) The natural control of animal populations. J. Anim. Ecol., 18(1): 1–35.

Sousa, A.L.V., Silva, D.B., Silva, G.G., Bento, J.M.S., Penãflor, M.F.G. and Souza, B. (2020). Behavioral response of the generalist predator *Orius insidiosus* to single and multiple herbivory by two cell content-feeding herbivores in rose plants. Biol. Control. Arthropod-plant interactions, 14: 227–236.

Souza, B. and Carvalho, C.F. (2002). Population dynamics and seasonal occurrence of adults of *Chrysoperla externa* (Hagen, 1861) (Neuroptera: Chrysopidae) in a citrus orchard in Southern Brazil. Acta. Zool. Academ. Sci. Hung., 48: 301–310.

Souza, B., Costa, R.I.F., Tanque, R.L., Oliveira, P.D.S. and Santos, F.A. (2008). Aspectos da predação entre larvas de *Chrysoperla externa* (Hagen, 1861) e *Ceraeochrysa cubana* (Hagen, 1861) (Neuroptera: Chrysopidae) em laboratório. Cienc. Agrotec., 32(3): 712–716.

Thomas, C.F.G., Holland, J.M. and Brown, N.J. (2002). The spatial distribution of carabid beetles in agricultural landscapes. In: Holland JM (ed) The agroecology of carabid beetles. Intercept, Andover, pp. 305–344.

Thomas, M.B., Wratten, S.D. and Sotherton, N.W. (1991). Creation of 'island" habitats in farmland to manipulate populations of beneficial arthropods: predator densities and emigration. J. Appl. Ecol. 28(3): 906–917.

Torres, J.B. and Boyd, D.W. (2009). Zoophytophagy in predatory Hemiptera. Braz. Arch. Biol. Technol. 52(5): 1199–1208.

Triltsch, H. (1997). Gut contents in field sampled adults of *Coccinella septempunctata* (Col: Coccinellidae). Entomophaga 42(1/2): 125–131.

Van Alebeek, F., Visser, A. and Van den Broek, R. (2007). Field margins as (winter) refuge for natural enemies. Entomolog. Ber. 67: 223–225.

Van Houten, Y.M., Hoogerbrugge, H., Lenferink, K.O., Knapp, M. and Bolckmans, K.J. (2016). Evaluation of *Euseius gallicus* as a biological control agent of western flower thrips and greenhouse whitefly in rose. Nihon Dani. Gakkai. Shi., 259(suppl): 147–159.

Venzon, M., Pallini, A. and Janssen, A. (2001) Interactions mediated by predators in arthropod food webs. Neotrop. Entomol., 30(1): 1–9.

Vieira, G.F., Bueno, V.H.P. and Auad, A.M. (1997). Resposta funcional de Scymnus (Pullus) argentinicus (Weise) (Coleoptera: Coccinellidae) a diferentes densidades do pulgão verde *Schizaphis graminum* (Rondani) (Homoptera: Aphididae). An. Soc. Entomol. Bras. 26(3): 495–502.

Wallin, H. (2002) Foreword. In: The agroecology of carabid beetles, Holland, J.M., (Ed.) Intercept, Andover, pp. xiii-xiv.

Wang, Y., Hou, Y.Y., Benelli, G., Desneux, N., Ali. A. and Zang, L.-S. (2022). *Trichogramma ostriniae* is more effective than *Trichogramma dendrolimi* as a biocontrol agent of the asian corn borer, *Ostrinia furnacalis*. Insects. 13: 70. https://doi.org/10.3390/insects13010070

Warren, J. and James, P. (2008). Do flowers wave to attract pollinators? A case study with Silene maritime. J. Evol. Biol., 21(4): 1024–1029.

Wasuwan, R., Phosrithong, N., Promdonkoy, B., Sangsrakru, D., Sonthirod, C. et al. (2022). The fungus *Metarhizium* sp. BCC 4849 is an effective and safe mycoinsecticide for the management of spider mites and other insect pests. Insects, 13: 42. https://doi.org/10.3390/insects13010

Weber, D.C. and Lundgren, J.G. (2009). Assessing the trophic ecology of the Coccinellidae: their roles as predators and as prey. Biol. Control. 51(2): 199–214.

Wissinger, S.A. (1997). Cyclic colonization in predictably ephemeral habitats: a template for biological control in annual crop systems. Biol. Control. 10(1): 4–15.

Xu, X. and Enkegaard, A. (2009). Prey preference of *Orius sauteri* between Western Flower Thrips and spider mites. Entomol. Exp. Appl., 132(1): 93–98.

6 Entomopathogenic Fungi: Bioweapons against Insect Pests

A.K. Hasith Priyashantha,[1] M.C.A. Galappaththi,[2] Samantha C. Karunarathna[3] and Saisamorn Lumyong[1*]

1. Introduction

Today, agriculture is often deemed the 'heart of a country's economy', particularly in developing countries that are largely dependent on the agricultural sector. We have witnessed the importance of a robust agricultural sector and the essentiality of food safety during the recent COVID-19 pandemic. Millions of people suffered from food insecurity during the pandemic, particularly in developing countries, with many facing a daily battle to find even a single bowl of rice (Rashid et al., 2020; Kakaei et al., 2022). As researchers, we must answer complex questions: What lessons have we learned from the COVID-19 pandemic about the importance of agriculture and food security, and what have we done since to prepare for such catastrophes in the future? Increasing production and minimizing harvest losses are essential to any agricultural system (Sharma et al., 2020). Today, pests, diseases, weeds, and hazardous climatic conditions are primary agricultural concerns (Tudi et al., 2021). The Food and Agriculture Organization of the United Nations (FAO) estimates that pests cause 20% to 40% of the world's harvests to be lost annually (Karar et al., 2021). Losses from invasive insect pests are about $70 billion annually (FAO, 2023).

To combat pests, in the late 1930s, synthetic pesticides were introduced. Since then, the application of these compounds has increased dramatically, over fifty times in magnitude so far (Hu, 2020). In 2021, 3.53 million metric tons of pesticides were used globally. Brazil consumed the most pesticides compared with other countries, with an estimated utilization of 719.51 thousand metric tons. With 457.39 thousand tons consumed, the United States came in second. The quantity of pesticides used worldwide rose 96% between 1990 and 2021 (www.statista.com). Farmers use numerous formulations of synthetic pesticides, sometimes neglecting the recommended application rate and dosage, which suggests that actual usage is perhaps more than the estimation (Tudi et al., 2021). Despite the overuse of pesticides, there has not been a significant reduction in annual crop losses due to pests; instead, there have been only slight variations up and down (Sharma et al., 2020). The catastrophic

[1] Department of Biology, Faculty of Science, Chiang Mai University, Chiang Mai 50200, Thailand.
[2] Harry Butler Institute, Murdoch University, Murdoch 6150 WA, Australia
[3] Center for Yunnan Plateau Biological Resources Protection and Utilization, College of Biological Resource and Food Engineering, Qujing Normal University, Qujing 655011, China.
Email: priyashanthahasith@gmail.com; mcagalappaththi@gmail.com; samanthakarunarathna@gmail.com
* Corresponding author: scboi009@gmail.com

nature of synthetic pesticides is well understood: soil degradation, underground and surface water pollution, threats to non-target organisms, and health concerns for humans and animals (Rani et al., 2021). The introduction of regulations on using some chemical products and the aforementioned adverse effects have compelled farmers to employ non-chemical means of pest control. Unfortunately, traditional methods such as mixed cropping, crop rotation, resistant cultivars/selective breeding, flooding, solarisation, steaming, pasteurization, hot water treatment, bio-fumigation, and the use of biocontrol agents have not shown satisfactory results, especially against insect pests (Priyashantha and Attanayake, 2021). This state of affairs has compelled researchers to develop effective and efficient alternatives to synthetic pesticides (Sharma et al., 2020). One such bio-control approach is the utilization of entomopathogenic fungi.

As their name implies, these fungi are pathogenic to insects (*ento*). They cause fatal diseases in the insects, generally causing death within a few days. They are not only infecting the insects, but also other arthropods, including spiders, mites, and ticks (Skinner et al., 2014; Mantzoukas et al., 2022). Entomopathogenic fungi, with their global presence, are not confined to specific regions. They thrive in diverse habitats such as tropical forests, agricultural lands, grasslands, urbanized areas, and even extreme environments like arid, desert, and arctic regions. Their adaptability is further demonstrated by their presence in aquatic habitats (Behie et al., 2015; Sharma et al., 2020). This widespread distribution underscores the importance of understanding their role in infecting a multitude of insect species.

According to the findings, out of the 31 insect orders, 20 are highly infected by entomopathogenic fungi. These fungi demonstrate their efficacy across all developmental stages, from eggs to adults, infecting a wide range of insects (Islam et al., 2021). The versatility of entomopathogenic fungi is evident in their ability to be host-specific or generalist. Host-specific fungi target a particular order of insects, while generalist strains can cause disease in multiple insect orders, thereby expanding their host range and enhancing their survivability by being able to move between several insect orders (Kidanu and Hagos, 2020). This adaptability is a key aspect of their potential in biocontrol, which is of great interest to researchers.

In this chapter, our focus is on the practical applications of entomopathogenic fungi in current agriculture systems, particularly in controlling a wide range of insect pests. We begin by introducing early studies and the phylogenetic placement of these fungi. We then delve into common infection mechanisms. Our discussion also includes laboratory experiments conducted to identify the most virulent strains and other crucial considerations for developing new biocontrol agents. We highlight suitable genera and species of entomopathogenic fungi that can be used to control the most troublesome insect pests. Finally, we provide insights into commercially available formulations of these fungi, underscoring their practicality and potential in the field of agriculture.

2. Early Studies and Phylogenetic Placement of Entomopathogenic Fungi

According to the keyword search 'entomopathogenic fungi' on the Scopus database (www.scopus.com), study findings on the subject started in 1966. Post-1978, documentation is in the double digits, and in 1990, for the first time, over 50 studies were reported. In 1993, publications exceeded 100. Twenty years later, in 2013, over 1,000 studies were published. This trend continues to the present; 2,263 publications were seen in 2020, and in 2023, 2,840 studies were published.

The earliest studies on entomopathogenic fungi were conducted in the 1800s and focused on developing control methods to manage diseases in the silkworm industry (Goettel et al., 2005). Agostino Bassi discovered and described the first entomopathogenic fungus in 1835, which caused white muscardine disease in insects (Mantzoukas et al., 2022). This strain was later named *Beauveria bassiana* (Hypocreales, Cordycepsaceae). A few years later, Elias Metschnikoff (1845–1916) discovered green muscardine, a fungal disease that attacks insects (*Anisoplia austriaca*) caused by *Metarhizium anisopliae* (Mantzoukas et al., 2022; Zimmermann et al., 1995). Thereafter, many studies have been conducted on *Beauveria* and *Metarhizium*, attempting to understand various aspects of the fungi, including the parasite-host relationship (Nel'zina et al., 1978), storage (Daoust and Roberts, 1983), formulation (Daoust et al., 1983), and characterization (Dorta et al., 1996). The foremost molecular-based studies conducted by Smithson et al. (1995); Bailey et al. (1996); Bogo et al. (1996), and Hegedus and Khachatourians (1996) provided further, in-depth understanding about fungi. Through years of taxonomical and diversity studies, today, over 1,000 entomopathogenic fungi species across 100 genera have been identified (Chen et al., 2021).

With continuous updates and changes in classification, there has been debate among researchers about placing entomopathogenic fungi into a particular taxon. However, today, there is overall agreement that entomopathogenic fungi are classified within seven distinct phyla: Ascomycota, Basidiomycota, Blastocladiomycota, Chytridiomycota, Entomophthoromycota, Microsporidia, and Zoopagomycota (Kaczmarek and Boguś, 2021). Initially, Oomycota was categorized as a fungus; it has now been recognized as a phylum of Chromista. Most importantly, even though they are morphologically similar to fungi, molecular phylogenetic results show that they are a distinct group. There are several species of entomopathogenic fungi classified into Oomycota, among them, *Lagenidium giganteum* is the most studied species (Sila et al., 2023), and *Crypticola clavulifera* has also been well-researched (Mendoza et al., 2018). According to the most recent classification, most entomopathogenic fungi species (over 40 genera) are found in the order Hypocreales. The most common entomopathogenic genus within the order is *Cordyceps*, which comprises over 400 species (Boomsma et al., 2014; Chandler, 2017).

3. Mechanism of Infection

The infection mechanism of entomopathogenic fungi begins through physical contact with the host. Fungi spores are the main causative agents; they may attach to the insect exoskeleton (cuticle), be ingested through the mouth, or pass through the spiracles before initiating infection. Fungi (e.g., *Beauveria bassiana*) intrusion through both oral and cuticle can overwhelm the insect's immune system (Mannino et al., 2019). Infection through the cuticle has been found to be the most common. The fungi generally enter the insect body via direct penetration by damaging the cuticle. After spore adhesion to the cuticle, it germinates and produces a germ tube. The germ tube then develops appressoria that can be seen at the tube tip (Bihal et al., 2023). The appressoria further develops into a penetration peg, and through mechanical force and extracellular enzymes (e.g., lipases, proteases, and chitinases), localized breakdown of the cuticle allows the penetration peg to enter the insect body. Notably, in addition to extracellular enzymes to soften the cuticle, appressorium also produces oxalic acid (Vidhate et al., 2023). An interesting characteristic is that when spores are in contact with an insect's wings, they will not germinate, as it is not a suitable location to initiate infection. Therefore, disruption of the fungi gene *MAD1* will reduce the adherence of spores to the wings and allow them to slip out for another place (Wang and St Leger, 2007).

After the penetration of fungal hyphae through the integument, it changes morphology into yeast-like hyphal bodies or blastospores (increasing surface area) inside the nutrient-rich hemocoel. These hyphal bodies multiply through the budding and freely circulate inside the hemocoel. This not only helps them absorb more nutrition but also to overcome, avoid, or suppress insect immune defenses (Boucias et al., 2016; Zhu et al., 2023). In the hemolymph, the yeast-like hyphal bodies proliferate exponentially, reaching densities that greatly exceed the number of circulating hemocytes. These hemolymph-borne cells revert synchronously to an apical growth process upon reaching a critical threshold density, generating the tissue-invasive mycelial cell phenotype. This subsequent mycelial (tissue-invasive) phase generates and secretes compounds that effectively degrade insect tissue and cause the rapid death of nutritionally deprived, weakened hosts (Boucias et al., 2016). The infection mechanism of entomopathogenic fungi is controlled by various genes, particularly virulence-related genes. In a comprehensive literature review, Shin et al. (2020) discussed the genes responsible for each infection process; for example, fungus penetration is related to the genes *chit1*, *chit3*, *chiti2*, *CYP52*, *CYP53*, *Pr1*, and *Pr2*. While blastospore formation is controlled by the *ATG7*, *Bbsnf1*, and *Fkh2* genes. The growth of the blastospore is manipulated by the gene *ATM1*. In addition, *Bbmpk1*, *MaMk1*, and *MaPls1* are responsible for the re-penetration of fungal hyphae. In addition to Shin et al. (2020), Vidhate et al. (2023) also discussed the infection mechanism and highlighted the role of associated fungus genes.

4. Entomopathogenic Fungi: Screening for Potential Biocontrol

Even though entomopathogenic fungi demonstrate the ability to control insects, screening and selection of species are necessary prior to their use as biocontrol agents. Among all the species of entomopathogenic fungi, the highest rates of infection and virulence are found among Entomophthorales, e.g., *Conidiobolus*, *Furia*, *Erynia*, and *Entomophaga*, though culturing difficulties restrict their utilization (Litwin et al., 2020). On the other hand, some of the entomopathogenic fungi produce fewer spores. As shown in Fig. 1, the species of *Cordyceps* and *Ophiocordyceps* are effective in killing insects; however, obtaining an adequate number of spores to prepare insecticidal formulations is difficult.

Fig. 1 (a) *Cordyceps ninchukispora* showing bright orange stromata emerging from insect pupa; (b) *Ophiocordyceps* sp. on an adult insect (photo credit: A.K. Hasith Priyashantha; Location: Chiang Mai, Northern Thailand).

Thus, high-sporulation strains, such as *Beauveria* sp. and *Metarhizium* sp., are of particular interest (Fig. 2).

It is interesting to discuss possible identification techniques for highly virulent strains of entomopathogenic fungi. Here, researchers first collect the fungi from natural habits, and species are identified through morphological and molecular techniques. Correctly identifying species is essential before employing them as biocontrol agents. After identification, initial bioassays are generally conducted to identify pathogenicity (Imoulan et al., 2017; Islam et al., 2023). However, here, it is not necessary to check initial pathogenicity against only targeted pests; it may be tested against any other insect. To confirm the pathogenicity of a strain of fungus, for instance, it is possible to first introduce the fungus to rice mealworm larvae (in PDA plates) and observe whether the larvae become infected. There may also be effects on non-insect pests. If successful, a major study can be initiated on the application against a targeted insect. It is necessary to recognize that fungi have the ability to infect non-targeted species. Vivekanandhan et al. (2022) tested the effects of *Metarhizium anisopliae* (ethyl acetate extracts) against several types of mosquitoes (larva, pupae, and adult), and found the fungus had lethal effects. The researchers further tested the fungus on *Artemia nauplii* and *Eudrilus eugeniae* (non-target hosts) and found effects on these species to be minimal, thus concluding the potential of *M. anisopliae* as a biocontrol agent against mosquitoes. Researchers sometimes determine a high-virulence

Fig. 2 (a) An insect infected with *Beauveria bassiana*; (b) *Metarhizium pinghaense* under compound microscope (photo credit: A.K. Hasith Priyashantha; Location: Chiang Mai, Northern Thailand).

strain using enzyme assays. Of these, chitinase, cellulase, protease, and lipase media can be prepared to test the degradation capacity of each fungus. Such media represent the components of an insect cuticle with the highest degradation (in Petri dishes), therefore indicating the most virulent strain or fungus (Stuart et al., 2020).

Determining biocontrol potential against a particular insect pest is intriguing. For example, conidia of a selected strain of entomopathogenic fungi can be harvested, and different concentrations (10^2 to 10^8 conidia/mL) of conidial suspension can be applied to chosen stages of the targeted host (mostly the larvae) or even another alternative host. Dipping larvae into each aqueous solution or spraying them are the general practice methods (Mantzoukas et al., 2020; Shahriari et al., 2021; Gao et al., 2022). The mortality of tested life stages is then counted after several days (e.g., 7 to 14 days), and lethal concentration 50 (LC50) values are calculated (Shahriari et al., 2021). Researchers may also count dead insects daily to better understand the mortality rate (Fergani and Refaei, 2021). Sometimes, researchers do not conduct initial pathogenicity tests (on non-targeted hosts), but do conduct the confirmation test. For instance, Fergani and Refaei (2021) demonstrated the effect of *Beauveria bassiana* against larval instars (L2:L5) of *Spodoptera littoralis*. They observed the death of host larvae after fungal treatment. They then collected the dead larvae and kept them in Petri dishes alongside sterilized moistened filter paper. They recorded fungus growth daily for mycosis testing to confirm mortality was due to *B. bassiana*.

Another important factor is the location where entomopathogenic fungi are applied for pest control. They may be used on crop farming lands or animal husbandry farms. In real-world applications, the fungus must be able to survive, reproduce (multiply), and disseminate. The efficacy of entomopathogenic fungi is highly dependent on environmental factors. Gindin et al. (2009) tested the applicability of a virulent *Metarhizium anisopliae*-K isolate in broiler houses against the lesser mealworm, *Alphitobius diaperinus*. They conducted a simulated poultry house bioassay since entomopathogenic fungi generally show lower survivability above 30°C, particularly *Metarhizium*, which sees optimal growth between 20 and 30°C (Seib et al., 2023). In the poultry industry, farmhouses may have temperatures higher than 30°C. The researchers evaluated the efficacy of the fungus in 28–32°C conditions, with 70–80% relative humidity, in a climate-controlled room. They found that the fungus showed considerable efficacy in controlling targeted pests, thus confirming its potential as a biocontrol agent (Gindin et al., 2009).

Another important factor to consider is effective formulation. Bait formulations have been used less frequently compared with wettable powders or oil dispersions, which are the most frequently used methods (Lei et al., 2023). Using bait formulations is significantly more challenging than direct application. Targeted pests must be attracted to the bait, by using attractants and phagostimulants, basically. Further, repeated visits and increased contact time between insects and bait are needed. Molasses, milk, and yeast have been used in early studies; however, today, sex pheromone and other food-based volatile organic compounds are also utilized (Baker et al., 2020). Bait may be either a solid or liquid-based formulation (Cheraghi et al., 2013).

Ángel-Sahagún et al. (2010) demonstrated the efficacy of various formulations for *Metarhizium anisopliae* and *Isaria fumosorosea*, namely tween (water, tween, and agricultural surfactant), citroline (water, tween, agricultural surfactant, and mineral oil-citroline), celite (celite-diatomaceous earth, water), and wheat bran. Among these, celite and wheat bran are recognized as better formulations (see Section 5.5). According to recent literature, oil-based formulations perform better than other types. For example, Lee et al. (2023) studied the best formulations for *Beauveria bassiana* (strain JN5R1W1), among several types of vegetable oils used as a main ingredient, including soybean, corn, and canola. Interestingly, all the oil-based formulations showed greater results in conidial survivability; overall, soybean oil indicated the highest conidial germination rate. The development of more complex formulations, e.g., oil-in-gum emulsions, is also reported. For instance, Cantú-Bernal et al. (2023) prepared an Acacia gum and a powdered vegetable oil-based formulation suitable for carrying the conidia of *Hirsutella citriformis*.

5. Entomopathogenic Fungi as Pest Killers

Thus far, we have discussed infection mechanisms and the selection of entomopathogenic fungi for pest control. Here, we further consider these fungi to understand the entomopathogenic effect on a wide range of insect groups. Entomopathogenic fungi in natural ecosystems are vital in controlling pest epidemics (Verma et al., 2020). The adaptability and plasticity of fungi increase their rates of survival (Gul et al., 2014). Further, several characteristics of these fungi, such as their ability to produce dormant or saprophytic stages in the short term and their minimal impact on alternative hosts, ensure their long-term survival without potential target organisms (Sharma et al., 2020). Soil-dwelling entomopathogenic fungi use other survival strategies. *Metarhizium robertsii* forms a root-rhizosphere association, transferring nitrogen from insect cadavers to plants and acquiring carbon in return, thus staying alive until finding a suitable host.

However, entomopathogenic fungi are susceptible to UV radiation, high temperatures, humidity, and pH levels, like many other fungi groups. According to experiments attempting to culture field-collected fungi, failure to culture in PDA sometimes occurred due to the destruction of germinability during transportation from field to laboratory. This mainly occurred due to prolonged UV exposure and storage at elevated temperatures. Besides abiotic factors, competition with other microbes and antagonistic enzymes and compounds in plants or insect hosts may also restrict growth (Sharma et al., 2020). Mycoparasitic fungi can also affect entomopathogenic fungi, e.g., *Syspastospora parasitica* attacks *Beauveria bassiana*.

Entomopathogenic fungi also use several mechanisms to infect large numbers of insects. For example, some species release infective spores at night. This helps protect the spores from heat and UV radiation. Further, many genera of insects are also more active at night than in the daytime, thus more susceptible to contamination by the released spores (Aak et al., 2018; Zulfitri et al., 2018). These fungi frequently lead to natural epizootics in nocturnal insect populations (Sharma et al., 2020). Another impressive characteristic of entomopathogenic fungi is their ability to control insect behaviour. *Cordyceps* (*Ophiocordyceps*) are well-known for controlling host behaviour, and, upon infection, their victims behave like zombies; thus, they are named "zombie fungi" (Sharma et al., 2023). After being infected, victim hosts exhibit constant, directionless movement. Before dying, they move away from their insect colony and seek elevated positions (summited behaviour). Later, the fruiting bodies of fungi emerge from the dead insects and eject spores. Releasing spores at elevated positions helps in dispersal by wind currents over greater distances, thereby achieving larger coverage areas (de Bekker et al., 2021). Another example of manipulating behaviour can be seen with the genus *Massospora*. To be successful, they consume the host (e.g., cicada) from within and rupture the abdomen. Despite a ruptured abdomen, the host remains alive and makes short flights, allowing the fungus to disperse its spores (Cooley et al., 2018; Boyce et al., 2019). In the following, we briefly discuss the most studied entomopathogenic fungi by targeted insect groups.

5.1 Coleoptera (Beetles and Weevils)

The coconut leaf beetle (*Brontispa longissima*) (Chrysomelidae) causes severe damage to coconut trees. It can be successfully controlled by the entomopathogenic fungus *Metarhizium anisopliae* (Liu et al., 1989). It has also been found that the coconut palm rhinoceros beetle (*Oryctes rhinoceros*) (Scarabaeidae) can be managed using *M. anisopliae*. Similarly, the red pumpkin beetle (*Aulacophora foveicollis*), which infects pumpkins, cucumbers, melons, and most fruits and vegetables belonging to Cucurbits, may be controlled through the application of *Beauveria bassiana* (Moorthi and Balasubramanian, 2016). Wood-boring red bay ambrosia beetle (*Xyleborus glabratus*) (a vector of *Raffaelea lauricola,* which causes laurel wilt in avocado) can also be minimized by *B. bassiana* (Carrillo et al., 2015). Further, a recent study tested controlling the red palm weevil (*Rhynchophorus ferrugineus*) using *B. bassiana* and *Isaria fumosorosea,* demonstrating promising results (Yang et al., 2023).

5.2 Diptera (Flies and Mosquitoes)

Mosquitoes (Culicidae) are mostly recognized for causing disease in humans. However, they also threaten livestock animals, including cattle, horses, pigs, sheep, working dogs, and so on. Not enough attention has been given to addressing this threat to animals (Folly et al., 2020; Kayedi et al., 2020). Entomopathogenic fungi have been found to be effective in controlling mosquitoes, benefiting both human well-being and animal husbandry. The two most widely recognized entomopathogenic fungi for controlling mosquitoes are *Metarhizium* spp. and *Beauveria* spp. (Accoti et al., 2021). Other fungi genera, such as *Coelomomyces, Culicinomyces,* and *Entomophthora*, have also shown the potential to decrease mosquito populations (Scholte et al., 2004). Initial isolation of *Lagenidium giganteum*, done by Couch (1935), reported the ability to infect *Culex* and *Anopheles* mosquito larvae in North Carolina, USA. Subsequent studies have shown their ability to infect *Aedes aegypti, A. mediovittatus, Culex quinquefasciatus*, *C. restuans*, *Mansonia* sp., *Ochlerotatus sollicitans*, *O. taeniorhynchus*, and *O. triseriatus* (Scholte et al., 2004).

Recent studies show that the fungus *Metarhizium anisopliae* holds promise as a biocontrol agent against *Anopheles gambiae,* a major malaria vector (Accoti et al., 2021; Cafarchia et al., 2022). Several fungi, including *M. anisopliae, M. brunneum, M. humberi, M. robertsii, Beauveria bassiana, B. brongniartii, Isaria fumosorosea,*

Mucor hiemalis, and *Fusarium oxysporum*, have been shown to be effective in controlling *A. aegypti* mosquitoes. These mosquitoes transmit diseases like dengue fever, Zika virus, and chikungunya fever (Cafarchia et al., 2022; Accoti et al., 2021; Cafarchia et al., 2022; Accoti et al., 2021; WHO, 2020). Interestingly, Deng et al. (2019) genetically modified *Beauveria bassiana* fungi by introducing the *Bacillus thuringiensis* (Bt) toxin gene Cyt2Ba to improve efficiency in killing mosquitoes.

The ability to control house flies (*Musca domestica*) using *B. bassiana, I. fumosorosea,* and *M. anisopliae* has been demonstrated. Larvae of houseflies create problems for livestock, the most prominent being that eggs laid in animal wounds hatch into larvae that feed on flesh, causing significant health problems for the animal. Among other recent studies, Baker et al. (2020) demonstrated the ability of *M. anisopliae* to manage houseflies and compared the efficacy of this approach with synthetic chemical applications. They recognized that a bait formulation of fungus-coated white rice with skim milk powder was most attractive to the flies, being even more effective than the conventional chemical bait test (Quickbayt®). The research, although the chemical bait method yielded quick results, killing 50% of tested flies within two days, showed a similar mortality rate when using the tested fungi within four to five days. Peach fruit fly (*Bactrocera zonata*) is another insect pest on fruits and vegetables, and it has been found that *B. bassiana* and *M. anisopliae* can be effectively used as biopesticides against them (Murtaza et al., 2022). The west Indian fruit fly *Anastrepha obliqua* is a prominent pest on guava and mango and also attacks several plant families, such as Annonaceae, Anacardiaceae, Fabaceae, Bignoniaceae, Rosaceae, and Myrtaceae (Lozano-Tovar et al., 2023).

5.3 Hemiptera (Aphids, Whiteflies, and Bugs)

The control of aphids is challenging due to their relatively higher resistance to chemical insecticides, shorter lifespans with large numbers of nymphs, and seasonal migration (Herron et al., 2020). Therefore, the application of entomopathogenic fungi is a beneficial alternative to controlling aphids; available data indicates positive results. The virus-carrying cotton aphid, *Aphis gossypii* (Aphididae), is a major agricultural pest in the cotton industry (Yi et al., 2023). Im et al. (2022) demonstrated the high virulence of *Beauveria bassiana* JEF-544 against *A. gossypii*. Abdel-Raheem et al. (2021) and Ullah et al. (2022) showed the potential of *B. bassiana* and *Metarhizium anisopliae* in controlling *A. craccivora* and *Myzus persicae*. Moreover, the study by Abdel-Raheem et al. (2021) showed the capability of *Verticillium lecanii* to manage *A. craccivora* populations. Whitefly is another virus-transmitting pest responsible for severe losses in almost all commercially cultured crops (Saurabh et al., 2021). Control of the soybean pest *Bemisia tabaci* is a topic of interest to many researchers. Zou et al. (2014) showed that the efficacy of *I. fumosorosea* (strain IfB01) can be increased by combining it with chemical insecticides such as spirotetramat, acetamiprid, imidacloprid, and thiamethoxam. Research largely suggests that combined application is far more effective than using either fungal or chemical treatment alone. It also indicates that, other than *I. fumosorosea, B. bassiana* is also suitable to control *B. tabaci*, with optimal control achieved through the combined application of synthetic insecticides such as acetamiprid, bifenthrin, dinotefuran, and pyriproxyfen (Iqbal et al., 2022).

5.4 Hymenoptera (Wasps)

Wasps belong to the family Vespidae. They are primarily found in tropical areas. They fulfil certain beneficial ecological roles for humans, such as helping control some pest species by killing them to feed their larvae and pollinating flowers while feeding on nectar. They also cause harmful effects, e.g., as a vector of disease-causing bacteria such as *Escherichia coli* and causing the loss of fruit crops such as grapes, plums, and pears (Szczepko et al., 2020). Therefore, they act as pests under certain circumstances. There are relatively few studies available on wasps. Among the few studies that have been conducted, an experiment by Harris et al. (2000) is notable. In their study, pathogenic activity on *Vespula vulgaris* by *Beauveria bassiana, Metarhizium anisopliae*, and *Aspergillus flavus* was reported. Wasps may also become invasive; the Asian paper wasp-*Polistes chinensis*, is a eusocial wasp species recognized as an invasive species in New Zealand (Beggs et al., 2011; Howse et al., 2022). Reason et al. (2022) have found that *Beauveria malawiensis* and *Ophiocordyceps humbertii* are two virulent fungi associated with *P. chinensis*. They estimate that about 3.3% of wild colonies of *P. chinensis* are infected with these fungi. Laboratory infection bioassays suggest that wasp nests treated with *B. malawiensis* have a considerable mortality rate of adult wasps, whereas *Ophiocordyceps* assays indicated no elevated infection compared to the control.

5.5 Ixodida (Ticks)

There are two families of ticks: soft ticks (family Argasidae) and hard ticks (family Ixodidae) (Greay et al., 2016). The parasitic behavior of ticks affects cattle, sheep, cats, horses, dogs, and other livestock animals globally. It has been reported that *Beauveria* and *Metarhizium* are the most recorded pathogenic genera of ticks (Ebani, and Mancianti, 2021; Cafarchia et al., 2022). In an early study, Kalsbeek et al. (1995) isolated *B. bassiana, B. brongniartii, Paecilomyces farinosus, P. fumosoroseus, Verticillium aranearum*, and *V. lecanii* from vegetation, small rodents, and deer. In later studies, researchers attempted to control tics with entomopathogenic fungi-based formulations. Ángel-Sahagún et al. (2010) investigated using fungi like *M. anisopliae* (33 isolates) and *Isaria fumosorosea* (20 isolates) to control *Rhipicephalus microplus* ticks. The experiment was conducted under both laboratory conditions and in the field, testing different formulations. The results of the laboratory experiment showed that *M. anisopliae* killed 2–100% of infected larvae, and *I. fumosorosea* (13 isolates) caused mortality between 7–94%. In the field experiment (14 days post-application), the best results were seen with celite and wheat bran formulations, causing 67.8 and 94.2% pest population reduction, respectively. In a similar study, Cruz-Avalos et al. (2015) found that, other than *M. anisopliae* and *I. fumosorosea*, *B. bassiana* is also effective in controlling *Rhipicephalus microplus*. Abdigoudarzi et al. (2009) recognized that *B. bassiana* is also a suitable candidate for controlling *Hyalomma anatolicum*. Later, Stafford and Allan (2010) highlighted the importance of *B. bassiana* in controlling *Ixodes scapularis*. Sullivan et al. (2020) recently demonstrated the effectiveness of *B. bassiana*, along with *M. anisopliae* and *M. brunneum*, against larval winter ticks, *Dermacentor albipictus*.

5.6 Lepidoptera (Moths)

In their study, Stuart et al. (2020) demonstrated the ability to control European pepper moths, *Duponchelia* fovealis, using fungal consortia—*Beauveria bassiana* (Bea 111, Bov 2 and Bov 3), one species of *Isaria javanica* (Isa 340), and one species of *Purpureocillium lilacinum*. Fitriana et al. (2021) and Gebreslasie et al. (2023) highlighted the effects of *B. bassiana* (*Aspergillus* spp.) and *Metarhizium rileyi* against the corn pest *Spodoptera litura*. Moreover, several other studies have shown the potential of *B. bassiana* in controlling sugarcane internode borer (*Chilo Sacchariphagus indices*) (Shahid et al., 2023), fall armyworm (*Spodoptera frugiperda*) (Noctuidae) in maize (Ramanujam et al., 2020), and *Rachiplusia nu* (Noctuidae) in soybean (Abalo et al., 2022).

5.7 Mesostigmata and Trombidiformes (Mites)

Varroa destructor (Mesostigmata) is a parasitic mite found in honeybee colonies. If there is no treatment, typically, colonies die after two years of initial infection (Kuster et al., 2014). Control of this mite was achieved by Hamiduzzaman et al. (2012). They tested the common pathogens—*Metarhizium anisopliae* and *Beauveria bassiana,* along with *Clonostachys rosea.* They found that for significant control of the mite (compared to non-inoculated mites), it takes about seven days to show significant mortality of *V. destructor*. Recently, Bava et al. (2022) summarized the ability of various entomopathogenic fungi, including species of *Beauveria, Hirsutella,* and *Metarhizium,* to control *V. destructor.*

According to the academic literature, *B. bassiana* is the most common pathogen used to control Trombidiformes (Al-Zahrani et al., 2023). The Two-Spotted Spider Mite (*Tetranychus urticae*) belongs to the Tetranychidae family and causes economic losses in 150 plant varieties. *T. urticae* is also reported to cause a 10-50% loss in tomato production. *Beauveria bassiana* controls two-spotted spider mites effectively (Al-Zahrani et al., 2023). Rasool et al. (2023) reported *M. brunneum* suppression and killing of the Two-Spotted Spider Mite on tomatoes. Raspberry Eriophyoid Mite (*Phyllocoptes gracilis*) is a pest found on raspberries (Gordon and Taylor, 1976), and both *B. bassiana* and *M. anisopliae* are recognized as mite killers for this species (Minguely et al., 2021). Table 1 accounts for the most recent studies; we have summarized some of the potential entomopathogenic fungi that can be used as biocontrol for various pests.

6. Commercial Practices

At the beginning of this chapter, we have highlighted the importance of bio-control agents against various pests, particularly those that have tremendous potential to replace chemical pesticides. We further emphasized the importance of entomopathogenic fungi in commercial formulations and their current utilization. As previously highlighted, chemical pesticides are still used most frequently due to their energy-saving aspects, efficiency, and ease of use. However, in 80% of cases, pesticide overuse leads to resistance to more than one class of pesticides. Along with sustainable agricultural practices, during the last two decades, the utilization of entomopathogenic fungi

Table 1 Entomopathogenic fungi were recently recognized as having significant potential as pest bio-control agents.

Entomopathogenic fungi	*Host species*	*Group of pests*	*Reference*
Aspergillaceae			
Aspergillus oryzae	*Spodoptera litura*	Moth	Fitriana et al., 2021
Penicillium citrinum	*Culex quinquefasciatus*	Mosquito	Accoti et al., 2021
Capnodiales			
Cladosporium oxysporum	*Aphis craccivora*	Aphid	Maina et al., 2018
Clavicipitaceae			
Aschersonia aleyrodis	*Bemisia tabaci*	White fly	Sani et al., 2020
Aschersonia placenta			
Metarhizium anisopilae	*Aedes aegypti*	Mosquito	Cafarchia et al., 2022
	Aedes albopictus		
	Amblyomma parvum	Tick	
	Anopheles gambiae	Mosquito	
	Anopheles stephensi		
	Aphis craccivora	Aphid	Maina et al., 2018
	Aphis gossypii		
	Bactrocera zonata	Fruit fly	Murtaza et al., 2022
	Bemisia tabaci	White fly	Sani et al., 2020
	Blattella germanica	Bug	Zhang et al., 2018
	Cnaphalocrocis medinalis	Moth	Hong et al., 2017
	Culex pipiens	Mosquito	Cafarchia et al., 2022
	Culex quinquefasciatus		
	Curculio chinensis	Weevil	Li et al., 2021
	Cylas formicarius		Dotaona et al., 2017
	Dermacentor albipictus	Tick	Cafarchia et al., 2022
	Haemaphysalis longicornis		
	Ixodes ricinus		
	Lipaphis erysimi	Aphid	Maina et al., 2018
	Locusta migratoria manilensis	Locust	Li et al., 2021
	Mahanarva fimbriolata	Bug	Mascarin et al., 2019
	Mahanarva spectabilis		
	Monochamus alternatus	Beetle	Li et al., 2021

Contd.

Table 1 *Contd.*

Entomopathogenic fungi	*Host species*	*Group of pests*	*Reference*
	Musca domestica	House fly	Farooq and Freed, 2016
	Oedaleus asiaticus	Locust	Li et al., 2021
	Plutella xylostella	Moth	
	Polyphylla laticollis	Beetle	
	Rhipicephalus decoloratus	Tick	Cafarchia et al., 2022
	Rhipicephalus microplus		
	Rhipicephalus sanguineus		
	Stomoxys calcitrans	Fly	Baleba et al., 2021
	Varroa destructor	Mite	Hamiduzzaman et al., 2012
	Vespula vulgaris	Wasp	Harris et al., 2000
Metarhizium brunneum	*Aedes aegypti*	Mosquito	Cafarchia et al., 2022
	Anopheles stephensi		
	Culex qinquefasciatus		
	Dermacentor albipictus	Tick	
	Ixodes scapularis		
	Rhipicephalus annulatus		
	Tetranychus urticae	Mite	Rasool et al., 2023
Metarhizium humberi	*Aedes aegypti*	Mosquito	Cafarchia et al., 2022
Metarhizium pemphigi	*Ixodes ricinus*	Tick	
Metarhizium rileyi	*Spodoptera litura*	Moth	Gebreslasie et al., 2023
Metarhizium robertsii	*Aedes aegypti*	Mosquito	Cafarchia et al., 2022
	Rhipicephalus microplus	Tick	
	Gonipterus scutellatus	Beetle	Mejía et al., 2024
Cordycipitaceae			
Beauveria bassiana	*Aedes aegypti*	Mosquito	Accoti et al., 2021
	Aedes albopictus		Deng et al., 2019
	Aphis gossypii	Aphid	Erol et al., 2020
	Anopheles coluzzii	Mosquito	Cafarchia et al., 2022
	Anopheles gambiae		
	Anopheles stephensi		
	Aphis craccivora	Aphid	Maina et al., 2018

Contd.

Table 1 *Contd.*

Entomopathogenic fungi	*Host species*	*Group of pests*	*Reference*
	Aulacophora foveicollis	Beetle	Moorthi and Balasubramanian, 2016
	Bactrocera zonata	Fruit fly	Murtaza et al., 2022
	Bemisia tabaci	Whitefly	Sani et al., 2020
	Brevicoryne brassicae	Aphid	Maina et al., 2018
	Chilo Sacchariphagus indices	Moth	Shahid et al., 2023
	Cosmopolites sordidus	Weevil	Mascarin et al., 2019
	Culex pipiens	Mosquito	Cafarchia et al., 2022
	Culex quinquefasciatus		
	Dermacentor albopictus	Tick	
	Gonipterus scutellatus	Beetle	Mejía et al., 2024
	Haemaphysalis longicornis	Tick	Cafarchia et al., 2022
	Haemaphysalis quinghaiensis		
	Hypothenemus hampei	Beetle	Mascarin et al., 2019
	Lipaphis erysimi	Aphid	Maina et al., 2018
	Musca domestica	House fly	Farooq and Freed, 2016
	Myzus percsicae	Aphid	Maina et al., 2018
	Rachiplusia nu	Moth	Abalo et al., 2022
	Rhipicephalus decoloratus	Tick	Cafarchia et al., 2022
	Rhipicephalus microplus		
	Rhipicephalus sanguineous		
	Rhopalosiphum padi	Aphid	Maina et al., 2018
	Rhynchophorus ferrugineus	Weevil	Yang et al., 2023
	Schizaphis graminum	Aphid	Maina et al., 2018
	Solenopsis invicta	Ant	Wei et al., 2021

Contd.

Table 1 *Contd.*

Entomopathogenic fungi	*Host species*	*Group of pests*	*Reference*
	Spodoptera frugiperda	Moth	Ramanujam et al., 2020
	Spodoptera litura		Fitriana et al., 2021; Shahid et al., 2023
	Tetranychus urticae	Mite	Al-Zahrani et al., 2023
	Varroa destructor		Hamiduzzaman et al., 2012
	Vespula vulgaris	Wasp	Harris et al., 2000
	Xyleborus glabratus	Beetle	Carrillo et al., 2015
Beauveria brongniartii	*Aedes aegypti*	Mosquito	Cafarchia et al., 2022
	Aedes albopictus		
Isaria spp.	*Bemisia tabaci*	Whitefly	Sani et al., 2020
Isaria fumosorosea	*Aedes aegypti*	Mosquito	Accoti et al., 2021
	Anopheles gambiae		
	Bemisia tabaci	Whitefly	Sani et al., 2020
	Lipaphis erysimi	Aphid	Maina et al., 2018
	Musca domestica	Housefly	Farooq and Freed, 2016
	Plutella xylostella	Moth	Maina et al., 2018
	Rhynchophorus ferrugineus	Weevil	Yang et al., 2023
	Spodoptera litura	Moth	Ullah et al., 2019
Lecanicillium lecanii	*Bemisia tabaci*	Whitefly	Sani et al., 2020
L. muscarium	*Bemisia tabaci*		
Davidiellaceae			
Cladosporium cladosporioides	*Aphis gossypii*	Aphid	Accoti et al., 2021
Dipodascaceae			
Geotrichum candidum	*Anophelus gambiae*	Mosquito	Accoti et al., 2021
Hypocreales			
Verticillium alfalfae	*Aphis gossypii*	Aphid	Erol et al., 2020
Verticillium lecanii	*Aphis craccivora*		Maina et al., 2018
	Lipaphis erysimi		
	Myzus persicae		
Purpureocillium lilacinum	*Lipaphis erysimi*		
Trichoderma viride	*Aphis gossypii*		Erol et al., 2020

Contd.

Table 1 *Contd.*

Entomopathogenic fungi	*Host species*	*Group of pests*	*Reference*
Mucoraceae			
Mucor hiemalis	*Aedes aegypti*	Mosquito	Accoti et al., 2021
Nectriaceae			
Fusarium oxysporum	*Aedes aegypti*	Mosquito	Accoti et al., 2021
	Anopheles stephensi		
	Culex quinquefasciatus		
Trichocomaceae			
Aspergillus flavus	*Vespula vulgaris*	Wasp	Harris et al., 2000

has received considerable attention (Bamisile et al., 2021; Bihal et al., 2023). As of April 2016, including entomopathogenic fungi, there are 1,401 available biopesticide products and 299 available biopesticide ingredients registered with the United States Environmental Protection Agency (Wang et al., 2021).

According to laboratory and field experiments, it is clear that myco-pesticides are the most auspicious and effective alternative to synthetic pesticides. Based on available evidence, most entomopathogenic fungi-based pesticides are considered safe to use and effective in reducing synthetic pesticide abuse (Bamisile et al., 2021). To improve the management of entomopathogenic fungi-based product use, it is recommended to consider factors such as identification of target pests, monitoring, application time, method of application, application equipment, phytosanitary management of nearby fields, product storage, and handling and transportation (Barra-Bucarei et al., 2019). When it comes to the commercial field, like academic-based studies, the most attention has been given to formulating *Beauveria bassiana* and *Metarhizium anisopliae*. Today, licensed commercial products derived from these two organisms are widely applied as conidial and/or mycelium-based formulations (Barra-Bucarei et al., 2019). These products have passed registration requirements and are now widely used for pest biocontrol in many countries. Similarly, *Lecanicillium lecanii* has been reported as a mycoinsecticide and was the first fungus to be developed as an inundating mycoinsecticide or used in medium- and large-scale greenhouse agriculture (Bamisile et al., 2021). In comprehensive studies, Bamisile et al. (2021) and Liu et al. (2023) have listed the number of commercially available biopesticides. Figure 3 illustrates some of these available biopesticides; this only represents a handful of available products.

In addition to the biocontrol of insect pests, entomopathogenic fungi also deliver several other benefits. They can also act as effective biofertilizers, particularly due to their endophytic lifecycle. These endophytic entomopathogenic fungi are considered an alternative to fertilizers and an effective and environmentally friendly way to ensure food security (Glick, 2014). In organic farming methods, the use of endophytic entomopathogenic fungi to increase yield and protect plants from damage is increasing. Endophytic entomopathogenic fungi also play a role as growth promoters for plants (Bamisile et al., 2021). These beneficial effects make entomopathogenic fungi an ideal addition to modern agriculture.

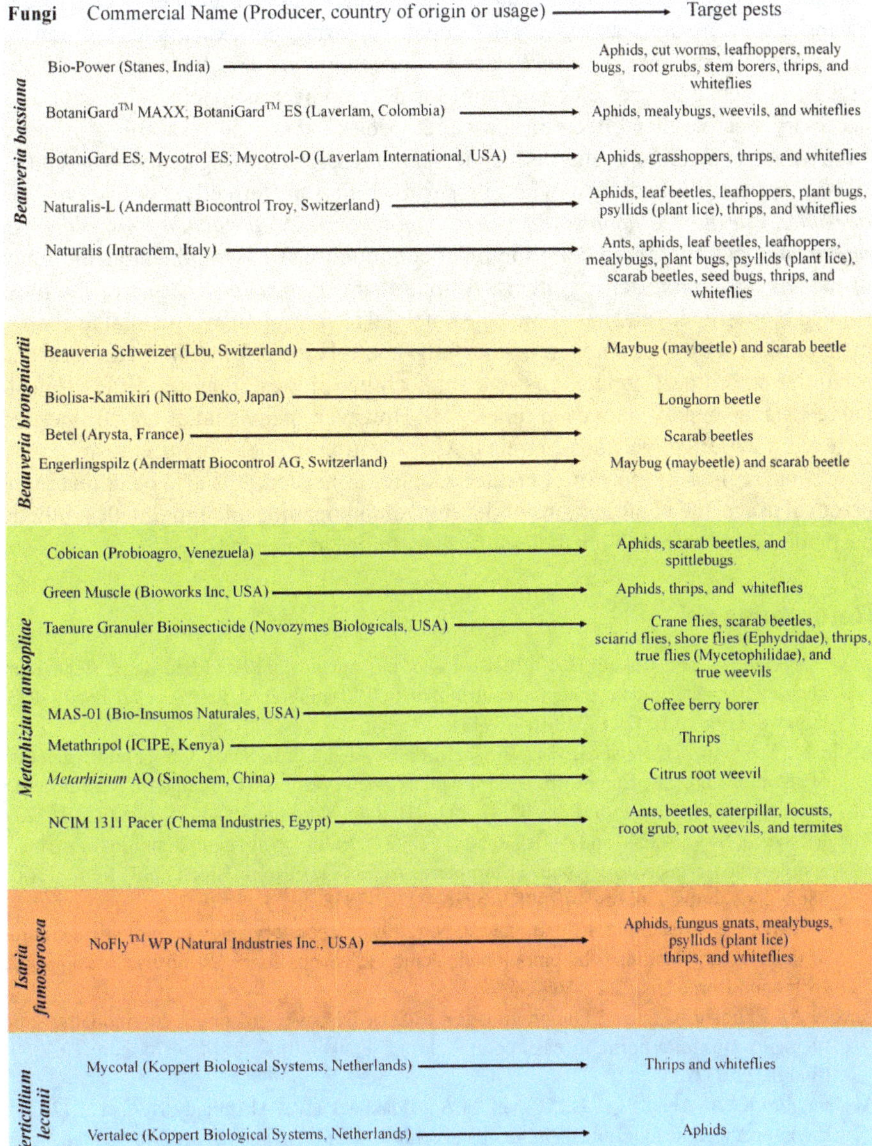

Fig. 3 Some common biopesticides that are designed and labelled as alternatives to chemical pesticides are based on recent literature. The bold letters represent the entomopathogenic fungi species used for the formulations. The commercial name of the product is shown at the rear end of the arrow, along with the producer's details and country of origin or applied region (within parentheses). The arrows point to the main targeted pests to be controlled.

Conclusion

Entomopathogenic fungi are widely recognized today, given their ability to control insect pests in cropland, thereby minimizing the necessity of hazardous synthetic pesticides. They have shown a peculiar way of infecting insects; unlike viruses and bacteria, they predominantly infect through rupturing the exocuticle. Laboratory experiments show that it is necessary to understand species and strains with the highest virulence before selecting fungi for development as biocontrol agents. It is necessary to screen their effects on non-target species and their survival ability in varied environmental conditions, such as high temperatures. Recognizing better formulations is vital, directly affecting the fungi's survivability and effectiveness. *Beauveria bassiana, Hirsutella thompsonii, Isaria fumosorosea, Metarhizium anisopliae, M. brunneum,* and *M. robertsii* are the most commonly used entomopathogenic fungi in today's agriculture farm systems. Among them, *B. bassiana* is one of the most studied biopesticides and has shown the potential to control many potential to control many insect groups. Entomopathogenic fungi control crop infestations in the field and minimize stored pest populations; the applicability of these fungi in different post-harvest stages makes them even more vital. However, many studies are still needed to reveal new effective fungal strains. As utilization of these fungi is restricted to only some regions of the world, greater adoption is needed. It is also clear that more practical usage and combinations of different entomopathogenic fungi with synthetic pesticides are valuable. Nonetheless, further studies are needed.

References

Aak, A., Hage, M. and Rukke, B.A. (2018). Insect pathogenic fungi and bed bugs: Behaviour, horizontal transfer and the potential contribution to IPM solutions. J. Pest Sci., 91(2): 823–835. https://doi.org/10.1007/s10340-017-0943-z

Abalo, M., Scorsetti, A.C., Vianna, M.F., Russo, M.L., De Abajo, J.M., et al. (2022). Field evaluation of entomopathogenic fungi formulations against *Rachiplusia nu* (Lepidoptera: Noctuidae) in soybean crop. J. Plant Prot. Res., 62(4): 403–410. https://doi.org/10.24425/jppr.2022.143232

Abdel-Raheem, Saad, A.F.A. and Abdel-Rahman. (2021). Entomopathogenic fungi on fabae bean aphid, *Aphis craccivora* (Koch) (Hemiptera: Aphididae). Rom. Biotechnol. Lett., 26(4): 2862–2868. https://doi.org/10.25083/rbl/26.4/2862-2868

Abdigoudarzi, M., Esmaeilnia, K. and Shariat, N. (2009). Laboratory study on biological control of ticks (Acari: Ixodidae) by entomopathogenic indigenous fungi (*Beauveria bassiana*). J. Arthropod Borne Dis., 3(2): 36–43.

Accoti, A., Engdahl, C.S. and Dimopoulos, G. (2021). Discovery of novel entomopathogenic fungi for mosquito-borne disease control. Front. Fungal Biol., 2: https://doi.org/10.3389/ffunb.2021.637234

Al-Zahrani, J.K., Al-Abdalall, A.H., Osman, M.A., Aldakheel, L.A., AlAhmady, N.A. et al. (2023). Entomopathogenic fungi and their biological control of *Tetranychus urticae*: Two-spotted spider mites. J. King Saud Univ. Sci., 35(8): 102910. https://doi.org/10.1016/j.jksus.2023.102910

Ángel-Sahagún, C.A., Lezama-Gutiérrez, R., Molina-Ochoa, J., Pescador-Rubio, A., Skoda, S.R. et al. (2010). Virulence of Mexican isolates of entomopathogenic fungi (Hypocreales: Clavicipitaceae) upon Rhipicephalus=*Boophilus microplus* (Acari: Ixodidae) larvae and the efficacy of conidia formulations to reduce larval tick density under field conditions. Vet. Parasitol., 170(3–4): 278–286. https://doi.org/10.1016/j.vetpar.2010.02.037

Bailey, A.M., Kershaw, M.J., Hunt, B.A., Paterson, I.C., Charnley, A.K. et al. (1996). Cloning and sequence analysis of an intron-containing domain from a peptide synthetase-encoding gene of the entomopathogenic fungus *Metarhizium anisopliae*. Gene, 173(2): 195–197. https://doi.org/10.1016/0378-1119(96)00212-0

Baker, D., Rice, S., Leemon, D., Godwin, R. and James, P. (2020). Development of a mycoinsecticide bait formulation for the control of house flies, *Musca domestica* L. Insects, 11(1): 47. https://doi.org/10.3390/insects11010047

Baleba, S.B.S., Agbessenou, A., Getahun, M.N., Akutse, K.S., Subramanian, S. et al. (2021). Infection of the stable fly, *Stomoxys calcitrans*, L. 1758 (Diptera: Muscidae) by the entomopathogenic fungi *Metarhizium anisopliae* (Hypocreales: Clavicipitaceae) negatively affects its survival, feeding propensity, fecundity, fertility, and fitness parameters. Front. Fungal Biol., 2: 637817. https://doi.org/10.3389/ffunb.2021.637817

Bamisile, B.S., Akutse, K.S., Siddiqui, J.A. and Xu, Y. (2021). Model application of entomopathogenic fungi as alternatives to chemical pesticides: Prospects, challenges, and insights for next-generation sustainable agriculture. Front. Plant Sci., 12: 741804. https://doi.org/10.3389/fpls.2021.741804

Barra-Bucarei, L., France Iglesias, A. and Pino Torres, C. (2019). Entomopathogenic fungi. *In* natural enemies of insect pests in neotropical agroecosystems (pp. 123–136). Cham: Springer International Publishing.

Bava, R., Castagna, F., Piras, C., Musolino, V., Lupia, C., Palma, E., Britti, D. and Musella, V. (2022). Entomopathogenic fungi for pests and predators control in beekeeping. Vet. Sci., 9(2): 95. https://doi.org/10.3390/vetsci9020095

Beggs, J. R., Brockerhoff, E. G., Corley, J. C., Kenis, M., Masciocchi, M. et al. (2011). Ecological effects and management of invasive alien Vespidae. BioControl, 56(4): 505–526. https://doi.org/10.1007/s10526-011-9389-z

Behie, S.W., Jones, S.J. and Bidochka, M.J. (2015). Plant tissue localization of the endophytic insect pathogenic fungi *Metarhizium* and *Beauveria*. Fungal Ecol., 13: 112–119. https://doi.org/10.1016/j.funeco.2014.08.001

Bihal, R., Al-Khayri, J.M., Banu, A.N., Kudesia, N., Ahmed, F.K., Sarkar, R., Arora, A. and Abd-Elsalam, K.A. (2023). Entomopathogenic fungi: An eco-friendly synthesis of sustainable nanoparticles and their nanopesticide properties. Microorganisms, 11(6): 1617. https://doi.org/10.3390/microorganisms11061617

Bogo, M.R., Vainstein, M.H., Aragão, F.J., Rech, E. and Schrank, A. (1996). High frequency gene conversion among benomyl resistant transformants in the entomopathogenic fungus *Metarhizium anisopliae*. FEMS Microbiol. Lett., 142(1): 123–127. https://doi.org/10.1016/0378-1097(96)00255-8

Boomsma, J.J., Jensen, A.B., Meyling, N.V. and Eilenberg, J. (2014). Evolutionary interaction networks of insect pathogenic fungi. Annu. Rev. Entomol., 59: 467–485, https://doi.org/10.1146/annurev-ento-011613-162054.

Boucias, D., Liu, S., Meagher, R. and Baniszewski, J. (2016). Fungal dimorphism in the entomopathogenic fungus *Metarhizium rileyi*: Detection of an in vivo quorum-sensing system. J. Invertebr. Pathol., 136: 100–108. https://doi.org/10.1016/j.jip.2016.03.013

Boyce, G.R., Gluck-Thaler, E., Slot, J.C., Stajich, J.E., Davis, W.J. et al. (2019). Psychoactive plant-and mushroom-associated alkaloids from two behavior modifying cicada pathogens. Fungal Ecol., 41: 147–164. https://doi.org/10.1016/j.funeco.2019.06.002

Cafarchia, C., Pellegrino, R., Romano, V., Friuli, M., Demitri, C. et al. (2022). Delivery and effectiveness of entomopathogenic fungi for mosquito and tick control: Current knowledge and research challenges. Acta Trop., 234: 106627. https://doi.org/10.1016/j.actatropica.2022.106627

Cantú-Bernal, S.H., Gomez-Flores, R., Flores-Villarreal, R.A., Orozco-Flores, A.A., Romo-Sáenz, C.I. et al. (2023). Adult *Diaphorina citri* biocontrol using *Hirsutella citriformis* strains and gum formulations. Plants, 12(18): 3184. https://doi.org/10.3390/plants12183184

Carrillo, D., Dunlap, C.A., Avery, P.B., Navarrete, J., Duncan, R.E. et al. (2015). Entomopathogenic fungi as biological control agents for the vector of the laurel wilt disease, the redbay ambrosia beetle, *Xyleborus glabratus* (Coleoptera: Curculionidae). Biol. Control, 81: 44–50. https://doi.org/10.1016/j.biocontrol.2014.10.009

Chandler, D. (2017). Basic and applied research on entomopathogenic fungi. In *Microbial Control of Insect and Mite Pests* (pp. 69–89). Elsevier.

Chen, W., Xie, W., Cai, W., Thaochan, N. and Hu, Q. (2021). Entomopathogenic fungi biodiversity in the soil of three provinces located in southwest China and first approach to evaluate their biocontrol potential. J. Fungi, 7(11): 984. https://doi.org/10.3390/jof7110984

Cheraghi, A., Habibpour, B. and Mossadegh, M.S. (2013). Application of bait treated with the entomopathogenic fungus *Metarhizium anisopliae* (Metsch.) Sorokin for the control of *Microcerotermes diversus* Silv. Psyche (Camb. Mass.), 2013: 1–5. https://doi.org/10.1155/2013/865102

Cooley, J.R., Marshall, D.C. and Hill, K.B.R. (2018). A specialized fungal parasite (*Massospora cicadina*) hijacks the sexual signals of periodical cicadas (Hemiptera: Cicadidae: Magicicada). Sci. Rep., 8(1): 1432. https://doi.org/10.1038/s41598-018-19813-0

Cruz-Avalos, A., Cruz-Vázquez, C., Lezama-Gutiérrez, R., Vitela-Mendoza, I. and Angel-Sahagún, C. (2015). Selección de aislados de hongos entomopatógenos para el control de *Rhipicephalus microplus* (Acari: Ixodidae). Trop. Subtrop. Agroecosyst. 18 (2): 175–180.

Daoust, R.A. and Roberts, D.W. (1983). Studies on the prolonged storage of *Metarhizium anisopliae* conidia: effect of temperature and relative humidity on conidial viability and virulence against mosquitoes. J. Invertebr. Pathol., 41(2): 143–150. https://doi.org/10.1016/0022-2011(83)90213-6

de Bekker, C., Beckerson, W.C., and Elya, C. (2021). Mechanisms behind the madness: How do zombie-making fungal entomopathogens affect host behavior to increase transmission? MBio, 12(5): e01872-21. https://doi.org/10.1128/mBio.01872-21

Deng, S.Q., Zou, W.-H., Li, D.L., Chen, J.T., Huang, Q. et al. (2019). Expression of *Bacillus thuringiensis* toxin Cyt2Ba in the entomopathogenic fungus *Beauveria bassiana* increases its virulence towards Aedes mosquitoes. PLOS Negl. Trop. Dis., 13(7): e0007590. https://doi.org/10.1371/journal.pntd.0007590

Dorta, B., Ertola, R.J. and Arcas, J. (1996). Characterization of growth and sporulation of *Metarhizium anisopliae* in solid-substrate fermentation. Enzyme Microb. Technol., 19(6): 434–439. https://doi.org/10.1016/s0141-0229(96)00017-8

Dotaona, R., Wilson, B.A.L., Ash, G.J., Holloway, J. and Stevens, M.M. (2017). Sweetpotato weevil, *Cylas formicarius* (Fab.) (Coleoptera: Brentidae) avoids its host plant when a virulent *Metarhizium anisopliae* isolate is present. J. Invertebr. Pathol., 148: 67–72. https://doi.org/10.1016/j.jip.2017.05.010

Ebani, V.V. and Mancianti, F. (2021). Entomopathogenic fungi and bacteria in a veterinary perspective. Biology, 10(6): 479. https://doi.org/10.3390/biology10060479

Erol, A., Abdelaziz, O., Birgücü, A.K., Senoussi, M.M., Oufroukh, A. and Karaca, İ. (2020). Effects of some entomopathogenic fungi on the aphid species, *Aphis gossypii* Glover (Hemiptera: Aphididae). Egypt. J. Biol. Pest Contr., 30(1): 108. https://doi.org/10.1186/s41938-020-00311-3

FAO (2023). Researchers Helping Protect Crops From Pests. Retrieved 2024 Jan 02 from https://www.nifa.usda.gov/about-nifa/blogs/researchers-helping-protect-crops-pests

Farooq, M. and Freed, S. (2016). Infectivity of housefly, *Musca domestica* (Diptera: Muscidae) to different entomopathogenic fungi. Braz. J. Microbiol., 47(4): 807–816. https://doi.org/10.1016/j.bjm.2016.06.002

Fergani, Y.A. and Refaei, E.A.E. (2021). Pathogenicity induced by indigenous *Beauveria bassiana* isolate in different life stages of the cotton leafworm, *Spodoptera littoralis* (Boisduval) (Lepidoptera: Noctuidae) under laboratory conditions. Egypt. J. Biol. Pest Contr., 31(1): 64. https://doi.org/10.1186/s41938-021-00411-8

Fitriana, Y., Suharjo, R., Swibawa, I.G., Semenguk, B., Pasaribu, L.T. et al. (2021). *Aspergillus oryzae* and *Beauveria bassiana* as entomopathogenic fungi of *Spodoptera litura* Fabricius (Lepidoptera: Noctuidae) infesting corn in Lampung, Indonesia. Egypt. J. Biol. Pest Control., 31(1): 127. https://doi.org/10.1186/s41938-021-00473-8

Folly, A.J., Dorey-Robinson, D., Hernández-Triana, L.M., Phipps, L.P. and Johnson, N. (2020). Emerging threats to animals in the United Kingdom by arthropod-borne diseases. Front. Vet. Sci., 7: 20. https://doi.org/10.3389/fvets.2020.00020

Gao, Y.-P., Luo, M., Wang, X.-Y., He, X.Z., Lu, W. and Zheng, X.-L. (2022). Pathogenicity of *Beauveria bassiana* PfBb and immune responses of a non-target host, *Spodoptera frugiperda* (Lepidoptera: Noctuidae). Insects, 13(10): 914. https://doi.org/10.3390/insects13100914

Gebreslasie, M.G., Nishi, O., Wasano, N. and Yasunaga-Aoki, C. (2023). Varied selectivity of caterpillar-specific *Metarhizium rileyi* and generalist entomopathogenic fungi against last instar larvae and pupae of common cutworm, *Spodoptera litura* (Lepidoptera: Noctuidae). Appl. Entomol. Zool. , 58(3): 219–228. https://doi.org/10.1007/s13355-023-00824-x

Gindin, G., Glazer, I., Mishoutchenko, A. and Samish, M. (2009). Entomopathogenic fungi as a potential control agent against the lesser mealworm, *Alphitobius diaperinus* in broiler houses. Biocontrol, 54(4): 549–558. https://doi.org/10.1007/s10526-008-9205-6

Glick, B.R. (2014). Bacteria with ACC deaminase can promote plant growth and help to feed the world. Microbiol Res., 169(1): 30–39. https://doi.org/10.1016/j.micres.2013.09.009

Goettel, M.S., Eilenberg, J. and Glare, T. (2005). Entomopathogenic fungi and their role in regulation of insect populations. *In* Comprehensive molecular insect science (pp. 361–405). Elsevier.

Gordon, S.C. andTaylor, C.E. (1976). Some aspects of the biology of the raspberry leaf and bud mite (Phyllocoptes (Eriophyes) Gracilis Nal.) eriophyidae in Scotland. J. Hortic. Sci., 51(4): 501–508. https://doi.org/10.1080/00221589.1976.11514719

Greay, T.L., Oskam, C.L., Gofton, A.W., Rees, R.L., Ryan, U.M. et al. (2016). A survey of ticks (Acari: Ixodidae) of companion animals in Australia. Parasit. Vectors 9(1): 207. https://doi.org/10.1186/s13071-016-1480-y

Gul, H.T., Saeed, S., and Khan, F.Z.A. (2014). Entomopathogenic fungi as effective insect pest management tactic: A review. App. Sci. Business Econ., 1(1): 10–18.

Hamiduzzaman, M.M., Sinia, A., Guzman-Novoa, E. and Goodwin, P.H. (2012). Entomopathogenic fungi as potential biocontrol agents of the ecto-parasitic mite, *Varroa destructor*, and their effect on the immune response of honey bees (*Apis mellifera* L.). J. Invertebr. Pathol., 111(3): 237–243. https://doi.org/https://doi.org/10.1016/j.jip.2012.09.001

Harris, R.J., Harcourt, S.J., Glare, T.R., Rose, E.A. and Nelson, T.J. (2000). Susceptibility of *Vespula vulgaris* (Hymenoptera: vespidae) to generalist entomopathogenic fungi and their potential for wasp control. J. Invertebr. Pathol., 75(4): 251–258. https://doi.org/10.1006/jipa.2000.4928

Hegedus, D.D. and Khachatourians, G.G. (1996). Identification and differentiation of the entomopathogenic fungus *Beauveria bassiana* using polymerase chain reaction and single-strand conformation polymorphism analysis. J. Invertebr. Pathol., 67(3): 289–299. https://doi.org/10.1006/jipa.1996.0044

Herron, G., Powis, K., and Rophail, J. (2000). Baseline studies and preliminary resistance survey of Australian populations of cotton aphid *Aphis gossypii* Glover (Hemiptera: Aphididae). Aust. J. Entomol., 39(1): 33–38. https://doi.org/10.1046/j.1440-6055.2000.00134.x

Hong, M., Peng, G., Keyhani, N.O. and Xia, Y. (2017). Application of the entomogenous fungus, *Metarhizium anisopliae*, for leafroller (*Cnaphalocrocis medinalis*) control and its effect on rice phyllosphere microbial diversity. Appl. Microbiol. Biotechnol., 101(17): 6793–6807. https://doi.org/10.1007/s00253-017-8390-6

Howse, M.W.F., McGruddy, R.A., Felden, A., Baty, J.W., Haywood, J. and Lester, P.J. (2022). The native and exotic prey community of two invasive paper wasps (Hymenoptera: Vespidae) in New Zealand as determined by DNA barcoding. Biol. Invasions, 24(6): 1797–1808. https://doi.org/10.1007/s10530-022-02739-0

Hu, Z. (2020). What socio-economic and political factors lead to global pesticide dependence? A critical review from a social science perspective. Int. J. Environ. Res. Public Health, 17(21): 8119. https://doi.org/10.3390/ijerph17218119

Im, Y., Park, S.E., Lee, S.Y., Kim, J.C. and Kim, J.S. (2022). Early-stage defense mechanism of the cotton aphid *Aphis gossypii* against infection with the insect-killing fungus *Beauveria bassiana* JEF-544. Front. Immunol., 13: 907088. https://doi.org/10.3389/fimmu.2022.907088

Imoulan, A., Hussain, M., Kirk, P.M., El Meziane, A. and Yao, Y.-J. (2017). Entomopathogenic fungus *Beauveria*: Host specificity, ecology and significance of morpho-molecular characterization in accurate taxonomic classification. J. Asia. Pac. Entomol., 20(4): 1204–1212. https://doi.org/10.1016/j.aspen.2017.08.015

Iqbal, M., Arif, M.J., Saeed, S., ul Hasan, M. and Javed, N. (2022). Biorational approach for management of whitefly, *Bemisia tabaci* (Gennadius) (Homoptera: Aleyrodidae), on cotton crop. Int. J. Trop. Insect Sci., 42(2): 1461–1469. https://doi.org/10.1007/s42690-021-00664-8

Islam, S.M.N., Chowdhury, M.Z.H., Mim, M.F., Momtaz, M.B. and Islam, T. (2023). Biocontrol potential of native isolates of *Beauveria bassiana* against cotton leafworm *Spodoptera litura* (Fabricius). Sci. Rep., 13: 8331. https://doi.org/10.1038/s41598-023-35415-x.

Islam, W., Adnan, M., Shabbir, A., Naveed, H., Abubakar, Y.S., et al. (2021). Insect-fungal-interactions: A detailed review on entomopathogenic fungi pathogenicity to combat insect pests. Microb. Pathog., 159: 105122. https://doi.org/10.1016/j.micpath.2021.105122

Kaczmarek, A. and Boguś, M.I. (2021). Fungi of entomopathogenic potential in Chytridiomycota and Blastocladiomycota, and in fungal allies of the Oomycota and Microsporidia. IMA fungus, 12(1): 29. https://doi.org/10.1186/s43008-021-00074-y

Kakaei, H., Nourmoradi, H., Bakhtiyari, S., Jalilian, M. and Mirzaei, A. (2022). Effect of COVID-19 on food security, hunger, and food crisis. *In* COVID-19 and the sustainable development goals (pp. 3–29). Elsevier.

Kalsbeek, V., Frandsen, F. and Steenberg, T. (1995). Entomopathogenic fungi associated with *Ixodes ricinus* ticks. Exp. Appl. Acarol., 19(1): 45–51. https://doi.org/10.1007/bf00051936

Karar, M.E., Alsunaydi, F., Albusaymi, S. and Alotaibi, S. (2021). A new mobile application of agricultural pests recognition using deep learning in cloud computing system. Alex. Eng. J., 60(5): 4423–4432. https://doi.org/10.1016/j.aej.2021.03.009

Kayedi, M.H., Sepahvand, F., Mostafavi, E., Chinikar, S., Mokhayeri, H. et al. (2020). Morphological and molecular identification of Culicidae mosquitoes (Diptera: Culicidae) in Lorestan province, Western Iran. Heliyon, 6(8): e04480. https://doi.org/10.1016/j.heliyon.2020.e04480

Kidanu, S. and Hagos, L. (2020). Research and Application of entomopathogenic fungi as pest management option: A review. Environ. Earth Sci., 10(3): 31–39. https://doi.org/10.7176/JEES/10-3-03

Kuster, R.D., Boncristiani, H.F. and Rueppell, O. (2014). Immunogene and viral transcript dynamics during parasitic Varroa destructor mite infection of developing honeybee (*Apis mellifera*) pupae. J. Exp. Biol., 217(10): 1710–1718. https://doi.org/10.1242/jeb.097766

Lee, J.Y., Mi Woo, R. and Dong Woo, S. (2023). Formulation of the entomopathogenic fungus *Beauveria bassiana* JN5R1W1 for the control of mosquito adults and evaluation of its novel applicability. J. Asia. Pac. Entomol., 26(2): 102056. https://doi.org/10.1016/j.aspen.2023.102056

Lei, C.J., Ahmad, R.H.I.R., Halim, N.A., Asib, N., Zakaria, A. and Azmi, W.A. (2023). Bioefficacy of an oil-emulsion formulation of entomopathogenic fungus, *Metarhizium anisopliae* against adult red palm weevil, *Rhynchophorus ferrugineus*. Insects, 14(5): 482. https://doi.org/10.3390/insects14050482

Li, S., Xu, C., Du, G., Wang, G., Tu, X. et al. (2021). Synergy in efficacy of *Artemisia sieversiana* crude extract and *Metarhizium anisopliae* on resistant *Oedaleus asiaticus*. Front Physiol., 12: 642893. https://doi.org/10.3389/fphys.2021.642893

Litwin, A., Nowak, M. and Różalska, S. (2020). Entomopathogenic fungi: Unconventional applications. Rev. Environ. Sci. Biotechnol. 19: 23–42.

Liu, D., Smagghe, G. and Liu, T.-X. (2023). Interactions between entomopathogenic fungi and insects and prospects with glycans. J. Fungus, 9(5): 575. https://doi.org/10.3390/jof9050575

Liu, S.D., Lin, S.C. and Shiau, J.F. (1989). Microbial control of coconut leaf beetle (*Brontispa longissima*) with green muscardine fungus, *Metarhizium anisopliae* var. anisopliae. J. Invertebr. Pathol., 53(3): 307–314. https://doi.org/10.1016/0022-2011(89)90094-3

Lozano-Tovar, M.D., Ballestas Álvarez, K.L., Sandoval-Lozano, L.A., Palma Mendez, G.M. and Barrera-Cubillos, G.P. (2023). Study on the insecticidal activity of entomopathogenic fungi for the control of the fruit fly (*Anastrepha obliqua*), the main pest in mango crop in Colombia. Arch. Microbiol., 205(3): 83. https://doi.org/10.1007/s00203-023-03405-2

Maina, U., Galadima, I., Gambo, F. and Zakaria, D. (2018). A review on the use of entomopathogenic fungi in the management of insect pests of field crops. J. Entomol. Zool. Stud, 6(1): 27–32.

Mannino, M.C., Huarte-Bonnet, C., Davyt-Colo, B. and Pedrini, N. (2019). Is the insect cuticle the only entry gate for fungal infection? Insights into alternative modes of action of entomopathogenic fungi. Fungi. J. Fungi, 5(2): 33. https://doi.org/10.3390/jof5020033

Mantzoukas, S., Kitsiou, F., Natsiopoulos, D. and Eliopoulos, P.A. (2022). Entomopathogenic fungi: Interactions and applications. Encyclopedia, 2(2): 646–656. https://doi.org/10.3390/encyclopedia2020044

Mantzoukas, S., Lagogiannis, I., Karmakolia, K., Rodi, A., Gazepi, M. and Eliopoulos, P. A. (2020). The effect of grain type on virulence of entomopathogenic fungi against stored product pests. Appl. Sci., 10(8): 2970. https://doi.org/10.3390/app10082970

Mascarin, G.M., Lopes, R.B., Delalibera, Í., Fernandes, É.K.K., Luz, C. et al. (2019). Current status and perspectives of fungal entomopathogens used for microbial control of arthropod pests in Brazil. J. Invertebr. Pathol., 165: 46–53. https://doi.org/10.1016/j.jip.2018.01.001

Mejía, C., Barrera, G., Pulgarín Díaz, J.A. and Espinel, C. (2024). The Eucalyptus snout beetle in Colombia: Selection and evaluation of entomopathogenic fungi as bioinsecticides against *Gonipterus platensis*. Biol. Control, 188: 105407. https://doi.org/10.1016/j.biocontrol.2023.105407

Mendoza, L., Vilela, R. and Humber, R.A. (2018). Taxonomic and phylogenetic analysis of the Oomycota mosquito larvae pathogen *Crypticola clavulifera*. Fungal Biol., 122(9): 847–855. https://doi.org/10.1016/j.funbio.2018.04.010

Minguely, C., Norgrove, L., Burren, A. and Christ, B. (2021). Biological control of the raspberry eriophyoid Mite *Phyllocoptes gracilis* using entomopathogenic fungi. Horticulturae, 7(3): 54. https://doi.org/10.3390/horticulturae7030054

Moorthi, P.V. and Balasubramanian, C. (2016). *Aulacophora foveicollis*, a natural diet to entomopathogenic fungus, *Beauveria bassiana*. J. Basic Appl. Zool., 73: 28–31. https://doi.org/10.1016/j.jobaz.2015.09.007

Murtaza, G., Naeem, M., Manzoor, S., Khan, H.A., Eed, E.M. et al. (2022). Biological control potential of entomopathogenic fungal strains against peach fruit fly, *Bactrocera zonata* (Saunders) (Diptera: Tephritidae). Peer J, 10: e13316. https://doi.org/10.7717/peerj.13316

Nayduch, D., Neupane, S., Pickens, V., Purvis, T. and Olds, C. (2023). House flies are underappreciated yet important reservoirs and vectors of microbial threats to animal and human health. Microorganisms, 11(3): 583. https://doi.org/10.3390/microorganisms11030583

Nel'zina, E.N., Mironov, N.P., Sorokina, L.I. and Bakhtinova, N.Z. (1978). Nature of the parasite-host relations of the entomopathogenic fungi, *Metarrhizium anisopliae* (Metsch.) Sorokin and *Beauveria bassiana* (Bals.) Vouill. (Fungi Imperfecti) to *Ceratophyllus fasciatus* Bosc. (Siphonaptera) fleas. Med. Parazitol. (Mosk.), 47(4): 86–89.

Priyashantha, A.K.H. and Attanayake, R.N. (2021). Can anaerobic soil disinfestation (ASD) be a game changer in tropical agriculture? Pathogens, 10(2): 133. https://doi.org/10.3390/pathogens10020133

Ramanujam, B., Poornesha, B. and Shylesha, A.N. (2020). Effect of entomopathogenic fungi against invasive pest *Spodoptera frugiperda* (J. E. Smith) (Lepidoptera: Noctuidae) in maize. Egypt. J. Biol. Pest Control., 30(1): 100. https://doi.org/10.1186/s41938-020-00291-4

Rani, L., Thapa, K., Kanojia, N., Sharma, N., Singh, S. et al. (2021). An extensive review on the consequences of chemical pesticides on human health and environment. J. Clean. Prod., 283: 124657. https://doi.org/10.1016/j.jclepro.2020.124657

Rashid, S.F., Theobald, S. and Ozano, K. (2020). Towards a socially just model: Balancing hunger and response to the COVID-19 pandemic in Bangladesh. BMJ Glob. Health, 5(6): e002715. https://doi.org/10.1136/bmjgh-2020-002715

Rasool, S., Markou, A., Hannula, S.E. and Biere, A. (2023). Effects of tomato inoculation with the entomopathogenic fungus *Metarhizium brunneum* on spider mite resistance and the rhizosphere microbial community. Front. microbiol., 14: 1197770. https://doi.org/10.3389/fmicb.2023.1197770

Reason, A., Bulgarella, M. and Lester, P.J. (2022). Identity, prevalence, and pathogenicity of entomopathogenic fungi infecting invasive Polistes (Vespidae: Polistinae) paper wasps in New Zealand. Insects, 13(10): 922. https://doi.org/10.3390/insects13100922

Sani, I., Ismail, S.I., Abdullah, S., Jalinas, J., Jamian, S. et al. (2020). A Review of the biology and control of whitefly, *Bemisia tabaci* (Hemiptera: Aleyrodidae), with special reference to biological control using entomopathogenic fungi. Insects, 11(9): 619. https://doi.org/10.3390/insects11090619

Saurabh, S., Mishra, M., Rai, P., Pandey, R., Singh, J. et al. (2021). Tiny flies: A mighty pest that threatens agricultural productivity-a case for next-generation control strategies of whiteflies. Insects, 12(7): 585. https://doi.org/10.3390/insects12070585

Scholte, E.J., Knols, B.G., Samson, R.A. and Takken, W. (2004). Entomopathogenic fungi for mosquito control: A review. J. Insect Sci., 4 (1):19. https://doi.org/10.1093/jis/4.1.19

Scopus. (2024). Retrieved January 22, 2024, from Scopus.com website: https://www.scopus.com/

Seib, T., Fischer, K., Sturm, A.M. and Stephan, D. (2023). Investigation on the influence of production and incubation temperature on the growth, virulence, germination, and conidial size of *Metarhizium brunneum* for granule development. J. Fungi, 9(6): 668. https://doi.org/10.3390/jof9060668

Shahid, M., Haq, E., Mohamed, A., Rizvi, P.Q. and Kolanthasamy, E. (2023). Entomopathogen-based biopesticides: Insights into unraveling their potential in insect pest management. Front. microbiol., 14: 1208237. https://doi.org/10.3389/fmicb.2023.1208237

Shahriari, M., Zibaee, A., Khodaparast, S.A. and Fazeli-Dinan, M. (2021). Screening and virulence of the entomopathogenic fungi associated with *Chilo suppressalis* Walker. J. Fungi, *7*(1): 34. https://doi.org/10.3390/jof7010034

Sharma, A., Srivastava, A., Shukla, A.K., Srivastava, K., Srivastava, A.K. et al. (2020). Entomopathogenic fungi: A potential source for biological control of insect pests. *In* phytobiomes: current insights and future vistas; Springer Singapore: Singapore, 2020; pp. 225–250 ISBN 9789811531507.

Sharma, H., Sharma, N. and An, S.S.A. (2023). Unique bioactives from zombie fungus (*Cordyceps*) as promising multitargeted neuroprotective agents. Nutrients, 16(1): 102. https://doi.org/10.3390/nu16010102

Shin, T.Y., Lee, M.R., Park, S.E., Lee, S.J., Kim, W.J. and Kim, J.S. (2020). Pathogenesis-related genes of entomopathogenic fungi. Arch. Insect Biochem. Physiol., 105(4): e21747. https://doi.org/10.1002/arch.21747

Sila, M.M., Musila, F.M., Wekesa, V.W. and Sangilu, I.S. (2023). Evaluation of pathogenicity of entomopathogenic oomycetes *Lagenidium giganteum* and *L. ajelloi* against Anopheles Mosquito Larvae. Psyche (Camb. Mass.), 2023: 1–10. https://doi.org/10.1155/2023/2806034

Skinner, M., Parker, B.L. and Kim, J.S. (2014). Role of entomopathogenic fungi in integrated pest management. *In* integrated pest management (pp. 169–191). Elsevier.

Smithson, S.L., Paterson, I.C., Bailey, A.M., Screen, S.E., Hunt, B.A. et al. (1995). Cloning and characterisation of a gene encoding a cuticle-degrading protease from the insect pathogenic fungus *Metarhizium anisopliae*. Gene, 166(1): 161–165. https://doi.org/10.1016/0378-1119(95)00609-3

Stafford, K.C. III and Allan, S.A. (2010). Field applications of entomopathogenic fungi *Beauveria bassiana* and *Metarhizium anisopliae* F52 (Hypocreales: Clavicipitaceae) for the control of *Ixodes scapularis* (Acari: Ixodidae). J. Med. Entomol., 47(6), 1107–1115. https://doi.org/10.1603/me10019

Statista. (2023). Pesticide consumption worldwide 2022.. Retrieved December 12, 2023, from Statista website: https://www.statista.com/statistics/1263069/global-pesticide-use-by-country/

Stuart, A.K.D.C., Furuie, J.L., Zawadneak, M.A.C. and Pimentel, I.C. (2020). Increased mortality of the European pepper moth *Duponchelia fovealis* (Lepidoptera:Crambidae) using entomopathogenic fungal consortia. J. Invertebr. Pathol., 177: 107503. https://doi.org/10.1016/j.jip.2020.107503

Sullivan, C.F., Parker, B.L., Davari, A., Lee, M.R., Kim, J.S. and Skinner, M. (2020). Evaluation of spray applications of *Metarhizium anisopliae*, *Metarhizium brunneum* and *Beauveria bassiana* against larval winter ticks, *Dermacentor albipictus*. Exp. Appl. Acarol, 82(4): 559–570. https://doi.org/10.1007/s10493-020-00547-6

Szczepko, K., Kruk, A. and Wiśniowski, B. (2020). Local habitat conditions shaping the assemblages of vespid wasps (Hymenoptera: Vespidae) in a post-agricultural landscape of the Kampinos National Park in Poland. Sci. Rep., 10: 1424. https://doi.org/10.1038/s41598-020-57426-8

Tudi, M., Daniel Ruan, H., Wang, L., Lyu, J., Sadler, R. et al. (2021). Agriculture development, pesticide application and its impact on the environment. Int. J. Environ. Res. Public Health, 18(3): 1112. https://doi.org/10.3390/ijerph18031112

Ullah, M.I., Altaf, N., Afzal, M., Arshad, M., Mehmood, N. et al. (2019). Effects of Entomopathogenic fungi on the biology of *Spodoptera litura* (Lepidoptera: Noctuidae) and its reduviid predator, *Rhynocoris marginatus* (Heteroptera: Reduviidae). Int. J. Insect Sci., 11: 1–7. https://doi.org/10.1177/1179543319867116

Ullah, S., Raza, A.B.M., Alkafafy, M., Sayed, S., Hamid, M.I. et al. (2022). Isolation, identification and virulence of indigenous entomopathogenic fungal strains against the peach-potato aphid, *Myzus persicae* Sulzer (Hemiptera: Aphididae), and the fall armyworm, *Spodoptera frugiperda* (J.E. Smith) (Lepidoptera: Noctuidae). Egypt. J. Biol. Pest Contr., 32:2. https://doi.org/10.1186/s41938-021-00500-8

Umaru, F.F. and Simarani, K. (2020). Evaluation of the potential of fungal biopesticides for the biological control of the seed bug, *Elasmolomus pallens* (Dallas) (Hemiptera: Rhyparochromidae). Insects, 11(5): 277. https://doi.org/10.3390/insects11050277

Verma, D., Banjo, T., Chawan, M., Teli, N. and Gavankar, R. (2020). Microbial control of pests and weeds. *In* natural remedies for pest, disease and weed control; Elsevier, 2020; pp. 119–126. https://doi.org/10.1016/B978-0-12-819304-4.00010-5.

Vidhate, R.P., Dawkar, V.V., Punekar, S.A. and Giri, A.P. (2023). Genomic determinants of entomopathogenic fungi and their involvement in pathogenesis. Microb. Ecol., 85(1): 49–60. https://doi.org/10.1007/s00248-021-01936-z

Vivekanandhan, P., Swathy, K., Murugan, A.C. and Krutmuang, P. (2022). Insecticidal efficacy of *Metarhizium anisopliae* derived chemical constituents against disease-vector mosquitoes. J. Fungi, 8(3): 300. https://doi.org/10.3390/jof8030300

Wang, C. and St Leger, R.J. (2007). The MAD1 adhesin of *Metarhizium anisopliae* links adhesion with blastospore production and virulence to insects, and the MAD2 adhesin enables attachment to plants. Eukaryot. Cell, 6(5): 808–816. https://doi.org/10.1128/ec.00409-06

Wang, H., Peng, H., Li, W., Cheng, P. and Gong, M. (2021). The toxins of *Beauveria bassiana* and the strategies to improve their virulence to insects. Front. Microbiol., 12: 705343. https://doi.org/10.3389/fmicb.2021.705343

Wang, X., Xu, J., Wang, X., Qiu, B., Cuthbertson, A.G. S. et al. (2019). *Isaria fumosorosea*-based zero-valent iron nanoparticles affect the growth and survival of sweet potato whitefly, *Bemisia tabaci* (Gennadius). Pest Manag. Sci., 75(8): 2174–2181. https://doi.org/10.1002/ps.5340

Wei, Z., Ortiz-Urquiza, A. and Keyhani, N.O. (2021). Altered expression of chemosensory and odorant binding proteins in response to fungal infection in the red imported fire ant, *Solenopsis invicta*. Front. Physiol., 12: 596571. https://doi.org/10.3389/fphys.2021.596571

WHO (2020). Vector-borne diseases. Retrieved January 20, 2024, from WHO.int website: https://www.who.int/news-room/fact-sheets/detail/vector-borne-diseases

Yang, T.H., Wu, L.H., Liao, C.T., Li, D., Young Shin, T. et al. (2023). Entomopathogenic fungi-mediated biological control of the red palm weevil *Rhynchophorus ferrugineus*. J. Asia. Pac. Entomol., 26(1): 102037. https://doi.org/10.1016/j.aspen.2023.102037

Yi, C., Teng, D., Xie, J., Tang, H., Zhao, D. et al. (2023). Volatiles from cotton aphid (*Aphis gossypii*) infested plants attract the natural enemy *Hippodamia variegata*. Front. Plant Sci., 14: 1326630. https://doi.org/10.3389/fpls.2023.1326630

Zhang, X.C., Li, X.X., Gong, Y.W., Li, Y.R., Zhang, K.L. et al. (2018). Isolation, identification, and virulence of a new *Metarhizium anisopliae* strain on the German cockroach. J. Econ. Entomol., 111(6): 2611–2616. https://doi.org/10.1093/jee/toy280

Zhu, G., Ding, W., Zhao, H., Xue, M., Chu, P. and Jiang, L. (2023). Effects of the entomopathogenic fungus *Mucor hiemalis* BO-1 on the physical functions and transcriptional signatures of *Bradysia odoriphaga* larvae. Insects, 14(2): https://doi.org/10.3390/insects14020162

Zimmermann, G., Papierok, B. and Glare, T. (1995). Elias Metschnikoff, Elie Metchnikoff or Ilya Ilich Mechnikov (1845-1916): A pioneer in insect pathology, the first describer of the entomopathogenic fungus *Metarhizium anisopliae* and how to translate a Russian name. Biocontrol Sci. Technol., 5(4): 527–530. https://doi.org/10.1080/09583159550039701

Zou, C., Li, L., Dong, T., Zhang, B. and Hu, Q. (2014). Joint action of the entomopathogenic fungus *Isaria fumosorosea* and four chemical insecticides against the whitefly *Bemisia tabaci*. Biocontrol Sci. Technol., 24(3): 315–324. https://doi.org/10.1080/09583157.2013.860427

Zulfitri, A., Lestari, A.S., Krishanti, N.P.R.A. and Zulfiana, D. (2018). Laboratory evaluation of the selected entomopathogenic fungi and bacteria against larval and pupal stages of *Spodoptera litura* L. IOP Conf. Ser. Earth Environ. Sci., 166: 012009. https://doi.org/10.1088/1755-1315/166/1/012009

7 Specificity and Biocontrol Potential of Entomopathogenic Fungi

P. Andrade-Hoyos,[1] L. Aguilar-Marcelino,[2] A. Luna-Cruz,[3] M.R. Vallejo-Pérez,[4] E. Molina-Gayosso,[5] M.N. Rivera-Jiménez[6] and G.S. Castañeda-Ramírez[7*]

1. Introduction

The term biological control was coined by a researcher in the entomology department at the University of California. However, the biological control researcher Harry Scott Smith (1919), 105 years ago, used the term biological control to refer to the use of natural enemies to control insects (Chant, 1967). Years later, Cisneros (1995) defined biological control as the repression of pests by their natural enemies, i.e., through the action of predators, parasitoids, and pathogens. However, the first experimental work with entomopathogenic fungi began 189 years ago. The first knowledge of fungi causing diseases in insects dates back to 1835, when Augostino Bass first demonstrated experimentally that the "sign disease" of the silkworm *Bombix mori* was caused by *Beauveria bassiana*.

Insect pest control began with studies in 1879 by Metchnikoff and in 1888 by Krassilstchick on the application of microorganisms with an insecticidal effect to control the beetle *Anisoplia austriaca* and the weevil *Cleonus pun*. 128 years ago, an experimental station for the production of *Beauveria bassiana* was established to regulate the leafhopper bug *Blissus leucopterus*. Results showed an effective reduction of *B. leucopterus*. In Florida, USA, between 1921 and 1924, microbial control was developed, and the fungus *Aschersonia aleyrodis* was released against the citrus whitefly *Dialeurodes citricola*. This was considered a successful action in the control of *D. citricola* (McCoy et al., 1988; Hernandez and Berlanga, 1996).

1 Instituto Nacional de Investigaciones Forestales Agrícolas y Pecuarias, INIFAP, Campo Experimental Zacatepec, Morelos, México.
2 Centro Nacional de Investigación Disciplinaria en Sanidad e Inocuidad Animal, INIFAP, Km 11 Carretera Federal Cuernavaca-Cuautla, No. 8534, C. P. 62550, Jiutepec, Estado de Morelos, México.
3 CONAHCYT-Instituto de Investigaciones Químico-Biológicas, Universidad Michoacana de San Nicol´as de Hidalgo, Morelia,58030 Michoacán, México.
4 CONAHCYT-Universidad Autónoma de San Luis Potosí, Coordinación para la Innovación y Aplicación de la Ciencia y la Tecnología, San Luis Potosí, C.P. 78210 S.L.P. México.
5 Universidad Politécnica de Puebla (UPP), Tercer Carril del Ejido, Serrano s/n, Cuanalá, 72640 Puebla, Pue.
6 Agrosis mg S.A.de C.V./ Biosoluciones Orgánico Minerales, Dirección de Investigación Agrícola, Villahermosa, Tabasco, México.
7 Centro de Investigación en Biotecnología (CEIB), Universidad Autónoma del Estado de Morelos (UAEM), Cuernavaca, Morelos, México. C.P, 62209.
* Corresponding author: sarahi.castaneda@uaem.mx; gs.castanedaramirez@gmail.com

It has been reported that the application of *A. aleyrodis* as a biofungicide is recent and not yet fully implemented, partly because it requires the selection of an isolate, as well as the production and formulation in large quantities (Baker and Cook, 1982). Biocontrol is carried out by a group of microorganisms, either naturally or by introduction, with antagonists (antibiosis), competition for nutrients, oxygen, or space, or hyperparasitism. This chapter deals with entomopathogenic fungal compatibility, host specificity, formulations, different approaches of evaluation, parasitism and pest management.

2. Entomopathogenic Fungi with Biocontrol Potential

There are many reports of entomopathogenic fungal species with over 700 species documented. However, few species have been studied with a focus on potential use as biocontrol of insect pests (Fig. 1) (Hajeck and Leger, 1994; López-Lastra et al., 2019; Castañeda-Ramirez et al., 2022). The most studied species are *Coelomyces, Leptolegnia, Lagenidium giganteum, Mucor, Entomophthorales* and *Ascosphaera*. Among these species, there are multiple reports of their entomopathogenic activity in laboratory tests, starting with the first report in 1879 of the fungus *M. anisopliae*, described by Metschnikoff (Zimmermann, et al., 1995). It has since been over 100 years since its discovery and multiple experiments around the world have been conducted. Currently, the most frequently utilized species globally are *M. anisopliae* (33.9%) and *Beauveria bassiana* (33.9%) (De Faria and Wraight, 2007). These species exhibit insecticidal effects ranging from 80 to 100% under various conditions.

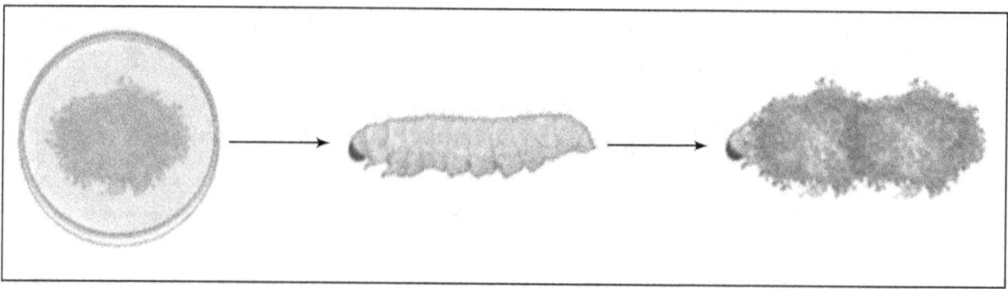

Fig. 1 Representation of a larva invaded by an entomopathogenic fungus (created with Biorender.com)

To determine that a fungus is entomopathogenic, it must have the following mechanisms of action: (a) the fungus spores attach to the body of the host; (b) it enters or breaks the cuticle of the insect; (c) the fungal hyphae invade the insect, weakening and causing death of the host; (d) conidiogenous cells are subsequently observed (Hyde et al., 2019). Some specificity has been observed at the family level. These characteristics allow these fungi to be considered as potential methods of pest control in agriculture. However, the use of these fungi as biocontrol is limited by some environmental factors, such as sun radiation, temperature, humidity, amount of yield, specificity, virulence with pathogens, and interactions with other compounds in the environment. When entomopathogenic fungi are tested under these conditions, few are able to move from *in-vitro* tests in the laboratory to *in-vivo* tests. Species that have been reported as commercially viable with biocontrol potential are: *V. lecanii*, *M. anisopliae*, *B. bassiana*, *B. brongniartii*, *M. flavoviridae*, *P. fumosoroseus* and *Lagenidium giganteum*.

3. Host Specificity

The different biological relationships in nature are the result of a co-evolutionary process (Roy et al., 2006). In parasitism, pathogens and hosts adapt to maximize their reproductive rate and ecological fitness through extremely diverse adaptive strategies, which are aimed at determining a certain level of host selection specificity by pathogens. In the parasitic relationship between entomopathogenic fungi and their hosts, this is denoted by the great taxonomic and behavioural diversity in both groups of organisms.

Fargues and Remaudiere (1977) recognized 4 categories of entomopathogenic fungi; the first two refer to accidental and occasional pathogens, which suggests chance events of encounter in nature and opportunism by the fungus due to injuries on the surface of insects; the third refers to facultative pathogens, fungi having the ability to develop by having a wide host range; the fourth category corresponds to obligate pathogens that can exhibit all degrees of specificity, from those that can attack insects belonging to several orders to those that only have a single host species. Onstad et al., (2006) mention that, in addition, facultative pathogens can multiply in the environment and in the host, in contrast to obligate pathogens, which can only multiply within the host body. It follows that the behaviour of entomopathogenic fungi in these categories is a function of a wide variety of adaptations, so that directly categorizing antagonistic behaviour under *in-vitro* conditions may overestimate the understanding of the parasitic relationship.

The specificity of entomopathogenic fungi is complex to determine as it is the result of the intricate ecological relationships between the fungus, the insect, the environment, and the timing of encounters between the entomopathogen and host insect populations. However, based on research to date, general relationships of specificity have been established between the different phylum to which different genera of entomopathogenic fungi belong and the insect groups they parasitize (Table 1).

Table 1 Phylum and orders of entomopathogenic fungi and the insect groups they parasitize

Phylum	*Group of insect*
Basidiomicota (orders Septobasidiales and Atheliales, 238 reported species)	Termite eggs (order Isoptera) and scale insects (family Diaspididae and order Hemiptera)
Entomophthoromycota (Order Entomophthorales, 474 reported species)	Lepidopteran larvae (order Lepidoptera) and aphid species (Family Aphididae)
Chytridiomycota (Orders Blastocladiales and Chytridiales, 65 reported species)	Flies and mosquitoes (order Hemiptera and Diptera)
Zygomycota (Class Trichomycetes and order Entomophthorales)	Trichomycetes grow in the guts of aquatic Order Diptera; Entomophthorales insects hosts are hemipterans, homopterans, lepidopterans, orthoptera, dipterans, etc.

Contd.

Table 1 *Contd.*

Phylum	*Group of insect*
Ascomycota (order Hypocreales, 476 reported species)	Order hemiptera, homoptera, lepidoptera, hymenoptera. orthoptera, diptera, coleoptera, Isoptera, etc
Fungus-like Oomycota (orders Lagenidiales, Leptomitales and Pythiales, 12 reported species)	Mosquitoe larvae (Diptera)

(*Source:* Lichtwardt 1986; Vega et al., 2012; Shukla and Afzal 2021; Quesada-Moraga et al., 2023)

The first contact between the conidium, or blastospore, of the fungus and the insect is considered the most important feature for the study of specificity. The conidium is adsorbed to the surface of the insect; shortly after germination, the appressorium is formed, or not, to initiate the penetration phase (Wang and St Leger, 2007, Pucheta Díaz et al., 2006). This process has been refined by the different adaptations that both entomopathogenic fungi and insects have undergone over time (Table 2).

Table 2 Characteristics related to specificity in the first stage of parasitism in entomopathogenic fungi-insects.

	Specific characteristics
Entomopathogenic fungus	Conidia or blastospores coated with hydrophobins and adhesion proteins that facilitate adhesion to the insect cuticle, e.g. aerial conidia can adhere to hydrophobic surfaces, blastospores to hydrophilic surfaces and submerged conidia to hydrophobic and hydrophilic surfaces (Holder and Keyhani, 2005; Holder, et al., 2007; Ortiz-Urquiza and Keyhani, 2013;).
	The surface of fungal cells contains various carbohydrate moieties, playing diverse roles in adhesion (Wang and St Leger ,2007).
	Modify surface components (carbohydrate moieties) that are typically recognized by the host immune system (Wanchoo, et al., 2009)
	Production of appressoria and subsequent production of a penetration tube that produces chitinases, lipases and proteases to facilitate the penetration process (Xiao et al., 2012; Skinner et al., 2014; Lacey et al., 2015), although certain species of entomopathogenic fungi do not form appressoria (Butt, et al., 1995; Hong, et al., 2023a).
Insect	Pre-contact strategy, insect odorant binding proteins react to the volatile organic compounds emitted by entomopathogenic fungi which activates host defence mechanisms (Zhang, et al., 2023a, 2023b).
	Hydrocarbons in the insect cuticle act as a carbon source for fungal adhesion and germination, insects can alter the hydrocarbon content and evade fungal infection (Pedrini, et al., 2013).
	Secondary metabolites secreted into the host lead to immunological alterations and a severe physiological disorder that induces a lethal pathological condition (Wang, et al., 2012; Gibson, et al., 2014)
	Carbohydrates moieties (in fungus) are compounds recognized by host for triggering immune signaling cascades (Wanchoo, et al., 2009; Brivio and Mastore, 2020)

Contd.

Table 2 *Contd.*

	Specific characteristics
	The cuticular integument, a robust protective barrier composed of diverse chemical components including n-alkane, fatty acids, chitin, and tanned proteins that some of them, exhibit antimicrobial properties (Brey, et al., 1993; Ashida and Brey, 1995; Ortiz-Urquiza and Keyhani, 2013)
	Release of substances such as formic acid, quinones, terpenes, and glucanases, as well as proteinase and chitinase inhibitors, from their epidermal, salivary or poison glands, and these secretions serve to hinder spore adhesion, germination, or appressorium formation (Bulmer, et al., 2009; Bulmer, 2004)
	The surface of insects is inhabited by diverse bacteria to form ectomicrobiomes, confer colonization resistance to hosts by producing antimicrobial compounds to inhibit the germination and growth of fungal conidia (Ortiz-Urquiza and Keyhani, 2013; Hong, et al., 2023b; Boucias, et al., 2018; Zhou, et al., 2019)
Environment	Spore germination on the insect cuticle at 20-30 °C (Skinner et al., 2014).

Understanding the specificity of different fungal species is a decisive factor in selecting the best candidate for application. However, by selecting an entomopathogenic fungus with low specificity and a wide host range, it can be assumed that it could survive in the environment on several insect species; conversely, the higher the specificity of the fungal agent, the less likely it is to persist in the environment. The former would affect pest insect communities but would also affect beneficial insects, while the latter would not affect beneficial insects such as pollinating insects (Goettel and Hajek, 2001). However, the choice of the former or the latter depends more on production and marketing considerations than on biological effectiveness; entomopathogens with the least specificity are the most widely produced and distributed in the world: *Beauveria bassiana* and *Metarhizium anisoplie* (Faria and Wraight, 2007).

4. Pest Management

4.1 Potential Uses of Entomopathogenic Fungi

The increasing world population necessitates a continuous supply of food in both sufficient quantity and quality. This demand has prompted the implementation of various strategies to enhance potential yields. However, various limiting factors must be overcome, and one of the most important is the presence of insect pests coexisting with crops. The application of insecticides is the most effective and commonly used strategy. Nevertheless, excessive use and inappropriate handling have resulted in numerous harmful effects on public health and the environment. Developing biopesticides with entomopathogens that target various economically important insect pests is crucial for promoting sustainable agricultural production and food security, as well as protecting human health and the environment (Sabbahi et al., 2022; Lacey et al., 2015). Fungi, bacteria, protozoa, viruses, and nematodes are useful as biopesticides. They are characterized by high specificity, require low product doses during phytosanitary management activities, and are harmless to the environment (Shahid et al., 2023).

The use of microbial biopesticides is one of the preventive strategies implemented in Integrated Pest Management (IPM) programs. Entomopathogenic fungi have particular advantages, as they possess a broader host range and can interact with plants, establishing symbiotic relationships and even stimulating plant growth. Sharma et al., (2023) propose the concept of myco-biocontrol, which involves the widespread application of fungal infective propagules to reduce pest insect outbreaks. This approach reduces crop losses by inducing direct insect mortality and indirectly through the production and release of metabolites or toxins in plants, such as destruxins, cyclosporines, beauvericin, cytochalasins, beauverolides, and others. When ingested by insects during their feeding, these substances affect their metabolism and biological cycle. Finally, under optimal environmental conditions, entomopathogenic fungi can complete their biological cycle and produce new propagules (spores) that perpetuate their presence in the agroecosystem. This leads to the production of new fungal cycles, facilitating the contagion of susceptible pest insects (Sabbahi et al., 2022).

Among the 171 entomopathogen-based products developed, those utilizing *B. bassiana* accounted for 33.9% of the total, while *M. anisopliae* products constitute another 33.9%. Additionally, *I. fumosorosea* and *B. brongniartii* products comprised 5.8% and 4.1%, respectively (Islam et al., 2021). Other fungal species used as formulated bioinsecticides have been discussed, but it is important to note that fungal susceptibility can be higher for nymphal and larval stages, while insect adults may exhibit different susceptibility. Certain fungi exhibit narrow host ranges; for example, *Aschersonia aleyrodis* exclusively infects whiteflies, and *Nomuraea rileyi* targets only lepidopteran larvae. Conversely, fungi such as *B. bassiana* and *M. anisopliae* can infect over 700 species, spanning various insect orders. Nevertheless, the data in these reports are primarily based on pathogenicity tests conducted on pests such as aphids, whiteflies, and thrips, with fewer assessments performed on lepidopteran and coleopteran insect pests (Maina et al., 2018).

The mechanism of fungal control relies on their ability to penetrate insect tissues at different developmental stages and the production of toxic metabolites during this process. It commences with conidia adhering to the epicuticle of the host insects, leading to germination and the formation of an appressorium. The mycelium then penetrates the procuticle, enters the hemolymph, and continues to proliferate (Shin et al., 2020). Subsequently, the fungus colonizes the interior of the infected insect´s body, and the formation of mycelium and conidia occurs. These conidia can later be dispersed by water or wind after emerging from the cadaver. This genetically mediated process can be highly specific, which is why it is used as a strategy within integrated pest management, in combination with cultural practices. Integrated Pest Management strategies (IPM) prioritize the utilization of biopesticides, resorting to chemical pesticides solely when deemed essential (Deguine et al., 2021). Furthermore, to enhance the dispersion capacity of entomopathogenic fungi, some authors have suggested the use of insect vectors that can transport the spores, such as pollinators (Baron et al., 2019).

Mutualistic relationships between entomopathogenic fungi and plants occur in the rhizosphere and on the phylloplane, where the fungi feed on nitrogen and sugars produced by their host. They can also behave as endophytes, taking advantage of being contained within the plant. When a chewing or sucking insect feeds, it can become infected with the fungus's propagules or ingest the toxins produced by the fungus, which can lead to its death (Quesada-Moraga 2020, Vega 2018). For instance, destruxin A, a toxin produced by *Metarhizium brunneum* (Ascomycota; Hypocreales), is detected after colonizing potato and melon plants, and it has control effects against *Bemicia tabaci* nymphs (Garrido-Jurado et al., 2017).

Other entomopathogens with the capacity to act as endophytes include *Beauveria bassiana*, *Lecanicillium lecani*, among others. These types of interactions can be intentionally promoted through direct spraying of the fungus on crops, seed inoculation, soil treatment, and even direct injection into the stem. Using endophytic fungi is safe and compatible with insect parasitoids and predators, potentially enhancing the plant´s resilience to various biotic and abiotic stresses. Additionally, they promote root growth, provide protection against phytopathogenic microorganisms, and enhance nutrition through interactions with other microorganisms in the rhizosphere. The mentioned attributes are advantageous for nursery production as they provide plant protection against potential insect pests (Quesada-Moraga 2020). However, a possible drawback is the environmental contamination caused by mycotoxins (aflatoxins, fumonisins and citrinins), which are produced by entomopathogenic fungi. Therefore, further investigation is necessary to quantify and mitigate these adverse impacts (Sinha et al., 2016).

4.2 *In vitro* Evaluation of Entomopathogenic Fungi

The *in-vitro* evaluation of entomopathogenic fungi contributes to comprehending the growth and capacity of the propagules to cause mycoses in different biological phases of susceptible insects. Based on the infection results, the isolates are selected for mass reproduction, application, introduction, and evaluation in the field, in such a way that strategies are generated that contribute to determining the agents that regulate pest populations. To ensure the efficacy of biocontrol against insect pests, it is essential to consider that microbial biological control relies on the virulence of the fungi, their ability to cause infection, induce disease, and efficiently spread in susceptible populations (Gallegos et al., 2003). Therefore, the selection of native or introduced fungal isolates that demonstrate adaptability capabilities in the agroecosystem becomes crucial. The utilization of biocontrol agents represents an efficient alternative for reducing production losses (Ni et al., 2022; Andrade-Hoyos et al., 2020).

In studies assessing the biological effectiveness of entomopathogenic fungi, standard measures include evaluating fungal growth in various media and temperatures, as well as examining their impact on germination, and signs of disease in insects. These measurements serve as benchmarks to evaluate fungal mechanisms and evaluations *in-vitro* in biological effectiveness tests against different insects (Jaronski and Mascarin, 2023; Yeo et al., 2003).

Currently, there is an ongoing development of rapid and efficient methods to measure the potential of entomopathogens, which are used as integral ecological management of pesticides or in combination with them. Techniques have evolved over time, and indirect measurements are being explored or implemented with microplate readers. This device employs an optical density of low volume cultures to monitor and estimate the mycelial growth of entomopathogens (Slowik et al., 2023).

The direct correlation in optical density with the biomass of fungal propagules, coupled with the direct comparison of results from conventional measurements on agar plates, indicates that nanospectrophotometers are accurate. In addition to these advancements, other complementary techniques involve the utilization of silver nanoparticles obtained from entomopathogens in the regulation of mosquitoes that transmit diseases to humans. These fungi have the ability to degrade the host using enzymes and toxins with bioinsecticidal activity, potentially yielding a satisfactory synergistic impact. Promising results have been observed in terms of mortality estimates.

The study areas of entomopathogens focus on the synthesis of silver microparticles with potential applications in the control of insects that affect humans. In agriculture, this approach could serve as an alternative for regulating insect populations. Therefore, it is crucial to expand toxicity studies to obtain a comprehensive understanding of potential effects (Santos et al., 2021).

For the management of agricultural and health-related pest insects, various species of entomopathogenic fungi have been studied for their potential as insecticides against certain dipteran species, including wild mosquitoes. For example, *Metarhizium anisopliae* and *Beauveria bassiana* have been employed to reduce populations of *Culicoide nubeculosus* (Ansari et al., 2011). Additionally, it has been determined that the genera *Purpureocillium* and *Fusarium* have insecticidal potential against mosquito larvae. This information is relevant because it enables the selection of fungal strains better adapted to the environment and with superior characteristics for their survival (Ni et al., 2022; Andrade-Hoyos et al., 2020; Ansari et al., 2011).

Humans are not exempt from the attack of pest insects; an example is the species *Forcipomyia taiwana* (Diptera); likewise, it is a pest that can be controlled with entomopathogenic fungi. In this regard, strains of *Fusarium verticillioides* and *Purpureocillium lilacinum* have been isolated from *F. taiwan* cadavers and tested against larvae, pupae, and adults of the same pest from which they were isolated. Two strains were tested, which showed 38 and 50% mortality with a concentration of 5×107 conidia/mL in fourth-stage larvae of *F. taiwana*; it also reduced egg hatching and adult emergence rate (Ni et al., 2022).

These results suggest that entomopathogenic fungi are a viable control option against this pest (Chen et al., 2021), as the use of chemical pesticides alone is not effective for controlling *F. taiwana* and only has a significant effect on adults. In the search for alternatives for pest control, other studies have utilized entomopathogenic filamentous fungi as mediators in biological synthesis methods of silver nanoparticles because they have the advantage of producing and secreting metabolites, many of which produce metallic salts and form easily manipulable nanoparticles. Several species have been used for silver nanoparticle synthesis, mainly for controlling microorganisms (Santos et al., 2021; Akter et al., 2018; Akther et al., 2020). However, despite the potential of these microorganisms for nanoparticle synthesis, they are still underutilized for this purpose. Their use is promising for the development of nanoparticles capable of controlling pest insects due to the production of enzymes and mycotoxins capable of causing mortality and may have a promising synergistic effect.

The application of entomopathogenic fungi in livestock focuses on tick control because they are one of the major economic threats to the global livestock industry, affecting productivity, health, and welfare. Furthermore, the prevalence of tick populations resistant to chemical pesticides encourages the use of alternative control methods to be applied in these populations and reduce the ecological consequences caused by the use of chemical components. Biocontrol using entomopathogenic fungi is a promising ecological alternative; therefore, the most studied fungal species both in laboratory and field settings are *Metarhizium anisopliae* and *B. bassiana*. In addition to agricultural pest control, these fungi have also been used for controlling the cattle tick *Rhipicephalus microplus*. However, it is necessary to study the effect of these entomopathogenic fungi on other species of livestock ticks, such as *Amblyomma mixtum* and *R. annulatus*. Despite the positive results obtained with these entomopathogenic fungi, a transdisciplinary approach is required to incorporate different types of tools, such as genomics, transcriptomics, and proteomics, to better understand the mechanism of pathogenicity and virulence of these fungi on ticks.

Laboratory trials have allowed for the measurement of the capacity of entomopathogenic fungi to control populations of pest insects, susceptible and resistant ticks, while field tests have demonstrated the control efficacy of *M. anisopliae* on different stages of *R. microplus* when applied both on pastures and livestock. It should be noted that epidemiological aspects of ticks and environmental factors are components that influence the acaricidal behavior of entomopathogenic fungi (Alonso-Díaz and Fernández-Salas, 2021).

Among the significant pests are grasshoppers, which cause significant damage to crops and pastures in the USA. The use of chemical pesticides has been the primary method of control; however, it has negative implications for the environment. In a study conducted by Dakhel et al., (2019), using *Metarhizium brunneum* strain F52 and *Paranosema locustae* against the migratory grasshopper *Melanoplus sanguinipes* under laboratory and greenhouse conditions, they found that when using third-stage grasshoppers reared in the laboratory, mortality caused by the two microorganisms combined at both high and low doses resulted in mortality rates of 75% and 77%, respectively. Under greenhouse conditions, a similar trend was observed, with the highest mortality (60%) recorded with the combination of *M. brunneum* strain F52 and *Paranosema locustae*.

Another control alternative is the use of chemical insecticides in combination with entomopathogenic fungi; in this regard, the efficacy of the combination of five synthetic pesticides, gamma-cyhalothrin, lambda-cyhalothrin, rynaxypyr, lufenuron, and methoxyfenozide, combined with various strains of *Beauveria bassiana*, *Metarhizium anisopliae*, and *Metarhizium robertsii*, the three fungi at concentrations of 1×10^8, 1×10^6, and 1×10^4 conidia/mL against the insect *Rachiplusia nu* (Guenée) (Lepidoptera: Noctuidae) was studied. *R. nu* is one of the main pests of defoliating lepidopterans of soybeans *Glycine max* Merrill; the study was conducted under laboratory conditions.

The results showed that the combination of entomopathogenic fungi with insecticides has a synergistic effect on larval mortality compared to any of the individual agents. Specifically, the combination of gamma-cyhalothrin and *B. bassiana* caused the highest mortality in *R. nu* larvae when using between 25 and

50% of the field doses recommended by the manufacturer. Additionally, the level of significance differed in the viability of conidia *in-vitro*, the mycelial growth, and the conidial production of the entomopathogenic fungal strains at different doses of synthetic insecticides (Pelizza et al., 2018).

The discovery and use of biological strategies as management alternatives began with observations of interactions between microbial communities in defined ecological niches (Bosch et al., 1982; Barratt et al., 2018). Later, the use of biological control increased; invasive pests cause losses by reducing crop yields with an approximate cost of 120 billion dollars per year (Pimentel, 2008); even more alarming are the populations resistant to pesticides that exist, which has led to the creation of organizations such as the Fungicide Resistance Action Committee (FRAC) (https://www.frac.info/) and the Insecticide Resistance Action Committee (IRAC) (https://irac-online.org/).

It is essential to recognize that microorganisms associated with plants are not necessarily pathogenic (Stark, 2010). Beneficial microorganisms are part of the comprehensive management strategies in phytosanitary protection and pest regulation (Meena et al., 2017).

Generally, the potential and effectiveness of entomopathogenic fungi reduce the use of insecticides and environmental impact, considering that they do not generate resistance in insects, in addition to effectively regulating pests (Bale et al., 2008; Benjamin and Wesseler, 2016; Barratt et al., 2018). However, biocontrol strategies have disadvantages against exotic or invasive microorganisms, insect resistance, specificity, incompatibility with insecticides, and risk assessment (Bale et al., 2008; Barratt et al., 2018; Köhl et al., 2019).

The complexity of the use of microorganisms proposes an approach that involves *in-vitro* and *in-situ* evaluations for the bioprospecting of microorganisms, using species of *Trichoderma* sp. for bioprocess model studies. This strategy integrates predictive biodesign, throughput analysis, and genomic engineering approaches for new bioproducts.

Overall, it can be observed that biological products, encompassing *in-vitro* and field-tested bioproducts such as fertilizers, stimulants, herbicides, and biocontrol products, constitute a flourishing agriculture sector and industry (Bale et al., 2008; Barratt et al., 2018; van Lenteren et al., 2018). This industry will continue to thrive year after year.

4.3 Classical and Modern Approaches

Environmental contamination, post-harvest chemical bioaccumulation, erosion of biodiversity, return of secondary pests, and declines in the natural predator community are all consequences of the widespread use of chemical insecticides in agriculture. Less use of conventional pesticides is closely correlated with an increase in the demand for biopesticides. Low toxicity, eco-friendliness, reversing pest resistance, blending with synthetic pesticides, target specificity, biodegradability, lack of post-harvest contamination, and alignment with integrated pest management are some of these benefits (Archana et al., 2022; Qadri et al., 2020).

The use of entomopathogens in biological insect control holds potential for lowering the environmental impact of synthetic insecticides and reducing pollution.

It has long been known that entomopathogenic fungi, like *Beauveria bassiana, Metarhizium anisopliae,* and *Lecanicillium* (=*Verticillium*) *lecanii*, are beneficial for controlling pests. Furthermore, a significant benefit of these fungi is that there hasn't been any documented development of resistance thus far (Lacey et al., 2015; Halder et al., 2012).

Programs for biocontrol frequently fall short of their potential efficacy for a variety of reasons. According to Leung et al., (2020), they include the use of microbial agents that are poorly suited to the local environment, unfavorable interactions between them and the local fauna, and evolutionary changes made by pest species after they are introduced. Another major obstacle is the high expense of manufacturing microbial biopesticides. This expense includes the costs of finding new biocontrol agents and developing them, as well as the costs associated with getting regulatory approval (Sabbahi et al., 2022).

There are no set standards for the acceptance or adoption of entomopathogens, but there is a case to be made for the use of bioproducts made from these microbes because of their many advantages. It takes work in the fields of policy and regulation as well as in lab and field trials to address the different underlying issues. To effectively use the potential applications of these microbes, economic, societal, and political restraints must be taken into account in addition to scientific factors (Bamisile et al., 2021).

4.4 Compatibility

Biopesticides are a possible substitute for chemical pesticides that open up new possibilities for managing insect pests in a way that is safe and sustainable for the environment. They have negligible interference with other organisms and a relative host specificity (Prithiva et al., 2018; Halder et al., 2021). Nevertheless, a variety of agricultural products, such as plant extracts, neem oil, other biological control agents, insecticides, fungicides, and conventional herbicides, may mix well with microbial biopesticides (Da Silva et al., 2013; Halder et al., 2021).

Since they frequently take longer to produce sufficient insect mortality than conventional approaches, entomopathogenic fungi are generally thought of as slow-acting. One possible answer to this problem could be to incorporate faster-acting compounds into a management approach along with entomopathogenic fungi (Nawaz et al., 2022).

A number of studies have looked into the simultaneous application of several species of entomopathogenic nematodes and fungi against different insect pests as a result of the recent trend of combining biocontrol agents to boost their efficiency. There have been a range of findings from these investigations, including antagonistic, additive, and synergistic effects. Furthermore, each pathogen's effects on development and reproduction vary from normal reproduction to exclusion. Overall, the pathogen's species, the host, and the degree and frequency of infection all affect how these interactions turn out (Půža V and Tarasco, 2023).

Parasites and pathogens employ various strategies to enhance their growth or reproduction in response to competition. Throughout their evolution, both entomopathogenic fungi and nematodes have developed multiple strategies to compete with each other (Mideo, 2009; Půža V and Tarasco, 2023).

Studies on the compatibility of different agrochemicals with biological control agents are essential for developing policies for their combined usage and figuring out when to apply them. Among these agrochemicals, fungicides have drawn attention because of the various ways in which they might harm entomopathogenic fungi. The type and quantity of the fungicide's active ingredient, the kind of enthomopatogen fungus, and environmental factors like temperature determine how harmful fungicides can be (Shoeb et al., 2021).

The efficacy of both fungicides and entomopathogens may be harmed by their combination. To counteract the harmful effects of fungicides and improve the pesticidal properties of entomopathogenic fungi, a number of strategies can be used. This involves mutagenesis, artificial selection, proper formulation procedures, endophytic entomopathogenic fungi, and the use of fungicides that are compatible with entomopathogenic fungi in pest management measures. Additionally, the field survival and efficacy of these fungi in fungicide-treated agroecosystems may be enhanced by the creation of aggressive and fungicide-tolerant strains of entomopathogenic fungi using traditional and/or genetic methods (Samal et al., 2023).

Regarding plant growth regulators (PGRs), it was discovered that mefluidide and flurprimidol were acceptable for use with *B. bassiana* because they did not significantly impede the fungus's germination and growth. However, when larvae of autumn armyworms (*Spodoptera frugiperda*) were exposed to conidia plus soil treated with paclobutrazol, the mortality resulting from *B. bassiana* treatment was greatly reduced (Storey and Gardner, 1986; Shoeb et al., 2021).

One potentially effective way to lessen the amount of chemicals needed for pest management is to apply entomopathogenic fungi in addition to chemical pesticides at sublethal concentrations. This strategy seeks to protect agricultural workers' health as well as the environment (Pelizza et al., 2018).

Using the entomopathogenic fungus *M. anisopliae* CG 168, Da Silva et al., (2013) carried out *in-vitro* investigations and found that fungicides like difenoconazole (69 mL/ha), propiconazole (75 mL/ha), trifloxystrobin (313 g/ha), and azoxystrobin (56 mL/ha) were harmful to conidial germination, mycelial growth, and sporulation. On the other hand, they discovered that some agrochemicals were safe to use with *M. anisopliae* CG 168. These included herbicides like glyphosate (1560 mL/ha), bentazon (720 mL/ha), and imazapic+imazapyr (84 g/ha), as well as insecticides like methyl parathion (240 mL/ha), thiamethoxam (31 g/ha), and lambda-cyhalothrin (6.3 mL/ha). These results are essential for advising farmers on the safe use of this entomopathogen in conjunction with pesticides, even if the *in-vitro* investigation did not take into account all factors related to pesticide use in the field. It is advised against using pesticides and herbicides in the same tank containing this fungus.

A suitable sublethal concentration of neem oil combined with fungi like *B. bassiana*, *M. anisopliae*, and *L. lecanii* has been successfully used to control a number of insect pests, thereby lowering the selection pressure on the target pests (Islam et al., 2010). However, literature on the compatibility of various entomopathogens and neem oil against major vegetable-sucking pests and their impact on beneficial fauna is lacking (Halder et al., 2021).

The use of substances presents in the agrochemical formulation or compounds released during agrochemical of plant insect metabolism may be the source of the stimulatory effect observed in some formulations on conidial germination or vegetative growth of fungal isolates. These substances can act as direct nutrients for the fungus. Furthermore, the fungus may attempt to reproduce, which would result in a higher generation of conidia. In contrast, formulations that are ionic or molecular have the potential to neutralize the electrostatic charge present on the surface and/or eliminate the mucous layer that covers conidia. This could interactions have an impact on the processes involved in substrate recognition and the transduction of signals that trigger germination (Moino and Alves, 1998; Shoeb et al., 2021).

5. Application of Commercial Formulations

More than 700 fungal species from 100 genera have been identified isolated from various sources, such as soil and insect carcasses (Shin et al., 2020). *Metarhizium acridum*, *M. brunneum*, and *M. anisopliae* (green muscardine fungus), *Lecanicillium lecanii* (white-halo fungus), *Isaria fumosorosea*, *B. bassiana*, and *Hirsutella thompsonii* are commercially supplied as bioinsecticides in various formulations worldwide (Dara, 2019). Commercial formulations of entomopathogenic fungi typically consist of several components, such as surfactants, carriers, UV protectants, stickers, and the active ingredient (conidia/dry spores). The effectiveness of the product relies primarily on the viability of conidia (Islam et al., 2021). Their application is usually through an inundative approach (Bamisile et al., 2021).

The formulations need to be specific to the host and safe for non-target species. Sometimes, the limited host range of entomopathogenic fungi can pose a challenge as it may lead to the emergence of secondary pests (Sinha et al., 2016). Most formulations have a shelf life ranging from three to six months, with fungal spore concentrations ranging from 10^9 to 10^{10} spores per gram. Optimal humidity and temperature vary depending on the species of entomopathogenic fungi; however, they generally require humidity levels of 75 to 100% humidity and a temperature range of 15 to 30°C (Islam et al., 2021). The application dose varies based on the formulation, severity of the infestation, insect pest biology, and predominant environmental conditions. This information must be available on the label of the container (Sharma et al., 2023). The reduced toxicity towards pollinators is an advantage in addressing social-political concerns related to ecology. Depending on the crop, insect pest, and environmental conditions, this kind of bioinsecticide has good compatibility and can be used alone, in rotation, or in combination with beneficial insects (e.g., predators and parasitoids), chemical insecticides, herbicides, or others. The multi-site action targeting an insect pest overcomes the possibility of pest populations developing resistance, a common occurrence with synthetic insecticides (Ruiu 2018).

The creation of tools for biological control relies on understanding the biology of insects and the abiotic factors influencing them (Lacey et al., 2015). The success of bioinsecticides in controlling insect pests is influenced by factors such as the exposure of fungal spores to unfavorable conditions of humidity, temperature, and solar radiation. These abiotic elements diminish the persistence and effectiveness of these fungi that are pathogenic to pest insects (Bamisile et al., 2021). The efficiency of the application method is crucial, and it should be carried out at the most appropriate time to ensure maximum effectiveness. This involves optimum humidity, light conditions, and temperature, as well as targeting the susceptible stage of the insect pest (Sharma et al., 2023). Other factors restricting their field use include inconsistent effectiveness in crop protection, diminishing market availability, and widespread rejection by farmers and distributors (Lacey et al., 2015). The fungal inoculum has a brief lifespan, but effective pest control requires a period of 2 to 3 weeks (Islam et al., 2021), which constitutes a major obstacle to their widespread use and market competitiveness (Kagimu et al., 2017). Additionally, the expensive cost of commercial formulations is prohibitive (Sinha et al., 2016).

However, the drastic decline in biodiversity is a consequence of the environmental damage resulting from the irrational use of chemicals. Consequently, it is imperative to undergo a shift in awareness to encourage the adoption of entomopathogenic fungal formulations. The development of highly effective formulations requires research focused on better understanding the interactions between entomopathogenic fungi, pest insects, crops, and favorable environmental conditions.

Some farmers have high expectations for bioinsecticides, frequently measuring their effectiveness with chemical insecticides. Nonetheless, their efficacy has been compromised due to haphazard handling and a lack of expertise, leading to unfulfilled expectations among farmers (Sharma et al., 2023). The use of biotechnology for crop improvement through the inoculation of plants with modified fungal strains would consequently reduce toxicity to humans, livestock, and the environment (Bamisile et al., 2021). Exploring the role of entomopathogenic fungi in the biocontrol of insects appears to be a growing research area, as evident from the high number of studies conducted (Deka et al., 2021). Further understanding regarding the impacts of climate change and agronomic practices on the bio-functionality of entomopathogenic fungi, offering diverse agroecosystem services, needs to be developed. There is a requirement for advanced technologies to enhance the production of entomopathogenic fungi and, most importantly, to improve the efficacy, stability, and reliability of the products under environmentally stressful conditions (Sabbahi et al., 2022). This shift towards biological control not only addresses environmental concerns but also aligns with the growing need for sustainable and eco-friendly agricultural practices.

6. Limitations of the Use of Entomopathogenic Fungi

Field utilization of biological control agents faces numerous challenges, including inconsistent effectiveness in crop protection, decreasing availability in the market, and resistance from both farmers and distributors (Lacey et al., 2015).

In order to use predators, parasitoids, and microorganisms, including bacteria, fungi, and viruses, as effective and sustainable pest management techniques, abiotic conditions are essential. Prominent strains of entomopathogenic fungi, including *Beauveria*, *Metarhizium*, *Isaria*, *Hirsutella*, and *Lecanicillium*, have been employed for over 150 years. Typically, fungal spores are applied by flooding them into the environment and subjecting them to adverse temperatures, humidity levels, and sun radiation. The persistence and efficacy of entomopathogenic fungi are naturally decreased by these abiotic conditions. More than 170 strains have been developed as mycopesticides and are marketed in contravention of these restrictions. The composition of the culture and storage media in bioformulation, as well as temperature management, are also important factors for their effectiveness (Bamisile et al., 2021; Sutanto et al., 2022).

One strategy to reduce abiotic stress is to create formulations that shield the propagule. For instance, conidia suspended in mineral oil exhibit a higher capacity to withstand heat stress in comparison to those suspended in water. As a result, using oil formulations to shield conidia from damage brought on by heat stress may be a useful strategy. Furthermore, augmenting media can function as a substitute strategy to lessen harm brought about by abiotic variables (Lovett and Leger, 2015; Paixão et al., 2017).

Despite increased research on novel entomopathogens for potential biopesticides, the identification and isolation of hemibiotrophic and endophytic entomopathogenic fungi remain challenging. As a result, there are limited quantities of microbial biopesticides being produced (Kumar et al., 2021). Entomopathogenic Hypocrealean fungi, belonging to the order Hypocreales within the phylum Ascomycota, employ a hemibiotrophic strategy, enabling them to alternate between biotrophic and saprotrophic phases. During the saprotrophic phase, these fungi colonize arthropod corpses. In the biotrophic phase, the infection process, known as parasitism, occurs. This process begins with an infectious spore adhering to the cuticle, followed by germination and penetration (Wang et al., 2019).

Entomopathogenic fungi are primarily isolated from insects but can also be found in the soil. Interestingly, they demonstrate versatility by colonizing plants either as endophytic fungi or as mycoparasites against phytopathogenic fungi. Studies have shown that certain fungal traits crucial for insect pathogenicity also contribute to the biocontrol of phytopathogens (Ownley et al., 2010; Ondráčková et al., 2019).

Accurate identification of microbial organisms is necessary for the integration of biological control techniques into pest management. Uncertainty about the fate of the inoculum and its effect on non-target insects gives rise to concerns about the discharge of bacteria, viruses, and fungal strains into the field. These worries are exacerbated by the challenges in identifying and tracking microbial organisms discharged into the field (Wang et al., 2004).

There is growing recognition of fungal endophytes as a distinct class of microbial biocontrol agents. As endophytes, entomopathogenic fungi can protect plants from diseases in addition to being effective against insect pests (Saidi et al., 2023).

A number of entomopathogenic fungi, including *Entomophthora muscae* (which kills flies), *Pandora neoaphidis* and *Neozygites fresenii* (both fungi control cotton aphids), and *Entomophaga maimaiga* (which works well against gypsy moths), are essential in naturally lowering insect pest populations through natural epizootics. Nevertheless, because they are hard to grow in artificial media, these entomopathogenic fungi have difficulties in culture despite their potential. As a result, as biopesticides, they presently lack commercial promise (Goettel et al., 2005).

A thorough understanding of numerous aspects is necessary to apply biopesticides based on entomopathogens for pest management in agriculture, forests, and urban environments. This entails figuring out the best dose for infecting hosts (insect pests) in various crops, taking ecological adaptations into account, assessing the host's range, and researching the dynamics of interactions between entomopathogens, insects, and plants. These realizations are essential to guaranteeing the effective and focused use of these biopesticides. It's also critical to understand that the use of microbes for controlling particular insect species may be restricted by the environmental circumstances in which insect populations finish their life cycle. The effective use of biopesticides is made more difficult by this factor (Brodeur, 2012; Sabbahi et al., 2022).

Testing for pathogenicity and toxicity in vertebrates as well as non-target organisms is essential for future registration of new fungal strains. It takes a complete strategy to reduce any possible dangers that come with using these fungal strains (Bamisile et al., 2021).

Conclusions

Entomopathogenic fungi have the potential to act as biocontrol agents against different insect pests. The advantage of using these fungi is that they are effective, do not pollute the environment, and can be isolated regionally. Due to a number of advantages, entomopathogenic fungal products are already available on the market. In addition, with scientific developments, entomopathogenic fungi are being studied with molecular and biochemical tools in order to better understand how they function.

References

Akter, M., Sikder, M.T., Rahman, M.M., Ullah, A.A., Hossain, K.F.B., Banik, S., Hosokawa, T., Saito, T. and Kurasaki, M. (2018). Systematic review on silver nanoparticles-induced cytotoxicity: Physicochemical properties and perspectives. J. Adv. Res 9:1–16. https://doi.org/10.1016/j.jare.2017.10.008

Akther, T., Khan, M.S. and Hemalatha, S. (2020). Biosynthesis of silver nanoparticles via fungal cell filtrate and their anti-quorum sensing against *Pseudomonas aeruginosa*. J. Environ. Chem. Eng. 8:104365. https://doi.org/10.1016/j.jece.2020.104365

Alonso-Díaz, M.A. and Fernández-Salas, A. (2021). Entomopathogenic Fungi for Tick Control in Cattle Livestock From Mexico. Front Fungal. Biol., 30: 657694. https://doi.org/10.3389/ffunb.2021.657694

Andrade-Hoyos, P., Silva-Rojas, H.V. and Romero-Arenas, O. (2020). Endophytic *Trichoderma* species isolated from *Persea Americana* and *Cinnamomum verum* Roots reduce symptoms caused by *Phytophthora cinnamomi* in avocado. Plants 9: 1220. https://doi.org/10.3390/plants9091220

Ansari, M.A., Pope, E.C., Carpenter, S., Scholte, E.J. and Butt, T.M. (2011). Entomopathogenic fungus as a biological control for an important vector of livestock disease: The *Culicoides* biting midge. PLoS One 6: e16108. https://doi.org/10.1371/journal.pone.0016108

Archana, H.R., Darshan, K., Amrutha, L.M., Ghoshal, T., Bashayal, B.M. and Aggarwal, R. (2022). Biopesticides: A key player in agro-environmental sustainability, *In*: Trends of Applied Microbiology for Sustainable Economy, Soni, R., Suyal, D.C., Yadav, A.N. and Goel, R. (Eds.), Academic Press, 613-653, https://doi.org/10.1016/B978-0-323-91595-3.00021-5

Ashida, M. and Brey, P.T. (1995). Role of the integument in insect defense: pro-phenol oxidase cascade in the cuticular matrix. Proc. Natl. Acad. Sci. USA 92:10698-702. https://doi.org/10.1073/pnas.92.23.10698

Baker, K.F. and Cook, R.J. (1982). Biological Control of Plant Pathogens. American Phytopathological Society. St. Paul, Minnesota, 433.

Bale, J.S., van Lenteren, J.C. and Bigler, F. (2008). Biological control and sustainable food production. Philos Trans R Soc Lond B Biol. 1492:761-76. https://doi.org/10.1098/rstb.2007.2182

Bamisile, B.S., Akutse, K.S., Siddiqui, J.A. and Xu, Y. (2021). Model application of entomopathogenic fungi as alternatives to chemical pesticides: Prospects, challenges, and insights for next-generation sustainable agriculture. Front. Plant Sci. 12:741804. https://doi.org/10.3389/fpls.2021.741804

Baron, N.C., Rigobelo, E.C. and Zied, D.C. (2019). Filamentous fungi in biological control: current status and future perspectives. Chil. J. Agric. Res. 79: 307–315. http://dx.doi.org/10.4067/S0718-58392019000200307

Barratt, B.I.P., Moran, V.C., Bigler F., Van Lenteren JC. (2018). The status of biological control and recomendations for improving uptake for the future. BioControl. 63: 155–16 https://doi.org/10.3389/fpls.2021.741804

Benjamin, E.O. and Wesseler, JHH. (2016). A socioeconomic analysis of biocontrol in integrated pest management: a review of the effects of uncertainty, irreversibility and flexibility. Wageningen J Life Sci 77:53–60. https://doi.org/10.1016/j.njas.2016.03.002

Bitsadze, N., Jaronski, S., Khasdan, V., Abashidze, E., Abashidze, M., Latchininsky, A., Samadashvili, D., Sokhadze, I., Rippa, M., Ishaaya, I. and Horowitz, A.R. (2013). Joint action of *Beauveria bassiana* and the insect growth regulators diflubenzuron and novaluron, on the migratory locust, *Locusta migratoria*. J. Pest. Sci. 86: 293-300. https://doi.org/10.1007/s10340-012-0476-4

Bosch, R., Messenger, P.S. and Gutierrez, A.P. (1982). An introduction to biological control. Plenum Press, N.Y. 247.

Boucias, D.G., Zhou, Y. and Huang, S. (2018). Microbiota in insect fungal pathology. Appl. Microbiol. Biotechnol. 102:5873–5888. https://doi.org/10.1007/s00253-018-9089-z

Brey, P.T., Lee, W.J., Yamakawa, M., Koizumi, Y., Perrot, S. and François, M. (1993). Role of the integument in insect immunity: epicuticular abrasion and induction of cecropin synthesis in cuticular epithelial cells. Proc. Natl. Acad. Sci. USA 90: 6275–6279. https://doi.org/10.1073/pnas.90.13.6275

Brivio, M.F. and Mastore, M. (2020). When appearance misleads: The role of the entomopathogen surface in the relationship with its host. Insect 11:387. https://doi.org/10.3390/insects11060387

Brodeur, J. (2012). Host specificity in biological control: insights from opportunistic pathogens, Evol. Appl., 5:470–480. https://doi.org/10.1111/j.1752-4571.2012.00273.x

Bulmer, M.S. (2004). Duplication and diversifying selection among termite antifungal peptides. Mol. Biol. Evol. 21(12):2256–2264. https://doi.org/10.1093/molbev/msh236

Bulmer, M.S., Bachelet, I., Raman, R., Rosengaus, R.B. and Sasisekharan, R. (2009). Targeting an antimicrobial effector function in insect immunity as a pest control strategy. Proc. Natl. Acad. Sci. USA 106(31):12652–12657. https://doi.org/10.1073/pnas.0904063106

Butt, T.M., Ibrahim, L., Clark, S.J. and Beckett, A. (1995). The germination behaviour of *Metarhizium anisopliae* on the surface of aphid and flea beetle cuticles. Mycol. Res. 99:945–950. https://doi.org/10.1016/S0953-7562(09)80754-5

Castañeda-Ramírez, G.S., Aguilar-Marcelino, L. and López-Guillen, G. (2022). Macroscopic and microscopic fungi with insecticidal activity. Chilean J. Agricul. Res 82:348-357. https://dx.doi.org/10.4067/S0718-58392022000200348

Chant, D.A. (1967). Harry Scott Smith Award: The Department of Biological Control, University of California. Can. Entomol., 99:779-779. https://doi.org/10.4039/Ent99779-7

Chen, M.E., Tsai, M.H., Huang, H.T., Tsai, C.C., Chen, M.J., Yang, D.S., Yang, T.Z., Wang, J. and Huang, R.N. (2021). Transcriptome profiling reveals the developmental regulation of NaCl-treated *Forcipomyia taiwana* eggs. BMC Genom. 22: 1–14. https://doi.org/10.1186/s12864-021-08096-x

Cisneros, V.F. (1995). Control de Plagas agrícolas. Agencia Peruana del. https://www.avocadosource.com/books/cisnerosfausto1995/cpa_toc.htm

Da Silva, R.A., Quintela, E.D., Mascarin, G.M., Barrigossi, J.A.F. and Lião, L.M. (2013). Compatibility of conventional agrochemicals used in rice crops with the entomopathogenic fungus *Metarhizium anisopliae.* Sci. Agric. 70:152–160. https://doi.org/10.1590/S0103-90162013000300003

Dakhel, W.H., Latchininsky, A.V. and Jaronski, S.T. (2019). Efficacy of Two Entomopathogenic Fungi, *Metarhizium brunneum,* Strain F52 Alone and Combined with *Paranosema locustae* against the Migratory Grasshopper, *Melanoplus sanguinipes*, under Laboratory and Greenhouse Conditions. Insects.10(4):94. https://doi.org/10.3390/insects10040094

Dara, S.K. (2019). Non-entomopathogenic roles of entomopathogenic fungi in promoting plant health and growth. Insects 10: 277. https://doi.org/10.3390/insects10090277

De Faria, M.R. and Wraight, S.P. (2007). Mycoinsecticides and Mycoacaricides: A comprehensive list with worldwide coverage and international classification of formulation types. Biol. Control., 43: 237–256. https://doi.org/10.1016/j.biocontrol.2007.08.001

Deguine, J.P., Aubertot, J.N., Flor, R.J., Lescourret, F., Wyckhuys, K.A. and Ratnadass, A. (2021). Integrated pest management: good intentions, hard realities. A review. Agron. Sustain. Dev., 41(3):38. https://link.springer.com/article/10.1007/s13593-021-00689-w

Deka, B., Baruah, C. and Babu, A. (2021). Entomopathogenic microorganisms: their role in insect pest management. Egypt. Biol. Pest Control, 31:121. https://doi.org/10.1186/s41938-021-00466-7

Fargues, J. and Remaudiere, G. (1977). Considerations on the specificity of entomopathogenic Fungi. Mycopathologia. 62: 31-37. https://link.springer.com/article/10.1007/BF00491993

Faria, M.R. and Wraight, S.P. (2007). Mycoinsecticides and Mycoacaricides: A comprehensive list with worldwide coverage and international classification of formulation types. Biol Control 43:237-256. https://doi.org/10.1016/j.biocontrol.2007.08.001

Gallegos, M.G., Cepeda, S.M. and Olayo, P.R.P. (2003). Entomopatógenos. Trillas, México, DF. 148 p

Garrido-Jurado, I., Resquín-Romero, G., Amarilla, S.P., Ríos-Moreno, A., Carrasco, L. and Quesada-Moraga, E. (2017). Transient endophytic colonization of melon plants by entomopathogenic fungi after foliar application for the control of *Bemisia tabaci* Gennadius (Hemiptera: Aleyrodidae). J. Pest. Sci. 90:319-330. https://doi.org/10.1007/s10340-016-0767-2

Gibson, D.M., Donzelli, B.G., Krasnoff, S.B. and Keyhani, N.O. (2014). Discovering the secondary metabolite potential encoded within entomopathogenic fungi. Nat. Prod. Rep. 31: 1287–1305. https://doi.org/10.1039/c4np00054d

Goettel, M.S. and Hajek, A.E. (2001). Evaluation of non-target effects of pathogens used for management of arthropods, In: Evaluating Indirect Ecological Effects of Biological Control, Wajnberg, E., Scott, J.K. and Quimby, P.C. (Eds) Wallingford UK: CAB International, pp. 81-97. https://doi.org/10.1079/9780851994536.0081

Goettel, M.S., Eilenberg, J. and Glare, T. (2005). Entomopathogenic fungi and their role in regulation of insect populations, In: Comprehensive Molecular Insect Science, Gilbert, L.I. (Ed.), Elsevier, Amsterdam, 361-405. https://10.1111/j.1365-2133.1975.tb06468.x

Hajek, A.E. and St. Leger, R.J. (1994). Interactions between fungal pathogens and insect hosts. Annu. Rev. Entomol. 39: 293-322. https://doi.org/10.1146/annurev.en.39.010194.001453

Halder, J., Divekar, P.A. and Rani, A.T. (2021). Compatibility of entomopathogenic fungi and botanicals against sucking pests of okra: an ecofriendly approach. Egypt. J. Biol. Pest. Control., 31:30. https://doi.org/10.1186/s41938-021-00378-6

Halder, J., Srivastava, C., Dhingra, S. and Dureja P. (2012). Effect of essential oils on feeding, survival, growth and development of third instar larvae of *Helicoverpa armigera* Hubner. Natl. Acad. Sci. Lett. 35: 271–276. https://doi.org/1df0.1007/s40009-012-0043-9

Hernandez, V.V.M. and Berlanga, P.A. (1996). Control microbiano con hongos entomopatógenos", en Memoria del II Curso de Actualización en Control Biológico, Colima, México, mayo de 1996, pp. 94–106.

Holder, D.J. and Keyhani, N.O. (2005). Adhesion of the entomopathogenic fungus *Beauveria (Cordyceps) bassiana* to substrata. Appl. Environ. Microbiol., 71: 5260–5266. https://doi.org/10.1128/aem.71.9.5260-5266.2005

Holder, D.J., Kirkland, B.H., Lewis, M.W. and Keyhani, N.O. (2007). Surface characteristics of the entomopathogenic fungus *Beauveria bassiana*. Microbiol., 153: 3448–3457. https://doi.org/10.1099/mic.0.2007/008524-0

Hong, S., Shang, J., Sun, Y., Tang, G. and Wang, C. (2023a). Fungal infection of insects: molecular insights and prospects. Trends Microbiol. https://doi.org/10.1016/j.tim.2023.09.005

Hong, S., Sun, Y., Chen, H., Zhao, P. and Wang C. (2023b). Fungus–insect interactions beyond bilateral regimes: the importance and strategy to outcompete host ectomicrobiomes by fungal parasites. Curr. Opin. Microbiol., 74:102336. https://doi.org/10.1016/j.mib.2023.102336.

Hyde, K.D., Xu, J., Rapior, S., Jeewon, R., Lumyong, S. and Niego, A.G.T. (2019). The amazing potential of fungi: 50 ways we can exploit fungi industrially. Fungal Diver., 97: 1–136. https://doi.org/10.1007/s13225-019-00430-9

Islam, M.T., Olleka, A. and Ren, S. (2010). Influence of neem on susceptibility of *Beauveria bassiana* and investigation of their combined efficacy against sweet potato whitefly, *Bemisia tabaci* on eggplant. Pestic. Biochem. Physiol. 98: 45–49. https://doi.org/10.1016/j.pestbp.2010.04.010

Islam, W., Adnan, M., Shabbir, A., Naveed, H., Abubakar, Y.S., Qasim, M., Tayyad, M., Noman, A., Shahid, N.M., Ali, K.K. and Ali, H. (2021). Insect-fungal-interactions: A detailed review on entomopathogenic fungi pathogenicity to combat insect pests. Microb Pathog 159: 105122. https://doi.org/10.1016/j.micpath.2021.105122

Jaronski, S.T. and Mascarin, G.M. (2023). Mass Production of Entomopathogenic Fungi—State of the Art. In : Mass production of beneficial organisms,2nd Edition Morales-Ramos, J.A., Guadalupe Rojas, M. and Shapiro-Ilan, D.I. (Eds.) pp. 317–357, Academic Press; Cambridge, MA, USA.

Kagimu, N., Ferreira, T. and Malan, A.P. (2017). The attributes of survival in the formulation of entomopathogenic nematodes utilised as insect biocontrol agents. Afr. Entomol., 25:275-291. https://hdl.handle.net/10520/EJC-9e1e16147

Köhl, J., Kolnaar, R. and Ravensberg, W. (2019). Mode of action of microbial biological control agents against plant diseases: Relevance beyond efficacy. Front. Plant Sci. 10: 1–19. https://doi.org/10.3389/fpls.2019.00845

Kumar, J., Ramlal, A., Mallick, D. and Mishra, V. (2021). An overview of some biopesticides and their importance in plant protection for commercial acceptance, Plants, 10: 1185. https://doi.org/10.3390/plants10061185

Lacey, L.A., Grzywacz, D., Shapiro-Ilan, D.I., Frutos, R., Brownbridge, M. and Goettel, M.S. (2015). Insect pathogens as biological control agents: back to the future. J. Invertebr. Pathol 132: 1–41. https://doi.org/10.1016/j.jip.2015.07.009

Leung, K., Ras, E., Ferguson, K.B., Ariëns, S. and Babendreier, D. (2020). Next-generation biological control: the need for integrating genetics and genomics. Biol. Rev. 95: 1838–1854. https://doi.org/10.1111/brv.12641

Lichtwardt, R.W. (1986). Host Specificity. In: The Trichomycetes. Springer, New York, NY. pp 38–42. https://doi.org/10.1007/978-1-4612-4890-3_6

Lopez-Lastra, C.C., Manfrino, R.G. and Toledo, A.V. (2019). Entomophthoromycota; Instituto Nacional de Tecnología Agropecuaria. 99-120. http://hdl.handle.net/11336/122212

Lovett, S.T.B. and Leger, R.J. (2015). Stress is the rule rather than the exception for *Metarhizium*. Curr. Genet. 61: 253–261. https://doi.org/10.1007/s00294-014-0447-9

Maina, U.M., Galadima, I.B., Gambo, F.M. and Zakaria, D. (2018). A review on the use of entomopathogenic fungi in the management of insect pests of field crops. J. Entomol. Zool. Stud., 6(1): 27–32.

McCoy, C.W., Samson, R.A. and Boucias, D.G. (1988). Entomogenous Fungi. In: Handbook of Natural Pesticides, Vol. V, Microbial Insecticides, Part A, Ignoffo, C.M. and Mandava, N.B., (Eds.) CRC Press, Boca Raton, 151–236.

Meena, V.S., Mishra, P.K., Bisht, J.K. and Pattanayak, A. (2017). Agriculturally important microbes for sustainable agriculture Vol. 2, pp. 374. https://doi. org/10.1007/978-981-10-5343-6

Mideo, N. (2009). Parasite Adaptations to Within-Host Competition. Trends Parasitol., 25:261–268. https://doi.org/10.1016/j.pt.2009.03.001

Moino, J.R. and Alves, A.S.B. (1998). Efeito de imicacloprid e fipronil sobre *Beauveria bassiana* (Bals.) Vuill.e *Metarhizium anisopliae* (Metsch.) Sorok.e no comportamento de limpeza de *Heterotermes tenuis* (Hagen). An. Soc. Ent. Bras., 27: 611–620. https://doi.org/10.1590/S0301-80591998000400014

Nawaz, A., Razzaq, F., Razza, A., Gogi, M.D. and Fernández-Grandon, G.M. (2022). Compatibility and synergistic interactions of fungi, *Metarhizium anisopliae*, and insecticide combinations against the cotton aphid, *Aphis gossypii* Glover (Hemiptera: Aphididae). Sci. Rep., 12: 4843. https://doi.org/10.1038/s41598-022-08841-6

Ni, N.T., Wu, S.S., Lia, K.M., Tu, W.C., Lin, C.F. and Nai, Y.S. (2022). Evaluation of the potential entomopathogenic fungi *Purpureocillium lilacinum* and *Fusarium verticillioides* for biological control of *Forcipomyia taiwana* (Shiraki). J. Fungi, 8: 861. https://doi.org/10.3390/jof8080861

Ondráčková, E., Seidenglanz, M. and Šafář, J. (2019). Effect of seventeen pesticides on mycelial growth of *Akanthomyces*, *Beauveria*, *Cordyceps* and *Purpureocillium* strains. Czech Mycol., 7: 123-135. https://doi.org/10.33585/cmy.71201

Onstad, D.W., Fuxa, J.R., Humber, R.A., Oestergaard, J., Shapiroilan, D.I. Gouli, V.V. (2006). An Abridged Glossary of Terms used in Invertebrate Pathology. 3rd Edition, Society for Invertebrate Pathology.

Ortiz-Urquiza, A. and Keyhani, N. (2013). Action on the surface: Entomopathogenic fungi versus the insect cuticle. Insects. 4:357–374. https://doi.org/10.3390/insects4030357

Ownley, B.H., Gwinn, K.D. and Vega, F.E. (2010). Endophytic fungal entomopathogens with activity against plant pathogens: ecology and evolution. BioControl 55: 113–128. https://doi.org/10.1007/s10526-009-9241-x

Paixão, F.R.S., Muniz, E.R., Barreto, L.P., Bernardo, C.C., Mascarin, G.M., Luz, C. and Fernandes, E.K.K. (2017). Increased heat tolerance afforded by oil-based conidial formulations of *Metarhizium anisopliae* and *Metarhizium robertsii*. Biocontrol. Sci. Technol. 27: 324–337. https://doi.org/10.1080/09583157.2017.1281380

Pedrini, N., Ortiz-Urquiza, A., Huarte-Bonnet, C., Zhang, S., Keyhani, N.O. (2013). Targeting of insect epicuticular lipids by the entomopathogenic fungus *Beauveria bassiana*: hydrocarbon oxidation within the context of a host-pathogen interaction. Front. Microbiol. 4:24. https://doi.org/10.3389/fmicb.2013.00024

Pelizza, S.A., Schalamuk, S., Simón, M.R., Stenglein, S.A., Pacheco-Marino, S.G. and Scorsetti, A.C. (2018). Compatibility of chemical insecticides and entomopathogenic fungi for control of soybean defoliating pest, Rachiplusia nu. Rev Argent. Microbiol. 50(2): 189–201. https://doi.org/10.1016/j.ram.2017.06.002

Pimentel, D. (2008). Invasive plants: Their Role in Species Extinctions and Economic Losses to Agriculture in the USA. Pp. 1–8. In: Drake J.A. 2009. Management of Invasive Weeds. (Eds). Springer Science.

Prithiva, J.N., Ganapathy, N., Jeyarani, S. and Ramaraju, K. (2018). Relative safety of *Beauveria bassiana* (Bb 112) oil formulation to *Cryptolaemus montrouzieri* Mulsant. J. Biol. Cont. 32:212–214. https://doi.org/10.18311/jbc/2018/17684

Pucheta-Díaz, M., Flores-Macías, A., Rodríguez-Navarro, S. and De La Torre, M. (2006). Mecanismo de acción de los hongos entomopatógenos. Interciencia, 31(12): 856–860.

Půža, V. and Tarasco, E. (2023). Interactions between entomopathogenic fungi and entomopathogenic nematodes. Microorganisms 8; 163. https://doi.org/10.3390/microorganisms11010163

Qadri, M., Short, S., Gast, K., Hernandez, J. and Wong ACN. (2020). Microbiome innovation in agriculture: development of microbial based tools for insect pest management. Front. Sustain. Food Syst. 4: 547751. https://www.frontiersin.org/article/10.3389/fsufs.2020.547751

Quesada-Moraga, E. (2020). Entomopathogenic fungi as endophytes: their broader contribution to IPM and crop production. Bio. Sci. Tec., 30(9), 864–877. https://doi.org/10.1080/09583157.2020.1771279

Quesada-Moraga, E., Garrido-Jurado, I., González-Mas, N. and Yousef-Yousef, M. (2023). Ecosystem services of entomopathogenic ascomycetes. J. Invertebr. Pathol., 201: 108015. https://doi.org/10.1016/j.jip.2023.108015.

Roy, H.E., Steinkraus, D.C., Eilenberg, J., Hajek, A.E. and Pell, J.K. (2006). Bizarre interactions and endgames: entomopathogenic fungi and their arthropod hosts. Annu. Rev. Entomol. 51: 331–357. https://doi.org/10.1146/annurev.ento.51.110104.150941

Ruiu, L. (2018). Microbial biopesticides in agroecosystems. Agronomy, 8(11): 235. https://doi.org/10.3390/agronomy8110235

Sabbahi, R., Hock, V., Azzaoui, K., Saoiabi, S. and Hammouti, B. (2022). A global perspective of entomopathogens as microbial biocontrol agents of insect pests. J. Agricul. Food. Res. 10:100376. https://doi.org/10.1016/j.jafr.2022.100376

Saidi, A., Mebdoua, S., Mecelem, D., Al-Hoshani, N., Sadrati, N., Boufahja, F. and Bendif, H. (2023). Dual biocontrol potential of the entomopathogenic fungus *Akanthomyces muscarius* against *Thaumetopoea pityocampa* and plant pathogenic fungi. Saudi J. Biol. Sci. 30:8. https://doi.org/10.1016/j.sjbs.2023.103719

Samal, I., Bhoi, T.K., Vyas, V., Majhi, P.K., Mahanta, D.K., Komal, J., Singh, S., Kumar, P.V. and Acharya, L.K. (2023). Resistance to fungicides in entomopathogenic fungi: Underlying mechanisms, consequences, and opportunities for progress. Trop. Plant Pathol. https://doi.org/10.1007/s40858-023-00585-6

Santos, T.S., Silva, T.M., Cardoso, J.C., Albuquerque-Júnior, R.L.C,, Zielinska, A., Souto, E.B., Severino, P. and Mendonça, M.D.C. (2021). Biosynthesis of silver nanoparticles mediated by entomopathogenic fungi: Antimicrobial resistance, nanopesticides, and toxicity. Antibiotics. 13:10(7): 852. https://doi.org/10.3390/antibiotics10070852

Shahid, M., Haq, E., Mohamed, A., Rizvi, P.Q. and Kolanthasamy, E. (2023). Entomopathogen-based biopesticides: insights into unraveling their potential in insect pest management. Front. Microbiol, 14:1208237. https://doi.org/10.3389/fmicb.2023.1208237

Sharma, A., Sharma, S. and Yadav, P.K. (2023). Entomopathogenic fungi and their relevance in sustainable agriculture: A review. Cogent Food Agric., 9(1): 2180857. https://doi.org/10.1080/23311932.2023.2180857

Shin, T.Y., Lee, M.R., Park, S.E., Lee, S.J., Kim, W.J, and Kim, J.S. (2020). Pathogenesis-related genes of entomopathogenic fungi. Arch. Insect Biochem. Physiol., 105: e21747. https://doi.org/10.1002/arch.21747

Shoeb, M., Solaiman, R., Abd-Elgyed, A. and Ahmed, M. (2021). Compatibility of entomopathogenic fungi, *Beauveria bassiana* (Bals. -Criv.) Vuill. and *Metarhizium anisopliae* (Metchn) Sorokin isolates with different agrochemicals commonly used in vineyards. Egypt. Acad. J. Biolog. Sci., 14:37-53. https://doi.org/10.21608/eajbsa.2021.147044

Shukla, C., Afzal, K. (2021). Entomopathogenic Fungi. In: Microbial Approaches for insect pest management. Omkar (Ed.). Springer. 316–334. https://doi.org/10.1007/978-981-16-3595-3

Sinha, K.K., Choudhary, A.K. and Kumari P. (2016). Entomopathogenic fungi. In Ecofriendly pest management for food security. Omkar (Ed.). Academic Press 475-505. https://doi.org/10.1016/B978-0-12-803265-7.00015-4

Skinner, M., Parker, B.L. and Kim, J.S. (2014). Role of entomopathogenic fungi. In: Integrated pest management. Abrol DP (Ed.) Academic Press, Cambridge, 169-191. https://doi.org/10.1016/B978-0-12-398529-3.00011-7

Slowik, A.R., Hesketh, H., Sait, S.M. and de Fine Licht, H.H. (2023). A Rapid Method for Measuring *In Vitro* Growth in Entomopathogenic Fungi. Insects. 14(8):703. https://doi.org/10.3390/insects14080703

Stark, L. A. (2010). Beneficial microorganisms: countering microbephobia. CBE Life Sci. Educ. 9: 387–389. https://doi.org/10.1187/cbe.10-09-0119

Storey, G.K. and Gardner, W. (1986). Sensitivity of the entomogenous fungus *Beauveria bassiana* to selected plant growth regulators and spray additives. Appl. Environ. Microbiol. 52: 1–3. https://doi.org/10.1128/aem.52.1.1-3.1986

Sutanto, K.D., Husain, M., Rasool, K.G., Malik, A.F., Al-Qahtani, W.H. and Aldawood, A.S. (2022). Persistency of indigenous and exotic entomopathogenic fungi isolates under ultraviolet B (UV-B) irradiation to enhance field application efficacy and obtain sustainable control of the red palm weevil. Insects 13:103. https://doi.org/10.3390/insects13010103

van Lenteren, J.C., Bolckmans, K., Köhl, J., Ravensberg, W.J., Urbaneja, A. (2018). Biological control using invertebrates and microorganisms: plenty of new opportunities. BioControl 63 39–59. https://doi.org/10.1007/s10526-017-9801-4

Vega, F.E. (2018). The use of fungal entomopathogens as endophytes in biological control: a review. Mycologia, 110: 4–30. https://doi.org/10.1080/00275514.2017.1418578

Vega, F.E., Meyling, N.V., Luangsa-ard, J.J. and Blackwellz, M. (2012). Fungal Entomopathogens. In Insect Pathology. 2nd Edition. Vega, F.E. and Kaya, H.K. (Eds). Elsevier 172–206. https://doi.org/10.1016/B978-0-12-384984-7.00006-3

Wanchoo, A., Lewis, M.W. and Keyhani, N.O. (2009). Lectin mapping reveals stage-specific display of surface carbohydrates *in vitro* and haemolymph-derived cells of the entomopathogenic fungus *Beauveria bassiana*. Microbiol., 155(9): 3121–3133. https://doi.org/10.1099/mic.0.029157-0

Wang, C. and St Leger, R.J. (2007). The MAD1 adhesion of *Metarhizium anisopliae* links adhesion with blastospore production and virulence to insects, and the MAD2 adhesin enables attachements to plants. Eukaryot Cell 6:808-816. https://doi.org/10.1128/EC.00409-06

Wang, C., Fan, M., Li, Z. and Butt, T.M. (2004). Molecular monitoring and evaluation of the application of the insect-pathogenic fungus *Beauveria bassiana* in southeast China. J. Appl. Microbiol, 96: 861–870. https://doi.org/10.1111/j.1365-2672.2004.02215.x

Wang, J., Lovett, S.B. and Leger, R.J. (2019). The secretome and chemistry of *Metarhizium*; a genus of entomopathogenic fungi. Fungal Ecol., 38:7–11. https://doi.org/10.1016/j.funeco.2018.04.001

Wang, L., Yang, M., Akinnagbe, A., Liang, H., Wang, J. and Ewald, D. (2012). *Bacillus thuringiensis* protein transfer between rootstock and scion of grafted poplars thuringiensis protein transfer between rootstock and scion of grafted poplar. Plant Biology. 14 (5): 1–6. https://doi:10.1111/j.1438-8677.2011.00555.x

Xiao, G., Ying, S.H. and Zheng, P. (2012). Genomic perspectives on the evolution of fungal entomopathogenic city in *Beauveria bassiana*. Sci. Rep., 2:483. https://doi.org/10.1038/srep00483

Yeo, H., Pell, J.K., Alderson, P.G., Clark, S.J. and Pye, B.J. (2003). Laboratory evaluation of temperature effects on the germination and growth of entomopathogenic fungi and on their pathogenicity to two aphid species. Pest Manag. Sci. Former. Pestic. Sci., 59: 156–165. https://doi.org/10.1002/ps.622

Zhang, W., Jia, C., Zang, L.S., Gu, M., Zhang, R. and Eleftherianos, I. (2023b). Entomopathogenic fungal-derived metabolites alter innate immunity and gut microbiota in the migratory locust. J. Pest. Sci. 97: 853–872. https://doi.org/10.1007/s10340-023-01685-7

Zhang, W., Xie, M., Eleftherianos, I., Mohamed, A., Cao, Y. and Song, B. (2023a). An odorant binding protein is involved in counteracting detection-avoidance and Toll-pathway innate immunity. J. Adv. Res. 48: 1–16. https://doi.org/10.1016/j.jare.2022.08.013

Zhou, F., Wu, X. and Xu, L. (2019). Repressed *Beauveria bassiana* infections in *Delia antiqua* due to associated microbiota. Pest Manag Sci. 75: 170–179. https://doi.org/10.1002/ps.5084

Zimmermann G, Papierok B, Glare T (1995). Elias Metschnikoff, Elie Metchnikoff or Ilya Ilich Mechnikov (1845–1916): a pioneer in insect pathology, the first describer of the entomopathogenic fungus *Metarhizium anisopliae* and how to translate a Russian name. Biocontrol Science and Technology 5: 527–530

8 Entomopathogenic Perspectives of *Nomuraea rileyi*

Mansi Kothari[1] and Sardul Singh Sandhu[2*]

1. Introduction

The global population is predicted to reach 10.12 billion by the year 2100. Due to the burgeoning population, agriculture today faces significant challenges in providing adequate food to meet the pressing demand. Improvements in crop varieties must be made to meet the increasing food demand. Therefore, crops should be resistant to insect pests and diseases, and their life cycles should be short. Loss of crop yields, 20 to 40% annually, is mainly due to insect pests and diseases. Synthetic chemicals are recklessly used on vegetables to control insect pests (Maina et al., 2018). This indiscriminate application of chemical pesticides results in insecticide resistance and health hazards (Namasivayam and Bharani, 2015). Across the world, fungal-based formulations, i.e., mycoinsecticides, have made an impact as a component of integrated pest management in field crops (Maina et al., 2018). Pests are multi-species and phytophagous. Crops that are grown under protected cultivation or in open fields may become seriously infected by pests (Pranab et al., 2014). One of the most abundant groups of living creatures, that also plays significant and varied roles in human life, are insects (Asokan, 2007).

In the post-green revolution era, it has been estimated that, on a global scale, 10.8% of crop losses are due to the hindrance caused by insect pests. Insecticides are employed to suppress infestations and reduce crop losses. Insecticides have shown high efficacy and require minimal application effort, thus yielding high profits (Sharma and Sharma, 2021). Since the discovery of synthetic pesticides, they have become predominant in insect pest management (Maina et al., 2018). Rachel Carson's book 'Silent Spring' (Lear, 1993) discusses the dangers posed by synthetic chemicals, providing a wake-up call to the world. Since then, greater priority has been given to alternative pest control products (Maina et al., 2018). The Central Insecticides Board and Registration Committee (CIBRC) of India governs the Insecticides Act of 1968 and the Insecticides Rules of 1971. Biopesticides are regulated by these bodies. The production, sale, distribution, and use of all insecticides, along with biopesticides, to ensure the safety of humans and animals were recommended by the board to central and state governments. Public as well as private institutions are involved in identifying and developing entomopathogens, such as microbial biopesticides. A huge number of

[1] Fungal Biotechnology and Invertebrate Pathology Laboratory, Department of Biological Sciences, Rani Durgavati University, Jabalpur 482001, Madhya Pradesh, India.

[2] Bio-Design Innovation Centre, Rani Durgavati University, Jabalpur 482001, Madhya Pradesh, India.

* Corresponding author: ssandhu@rediffmail.com

wettable powders of bio-fungicides are registered and have been sold in India (Kumar et al., 2019). This chapter deals with the entomopathogenic perspectives of *Nomuraea rileyi* in the field of agriculture.

2. Microbial Biopesticides

According to the United States Environmental Protection Agency (EPA), a wide range of bio-based substances, capable of acting against pests through distinct mechanisms, are classified as biopesticides or biological pesticides. Based on this definition, biopesticides are further categorized into three main classes; plant-incorporated protectants, naturally occurring biochemicals, and microbial entomopathogens. The general hypothesis behind the interaction of living entities with natural products is to enhance ecosystem functions by reducing the reproductive potential of pests. To restore lost ecological balance in agricultural ecosystems, the application of natural enemies as biocontrol agents should be used to combat increased populations of harmful insects and invertebrates. Baculoviruses, fungi, nematodes, and bacteria are several microbial agents that act as invertebrate pathogens (Ruiu, 2018).

Commercially, several biopesticides are available to farmers. Globally, there are 700 products and approximately 175 registered biopesticide active ingredients available. Presently, only 12 biopesticides have been registered in India, of which three are fungal biopesticide products. *Nomuraea* is among the popular biopesticides used for plant protection (Ranga et al., 2007).

2.1 Fungi

Conservatively, about 1.5 million species of fungi are estimated to exist on Earth. The fungal species number was hypothesized by extrapolating data of well-studied fungi from plant hosts and known fungi from different regions. The range of known fungal species is approximately 72,000-100,000 and although it is an uncertain number (Hawksworth and Rossman, 1997).

The Kingdom Fungi incorporates an abundance of eukaryotic species such as lichens, molds, yeasts, rusts, mushrooms, and smuts (Stajich et al., 2009). These species proliferate in diverse environments (Corbu et al., 2023). They make crucial contributions to human industry, research, the biosphere, and medicine (Stajich et al., 2009). As components of the microbiota, fungi also have essential roles where they act as parasites, symbionts, saprotrophs, or endophytes. Based on their size, fungi are generally classified as macroscopic, such as filamentous fungi, or microscopic, such as yeast. In filamentous fungi, a thread or thin filament-like structure known as hyphae is generally 2-10 μm in length. Hyphae form complex networking structures known as mycelium, generally on a centimeter-to-meter scale, thus visible to the naked eye. Fungi host their microbiota that is adhered to the hyphal surface by forming pseudo-tissues that are produced by hyphal aggregation. All ecosystems are inhabited by species from the Kingdom of Fungi. They have a remarkable ability to adapt to different environments, being highly durable in stressful conditions. In ecology, they can act as symbionts as well as decomposers, and they exist across mediums from soil to water; they can persist in all kinds of environments. Fungi can exist in various morphologies; they can be either unicellular or multicellular. In nature, they can be found as either free-living organisms or in symbiotic associations. For centuries, fungi have had various roles in medicine and sustenance. With numerous applications across several sectors, fungi have emerged as a valuable and sustainable resource in modern biotechnology (Corbu et al., 2023). Generally, fungi release hydrolytic enzymes into their surroundings, thus digesting nutrients externally.

Fungi are categorized as heterotrophic organisms. They can cause allergic, systemic, cutaneous, superficial, and subcutaneous diseases. Fungi possess cell walls, plasma membranes with sterol ergosterol, 80S rRNA, and microtubules that consist of tubulin. Fungi can synthesize lysine through the L-α-adipic acid pathway. For the synthesis of carbohydrates, lipids, nucleic acids, and proteins, fungi utilize several carbon sources to meet their requirements. Energy is sourced through the oxidation of sugars, alcohols, proteins, lipids, and polysaccharides. The general differentiation of yeasts is done through morphological observation. Besides this, there are differences in the ability to utilize different carbon sources, such as simple sugars, sugar acids, and sugar alcohols. For the synthesis of amino acids in proteins, purines, pyrimidines, glucosamine, and vitamins, fungi require a source of nitrogen. Fungi are unable to synthesize nitrogen but can obtain it from ammonium, nitrate, nitrite, or organic nitrogen. Different forms of nitrogen exist in the environment, and it depends on the ability of fungi to determine which form of nitrogen they are likely to consume. Primarily, fungi utilize nitrate (a form of nitrogen), which is first reduced to nitrite in the presence of nitrate reductase and then to ammonia (McGinnis and Tyring, 1996). Fungi cells can be multinucleated without cross-walls, which are coenocytic in nature, or they can produce hyphae with cross-walls, which means that fungi have septa that help in forming uninucleate and multinucleate compartments. The septum provides a path for cytoplasmic communication, including the intercellular transfer of nuclei (Cole, 1996).

2.2 Entomopathogenic Fungi

The term entomogenous is a Greek word that is derived from "entomon," meaning, to interpret insects, and "genes," meaning, to arise in. Thus, an entomogenous microorganism refers to "microorganisms that arise in insects." In recent years, entomogenous microorganisms have attracted the attention of several microbiologists, molecular biologists, and entomologists, as microbial control of insect pests can directly impact human welfare. Effective long-term and short-term control can be inundatively introduced into a variety of habitats using several entomopathogens (Sandhu et al., 2012).

Organisms that cause infection in insects are called entomopathogens (Altinok et al., 2019). Entomopathogens, such as bacteria, fungi, nematodes, protozoans, and viruses, can infect various types of insects (Rajan et al., 2009). The action of eliminating harmful insects with the help of microorganisms such as fungi, bacteria, viruses, rickettsia, and nematodes is known as biological control (Altinok et al., 2019). Biological control is a chemical-free method in which living entities, such as microorganisms, are applied for the management of plant diseases. The microbial inoculants potentially replacing harmful pesticides are known as biocontrol agents. For sustainable agriculture, microbial inoculants are considered a promising tool, as microorganisms help enhance plant growth by promoting nutrient availability and uptake and simultaneously supporting plant health (Pirttilä et al., 2021). Regulation of pests, exclusionary systems of protection, and systems of self-defence are the main approaches for categorizing biological control. There are many qualities of biocontrol agents, such as protecting the crop throughout the crop period, being non-toxic to plants, easily combined with biofertilizers, the ability to easily multiply in soil, and not causing residual problems. These agents are also safe for the environment (Sharma et al., 2013).

Entomopathogenic biocontrol agents are eco-friendly alternatives to pesticides for managing crop pests. With the prominence of environmental protection and sustainable agriculture, biopesticides have become essential alternatives (Grewal and Joshi, 2022). The potential of fungi to control insects was first recognized by Louis Pasteur (Revathi et al., 2011). The first organism that was used as a biocontrol agent for pests was fungus. The group of fungi that attack and infect an insect host, ultimately causing death, is known as entomopathogenic fungi (Kidanu and Hagos, 2020). Many species of the Entomophthorales order are obligate pathogens, making them very difficult to mass produce in culture as they have specific adaptations to the life cycles of their hosts. Ascomycotina consists of many species that show an anamorphic nature, i.e., the sexual phase is absent in these fungi. The hypocrealean fungi *Cordyceps* and *Torrubiella* reproduce sexually (Bahadur, 2018). Entomophagous fungi are those that get their nutrition from living insects. Entomopathogenic fungi are classified as a group of phylogenetically diverse, eukaryotic, heterotrophic, unicellular, or multicellular microorganisms that either reproduce via sexual or asexual spores, or both. These fungi are usually non-mobile, and their cells are chitinized in nature (Mora et al., 2018). In tropical and subtropical agroecosystems, entomopathogenic fungi are widely distributed, especially in forest, agricultural, pasture, desert, and urban areas. Insect-pest populations in tropical and temperate habitats can potentially be easily regulated by implementing entomopathogenic fungi. Protection from UV radiation and other abiotic and biotic stresses can be provided by the soil, and, thus, the soil is considered an excellent environmental shelter for entomopathogenic fungi. Temperature, humidity, and rainfall are some of the weather parameters that play a significant role in the distribution, prevalence, and antagonistic efficacy of entomopathogenic fungi (Moanaro et al., 2017). Entomopathogenic fungi specifically infect insects, and they are derived from nature; therefore, they do not have any harmful effects on the environment (Rajan et al., 2009).

In around 100 genera, approximately 750 to 1000 fungal entomopathogens have been identified, and the estimated number of fungi that exist in the world ranges from 1.5 to 5.1 million. For fungi to invade an insect host, certain secondary metabolites and enzymes are required. It has been reported that hundreds of entomopathogenic fungi are responsible for the production of thousands of secondary metabolites, yet their functions are unclear in the host infection process. Secondary metabolites act as immune-suppressive agents (Paschapur et al., 2021).

3. Arthropods and Fungal Interaction

Arthropods are among the most successful species that can be found in all types of environments on the planet (Ten et al., 2022). Insects are members of the phylum Arthropoda (Fish et al., 1989). Soil arthropods are important biotic components of soil, and they act as potential hosts for fungi (Tkaczuk et al., 2012). In a variety of ways, arthropods harm crops. They harm developing plants by laying eggs in parts of the plant; chewing bark, leaves, fruits, stems, or buds; tunnelling or boring through stems, seeds, bark, nuts, or twigs, transmitting plant pathogens; within the plant, they live by causing cancer-like growths; and constructing nests or shelters by taking plant parts. Arthropods contaminate plant-based products with their secretions, laying eggs and depositing faecal material. They also consume plant products as food. These are a few ways that arthropods harm the value of stored plant products, creating problems and additional expense and labour requirements in sorting, packing, and preserving foods. Comparing modern and traditional agricultural practices, it has been reported that crop losses due to insect pests were found to be greater in modern practices (Culliney, 2014). One of the most important priorities set by the European Union in matters relating to environmental protection was the protection of biodiversity in rural areas (Tkaczuk et al., 2012).

Entomopathogenic fungi contribute to the biological control of arthropods, as they are found worldwide and can infect insect hosts that damage several economically important crops (Sabbour and Abdel-Rahman, 2013). Arthropods are initially infected by the fungal spore; once the spore is established in the insect cuticle, a cascade of events is initiated, ultimately leading to the host's death (Litwin et al., 2020). Entomopathogenic fungi act as parasitic microorganisms. Entomopathogenic fungi are a safe alternative to toxic chemical insecticides, which are capable of secreting several bioactive compounds to kill the host insects (Keppanan et al., 2019). Entomopathogenic fungi are heterogeneous organisms; they can infect a wide variety of arthropod species as they have a wide spectrum of activity and play various ecological roles. Some of these fungi are utilized in Chinese medicine (Litwin et al., 2020).

Arthropods in terrestrial and aquatic habitats interact with entomopathogenic fungi. In natural ecosystems, there are more than 100 genera, which consist of between 750 and 1,000 entomopathogenic fungi that play a major role in the dynamics of arthropod populations. A wide range of genera and species with high ecological, morphological, and phylogenetic diversity are comprised of entomopathogenic fungi. Arthropods are used as substrates by entomopathogenic fungi for their reproduction (Alonso-Díaz and Fernández-Salas, 2021); the fungi do not infect host plants, as they are very specific to insects (Namasivayam and Vidyasankar, 2014). It is estimated that over 700 species of fungi are pathogenic to insects. Certain insect pathogenic fungi have a restricted host range, while other fungal species have a wide host range (Ibrahim et al., 2012). Once insects are infected by the fungi, a series of pathogenic processes begin, resulting in symptoms. This is due to certain metabolites and depsipeptides synthesized by entomopathogenic fungi that act as toxins (Sánchez-Pérez et al., 2014). The first enzyme that was synthesized by the entomopathogenic fungi to break down the insect's integument was lipase. Enzymes are synthesized by entomopathogenic fungi through enzymatic activity. These enzymes are responsible for pathogenic activity as well as for the breakdown of the insect's integument (Maravi et al., 2018). Susceptibility of the insect and virulence of the fungus determine the effectiveness of myco-biocontrol agents; selection of a stable strain with efficacy specific to the target hosts determines fungus virulence (Kidanu and Hagos, 2020). Entomopathogenic fungus inoculum can be recycled as it can develop on cadavers, allowing the fungus to persist in the environment. To manage pest populations, entomopathogenic fungi act as a natural epizootic control agent. Fungal infections are associated with about 60% of insect diseases (Mantzoukas et al., 2022).

4. *Nomuraea rileyi* as an Entomopathogenic Fungus

Nomuraea is a genus of highly abundant fungi with multiple strains (St. Leger and Wang, 2020). In 1883, *M. rileyi* was classified as *Botrytis rileyi*. Later classified as *Spicaria rileyi* by Charles (Charles, 1936), ultimately being reclassified under the genus *Nomuraea* by Kish (Kish et al., 1974). *N. rileyi* isolates have been collected from all over the world. It is a dimorphic fungus with hyphal bodies and a true filamentous growth phase (Fronza et al., 2017). *Nomuraea* sp. can infect 32 species of insects (mainly lepidopterans) (Perinotto et al., 2012).

Nomuraea rileyi is an entomopathogenic fungus that mainly attacks prominent caterpillar pests of soybeans (Pavone et al., 2009). The velvet bean caterpillar (VBC), *Anticarsia gemmatalis*, is a major soybean pest. Globally, soybean cultivation is over 14 million ha, mainly under humid weather conditions. The fungus *Nomuraea rileyi* acts as a natural enemy of soybean caterpillars. From the mid-to-late vegetative stage, i.e., the middle of December, the velvet bean caterpillar usually occurs in high numbers, although populations can be detected within a few days after plant emergence. *Nomuraea rileyi* helps maintain the pest population below levels of significant economic loss, even in high humidity (Sosa-Gómez et al., 2003). Specific requirements for the fast growth of fungi include temperature, humidity, and aeration. The virulence of entomopathogenic fungi isolates varies according to the target insects. The time of conidial germination and mortality of insects are the parameters for assessing virulence.

From an environmental point of view, *N. rileyi* can be an excellent alternative, as it is reportedly safe for human beings as well as other non-target organisms (Stefanelli et al., 2021). In a variety of insects, the natural occurrence of this fungus has been reported in India (Table 1). *N. rileyi* prefers maltose as a carbon source (Tincilley et al., 2004). The potential of *N. rileyi* as an entomopathogen is undeniable. During growth, *N. rileyi* produces large amounts of extracellular polysaccharides. An important role is played by the polysaccharides in adhesion to the host cuticle and in the possible mummification of infected larvae (Fronza et al., 2017). Some active metabolites have reportedly been produced by *N. rileyi* against the larvae of *H. virescens*, *Bombyx mori*, and *Heliohis zea*. The biosynthesis of bioactive compounds produced by *N. rileyi* could be triggered by the insect-derived material added (Marcinkevicius et al., 2017).

The United States, Brazil, Argentina, China, India, Paraguay, Canada, and Uruguay are some of the top soybean-producing countries worldwide. Annually, 89.55 million ha of soybean have been planted by the countries in South and North America, i.e., approximately 76% of the total global soybean cultivation area. Around the world, more than 180 invertebrate species attack soybean crops. *Thysanoplusia orichalcea*, *Chrysodeixis acuta*, *H. armigera*, and *Spodoptera litura* are some of the most prominent lepidopteran soybean pests found in India. *Omiodes indicata*, *Spilosoma obliqua*, and *S. exigua* are other regionally prominent lepidopteran pests (Sosa-Gómez, 2017). Farlow (Farlow, 1883) reported and described the fungus *N.*

rileyi as *Botrytis rileyi*, although the systematic documentation on the occurrence of the fungus was traced back to 1915. Around the world, serious attempts were made to harness the potential of *N. rileyi* as a bio-suppression agent after 1955.

Table 1 Epizootic occurrence of *Nomuraea rileyi* in India.

Pest	*Plant host*	*State*	*Reference*
Spodoptera litura	Tobacco	Pune	Patil and Abhilash, 2014
S. podoptera exigua	Black gram, and bajra		
S. litura	Groundnut, soybean, potato, and cabbage	Dharwad	
Helicoverpa. armigera and *Plusia*	Groundnut, soybean, lucerne, niger, sunflower, cotton, and sorghum		
H. armigera	Tomato, field beans, and pigeon pea	Bangalore	
S. litura	Groundnut	Bapatla	
H. armigera, S. litura and *Plusia*	Groundnut and cotton	Andra Pradesh	Bhargavi et al., 2018.

Microbial agents can be used to control most of the lepidopteran species that are susceptible to diseases (Sosa-Gómez, 2017). All over the world, *N. rileyi* is considered a prominent mortality factor for many lepidopteran insects. Under favourable environmental conditions, *N. rileyi* has shown the potential to cause natural epizootics. *Nomuraea rileyi* is an entomopathogenic fungus found in several countries and exhibits an anamorphic mode of reproduction (Bhargavi et al., 2018). *Chrysodeixis* (formerly *Pseudoplusia*), *Spodoptera*, *Anticarsia*, *Heliothis*, and *Rachiplusia* are certain genera of *Noctuid* caterpillars that *N. rileyi* attacks. Even among the same genus, the susceptibility of host species varies (Fronza et al., 2017).

All around the world, *Spodoptera litura* (Lepidoptera: Noctuidae) is a pest defoliator that causes yield losses of high magnitude. Throughout Asia and the Pacific Islands, this pest is widely distributed. *Spodoptera litura* (Lepidoptera: Noctuidae) is a pest species that is highly polyphagous. It has been reported that this pest feeds on plants that belong to 44 different families consisting of 112 species (Namasivayam and Bharani, 2015). The tobacco budworm, oriental leafworm moth, tropical armyworm, and taro caterpillar are some other known species of these insects. Over 120 plant species, such as vegetables, fruits, and ornamental crops, are susceptible to this insect, which causes extensive losses (Krutmuang and Thungrabeab, 2017). It has been reported that, mostly in humid weather conditions, more than 30 species of lepidopteran larvae are susceptible to *Nomuraea rileyi,* which acts as a pathogen (Table 2). *Spodoptera frugiperda* is a very prominent phytophagous species in maize fields and a well-suited host for fungi. Entomopathogenic fungi can be easily isolated from cultivated soils and dead insects (Marcinkevicius et al., 2017).

4.1 Occurrence and Isolation of Nomuraea rileyi

Infected insect hosts are the natural source for obtaining entomopathogenic fungi. For further study of the fungus, the specimen can be collected from the field and further

Table 2 Reports of *Nomuraea rileyi* morbific on different insect species in India.

Insect	*Plant*	*Occurrence*	*Reference*
Acontia graellsii	Soybean	Natural	Devi and Duraimurugan, 2013
Achaea Janata	Castor	Laboratory	Phadke and Rao, 1978; Kamat et al., 1978
Acrocercops tenera	Forest	Natural	Sandhu et al., 1993
Agrotis ipsilon	Various crops and forest nurseries	Natural	Devi and Prasad, 2001
Amsacta moorei	Various crops and forest nurseries	Natural	Devi and Prasad, 2001
Atteva fabriciella	Forest	Natural	Sandhu et al., 1993
Chrysodeixis acuta	Soybean	Natural	Devi and Duraimurugan, 2013
Diacrisa obliqua	Soybean	Natural	Singh and Gangrade, 1975
Eutectona machaeralis	Teak	Natural	Sandhu et al., 1993
Helicoverpa armigera	Cotton, tomato, pigeon pea, and field beans	Natural	Devi and Prasad, 2001
Hyblea puera	Teak	Natural	Sandhu et al., 1993
Hypocala rostrata	Various crops and forest nurseries	Natural	Devi and Prasad, 2001
Junonia orithiya	Justicia gendarussa	Natural	Devi and Prasad, 2001
Lamprosema indicata	Various crops and forest nurseries	Natural	Devi and Prasad, 2001
Mocis undata	Various crops and forest nurseries	Natural	Devi and Prasad, 2001
Plusia orichalcea	Various crops and forest nurseries	Natural	Devi and Prasad, 2001
Spodoptera exigua	Gram and millet	Natural	Devi and Prasad, 2001
Spodoptera litura	Tobacco, castor and groundnut	Natural	Devi and Prasad, 2001

cultivated in laboratory conditions (Shahid et al., 2012). If the fungus has already sporulated, then the pathogens can be directly retrieved from the surface of cadavers. Fungus can be directly scraped off the cadavers. Infected cadavers are kept in such a way that they can directly contact the nutrient surface. To encourage sporulation, cadavers should be kept in a petri dish that is lined with moist filter paper. The petri dish then acts as a humid chamber where the cadavers can incubate. To isolate the pathogen, sporulating cadavers are dabbed over the selected media (Butt and Goettel, 2000). The insect bait method is a more efficient way of isolating entomopathogenic fungi compared to direct isolation from soil (Chang et al., 2021).

Other methods for isolating fungi include discontinuous density gradients or aqueous solutions that can be extracted by conjugating directly with a selective medium. Insect live baiting is an indirect way of isolating pathogens from the soil.

Selective media are mainly supplemented with antibiotics or fungicides that help enhance the growth of entomogenous fungi and suppress the growth of bacteria and saprophytic fungi. On several media, isolated fungi can be maintained in vitro. Sterile glass ampoules can be used for storing freeze-dried fungal mycelium, or cryovials can be used for storing fungal mycelium under liquid nitrogen. A desiccator can be used for storing freshly harvested conidia (Butt and Goettel, 2000).

4.2 Morphology of *Nomuraea rileyi*

Presently, the *N. rileyi* fungus could be identified by looking for malachite-green coloration on the insect body surface (Vimala, 2018). *N. rileyi* is an anamorphic fungus morphologically, and it is classified as an imperfect fungus. According to colony progress, its color ranges from white-green-pale to green-intense. *N. rileyi* is septate; the diameter of the hyphae ranges from 2-3 mm and is hyaline to slightly pigmented; the conidiophores are erect and septate; the conidia form divergent chains and are smooth, ellipsoidal, and sometimes cylindrical with a pale-green color (Álvarez et al., 2018); the size of the conidiospores ranges from 3.5–4.5 × 2.0–3.1 μm (Vimala, 2018). Initially, *N. rileyi* reproduces in the host's hemolymph-like hyphal bodies like yeast that differentiate into elongate mycelia, which later results in producing external septate conidiophore structures. Lipid bodies and various membrane-bound organelles are present in all stages of *N. rieyi* (Boucias et al., 1984).

Fatty acids mainly consist of palmitic, oleic, and linoleic neutral lipids, such as triglycerides and sterols, and polar lipids such as phosphatidylethanolamine, phosphatidylinositol, and phosphatidylserine; these comprise the basic lipid composition of *N. rileyi*. Identification of only a few proteins in *N. rileyi* spores has been achieved. Even by two-dimensional electrophoresis, 252 proteins were separated and then analyzed by mass spectrometry (MS) analysis. As a result, it was found that only 121 proteins produced a good MS signal; attributed to the lack of genomic information about *N. rileyi* (Fronza et al., 2017). Sabourd's agar medium, fortified with yeast extract, is the best growth medium for the multiplication of the *Nomuraea rileyi* fungus. For the long-term preservation of the fungus, it is recommended that the culture be stored under 10% glycerol in liquid nitrogen (–196°C), which does not cause any genetic change. Without any loss of viability in fungus slants overlaid with sterile mineral oil, they can be held at 25°C for 6–12 months, but for the maintenance of virulence, regular passage through host insects is essential (Vimala, 2018).

4.3 Life Cycle of *Nomuraea rileyi*

Insects' epicuticles are hydrophobic. Conidiospores, which are passively dispersed, are covered in rodlet bundles embodied in hydrophobins. In the first stage of the infection, these hydrophobins bind to the hydrophobic surface of the insect epicuticle (Fig. 1). Hydrolytic enzymes are impregnated within the conidial surface coat and help in actively degrading the epicuticle under moist conditions, ultimately producing substances that help initiate the germination process. At this stage, the production of apical-growing penetrant hyphae starts, which releases a cocktail of cuticle-degrading meta that helps germ tubes enter and multilaminate insect cuticles. As soon as the germ tube reaches the nutrient-rich hemolymph, the formation of freely circulating hyphal bodies starts as the apical hyphae transit into a budding growth phase. In the nutrient-rich hemolymph, yeast-like hyphal bodies grow exponentially and reach a density that exceeds that of circulating hemocytes. The hemolymph-borne cells revert synchronously to an apical growth process when a critical threshold density is achieved, thus forming the tissue-invasive mycelial cell phenotype. Insect tissues are covered by basement membranes, which are attached by laminin-binding materials that encase the hyphal tips of the cells. Finally, a suite of metabolites that are produced and secreted by the ensuing mycelium modulates host development and ultimately kills the host. Infected tissues are efficiently digested, thus leading to the mummification of infected larvae (Boucias et al., 2016).

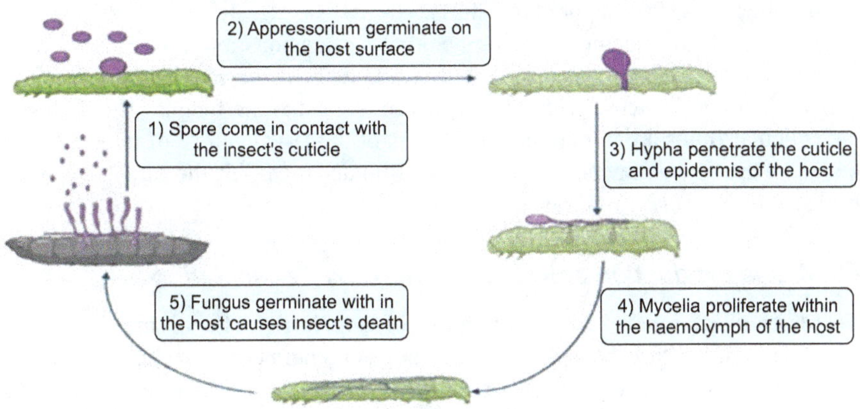

Fig. 1 Life cycle of *Nomuraea rileyi*.

A variety of fungal structures and cell types are involved in the natural life cycle of entomopathogenic fungi. In the environment, fungal conidia disseminate and thus get transmitted to new insects, soil niches, and plants. When fungi proliferate inside the host hemolymph, the formation of blastospores occurs naturally (Gotti et al., 2023). The optimum temperature is 20 to 30°C for *N. rileyi* conidia to germinate and penetrate a noctuid insect body, and the optimum humidity is 95 to 100% for conidial germination, infection, and sporulation (Padanad and Krishnaraj, 2009).

4.4 Action of Enzymes of *Nomuraea rileyi*

The tegument is present in most insects and is a segmented cylindrical structure with the presence of a cuticle, epidermis, and basal membrane. These three layers are responsible for the rigidity of insects. Chitin is a polysaccharide that is similar to cellulose and is considered an essential component of the cuticle. Chitin is a structure that is formed by crystalline chitin nanofibers that exist inside a protein, polyphenol, and lipid matrix. The polysaccharide mainly provides protection, acting as a barrier against parasites and diseases (Dar et al., 2017). The presence of cuticles on host species influences the production, expression, and timing of the extracellular enzyme (Fronza et al., 2017).

N. rileyi mainly secretes chitinase enzymes at the time of invasion. β-1,4 bonds of the chitin polymer get hydrolyzed by the action of chitinases that result in the production of dominant N-N′-diacetyl chitobiose. Chitobiose is responsible for the breakdown of the monomer N-acetyl glucosamine (GlcNac). Insect cuticles get degraded when protease enzymes collaborate with chitinase enzymes. Chitinases have varied roles in different stages of the life cycle: nutrition and defence against competitors, hyphal growth, germination, and morphogenesis of an insect. Chitinases and N-acetylglucosaminidases are chitinolytic enzymes that differ in their breakdown patterns. In the early stage, terminal non-reducing N-acetylglucosamine (GlcNac) residues get catalysed from the chitin, and at the latter stage, β-1,4 linkages of chitin and chitooligomers get hydrolyzed, ultimately liberating a short chain of monomers or chitooligomers (Paschapur et al., 2021).

Larval mortality achieved using conidial suspensions alone was found to be lesser compared to larval mortality when treated with *N. rileyi*, which was 2.7 times higher when administered with purified lipase alongside the fungus. This indicates that cuticular lipids are important constituents of the larval cuticle of insects, where the role of the lipase enzyme has been underestimated regarding the pathogenesis and virulence of *N. rileyi* (Fronza et al., 2017).

4.5 Mass Production of *Nomuraea rileyi* by Fermentation

Elie Metchnikoff initiated the mass culturing of fungi in 1883. For his first experiment on mass fungus culture, he used two beetle pests (Jitendra et al., 2012). In biocontrol strategies, the hyphae and conidial biomass of fungi are the main infective fungal structures. Fermentation in standard media for the mass production of biocontrol fungi is the most commercially used method. Low-cost agricultural by-products are prominent mediums for fermentation on solid substrates. Low demand for water, easy aeration using small batches, simulation of the natural environment, the use of solid support for microorganisms, high productivity, and lower sterility requirements are some of the important features of using solid substrates for multiplication (Bich et al., 2018). In solid culture, *N. rileyi* conidiates abundantly. Conidia is the end product, which is the advantage of using solid substrates (Hall and Papierok, 1982).

One of the essential components of the biocontrol programme is the production of the required quantities of inoculum. Based on the quantity of the product required, the production of entomopathogens will be undertaken by several methods. During the development of mycopesticides such as *N. rileyi*, relatively small quantities of the inoculum were used for laboratory experiments and field trials. The basic multiplication procedures for the development of a simple and reliable production system lie in the fact that for the production of aerial conidia, solid-state fermentation is performed, and for the production of blastospores, submerged liquid fermentation is performed. Blastospores are short-lived and hydrophilic. In the diphasic strategy, the fungal inoculum is initially germinated in a liquid culture, which is later used for the inoculation and production of conidia in solid-state fermentation (Sahayaraj and Namasivayam, 2008).

Solid and submerged cultures are the two methods that are employed for fungal growth. There are many technical and economic limitations to the solid culture method, but this method is easy to achieve in laboratory conditions. Most fungi sporulate on solid media easily; hence, this method is more favorable. In another method for obtaining desirable strain growth, certain criteria must be maintained, which include an optimal growth environment, reduced risk of contamination, carefully controlled initial pH, and less labour and dissolved oxygen (DO) concentration. Some studies have reported that blastospores are more virulent than conidia and that they can easily germinate in submerged media; this requires only a few days, whereas most entomopathogenic fungi take a longer time to germinate conidia on solid substrates. Significant differences in pathogenicity can be obtained from the presence of minute variations in host-pathogen interactions. Initially, the conidia successfully adhere and then germinate, followed by differentiation and ultimately the penetration of fungal hyphae. These are the unique pathogenic processes through which the conidia of entomopathogenic fungi often infect their hosts. Copious mucilage can be produced by the blastospore, which does not resemble conidia and is readily adhered to the cuticle surface. Adherence to spores determines how successful an infection is. Afterwards, spores invade through the gut following ingestion (Zhao et al., 2023).

For the mass production of most entomopathogenic fungi, information on growth requirements is essential. Unfortunately, information availability is poor. Nutritional requirements must be chosen according to the desired fungus. Entomopathogenic fungi generally require water, minerals, and organic and inorganic sources for their growth. An essential component of the biocontrol programme is the production of good-quality inoculum in adequate quantities. For the mass production of entomopathogenic fungi, a wide variety of organic materials have been evaluated as substrates. Numerous researchers from different countries are trying to evaluate suitable substrates for the mass production of entomopathogenic fungi. They are continuously putting their efforts into identifying low-cost agricultural materials, mainly based on agricultural waste products and byproducts. Some methods of mass production are liquid fermentation and a biphasic culture system subdivided into two types of fermentation systems, i.e., solid-state fermentation, and submerged-state fermentation. Among these, solid-state fermentation has emerged as a highly appropriate method. In the tropics and the Northern Hemisphere, the major substrates that are used for fungal production are barley and rice (Ranadev et al., 2023).

As suitable solid substrates, groundnut cake, finger millet, and rice agricultural products and byproducts have been evaluated by researchers. Molasses and yeast extract are the components of a liquid medium for submerged fermentation that was developed to produce inoculum for rice gruel semisolid substrate or for developing still culture. A plastic bag with autoclaved mijo grains that were inoculated with 8-day-old fragments of agar grown with the sporulated fungus was devised for mass production in Colombia (Jaronski., 2014). Insect pests can be infected even in unfavorable conditions by using the oil-based formulation method. Stable and viable inoculum production is critical in the utilization of entomopathogenic fungi (Montecalvo et al., 2023).

Hardened masses of pigmented hyphal aggregates that are approximately 50–600 μm are known as fungal microsclerotia. These hardened masses were first identified in the entomopathogenic fungus species *Metarhizium*. These masses serve as survival structures. The microsclerotia were initially observed when grown dimorphically in submerged liquid cultures. Initially, conidia germinate and form mycelium, then a compact mass of mycelia leads to the development of microsclerotia. This dimorphism is observed in *Metarhizium*. These structures utilize their endogenous reserves as a source of carbon, and thus they are capable of producing infective conidia and do not require any exogenous carbon sources. These fungal structures can survive in high desiccation conditions, and thus the fungal species can be applied as mycoinsecticide biological control agents (Paixão et al., 2021). A pigmented rind, a thin-walled cortex, and a large central medulla are the three distinct layers that a mature microsclerotia exhibits. The hyphae swell and aggregate during the initial stages of development. Pigment emerges as soon as aggregations enlarge. Within the cell wall, the pigment is deposited; this happens in the last phase (Song et al., 2013). Germination potential and conidial yield may be influenced by the storage conditions of microsclerotia. Microsclerotia may also exert an influence on fungal strain, formulation, and packaging (Yousef-Yousef et al., 2022).

For developing sustainable agriculture, generally, naturally derived pesticides are preferred over synthetic pesticides. *N. rileyi* is a natural mycopesticide that is utilized in integrated pest management. The development of inoculum in an adequate amount for field application determines the success of any microbial control programme. For commercializing entomopathogenic fungi, various laboratory-efficient production technologies have been developed (Rajan et al., 2009).

4.6 Mass Production of Nomuraea rileyi on Natural Substrates

It has been over 100 years since the fungus *N. rileyi* was first identified, but not a single attempt was made to mass cultivate and apply it as a biological control agent until 1995 (Elanchezhyan, 2007). Sorghum and barley can be used as natural substrates for the mass production of *N. rileyi* (Vimala, 2018). In the laboratory, *N. rileyi* strains are generally maintained on media, with each successive transfer of strains resulting in a reduction of pathogenicity or virulence. Through a host insect, virulence can be restored to the fungus. The number of generations produced on the culture medium is also affected by the ability of the fungus to sporulate on cadavers. Humidity, light, and temperature are some of the environmental conditions that influence the germination and sporulation of *N. rileyi* in the laboratory (Fronza et al., 2017) (Fig. 2).

5. Formulation and Commercialization of Mycopesticides

To maintain pest control efficacy, various genera of hypocrealean fungi have been recognized as being effective against numerous species of pests. The fungi of these genera are evaluated as integral components of integrated pest management (IPM) strategies. These fungi offer environmentally sustainable pest suppression by diminishing the risk of inorganic pesticide resistance. To safeguard the environment, many entomopathogen-based biopesticides have been formulated (Bamisile et al., 2021). Stabilizing and improving the efficacy of product formulations is required. Fungus modes of action should determine formulation. The storage period of bioinsecticides should be 18 months until they reach the farmer's market (Hall and Papierok, 1982).

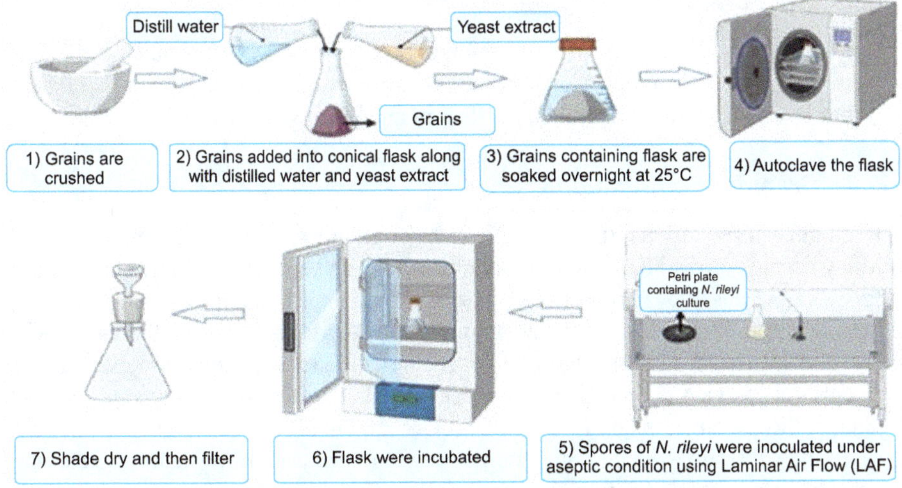

Fig. 2 Mass production of *Nomuraea rileyi* under laboratory conditions (Devi and Duraimurugan, 2013).

Every year, the application of biopesticides increases by almost 10% on a global scale. In the global market, consumption of these pesticides is expected to increase in the future. Over-reliance on chemical pesticides can be reduced with a synchronous increase in utilization of biopesticides (Paschapur et al., 2021). The global biopesticides market is growing at a five-year Compound Annual Growth Rate (CAGR) of 14.1%, and it is predicted that in 2021 it will reach about $7.7 billion (Ruiu, 2018).

Regulations used to assess the use of synthetic active substances were kept the same for biopesticides in the EU. Prospective biopesticide product registration is facilitated by the preparation of new guidelines. Compared to the USA, India, Brazil, or China, it is documented that there are fewer active substances in biopesticides registered in the EU. By the early 2050s, it is expected that the use of biopesticides will equal that of synthetic pesticides in terms of market capacity, but major uncertainties regarding the rate of consumption were accounted for especially in areas such as Africa and Southeast Asia (Paschapur et al., 2021).

In achieving sustainable agriculture, biocontrol agent formulations are powerful tools (Pavone et al., 2009). Optimising the formulation of biological control agents improves product viability and stability.

5.1 Oil Formulation

Oil-formulated entomopathogenic fungi efficiency is better in comparison with non-oil formulations (Bhargavi et al., 2018). Fungus field performance is enhanced when oil is added to liquid formulations. UV filters are chemical additives that when mixed with formulating agents they, protect entomopathogenic fungi against UV radiation (Pavone et al., 2009).

5.2 Wet Formulation

Formulating agents are mixed with water, sprayed over the plant tissues, and act as wetting agents. Powdered casein, saponins, soaps, dried milk, gelatine, and oils are examples of wet formulating agents (Bharani and Namasivayam, 2016).

5.3 Granular Formulations

In granular formulations, small particle sizes of granules, 0.3-1 mm, are formulated. Many species of conidia have hydrophobic walls, and if they are formulated in the form of granules, they remain stable under different environmental conditions. In the field, these granules act as a solid substrate for fungal growth or as an infectious agent. As soon as insects are attracted to these granules, contact is established, and thus the insects get infected. These formulated granules can be utilized as mycoherbicides, plant pathogen antagonistic fungi, insecticides, and nematicides. These granules act as a solid substrate, which allows fungal growth to occur in the agriculture field; the granules are a vehicle for fungal infection (Pavone et al., 2009).

For the successful utilization of myco-insecticides as pest control, it is essential to develop suitable formulations. During storage, packaging, and prior to application in the field, the viability of the spore, that is, the infective unit, should be maintained to ensure high virulence against the target pest. During the production and storage of entomopathogenic fungi, factors such as humidity, temperature, strain selection, and light exposure affect the stability of conidia.

For commercial production, rice is considered a promising alternative to other solid substrates as it is viable on a large scale and is easily available (Fronza et al., 2017). Entomopathogenic fungi, which are mainly based on the suspensions of conidia (Ruiu, 2018), have been used for the development of commercial products (Sabbour and Abdel-Rahman, 2013). Based on the 12 entomopathogenic fungal species, about 170 pest control agents have been produced so far and marketed commercially (Singh et al., 2016).

There are more than 100 commercial products based on entomopathogenic fungi. Currently, the U.S. Environmental Protection Agency has registered nine mycoinsecticides in the United States. The Organisation for Economic Cooperation and Development (OECD) has registered 21 different fungi in the European Union (EU) (Jaronski, 2014). One commercially available mycoinsecticide (*N. rileyi*) is produced in Colombia (Fig. 3) (Maina et al., 2018). These fungal agents can be applied in the field in different ways. Some field application methods are shown in Figure 4 (Devi and Duraimurugan, 2013).

For the maintenance and bioassay production of strains, the potential loss of virulence is very relevant. Before large-scale fermentation, the commercialized strains must undergo several routine conidium-to-conidium cycles that are particularly critical. Virulence loss potential and the quality of the product can be implicated in the results of a particular fermentation batch. To ensure the consistent quality of the marketed material, it is essential to maintain its virulence throughout the mass-production process. Among different isolates, and species, the morphological characteristics and virulence of entomopathogenic fungi appear to vary considerably due to the effects of repeated in vitro subcultures. It has been reported that after

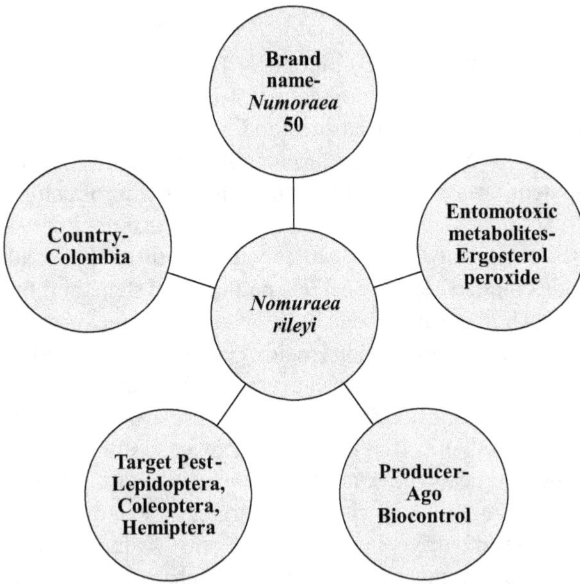

Fig. 3 Commercially available pesticides for *Nomuraea rileyi*.

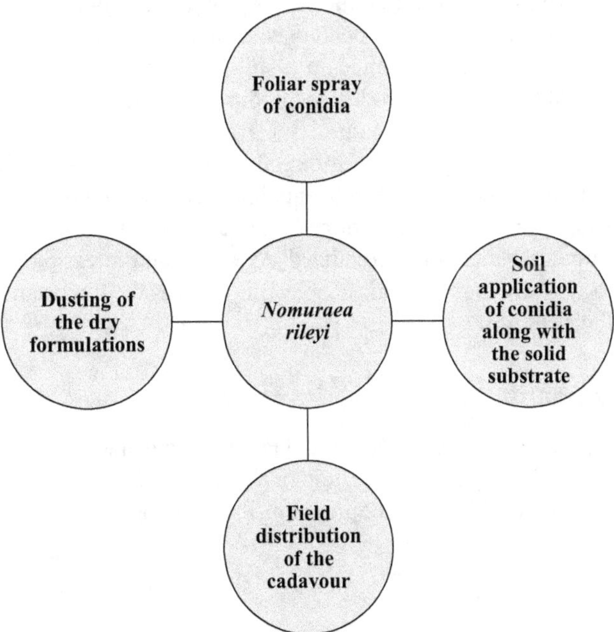

Fig. 4 Field application of mycopesticides of *Nomuraea rileyi*.

repeated transfer of *Nomuraea rileyi* isolates, the virulence of the fungi remains, and some have reported that loss of pathogenicity in isolates takes place (Brownbridge et al., 2001). Currently, commercial products based on entomopathogenic fungi, including *N. rileyi*, are either under development or in use (Edelstein et al., 2004).

Conclusion

The world's plant resources are continuously being harmed by pests, especially arthropods. For a growing human population, it is necessary to protect crops using effective and viable methods that ensure food supply. A relevant approach should be applied to reduce crop losses for sustainable production. By focusing on the entire agroecosystem, it is necessary to approach pest control comprehensively. To develop a healthy crop environment, it is necessary to acquire knowledge of factors that help enhance plant growth and enable them to withstand pest attack. Synthetic chemical pesticides induce negative effects on the ecosystem, and thus eco-friendly pest management techniques have been developed. To enhance crop productivity and suppress insect pest infestations, biopesticides can be employed. Biopesticides are safe for the environment. Biopesticides have no residual effects on the environment and thus can be used as an alternative to chemical pesticides. In various parts of the world, entomopathogenic fungi have been extensively used as biopesticides. Entomopathogenic fungi can attack and infect the insect host, ultimately resulting in host death. Arthropods are mainly used as substrates by entomopathogenic fungi for their reproduction. Various mycoinsecticides function as biocontrol agents. Hence, mycoinsecticides offer an alternative approach to chemical pesticides. One of the important mycoinsecticides is *N. rileyi*, which is increasingly utilized worldwide for pest management. *N. rileyi* helps maintain the pest population below levels of economic loss. For sustainable agriculture, mass produced, commercially shelf-stable, and efficient formulations of entomopathogenic fungi are required. Mass culturing of fungi (*N. rileyi*) can be performed in a solid state as well as in a submerged state. In solid culture, *N. rileyi* produces conidia abundantly. For mass cultivation, semi-synthetic media as well as natural substrates can be used. Another way to cultivate *N. rileyi* is to mass produce it directly using insect pests or bait methods. The formulation of biological control agents improves the viability and stability of the product. Thus, *N. rileyi* is an entomopathogenic fungus that can be easily isolated, cultivated, formulated, and commercialized. *N. rileyi* can protect plants from insect pests in an eco-friendly manner. This chapter emphasizes the importance of *N. rileyi* in the field of agriculture.

Acknowledgments

The authors wish to express gratitude to the Fungal Biotechnology and Invertebrate Pathology Laboratory, Department of Biological Sciences, Rani Durgavati University, Jabalpur (Madhya Pradesh), India, for providing their support and guidance throughout my work.

References

Alonso-Díaz, M.A. and Fernández-Salas, A. (2021). Entomopathogenic fungi for tick control in cattle livestock from Mexico. Frontiers in Fungal Biology., 2: 657–694.

Altinok HH, Altinok MA, Koca AS. (2019). Modes of action of entomopathogenic fungi. Curr. Trends Nat. Sci., 8(16): 117–124.

Álvarez, S.P., Guerrero, A.M., Duarte, B.N.D., Tapia, M.A.M., Medina, J.A.C. et al. (2018). First report of a new isolate of *Metarhizium rileyi* from maize fields of Quivican, Cuba. Indian journal of microbiology., 58(2): 222–226.

Asokan, R. (2007). Genetic engineering of insects. Resonance., 12: 47–56.

Bahadur, A.B. (2018). Entomopathogens: role of insect pest management in crops. Trends in Horticulture., 1(1).

Bamisile, B.S., Akutse, K.S., Siddiqui, J.A. and Xu, Y. (2021). Model application of entomopathogenic fungi as alternatives to chemical pesticides: Prospects, challenges, and insights for next-generation sustainable agriculture. Frontiers in Plant Science., 12: 741–804.

Bharani, R.A. and Namasivayam, S.K.R. (2016). Evaluation of Persistence and Compatibility with Synthetic Chemical Pesticides of Formulated Entomopathogenic Fungi *Metarhizium anisopliae (M.)* Sorokin. Biosciences Biotechnology Research Asia., 13(3): 1617–1621.

Bhargavi, G.B., Manjula, K., Rao, A.R. and Reddy, B.R. (2018). Viability of *Nomuraea rileyi* Conidia in Vegetable and Mineral Oil Based Formulations. Int. J. Pure App. Biosci., 6(4): 751–755.

Bhargavi, G.B., Manjula, K., Rao, R. and Reddy, B.R. (2018). Efficacy of oil-based formulations of *Nomuraea rileyi* (*Farlow*) samson against *Spodoptera litura* in vitro. International Journal of Current Microbiology and Applied Science., 7(10): 3413–3422.

Bich, G.A., Castrillo, M.L., Villalba, L.L. and Zapata, P.D. (2018). Evaluation of rice by-products, incubation time, and photoperiod for solid state mass multiplication of the biocontrol agents *Beauveria bassiana* and *Metarhizium anisopliae*.

Boucias, D., Liu, S., Meagher, R. and Baniszewski, J. (2016). Fungal dimorphism in the entomopathogenic fungus *Metarhizium rileyi*: Detection of an in vivo quorum-sensing system. Journal of invertebrate pathology., 136: 100–108.

Boucias, D.G., Brasaemle, D.L. and Nation, J.L. (1984). Lipid composition of the entomopathogenic fungus *Nomuraea rileyi*. Journal of invertebrate pathology., 43(2): 254–258.

Brownbridge, M., Costa, S. and Jaronski, S.T. (2001). Effects of in vitro passage of *Beauveria bassiana* on virulence to *Bemisia argentifolii*. Journal of Invertebrate Pathology., 77(4): 280–283.

Butt, T.M. and Goettel, M.S. (2000). Bioassays of entomogenous fungi. Bioassays of entomopathogenic microbes and nematodes., 141–195.

Chang, J.C., Wu, S.S., Liu, Y.C., Yang, Y.H., Tsai, Y.F. et al. (2021). Construction and selection of an entomopathogenic fungal library from soil samples for controlling *Spodoptera litura*. Frontiers in Sustainable Food Systems., 5, p.596316.

Charles, V.K. (1936). The synonymy of *Botrytis rileyi* Farlow., 397–398.

Cole, G.T. (1996). Basic biology of fungi. Medical Microbiology. 4th edition.

Corbu, V.M., Gheorghe-Barbu, I., Dumbravă, A.Ş., Vrâncianu, C.O. and Şesan, T.E. (2023). Current Insights in Fungal Importance—A Comprehensive Review. Microorganisms., 11(6): 1384.

Culliney, T.W. (2014). Crop losses to arthropods. Integrated Pest Management: Pesticide Problems., 3: 201–225.

Dar, S.A., Rather, B.A. and Kandoo, A.A. (2017). Insect pest management by entomopathogenic fungi. J. Entomol. Zool. Stud., 5: 1185–1190.

Devi, P.V. and Duraimurugan, P. (2013). Exploitation of *Nomuraea rileyi* and *Beauveria bassiana* for the management of lepidopteran pests.

Devi, P.V. and Prasad, Y.G. (2001). *Nomuraea rileyi*—a potential mycoinsecticide. In Biocontrol Potential and its Exploitation in Sustainable Agriculture., Insect Pests. Boston, MA: Springer US., 2: 23–38.

Edelstein, J.D., Lecuona, R.E. and Trumper, E.V. (2004). Selection of culture media and in vitro assessment of temperature-dependent development of *Nomuraea rileyi*. Neotropical Entomology., 33: 737–742.

Elanchezhyan, K. (2007). Effect of combination of grains in media on the sporulation of *Nomuraea rileyi* (*Farlow*) Samson.

Farlow, W. (1883). *Botrytis rileyi*. Rep. US Comm. Agric., 120–121.

Fish, J.D., Fish, S., Fish, J.D. and Fish, S. (1989). Arthropoda. A Student's Guide to the Seashore, 282–350.

Fronza, E., Specht, A., Heinzen, H. and de Barros, N.M. (2017). *Metarhizium* (*Nomuraea*) *rileyi* as biological control agent. Biocontrol Science and Technology., 27(11): 1243–1264.

Gotti, I.A., Moreira, C.C., Delalibera Jr, I. and De Fine Licht, H.H. (2023). Blastospores from *Metarhizium anisopliae* and *Metarhizium rileyi* Are Not Always as Virulent as Conidia Are towards *Spodoptera frugiperda* Caterpillars and Use Different Infection Mechanisms. Microorganisms., 11(6): 1594.

Grewal, G.K. and Joshi, N. (2022). Evaluation of adjuvants on growth and virulence of *Metarhizium rileyi* against *Spodoptera litura* (*F.*). Indian Journal of Entomology., 1–4.

Hall, R.A. and Papierok, B. (1982). Fungi as biological control agents of arthropods of agricultural and medical importance. Parasitology., 84(4): 205–240.

Hawksworth, D.L. and Rossman, A.Y. (1997). Where are all the undescribed fungi? Phytopathology., 87(9): 888–891.

Ibrahim, A.A., Haroun, B.M., El-Fekky, F.A. and Bekhiet, H.K. (2012). Isolation and identification of three entomopathogenic fungi. Egyptian Journal of Agricultural Research., 90(2): 558–574.

Jaronski, S.T. (2014). Mass production of entomopathogenic fungi: state of the art. Mass production of beneficial organisms.

Jitendra, M., Kiran, D., Ambika, K., Priya, S., Neha, K. et al. (2012). Biomass production of entomopathogenic fungi using various agro products in Kota region, India. International Research Journal of Biological Sciences., 1(4): 12–16.

Kamat, M.N., Bagal, S.R., Thobbi, V.V., Rao, V.G. and Phadke, C.H. (1978). Biological control of castor semi-looper through the use of entomogenous fungus *Nomuraea rileyi*. Indian Journal of Botany., 1(1/2): 69–74.

Keppanan, R., Krutmuang, P., Sivaperumal, S., Hussain, M., Bamisile, B.S., et al. (2019). Synthesis of mycotoxin protein IF8 by the entomopathogenic fungus *Isaria fumosorosea* and its toxic effect against adult *Diaphorina citri*. International journal of biological macromolecules, 125: 1203–1211.

Kidanu, S. and Hagos, L. (2020). Entomopathogenic fungi as a biological pest management option: A review. Int. J. Res. Stud. Agric. Sci., 6: 1–10.

Kish, L.P., Samson, R.A. and Allen, G.E. (1974). The genus *Nomuraea maublanc*. Journal of invertebrate pathology, 24(2), pp.154–158.

Krutmuang P, Thungrabeab M., (2017). Efficacy of the entomopathogenic fungus *Nomuraea rileyi* in the biological control of vegetable pest *Spodoptera litura* (Lepidoptera: Noctuidae). International Journal of Environmental and Rural Development., 8(2): 8–12.

Kumar, K.K., Sridhar, J., Murali-Baskaran, R.K., Senthil-Nathan, S., Kaushal, P., et al. (2019). Microbial biopesticides for insect pest management in India: Current status and future prospects. Journal of invertebrate pathology., 165: 74–81.

Lear, L.J. (1993). Rachel Carson's Silent Spring. Environmental history review, 17(2), pp.23–48.

Litwin, A., Nowak, M. and Różalska, S. (2020). Entomopathogenic fungi: unconventional applications. Reviews in Environmental Science and Bio/Technology., 19(1): 23–42.

Maina UM, Galadima IB, Gambo FM, Zakaria D. (2018). A review on the use of entomopathogenic fungi in the management of insect pests of field crops. J. Entomol. Zool. Stud., 6(1): 27–32.

Mantzoukas, S., Kitsiou, F., Natsiopoulos, D. and Eliopoulos, P.A. (2022). Entomopathogenic fungi: interactions and applications. Encyclopedia., 2(2): 646–656.

Maravi M.K., Rai S., and Sandhu S.S., (2018). Entomopathogenic fungi: Bio-Resource; Boon potentially with special focal point as biopesticide, International Journal of Pharmacy and Biological Sciences., 8(4): 817–825.

Marcinkevicius, K., Salvatore, S.A., Bardon, A.D.V., Cartagena, E., Arena, M.E. et al. (2017). Insecticidal activities of diketopiperazines of *Nomuraea rileyi* entomopathogenic fungus.

McGinnis, M.R. and Tyring, S.K. (1996). Introduction to mycology. Methods Gen. Mol. Microbiol., 925–928.

Moanaro, K.A., Choudhary, J.S., Pan, R.S. and Maurya, S. (2017). Natural incidence of *Nomuraea rileyi*, an entomopathogenic fungus on *Spodoptera litura* infesting groundnut in eastern region of India. The Bioscan., 12(2): 843–846.

Montecalvo, M.P., Macaraig, J.S.T., Navasero, M.M., Navasero, M.V. and Navasero, J.M.M. (2023). Effect of emulsifiable concentrate of *Metarhizium rileyi* (*Farl.*) Kepler, SA Rehner & Humber to third larval instar of fall armyworm, *Spodoptera frugiperda* (JE Smith) (*Lepidoptera: Noctuidae*).

Mora, M.A.E., Castilho, A.M.C. and Fraga, M.E. (2018). Classification and infection mechanism of entomopathogenic fungi. Arquivos do Instituto Biológico, 84.

Namasivayam, S.K.R. and Bharani, A.R.S. (2015). Biocontrol potential of entomopathogenic fungi *Nomuraea rileyi* (*F.*) Samson against major groundnut defoliator *Spodoptera litura* (*Fab.*) Lepidoptera; Noctuidae. Adv Plants Agric Res., 2(5): 221–225.

Namasivayam, S.K.R. and Vidyasankar, A. (2014). Biocompatible formulation of potential fungal biopesticide *Nomuraea rileyi* (*f.*) Samson for the improved post-treatment persistence and biocontrol potential. Nature Environment and Pollution Technology., 13(4): 835.

Padanad, M.S. and Krishnaraj, P.U. (2009). Pathogenicity of native entomopathogenic fungus *Nomuraea rileyi* against *Spodoptera litura*. Plant Health Progress., 10(1): 11.

Paixão, F.R., Huarte-Bonnet, C., Ribeiro-Silva, C.D.S., Mascarin, G.M., Fernandes, É.K. et al. (2021). Tolerance to abiotic factors of microsclerotia and mycelial pellets from *Metarhizium robertsii*, and molecular and ultrastructural changes during microsclerotial differentiation. Frontiers in Fungal Biology., 2: 654737.

Paschapur, A., Subbanna, A.R.N.S., Singh, A.K., Jeevan, B., Stanley, J., et al. (2021). Unraveling the importance of metabolites from entomopathogenic fungi in insect pest management. Microbes for Sustainable Insect Pest Management: Hydrolytic Enzyme & Secondary Metabolite., 2: 89–120.

Patil, R.H. and Abhilash, C., (2014). *Nomuraea rileyi* (*Farlow*) Samson: a bio-pesticide IPM component for the management of leaf eating caterpillars in soybean ecosystem. In International conference on Biological, Civil and Environmental Engineering (BCEE-2014), Dubai., 131–132.

Pavone, D., Díaz, M., Trujillo, L. and Dorta, B. (2009). A granular formulation of *Nomuraea rileyi Farlow* (Samson) for the control of *Spodoptera frugiperda* (Lepidoptera: Noctuidae). Interciencia., 34(2): 130–134.

Perinotto, W.M.S., Terra, A.L.M., Angelo, I.C., Fernandes, E.K.K., Golo, P.S., et al. (2012). *Nomuraea rileyi* as biological control agents of *Rhipicephalus microplus* tick. Parasitology research., 111: 1743–1748.

Phadke, C.H. and Rao, V.G., (1978). Studies on the entomogenous fungus *Nomuraea rileyi* (*Farlow*) Samson I. Current Science., 47(14): 511–512.

Pirttilä, A.M., Mohammad Parast Tabas, H., Baruah, N. and Koskimäki, J.J. (2021). Biofertilizers and biocontrol agents for agriculture: How to identify and develop new potent microbial strains and traits. Microorganisms., 9(4): 817.

Pranab, D., Patgiri, P., Pegu, J., Himadri, K. and Boruah, S., (2014). First record of *Nomuraea rileyi* (*Farlow*) Samson on *Spodoptera litura Fabricius* (Lepidoptera: Noctuidae) from Assam, India. Current Biotica., 8(2): 187–190.

Rajan, T.S., Muthukrishnan, N. and Thangasamy, P. (2009). Growth and Sporulation of *Nomuraea rileyi* Isolates on Nitrogen Based Media. Madras Agricultural Journal., 96(1/6): 189–193.

Ranadev, P., Nagaraju, K., Kumari, R.V. and Muthuraju, R. (2023). Evaluating the Stability of Post Mushroom Substrate (PMS) and Other Agro-wastes for Mass Production of Entomopathogenic Fungi. Current Journal of Applied Science and Technology., 42(13): 1–9.

Ranga Rao, G.V., Rupela, O.P., Rao, V.R. and Reddy, Y.V.R., (2007). Role of biopesticides in crop protection: present status and future prospects. Indian journal of plant protection., 35(1): 1–9.

Revathi, N., Ravikumar, G., Kalaiselvi, M., Gomathi, D. and Uma, C. (2011). Pathogenicity of three entomopathogenic fungi against *Helicoverpa armigera*. J. Plant Pathol. Microbiol, 2(4): 1–4.

Ruiu, L. (2018). Microbial biopesticides in agroecosystems. Agronomy., 8(11): 235.

Sabbour, M.M. and Abdel-Rahman, A., (2013). Efficacy of isolated *Nomuraea rileyi* and Spinosad against corn pests under laboratory and field conditions in Egypt. Annual Research & Review in Biology., 903–912.

Sahayaraj, K. and Namasivayam, S.K.R. (2008). Mass production of entomopathogenic fungi using agricultural products and by-products. African Journal of Biotechnology., 7(12).

Sánchez-Pérez, L.D.C., Barranco-Florido, J.E., Rodríguez-Navarro, S., Cervantes-Mayagoitia, J.F. and Ramos-López, M.Á., (2014). Enzymes of entomopathogenic fungi, advances and insights. Advances in Enzyme Research., 2(02): 65–76.

Sandhu, S.S., Rajak, R.C. and Agarwal, G.P. (1993). Microbial control agents of forest pests at Jabalpur. Annals of Forestry., 1(2): 136–140.

Sandhu, S.S., Sharma, A.K., Beniwal, V., Goel, G., Batra, P. et al. (2012). Myco-biocontrol of insect pests: factors involved, mechanism, and regulation. Journal of pathogens.

Shahid, A.A., Rao, Q.A., Bakhsh, A. and Husnain, T. (2012). Entomopathogenic fungi as biological controllers: new insights into their virulence and pathogenicity. Archives of Biological Sciences., 64(1): 21–42.

Sharma R, Sharma P. (2021). Fungal entomopathogens: a systematic review. Egyptian Journal of Biological Pest Control., 31(1): 1–13.

Sharma, A., Diwevidi, V.D., Singh, S., Pawar, K.K., Jerman, M., et al. (2013). Biological control and its important in agriculture. International Journal of Biotechnology and Bioengineering Research., 4(3):175–180.

Singh, D., Son, S.Y. and Lee, C.H. (2016). Perplexing metabolomes in fungal-insect trophic interactions: A Terra incognita of mycobiocontrol mechanisms. Frontiers in Microbiology., 7: 1678.

Singh, O.P. and Gangrade, G.A. (1975). Parasites, predators and diseases of larvae of *Diacrisia obliqua* Walker (Lepidoptera: Arctiidae) on soybean. Current science.

Song, Z., Yin, Y., Jiang, S., Liu, J., Chen, H. et al. (2013). Comparative transcriptome analysis of microsclerotia development in *Nomuraea rileyi*. BMC genomics., 14: 1–9.

Sosa-Gómez, D.R. (2017). Microbial control of soybean pest insects and mites. In Microbial control of insect and mite pests. Academic Press., 199–208.

Sosa-Gómez, D.R., Delpin, K.E., Moscardi, F. and Nozaki, M.D.H. (2003). The impact of fungicides on *Nomuraea rileyi* (*Farlow*) Samson epizootics and on populations of *Anticarsia gemmatalis* Hübner (Lepidoptera: Noctuidae), on soybean. Neotropical Entomology., 32: 287–291.

St. Leger, R.J. and Wang, J.B. (2020). *Metarhizium*: Jack of all trades, master of many. Open biology., 10(12): 200–307.

Stajich, J.E., Berbee, M.L., Blackwell, M., Hibbett, D.S., James, T.Y., et al. (2009). The fungi. Current Biology., 19(18): 840–R845.

Stefanelli L.E.P., Garcia R.D.M., Filho T.M.M.M., Camargo R.D.S., Nakai M., et al. (2021). Pathogenicity of Entomopathogenic Fungi *Metharhizium rileyi* (*Farlow*) to *Spodoptera litura* (*Fabricius*) (Lepidoptera: Noctuidae), International Journal of Agriculture Innovations and Research., 9(5): 358–365.

Ten Hagen, K.G., Nakato, H., Tiemeyer, M. and Esko, J.D. (2022). Arthropoda. Essentials of Glycobiology [Internet]. 4th edition.

Tincilley, A., Easwaramoorthy, S. and Santhalakshmi, G. (2004). Attempts on mass production of *Nomuraea rileyi* on various agricultural products and by-products. Journal of Biological Control., 18(1): 35–40.

Tkaczuk, C., Krzyczkowski, T. and Wegensteiner, R. (2012). The occurrence of entomopathogenic fungi in soils from mid-field woodlots and adjacent small-scale arable fields. Acta Mycologica., 47(2).

Vimala Devi, P.S. (2018), April. Mass production of *Bacillus thuringiensis* and *Nomuraea rileyi.* ICAR.

Yousef-Yousef, M., Romero-Conde, A., Quesada-Moraga, E. and Garrido-Jurado, I. (2022). Production of Microsclerotia by *Metarhizium sp*., and Factors Affecting Their Survival, Germination, and Conidial Yield. Journal of Fungi., 8(4): 402.

Zhao, X., Chai, J., Wang, F. and Jia, Y. (2023). Optimization of Submerged Culture Parameters of the Aphid Pathogenic Fungus *Fusarium equiseti* Based on Sporulation and Mycelial Biomass. Microorganisms., 11(1): 190.

9 Potential of *Trichoderma* in Combating Insect Pests

Parthasarathy Seethapathy[1*]

1. Introduction

In agricultural systems, insect pests pose a significant threat, reducing crop yields and storage losses. In tropical regions, the impact of insects on agricultural production can be substantial, with losses, particularly in stored products, reaching as high as 70%. Such agricultural losses can significantly impact global food production. Studies have shown that average worldwide losses range from 18% to 25%. These losses are exceptionally high in food-insecure countries with rapidly rising populations, where emerging and invasive pests pose a constant threat (Savary et al., 2019). Various crops face unique challenges from a range of pest insects. These pests can wreak havoc on specific crops, causing significant damage. One such example is the lepidopteran *Helicoverpa armigera*, which poses a threat to crops like chickpea, cotton, pigeon pea, and tomato. With the detrimental impact of insect pests on agriculture and the environmental harm caused by the corresponding excessive use of chemical insecticides, there is a pressing need to explore alternative pest control methods. In recent years, researchers have made significant progress in developing practical and eco-friendly alternatives, including the use of fungal biopesticides.

Trichoderma is a well-known fungal genus within the order Hypocreales, renowned for its ecological importance and utilization in multiple domains. The genus is mostly found in soil and root systems, where it is recognized for its capacity to inhabit the rhizosphere and establish mutually beneficial relationships with plants. In 1932, Weindling identified the genus *Trichoderma* as having the ability to manage plant diseases through biological mechanisms. The genus is omnipresent, inhabiting both above and belowground plant anatomies, developing inside plants as endophytes, and is a main element of the fungal community in soil. The capacity of *Trichoderma* to recognize, penetrate, and combat harmful organisms is a crucial factor behind its commercial success as a biopesticide (Sharma et al., 2023). *Trichoderma* has gained recognition for its multifaceted role in agriculture, particularly its efficacy in controlling insect pests. *Trichoderma* species are renowned for their mycoparasitic, biocontrol, and plant growth-promotion properties (Narayanan et al., 2016). This chapter provides a comprehensive overview of the biological potential and modes of action of *Trichoderma* spp. in combating insect pests. In the following, we delve into the specific modes of action employed by *Trichoderma* in insect

[1] Department of Plant Pathology, Amrita School of Agricultural Sciences, Amrita Vishwa Vidyapeetham, Coimbatore, Tamil Nadu, India

* Corresponding author: spsarathyagri@gmail.com

control, including parasitism, the production of secondary metabolites, the induction of green leaf volatiles, competition for resources, and interference with pest life cycles. Furthermore, the compatibility of *Trichoderma*-based biocontrol agents with other biological control methods is explored, highlighting the potential for integrated pest management (IPM) strategies that maximize pest suppression while minimizing environmental impact (Woo et al., 2023). We discuss the mechanisms by which *Trichoderma* enhances plant resilience, potentially reducing the susceptibility of crops to insect infestations (Poveda, 2021).

2. *Trichoderma*: A Versatile Multifaceted Fungi

Trichoderma species play a significant role in agriculture by improving crop yield and offering defense against plant diseases (Woo et al., 2023). *Trichoderma* stands out for its ability to function as a biocontrol agent. *Trichoderma* protects plants from stress by preventing or restricting the growth of harmful organisms. It does this in several ways, one of which is by parasitizing on pests and pathogens. This makes it a sustainable substitute for chemical pesticides in managing crop diseases (Sood et al., 2020). In addition to its agricultural functions, *Trichoderma* plays distinctive roles in other industrial sectors, notably in synthesizing enzymes and biofuels. It is known for its capacity to generate a diverse array of extracellular enzymes, such as cellulases and hemicellulases, that play actively in the decomposition of plant biomass (Manzar et al., 2022).

Due to its enzymatic potential, *Trichoderma* is valuable across biotechnology, since it finds applications in several industries, such as textile, paper, and biofuel. In addition, *Trichoderma* species are employed in remediating polluted soils due to their capacity to break down diverse contaminants. Recent research investigates the potential of *Trichoderma* in various domains, such as its capacity to improve plant development and increase resistance to stress, its use in sustainable agriculture, and its efficacy in industrial processes (Woo et al., 2023). *Trichoderma* is a very flexible and essential genus that is gaining attention due to its ability to provide sustainable and ecologically friendly solutions and its instrumentality in both biological and industrial fields (Dutta et al., 2022).

3. Biology of *Trichoderma*

Trichoderma, a fungus initially identified and named by Persoon in 1794, is a complex group of wood-decaying filamentous fungi essential in several ecological and commercial environments. The fungus has been reclassified as Phylum Ascomycota, Order Hypocreales, and Family Hypocreaceae. *Trichoderma*, an anamorphic or mitosporic genera, exhibits pleomorphism, which encompasses teleomorphs in genera such as *Hypocrea*, *Podostroma*, and *Sarawakus* (Guzmán-Guzmán et al., 2023), resulting in the adoption of a dual nomenclature. The *Hypocrea* (teleomorph) is infrequently encountered in nature, thriving only on plant detritus and particular basidiomycete fungi found in rotting bark. In contrast, *Trichoderma* (anamorph) is common (Woo et al., 2023). As per the International Commission on Taxonomy of *Trichoderma* (www.trichoderma.info), *Trichoderma* was preferentially selected for use, followed by *Hypocrea*. *Trichoderma* spp. that have lost the ability to reproduce sexually are referred to as "agamospecies," but the majority of genetic variations in the genus are comprised of sexual forms. *Trichoderma* was initially considered a monotypic genus, with most isolates identified as *T. viride*.

The genus originated approximately 66 million years ago, followed by the subsequent development of the sections *Trichoderma* and *Longibrachiatum*, as well as the clade *Virens* and *Harzianum* (Kubicek et al., 2019). However, detailed studies have revealed a much broader species diversity. *Trichoderma* species, historically considered morphologically similar and often categorized as *T. viride*, are now recognized for their diversity. Species such as *T. viride, T. harzianum, T. asperellum, T. atroviride, T. longibrachiatum, T. citrinoviride, T. gamssi, T. koningii, T. parareesei, T. polysporum, T. pseudokoningii, T. reesei,* and *T. virens* are widely studied due to their beneficial effect on plants and their natural products with potential application in drug and farming sectors (Guzmán-Guzmán et al., 2023). Most biocontrol strains do not have a reported sexual stage. Asexual fungi are organisms that reproduce without sexual reproduction. They consist of clonal individuals and communities that often have different genetic material within their cells. These organisms probably evolved independently during their asexual stage (Morán-Diez et al., 2021). They possess significant genetic variation and can produce a wide range of commercially and ecologically valuable products. They are highly productive in synthesizing extracellular proteins, particularly enzymes that facilitate the breakdown of cellulose and chitin. On average, there are 10^1–10^3 *Trichoderma* spp. that may be cultured per gram of temperate and tropical soils (Abbas et al., 2022).

These species are distinguished by their fast growth and abundant spore generation. They create a well-developed, branched mycelium composed of clear, segmented hyphae. *Trichoderma* exhibits a unique conidiation pattern, wherein it produces green asexual conidia at the ends of extensively branched conidiophores. Conidia are the predominant cellular form generated through asexual reproduction, facilitating spread. Profusion of conidiophores culminates in phialides, which are pyramidal in shape and whose branches develop in pairs (Guzmán-Guzmán et al., 2023). Certain species also produce chlamydospores as structures for survival. These species thrive best at temperatures wavering between 25 to 30 °C and are commonly found in diverse settings, particularly those with decaying organic material (Morán-Diez et al., 2021).

Trichoderma exhibits rapid growth and forms colonies that have a cottony, fluffy, or granular appearance, which can range in color from white to green. Under microscopic examination, these organisms can be recognized based on the specific morphology and organization of conidia and phialides, as well as the branching structure of conidiophores. However, the presence of substantial genetic variation within the genus frequently necessitates the use of molecular methods for accurate identification (Schmoll and Schuster, 2010). The genus comprises many species that have evolved to thrive in distinct ecological conditions (Kubicek et al., 2019). The primary emphasis on identifying *Trichoderma* species lies in their physical characteristics, which can occasionally be ambiguous. Conidiophores are clustered together in fascicles or pustules, and soluble pigments aid in distinguishing between different species. *Longibrachiatum* strains frequently generate vivid greenish-yellow colours. *T. aureoviride* exhibits a distinctive characteristic of crystal production in the media. *T. viride* and occasionally *T. atroviride* are characterized by different aromatic odours, such as coconut. The shape, size, and pattern of phialides exhibit variation throughout different sections, hence aiding in the distinction of species. Distinguishing characteristics of conidia, such as their shape, size, and external adornment, can be practically achieved using scanning electron microscopy (SEM) (Schmoll and Schuster, 2010). As an example, the conidia of *T. viride* have a rough or verrucose surface, but *T. saturnisporum* and *T. ghanense* have distinct protrusions on their outer walls. Berkeley first proposed the connection between the genus and *Hypocrea* in 1860, and Tulasne and Tulasne later confirmed it in 1865 by observing the presence of phialides on the conidiophore. Brefeld and von Tavel made an essential advancement in the connection between *H. rufa* and *T. viride* in 1891 by studying sole ascospore cultures. The genus *Trichoderma* is exclusive as many of its species can parasite and saprophyte even on Ascomycetes and also on phylogenetically related species (Kubicek et al., 2019).

4. Genomics of Antagonistic Potential in *Trichoderma*

The genome of *Trichoderma*, as observed in species such as *T. reesei*, contains a significant number of genes responsible for breaking down cellulose and hemicellulose, highlighting their effectiveness in decomposing plant matter (Schmoll and Schuster, 2010). In addition, they possess genes that facilitate their interaction with plants and their pathogens. Until today, the species-level report of *Trichoderma* was determined using a blend of multi-gene phylogeny and geomorphologic traits (Schmoll et al., 2016). The identification of *Trichoderma* spp. strains involved morphological and molecular analytical techniques, which utilized genes including *act*, *cal*, *rpb2*, *tef1*, and *ITS* (Abbas et al., 2022). The NCBI taxonomy browser (accessed on January 31, 2024) lists over 450 *Trichoderma* species and, moreover, 2000 unsystematic *Trichoderma* species. New genomic methods and physiological tests were also used to find the different purposeful groups in *Trichoderma* spp. that are significant for synthesizing secondary metabolites for commercial utilization.

An examination of the genomes of three *Trichoderma* species (*T. atroviride, T. reesei,* and *T. virens*) indicated that mycoparasitism is an inherent trait of *Trichoderma.* However, these species also possess significant nutritional adaptability (Kubicek et al., 2019). Significant progress has been made in the classification of *Trichoderma.* However, there are still challenges in distinguishing species within the genus. This is because most *Trichoderma* species are not linked to their teleomorphic state and are thus treated as mitotic and monoclonal. Recent efforts to enhance their categorization using barcode oligonucleotides involve online DNA barcoding marker tools like TrichoKEY and TrichoMARK. TrichoMARK helps to trim the genomic DNA sequences of the *ITS*, *rpb2*, and *tef1* DNA barcoding markers used for molecular identification of *Trichoderma.* TrichoBLAST searches use these markers, such as internal transcribed sequences (ITS), *rpb2*, and *tef1* genes, to do targeted BLAST-type searches with the reference sequences of 357 *Trichoderma* species. Dou and colleagues have recently introduced an online Multilocus Identification System for *Trichoderma* (MIST) for the automated identification of *Trichoderma* and *Hypocrea* species (http://mmit.china-cctc.org/). This system can detect 349 potential *Trichoderma* species by utilizing a set of three DNA barcodes (Dou et al., 2020).

5. Multiplication of *Trichoderma*

Asexual reproduction is frequently observed and plays a dynamic role in the rapid multiplication and distribution of *Trichoderma.* The primary mode of reproduction is mainly achieved through the production of conidia. The conidia, which can be green but differ in color depending on the species, are released abundantly into the environment. Under favorable conditions, they initiate germination to establish fresh mycelial colonies. In addition, certain *Trichoderma* species produce chlamydospores, which are robust dormant spores specifically adapted to withstand unfavorable climatic conditions, thus guaranteeing the species survival and persistence. The occurrence of sexual reproduction in *Trichoderma,* though less common, is of utmost importance in shaping the genetic diversity and evolution of the species (Kubicek et al., 2019).

Sexual reproduction involves the formation of fruiting bodies called ascomata, which are generally perithecial in structure. These ascomata within the asci are where the sexual spores, or ascospores, form. Sexual reproduction encompasses an intricate series of events, including plasmogamy, karyogamy, and meiosis, ultimately resulting in genetic recombination. This procedure necessitates the interaction of compatible mating types and leads to the formation of haploid ascospores during meiosis and subsequent mitotic division. Upon reaching maturity, the ascomata releases the ascospores into the surrounding environment through rupture. Similar to asexual spores, these ascospores undergo germination under favorable conditions, resulting in new progeny. The dual reproduction strategy of *Trichoderma* is crucial for its adaptability, ecological success, and genetic diversity. Asexual reproduction by conidia is the primary method used, which allows for fast growth and spreading. This is crucial for colonizing new substrates and surroundings. On the other hand, sexual reproduction, although less frequent, is essential for producing genetic diversity and enhancing species adaptability and evolution (Schmoll and Schuster, 2010).

The study of these reproductive systems at the molecular and genetic level has uncovered regulatory pathways and genes that play a role in spore formation and germination (Schmoll et al., 2016). Environmental factors like temperature, humidity, and nutrition availability significantly impact *Trichoderma* reproductive cycles (Kubicek et al., 2019). This emphasizes the intricate nature of these interactions. According to Woo et al. (2023), recent biotechnological advancements have facilitated the genetic modification of *Trichoderma* strains. This modification has resulted in enhanced spore output and improved tolerance to stress, which are crucial for *Trichoderma* applications.

6. Biocontrol Potential of *Trichoderma*

Recently, there has been growing apprehension about using chemical disease management measures, such as fungicides and pesticides, because they leave behind hazardous residues in agricultural products, soil, and water sources, contributing to the emergence of pathogen resistance. Except for creating disease-resistant plant types, contemporary crop protection methods have had detrimental impacts on the environment and soil fertility (Balakrishnan et al., 2017a). *Trichoderma* fungi are found in various habitats across various climatic zones, from polar to equatorial latitudes. As a result, they are among the most geographically distinct fungi in soil ecosystems. *Trichoderma* species are widely distributed and have a remarkable ability to colonize cellulosic materials. They can be found in various environments, including areas with decaying plant material and within the rhizosphere of plants (Poveda, 2021). *Trichoderma* spp. is an essential partner in combating diseases caused by many plant pathogens that reside in the soil, including *Fusarium, Phytophthora*, and *Sclerotium* (Balakrishnan et al., 2017b). These diseases present a substantial menace to various crops, such as cereals, cash crops, horticultural crops, millets, oilseeds, pulses, spices, and vegetables, both in India and worldwide.

The role of *Trichoderma* in disease control becomes evident in diverse agricultural contexts. In rice farming, treatment with *T. harzianum* effectively controls the debilitating sheath blight produced by *Rhizoctonia solani*. In various countries, *T. harzianum* provides disease-free conditions for many crops and is effective against root rot caused by *Sclerotium rolfsii* and *Fusarium oxysporum* (Narayanan et al., 2016). Similarly, the problem of charcoal rot caused by *Macrophomina phaseolina* in maize can be reduced by applying *Trichoderma* spp. to the soil or by treating the seeds before planting. In field circumstances, *Trichoderma* spp. has shown its efficacy in managing various diseases, such as *Fusarium, Rhizoctonia, Macrophomina*, and other pathogens (Balakrishnan et al., 2017b). A wide range of vegetables, including chilies, cucurbits, brinjal, peas, potatoes, and others, are prone to many soil- and root-borne diseases, such as charcoal rot, damping-off, root rot, seedling blight, seed rot, and wilt diseases. *Trichoderma* spp. provides practical remedies through formulations, soil inoculants, or commercial products. These applications have demonstrated efficacy in controlled greenhouse and open-field environments. *Trichoderma*, widely distributed in numerous environments, constitutes a significant proportion of the fungus community in forests and different soil types. The organism competes with soil-borne and plant-pathogenic fungi for vital exudates that promote the growth of pathogens in the soil. *Trichoderma* species have been acknowledged for their antagonistic characteristics for more than seventy years, and multiple studies have further established their reputation as effective biological control agents.

7. Mechanism of *Trichoderma* against Insect Pests

Entomopathogens refer to parasitic microorganisms that possess the capability to colonize and destroy arthropods. The genera that receive the most attention and are commonly utilized in agriculture are *Beauveria, Metarhizium, Isaria, Lecanicillium, Pochonia, Paecilomyces, Hirsutella, Ophiocordyceps, Cordyceps,* and *Torubiella*. Although *Trichoderma* spp. are widely recognized as biocontrol agents, research is scarce about their impact on insect herbivory. *Trichoderma* also possesses the ability to suppress arthropods and functions as a symbiotic organism that can inhabit roots and engage in intricate molecular communication with the plant host (Manzar et al., 2022).

The relationship between this genus, its ability to adapt to different climatic and soil conditions, and its rapid growth contribute to its superiority over other filamentous fungi and establish it as a promising candidate for biological control agents. *Trichoderma* is known to exhibit various mechanisms of action that make it an effective biological control agent. A possible mechanism for entomopathogenic strains of *Trichoderma* is that they typically infect insects by directly penetrating their cuticles. To accomplish this, they rely on adhesion molecules and lytic proteins such as chitinases, proteases, and lipases. After penetration, the fungus successfully suppresses the insect's immunity and establishes itself in the insect's body, ultimately generating and disseminating new spores from the deceased host (Fig. 1). Entomopathogenic fungi must synthesize diverse insecticidal secondary metabolites during their life period (Poveda, 2021).

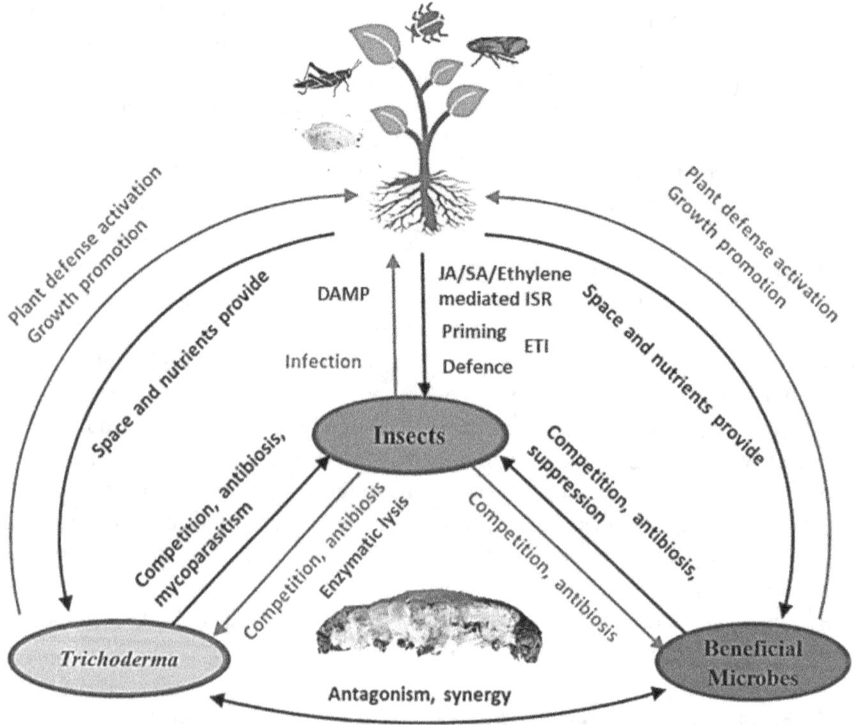

Fig. 1 Biocontrol mechanisms of *Trichoderma* spp. against insect pests.

Multiple studies have shown that *Trichoderma* spp. employ a significant variety of mechanisms for biocontrol. These mechanisms include mycoparasitism, priming effect, competition for nutrients and space, antibiosis, plant growth promotion, tolerance to stress by promoting root and plant development, solubilization and sequestration of inorganic nutrients, induced systemic resistance (ISR), enzyme inactivation, and stimulation of plant defences against invading pests and parasites (Table 1). *Trichoderma* spp. are the dominant commercially available biological control agents (BCAs) in the global market due to their versatile role and potential (Manzar et al., 2022).

Table 1 Antagonistic potential against insect pests by *Trichoderma* species.

Trichoderma	*Insect species*	*Mechanism*	*Reference*
T. album	*Rhyzopertha dominica*	Insect parasitism	(Mohamed and Taha, 2017)
T. asperellum	*Tetranychus urticae*	6-pentyl α-pyrone toxicity	(Salwa and Kottb, 2017)
T. asperellum, T. atroviride	*Thrips tabaci*	Induced systemic resistance	(Muvea et al., 2014)
T. atroviride	*Spodoptera littoralis*	JA-mediated systemic resistance	(Coppola et al., 2017)
T. atroviride	*Spodoptera frugiperda*	JA-mediated systemic resistance	(Contreras-Cornejo et al., 2018)
T. atroviride	*Drosophila melanogaster*	Secondary metabolite toxicity	(Atriztán-Hernández et al., 2019)
T. atroviride, T. koningiopsis, T. gamsii	*Delia radicum*	Insect parasitism	(Razinger et al., 2014)
T. citrinoviride	*Schizaphis graminum*	Secondary metabolite toxicity, antifeedant compounds	(Ganassi et al., 2001)
T. gamsii	*Trichoplusia ni*	JA-mediated systemic resistance	(Zhou et al., 2018)
T. gamsii	*Auchenorrhyncha* sp.	JA-mediated systemic resistance	(Parrilli et al., 2019)
T. harzianum	*Periplaneta americana*	Secondary metabolite toxicity	(Omar and Abdul-Wahid, 2012)
T. harzianum	*Aedes aegypti*	Secondary metabolite toxicity	(Sundaravadivelan and Padmanabhan, 2014)
T. harzianum	*Earias insulana*	Secondary metabolite toxicity	(Vijayakumar and Alagar, 2017)
T. harzianum	*Cimex hemipterus*	Insect parasitism	(Zahran et al., 2017)
T. harzianum	*Pectinophora gossypiella*	Secondary metabolite toxicity	(El-Massry et al., 2016)
T. harzianum	*Diuraphis noxia, Tribolium castaneum*	Secondary metabolite toxicity	(Rahim and Iqbal, 2019)
T. harzianum	*Bemisia tabaci*	SA-mediated systemic resistance	(Jafarbeigi et al., 2020)
T. harzianum, T. asperellum, T. atroviride	*Xylosandrus germanus*	Insect parasitism	(Kushiyev et al., 2021)
T. harzianum, T. atroviride, T. citrinoviride, T. gamsii	*Acanthoscelides obtectus, Xylotrechus arvicola*	Insect parasitism	(Rodríguez-González et al., 2017)

Contd.

Table 1 *Contd.*

Trichoderma	*Insect species*	*Mechanism*	*Reference*
T. longibrachiatum	*Macrosiphum euphorbiae*	Induced systemic resistance	(Battaglia et al., 2013)
T. longibrachiatum	*Bemisia tabaci*	Insect parasitism	(Anwar et al., 2016)
T. longibrachiatum	*Leucinodes orbonalis*	Insect parasitism	(Ghosh and Pal, 2016)
T. viride	*Culex quinquefasciatus*	Secondary metabolite toxicity	(Geetha et al., 2003)
T. viride	*Corcyra cephalonica*	Antifeedant compounds	(Berini et al., 2016)
T. viride	*Helicoverpa armigera*	Antifeedant compounds	(Chinnaperumal et al., 2018)
Trichoderma sp.	*Oryctes rhinoceros*	Insect parasitism	(Nasution et al., 2018)
Trichoderma sp.	*Unaspis mabilis*	JA-mediated systemic resistance	(Silva et al., 2019)
Trichoderma sp.	*Myzus persicae*	Antifeedant compounds	(Kaushik et al., 2020)
Trichoderma sp.	*Locusta migratoria*	Secondary metabolite toxicity	(Laib et al., 2020)
Trichoderma sp.	*Amrasca bigutulla, Spodoptera litura*	Secondary metabolite toxicity	(Nawaz et al., 2020)
Trichoderma sp.	*Aphis gossypii*	Secondary metabolite toxicity	(Nawaz et al., 2020)

7.1 Mycoparasitism

Mycoparasitism is a process that consists of four consecutive stages: chemotaxis, recognition purposes, dependency, and encircling. Additionally, it entails the entry of the parasite cell wall and the digestion of the host. Similar to other entomopathogenic fungi, *Trichoderma* has the ability to actively parasitize insects on buccal parts and intersegmental membranes, or through the spiracles, utilizing them as a nutritional basis to produce new conidiospores. Most early studies have focused on delivering fungal spores externally. in laboratory environments. These studies have reported varying fatality rates depending on the insect pest and the duration of the study. *Trichoderma* spp. strains used in insect biocontrol have been effective in combating various insect pests in both lab and field conditions, as revealed in the literature (Rodríguez-González et al., 2020), including *Delia radicum* (Diptera), *Leucinodes orbonalis* (Lepidoptera), *Lucanus cervus* (Coleoptera), *Thrips tabaci* (Thysanoptera), as well as the immature stages of *Acanthoscelides obtectus* and *Xylotrechus arvicola* (Coleoptera) and others.

The strains of *T. harzianum* and *T. citrinoviride* displayed significant potency against adult *A. obtectus* insects that infest *Phaseolus vulgaris*. The spores of the *Trichoderma* isolate exhibited greater attractiveness to both males and females, and resulted in a higher mortality among pests that attacked treated beans (Rodríguez-González et al., 2018). *Trichoderma* species are well-known for their capacity to parasitize other chitin-walled organisms through a process known as mycoparasitism. Using *Trichoderma* species for their mycoparasitic ability is a sophisticated and remarkably efficient biological control method to combat several plant pests and diseases. *Trichoderma* identifies distinct molecules present on the surface of the pest or pathogen. Receptors on the cellular surface of pests and pathogens assist in this recognition process. After acquiring an appropriate target, *Trichoderma* adheres to the host. *Trichoderma* wraps itself around the pest and pathogen, creating appressoria structures that help it penetrate the host. The physical interaction has a vigorous protagonist in the succeeding stages of mycoparasitism. *Trichoderma* absorbs the nutrients that are released when cell walls degrade. This not only helps *Trichoderma* grow, but also weakens or kills the pests and fungal pathogens.

In addition to direct parasitism, *Trichoderma* can synthesize secondary metabolites with antifungal qualities, further suppressing the growth of pests and pathogens. Investigations of *Trichoderma*-insect interactions suggest that *Trichoderma* spp. may encounter insect attacks in their natural environment. *Trichoderma* has the advantage of being able to counteract colonies of the fungus Termitomyces, which is a symbiotic organism of *Macrotermes natalensis* (termites). Due to this hostility, termites generate antifungal substances that hinder the growth of *Trichoderma* (Atriztán-Hernández et al., 2019). In addition, leaf-cutting ants *Atta sexdens rubropilosa* utilize *Trichoderma* as plant endophytes to nourish their symbiont, *Leugoagaricus*. Research suggests that ants can function as a Trojan horse for *Trichoderma* spp (Atriztán-Hernández et al., 2019).

T. longibrachiatum was obtained from cotton mealybugs, and its ability to cause sickness was assessed in a lab setting on *B. tabaci*. The entomopathogenic capability of *T. longibrachiatum* was evaluated in both the nymph and mature stages of *B. tabaci*. The entomopathogenic action of *T. longibrachiatum* was more significant in the nymph stage of *B. tabaci* than in the adult stage, as measured by the germination percentage of conidia on the external surface of the nymph and adult stages (Anwar et al., 2016). Similarly, the spore solution of *T. longibrachiatum* proved to be highly effective in combating *Leucinodes orbonalis*, a significant brinjal pest. The lethal dose (LD50) and lethal time (LT50) of *T. longibrachiatum* were determined to be 2.87×10^7 spores/ml and 11.7 days, respectively. Field application of a prepared product containing *T. longibrachiatum* at 10^8 spores/ml, applied every 15 days, resulted in a 56.02% increase in crop production. This therapy exhibited comparable efficacy to the chemical malathion in managing the insect pest *L. orbonalis* in brinjal crops (Ghosh and Pal, 2016).

The impact of *T. afroharzianum* on the lepidopteran pest *Spodoptera littoralis* is found to be significant when smeared as a seed treatment on tomatoes, as it disrupts the gut microbiota and its role in larval nutrition (Di Lelio et al., 2023). Pathogenicity bioassays were conducted by immersing mature meadow spittlebugs (*Philaenus spumarius*) in a solution containing *Trichoderma chlorosporum* cell suspension (120 mg/mL) or the cell-free culture supernatant. The results demonstrated a dose-dependent impact, with 97% and 87% mortality rates seen within 24 hours, respectively. The *T. chlorosporum* examined in this study demonstrates promising potential for the creation of efficient, environmentally friendly management methods for *P. spumarius*, consequently mitigating the olive tree fast decline syndrome (Ganassi et al., 2023).

7.2 Enzymatic Degradation

Trichoderma is known for its capacity to synthesize a diverse range of hydrolytic enzymes, such as glucanases, chitinases, and proteases. These enzymes play a vital role in the degradation of the cell wall constituents of pathogenic fungi. Chitinases and glucanases have a critical function in this process by explicitly targeting chitin and β-glucans, key mechanisms of fungal cell walls (Kubicek et al., 2019). *Trichoderma*, through colonizing soil and producing enzymes that are released, has shown the ability to directly regulate insects, specifically plant-eating Lepidoptera *Bombyx mori*, by interfering with the construction and porosity of the protective layer in their digestive system called the midgut peritrophic tissue. A bioassay confirmed these findings and showed that ingestion of combined exo- and endo-enzymes leads to changes in the peritrophic matrix, resulting in detrimental effects on the growth and maturation of larvae. This also has a negative impact on the weight of pupae and can even cause fatalities (Berini et al., 2016).

Trichoderma is also regarded as an entomopathogenic fungus due to its ability to utilize hydrolytic enzymes such as chitinases and proteases. These enzymes function as primary weapons for devouring insects. It has developed the ability to assimilate a diverse array of genes involved in the production of different enzymes and secondary metabolites, owing to several evolutionary developments. The phenomenon is apparent in the genetic codes of several *Trichoderma* species (accessible at https://mycocosm.jgi.doe.gov/Trichoderma400_project) (Woo et al., 2023). Chitinase-producing fungi are currently receiving significant interest as potential agents for controlling insect infestations. The research evaluated the potential of the Chit1 gene (chitinase) isolated from *T. longibrachiatum* against *Aphis gossypii*. They introduced Chit1 into cotton through a Geminivirus vector resulted in enhanced resistance, leading to 38.75% and 21.67% mortality rates were recorded in nymphs and adults of *A. gossypii*. These findings highlight the potential of *Trichoderma*-derived chitinase for pest management applications (Anwar et al., 2023).

7.3 Antibiosis

Trichoderma functions as an entomopathogen by directly parasitizing insects and producing insecticidal secondary metabolites, anti-feedant chemicals (methyl linoleate, linoleic acid, bisorbicillinoids, citrantifidiene, and citrantifidiol), and repelling metabolites. *Trichoderma* species are exceptionally effective at producing a diverse range of secondary metabolites, such as antibiotics and other bioactive substances. These compounds are essential for protecting plants and managing diseases. These metabolites not only hinder the proliferation of plant diseases but also enhance plant well-being through multiple mechanisms. *Trichoderma* produces a variety of secondary metabolites, such as polyketides, pyrones, non-ribosomal peptides, and terpenoids. These compounds display a variety of bioactivities, including antifungal, antibacterial, and antiviral effects (Mendoza-Mendoza et al., 2018).

Trichoderma spp. generates various secondary metabolites, including antibacterial and antifungal antibiotics. Several species within the *Trichoderma* genus have the ability to produce peptidols, which are non-proteinogenic amino acids. One example is α-aminoisobutyric acid, a type of polypeptide antibiotic. The N-terminus of these molecules undergoes acetylation, while the C-terminus contains an amino-alcohol. Peptaibols, such as short peptides, can break the cell membranes of harmful fungi. Specific metabolites derived from *Trichoderma* display antibacterial properties, which are advantageous for managing bacterial plant diseases (Sharma et al., 2023). These chemicals, such as polyketides, can disrupt bacterial cell processes. Various strains of *Trichoderma* secrete distinct antibiotics. For instance, *T. viride* synthesizes mucorin, mucortoxins A and B, and trichophyton. Furthermore, *T. mucorin* was the source of isolation for both mucorin A and B. *T. harzianum* synthesizes tricholongins BI and BII, whereas trichokonins and longibrachins were isolated from *T. koningii.* Atroviridines A-C and neoatroviridines A-D have been isolated from the culture of *T. atroviride*. Also, other prominent metabolites consist of both volatile and nonvolatile toxic compounds such as acetaldehydes, gliotoxin, alamethicins, formic aldehydes, gliovirin, glisoprenins, harmonic acid, harzianopyridone, harziandione, heptelidic acid, massoilactone, terpenoids, trichodermin, tricholin, viridian, viridin, and others (Abbas et al., 2022; Dutta et al., 2022; Guzmán-Guzmán et al., 2023). Antifungal chemicals are among the most notable metabolites generated by *Trichoderma*. *Trichoderma* also synthesizes metabolites that enhance plant growth and well-being, such as auxin-like chemicals that can impact root structure and plant growth (Contreras-Cornejo et al., 2018). *Trichoderma* can neutralize the virulence factors or toxins produced by pests, reducing their detrimental impact on plants (Li et al., 2015; Shakeri and Foster, 2007).

The plant root zone perceives chemical substances from microbes as auxin-like metabolites, effector molecules, and volatile organic compounds (VOCs). These substances, in turn, govern developmental processes or trigger effective defense responses against various aggressors. The use of root endophytic *Trichoderma* for insect suppression was related to an upsurge in the release of volatile terpenes and the buildup of jasmonic acid, which stimulates the activation of defence mechanisms against herbivores. *T. atroviride* was found to generate the volatile compounds 1-octen-3-ol and 6-pentyl-2H-pyran-2-one, as determined by chemical studies (Contreras-Cornejo et al., 2021). The scientific research demonstrated that both substances decreased the intake of leaf tissue and modified the eating behaviour of *S. frugiperda* due to VOCs from *T. atroviride*. Also found were the induction of terpenes such as α-humulene, β-linalool, β-myrcene, γ-terpineol, and (E)-β-caryophyllene in maize hosts against *S. frugiperda* (Contreras-Cornejo et al., 2021). The odorscape of the maize field during the oviposition cycle of the European corn borer showed a comparable composition of monoterpenes. This occurred during a crucial phase before herbivory, likely as a preventive defence strategy (Leppik and Frérot, 2014).

7.4 Competition for Nutrients and Space

Trichoderma prevails in several biological habitats, particularly in roots, and prevents the occurrence of plant diseases by outcompeting them for resources and territory. This mechanism operates through multiple facets. *Trichoderma* species are renowned for their fast colonization of substances and abundant reproduction rates. This allows them to quickly colonize available habitats and exploit accessible resources, outcompeting slow-growing pathogens (Affokpon et al., 2011). *Trichoderma* alters the rhizosphere habitat to establish conditions conducive to its growth and unfavorable for pathogenic fungi. This encompasses changes in pH, availability of nutrients, and the production of specific chemicals that can hinder the development of competitors. *Trichoderma* has highly efficient mechanisms for assimilating essential nutrients, specifically phosphorus and nitrogen, which are vital resources.

The optimal assimilation of nutrients may lead to the depletion of critical resources for other soil microorganisms, particularly pathogenic organisms. Field inoculation with *T. harzianum* in maize roots during the initial plant growth stage had a significant impact on the arthropod population associated with the leaves (Mendoza-Mendoza et al., 2018). It resulted in a decline in the number of chewing herbivores and a reduction in the quantity of piercing-sucking pests. Additionally, significant alterations were observed in the concentrations of sucrose, jasmonic acid, and (Z)-3-hexen-1-ol (Contreras-Cornejo et al., 2021).

7.5 Induced Systemic Resistance

Beyond direct antimicrobial activity, certain *Trichoderma*-derived metabolites can induce systemic resistance in plants, priming their defense mechanisms to respond more effectively to insect attacks. In most cases, components from beneficial microbes stimulate the host immune response through the induced systemic response (ISR), which is controlled by the jasmonic acid (JA) and ethylene (ET) pathways. On the other hand, substances originating from pathogenic microbes trigger systemic acquired resistance (SAR), which is regulated by salicylic acid (SA). The initiation of this ISR is facilitated by the activation of the JA and ethylene signaling pathways, which are also implicated in the plant's defense mechanisms against insects. The effect of *Trichoderma* on stimulating host immunity has been examined separately, both at the local and systemic levels (Salwan et al., 2022). Sucrose in root exudates has been shown to play a significant role in attracting *Trichoderma* fungi during root contact. Hydrophobin-like proteins facilitate the connection to roots, while the internalization of fungal hyphae is aided by the production of swollenins and plant cell wall disintegrating enzymes (Sood et al., 2020).

At the beginning of the invasion process, it is crucial to suppress the host defense mechanisms to colonize the roots successfully (Anwar et al., 2023). This suppression is facilitated by the secretion of modest, soluble protein molecules rich in cysteine, known as effector-like proteins. At this stage, *Trichoderma* exhibits behavior that is comparable to that of a plant disease infiltrating root system. Still, the fungus can be distinguished from pathogens due to occurrences such as oxidative bursts, the production of SA by plants, and the release of elicitor-like proteins. These activities stimulate the development of immunity in plants, which aids in defending against future attacks by pathogens and insects (Coppola et al., 2017; Mendoza-Mendoza et al., 2018).

Trichoderma first establishes itself in the root system of the plant. The colonization process involves exchanging specialized signaling molecules between the plant and fungus, enabling a mutually advantageous connection. *Trichoderma* elicits a defense reaction in the plant upon invasion. *Trichoderma* produces tiny bioactive chemicals that function as elicitors to do this. The plant's immune system detects these chemicals, triggering defense reactions. An essential component of ISR is activating the plant immune system through a process known as 'priming.' Priming enhances the plant's ability to promptly and efficiently react to future disease invasions (Abbas et al., 2022; Coppola et al., 2017; Mendoza-Mendoza et al., 2018; Morán-Diez et al., 2021). Consequently, the plant's defense mechanisms remain ready and can be promptly triggered. *Trichoderma*-induced ISR can produce several antifungal chemicals in plants, including phytoalexins, which possess toxicity against numerous diseases. The ISR also entails fortifying physical barriers, such as cell walls, by depositing lignin and other reinforcing chemicals, thus increasing resistance against pest infestation. *Trichoderma* can stimulate the activation of many plant defense genes, including those responsible for synthesizing pathogenesis-related proteins, signaling molecules such as SA, JA, and ethylene, and enzymes like peroxidases and chitinases (Narayanan et al., 2016; Salwan et al., 2022).

Trichoderma can alter the soil microbial community composition, consequently influencing insect pest populations. These modifications can affect the presence of specific root exudates or change the environment in ways that are less advantageous for insect pests. Certain *Trichoderma* species emit volatile chemical substances that can deter insect pests or attract their natural predators. These VOCs provide additional protection by indirectly influencing insect behavior and survival (Coppola et al., 2017; Leppik and Frérot, 2014). The interface between *Trichoderma* and the plant root system can modify the rhizosphere, rendering it less appealing or conducive to soil-dwelling insect pests. This phenomenon can diminish the probability of root impairment caused by insects such as rootworms or nematodes. *Trichoderma* can, in certain instances, exhibit synergistic effects when combined with other biocontrol agents, such as entomopathogenic fungi or bacteria, resulting in enhanced control of insect pests. This collaboration can result in the development of more holistic pest control solutions.

B. tabaci biotype B exhibited reduced efficacy on tomato plants with elevated levels of phenolic content, namely those treated with root applications of β-aminobutyric acid + *Trichoderma*, salicylic acid + *Trichoderma*, and *Trichoderma* alone. The nymphs exhibited slower development when nourished on tomato plants treated with plant sap. Additionally, there was an increase in mortality among the nymphs on this treated vegetation, and they were found to be less appealing to *B. tabaci* females (Jafarbeigi et al., 2020). The study evaluated the impact of *T. harzianum* on metabolic alterations in tomatoes, and changes affected the aphid *Macrosiphum euphorbiae* and the parasitoid *Aphidius ervi*, which are part of the same ecosystem. Treatments with *T. harzianum* produce a condition of priming that, when aphids attack, results in a more significant attraction of aphid parasitoids. This attraction is facilitated by the increased synthesis of VOCs, which are known to stimulate the flight of *A. ervi* (Battaglia et al., 2013).

Sequencing the transcriptome of plants treated with *T. harzianum* and infested by aphids revealed a noteworthy growth in the expression of genes for terpenoids and the salicylic acid pathway. These findings align with the observed action of *A. ervi* and the VOC profile that drives this behavioral reaction (Coppola et al., 2017). A study investigated the enzymatic antioxidant activity, mortality, and gene regulation of cabbage aphids in response to *T. longibrachiatum* at various time intervals ranging from 1-7 days following inoculation. The study found that the most significant number of deaths of *Brevicoryne brassicae* occurred on Day 7 when exposed to a concentration of 1×10^8 spores/ml (73%) of *T. longibrachiatum*, compared to the control group on the 7th day (11%). The activities of the enzymes ascorbate peroxidase (APX), glutathione peroxidase (GPX), glutathione S-transferase (GST), peroxidase (POD), superoxide dismutase (SOD), and catalase (CAT), were all increased in the host against *B. brassicae* (Inayat et al., 2022).

8. Mechanisms of *Trichoderma* against Plant Parasitic Nematodes

In recent times, scientists have been fascinated with the biological control agents of Plant Parasitic Nematodes (PPNs). *Trichoderma* species have been extensively studied for their potential as biological control agents against root-knot nematodes (RKNs). *Trichoderma* has been tested under different experimental conditions on various crops, such as *M. javanica* and *M. incognita*. The underlying mechanism has yet to be extensively investigated. A review explained the mechanisms by citing the following study on *T. harzianum*, which examined the expression profiles of proteases. The analysis revealed that 13 genes for programming peptidases, such as PRA1, P5216, P6281, P7455, and P9438, were expressed during *in-vitro* nematode egg colonization. This finding suggests that these genes may have essential functions in infecting PPN eggs. *Trichoderma* can colonize nematode eggs and juveniles by releasing certain extracellular hydrolytic enzymes. These include the chitinolytic enzymes chi18-5 and chi18-12 from multiple *Trichoderma* species, the serine protease SprT from *T. pseudokoningii*, and the trypsin-like protease PRA1 from *T. harzianum*. In addition, *Trichoderma* spp. have yielded various nematicidal secondary metabolites, including β-vinyl cyclopentane-1α, 3α-diol, 6-pentyl-2H-pyran-2-one, 4-(2-hydroxyethyl) phenol, and trichodermin (Li et al., 2015).

A study (TariqJaveed et al., 2021), revealed the significant potential of various isolates of *Trichoderma* in combating root-knot nematodes. They found that fungal hyphae, spores, and metabolites produced by these isolates exhibited multi-mechanism action action against nematodes. In addition, *T. harzianum* (T-79) exhibits early activation of SA and JA-related responses in tomatoes that are infested with RKNs. According to a study conducted by Martinez et al. (2023), it was found that the plant defence response was enhanced, leading to the suppression of the growth and reproduction of existing nematodes. *T. longibrachiatum* has been found to exhibit antagonistic activity against *Heterodera avenae*. Through meticulous examination, utilizing microscopic observation and bioassays, it has been determined that *T. longibrachiatum* exhibits promising characteristics as a biological control agent for *H. avenae*. This particular organism can directly parasitize and inflict fatal effects on the eggs and J2 stages. In their study, Khan et al. (2020) used inducers to stimulate the production of *T. koningiopsis* in a fermentation liquid that contained a significant quantity of chitinase. They used the fermentation liquid against *M. incognita* juveniles (J2), and found 90.4% and 63.2% of mortality and egg hatching inhibition respectively.

The increased level of chi18-5 and chi18-12 (chitinase) gene expression in *T. harzianum* FB10 was recorded during the parasitism on second-stage juveniles of

M. incognita and eggs (Baazeem et al., 2021). This suggests an increase in chitinase content, which could create favorable conditions for egg cleavage. The second-stage juvenile (J2) population in tomato roots was significantly reduced by up to 80% when exposed to *T. asperellum* T-16. Similarly, *T. brevicompactum* T-3 exhibited a remarkable suppression of egg production, with up to 86% reductions. The use of fungal biocontrol agents, particularly *T. asperellum* T-16, resulted in a significant upsurge of over 30% in tomato yields. While there were no notable impacts on carrot galling and yield, the presence of *T. asperellum* T-12 in the treated plots led to a significant reduction in soil J2 densities, with suppression levels reaching as high as 94% compared to the non-fungal controls (Affokpon et al., 2011). The bio-control strain FbMi6 of *T. asperellum* exhibited significant nematicidal activity in controlled laboratory conditions, resulting in a remarkable suppression of egg hatch (96%) and mortality of juvenile *M. incognita* (90%). The fungus *T. asperellum* was analyzed in both pot and field conditions following the enrichment of neem cake. The fungus was tested individually and in combination with other factors, and the results were compared to control samples. Using *T. asperellum* enriched neem cake resulted in a significant increase of 28% in bhendi yield. It also led to a notable decrease of 57% in the nematode population compared to the control group. The polyphenol content in bhendi was found to be higher when treated with *T. asperellum* FbMi6 enriched neem cake (Saharan et al., 2023).

9. Market Status of *Trichoderma*-Based Products

Trichoderma-based products, such as biopesticides and biofungicides, have gained attention in agriculture for their environmentally friendly and sustainable characteristics. The estimated value of the global *Trichoderma* products market is approximately $1 billion, with an expected compound annual growth rate (CAGR) of 10% over the next five years. The primary catalysts for this expansion are the escalating need for environmentally friendly farming methods, the rising recognition of the detrimental consequences of chemical pesticides, and the surging use of *Trichoderma* products in emerging nations. *Trichoderma* products are utilized in several crops, including cereals, fruits, vegetables, and ornamental plants. *T. harzianum* is the predominant species of *Trichoderma* utilized in the global market for biological goods. Approximately 60% of fungal-based Biological Control Agents (BCAs) are attributed to biopesticides containing *Trichoderma*. These are predominantly utilized in Asia as a BCA, with *T. harzianum*-based formulations accounting for 70% of the available products, especially in India.

Trichoderma products are usually considered safe and do not pose substantial hazards to well-being or the environment. These biopesticides are available in various formulations. In addition, dry formulations consist of six formats: tablets, pellets, dust, wettable powder, granules, and powders for seed dressing. On the other hand, liquid-state formulations include suspension concentrates, oil dispersions, suspo-emulsions, emulsions, oil-in-water, water-in-oil, capsule suspensions, and ultralow volume formulations. The market for *Trichoderma*-based biofungicides is projected to experience a growth rate of 4.80% during the forecast time frame of 2021-2028, ultimately reaching a value of USD 1.07 billion by 2028. The study by Data Bridge

Market Research on the *Trichoderma* biofungicides market offers analysis and insights into the elements that are projected to be present during the entire forecast period and their effects on the market's growth. The global increase in environmental concerns is driving the development of the *Trichoderma* biologicals market.

Conclusion

Due to its diverse functions in agriculture and other domains, *Trichoderma* is an extraordinary antagonistic fungus with extensive importance. Within agriculture, it functions as a potent ally in managing diseases, as well as insects and nematodes, providing sustainable and ecologically sound methods to counteract a diverse range of plant infections. *Trichoderma* improves crop health and yields by preventing soilborne infections of pests, pathogens, and parasites, suppressing foliar infections, and extending the postharvest life of fruits and vegetables. It reduces the need for chemical pesticides. Its compatibility with organic farming practices highlights its crucial role in supporting environmentally responsible agriculture. In addition to agriculture, *Trichoderma* has applications in several fields. Bioremediation involves actively engaging in the process of purifying polluted soils, thereby playing a significant role in the restoration of the environment. They are hosts for expressing therapeutic proteins in the production of biofuels, enzymes, biotechnology, and medicines. *Trichoderma* exemplifies controlling pests in agriculture, and its versatile uses in biotechnology and environmental cleanup make it a symbol of sustainable advancement and a demonstration of the potential of helpful microorganisms in tackling important issues.

References

Abbas, A., Mubeen, M., Zheng, H., Sohail, M.A., Shakeel, Q., et al. (2022). *Trichoderma* spp. genes involved in the biocontrol activity against *Rhizoctonia solani.* Front. Microbiol., 13. https://doi.org/10.3389/fmicb.2022.884469

Affokpon, A., Coyne, D.L., Htay, C.C., Agbèdè, R.D., Lawouin, L. et al. (2011). Biocontrol potential of native *Trichoderma* isolates against root-knot nematodes in West African vegetable production systems. Soil Biol. Biochem., 43(3): 600–608. https://doi.org/10.1016/j.soilbio.2010.11.029

Anwar, W., Amin, H., Khan, H.A.A., Akhter, A., Bashir, U., et al. (2023). Chitinase of *Trichoderma longibrachiatum* for control of *Aphis gossypii* in cotton plants. Sci. Rep., 13(1): 1–9. https://doi.org/10.1038/s41598-023-39965-y

Anwar, W., Subhani, M.N., Haider, M.S., Shahid, A.A., Mushtaq, H., et al. (2016). First record of *Trichoderma longibrachiatum* as entomopathogenic fungi against *Bemisia tabaci* in Pakistan. Pak. J. Phytopathol., 28(02): 287–294.

Atriztán-Hernández, K., Moreno-Pedraza, A., Winkler, R., Markow, T. and Herrera-Estrella, A. (2019). *Trichoderma atroviride* from predator to prey: Role of the mitogen-activated protein kinase tmk3 in fungal chemical defense against fungivory by drosophila melanogaster larvae. Appl. Environ. Microbiol., 85(2): 1–15. https://doi.org/10.1128/AEM.01825-18

Baazeem, A., Almanea, A., Manikandan, P., Alorabi, M., Vijayaraghavan, P. et al. (2021). In vitro antibacterial, antifungal, nematocidal and growth promoting activities of *Trichoderma hamatum* fb10 and its secondary metabolites. J. Fungi., 7(5): 1–13. https://doi.org/10.3390/jof7050331

Balakrishnan, S., Gopalakrishnan, C., Kamalakannan, A., Senthil, K. and Parthasarathy, S. (2017a). Evaluation of antagonistic activity and plant growth promotion by paste formulation of *Trichoderma harzianum*. J. Pharmacogn. Phytochem., 6(6): 355–360.

Balakrishnan, S., Parthasarathy, S., Kamalakannan, A., Senthil Kuppusamy, K. and Gopalakrishnan, C. (2017b). A noval method to develop paste formulation of *Trichoderma harzianum*. Int. J. Curr. Microbiol. Appl. Sci., 6(11): 3286–3293. https://doi.org/10.20546/ijcmas.2017.611.385

Battaglia, D., Bossi, S., Cascone, P., Digilio, M.C., Prieto, J.D., et al. (2013). Tomato below ground-above ground interactions: *Trichoderma longibrachiatum* affects the performance of *Macrosiphum euphorbiae* and its natural antagonists. MPMI, 26(10): 1249–1256. https://doi.org/10.1094/MPMI-02-13-0059-R

Berini, F., Caccia, S., Franzetti, E., Congiu, T., Marinelli, F., et al. (2016). Effects of *Trichoderma viride* chitinases on the peritrophic matrix of Lepidoptera. Pest Manag. Sci., 72(5): 980–989. https://doi.org/10.1002/ps.4078

Chinnaperumal, K., Govindasamy, B., Paramasivam, D., Dilipkumar, A., Dhayalan, A., et al. (2018). Bio-pesticidal effects of *Trichoderma viride* formulated titanium dioxide nanoparticle and their physiological and biochemical changes on Helicoverpa armigera (Hub.). Pestic. Biochem. Physiol., 149: 26–36. https://doi.org/10.1016/j.pestbp.2018.05.005

Contreras-Cornejo, H.A., Macías-Rodríguez, L., del-Val, E. and Larsen, J. (2018). The root endophytic fungus *Trichoderma atroviride* induces foliar herbivory resistance in maize plants. Appl. Soil Ecol., 124: 45–53. https://doi.org/10.1016/j.apsoil.2017.10.004

Contreras-Cornejo, H.A., Viveros-Bremauntz, F., del-Val, E., Macías-Rodríguez, L., López-Carmona, D.A., et al. (2021). Alterations of foliar arthropod communities in a maize agroecosystem induced by the root-associated fungus *Trichoderma harzianum*. J. Pest Sci., 94(2): 363–374. https://doi.org/10.1007/s10340-020-01261-3

Coppola, M., Cascone, P., Chiusano, M.L., Colantuono, C., Lorito, M., et al. (2017). *Trichoderma harzianum* enhances tomato indirect defense against aphids. Insect Sci., 24(6): 1025–1033. https://doi.org/10.1111/1744-7917.12475

Di Lelio, I., Forni, G., Magoga, G., Brunetti, M., Bruno, D., et al. (2023). A soil fungus confers plant resistance against a phytophagous insect by disrupting the symbiotic role of its gut microbiota. PNAS USA., 120(10). https://doi.org/10.1073/pnas.2216922120

Dou, K., Lu, Z., Wu, Q., Ni, M., Yu, C., Wang, M., et al. (2020). MIST: A multilocus identification system for *Trichoderma*. Appl. Environ. Microbiol., 86(18). https://doi.org/10.1128/AEM.01532-20

Dutta, P., Deb, L. and Pandey, A.K. (2022). *Trichoderma*- from lab bench to field application: Looking back over 50 years. Front. Agron., 4: 1–25. https://doi.org/10.3389/fagro.2022.932839

El-Massry, S., Shokry, H. and Hegab, M. (2016). Efficiency of *Trichoderma harzianum* and some organic acids on the cotton bollworms, *Earias insulana* and *Pectinophora gossypiella*. J. Plant Prot. Pathol., 7(2): 143–148. https://doi.org/10.21608/jppp.2016.50106

Ganassi, S., Domenico, C. Di, Altomare, C., Samuels, G. J., Grazioso, P., et al. (2023). Potential of fungi of the genus *Trichoderma* for biocontrol of *Philaenus spumarius*, the insect vector for the quarantine bacterium *Xylella fastidosa*. Pest Manag. Sci., 79(2): 719–728. https://doi.org/10.1002/ps.7240

Ganassi, S., Moretti, A., Stornelli, C., Fratello, B., Bonvicini Pagliai, A.M., Logrieco, A. and Sabatini, M.A. (2001). Effect of *Fusarium*, *Paecilomyces* and *Trichoderma* formulations against aphid Schizaphis graminum. Mycopathologia, 151(3): 131–138. https://doi.org/10.1023/A:1017940604692

Geetha, I., Paily, K.P., Padmanaban, V. and Balaraman, K. (2003). Oviposition response of the mosquito, *Culex quinquefasciatus* to the secondary metabolite(s) of the fungus, *Trichoderma viride*. Mem. Inst. Oswaldo Cruz., 98(2): 223–226. https://doi.org/10.1590/S0074-02762003000200010

Ghosh, S. K. and Pal, S. (2016). Entomopathogenic potential of *Trichoderma longibrachiatum* and its comparative evaluation with malathion against the insect pest *Leucinodes orbonalis*. Environ. Monit. Assess., 188(1): 1–7. https://doi.org/10.1007/s10661-015-5053-x

Guzmán-Guzmán, P., Kumar, A., de los Santos-Villalobos, S., Parra-Cota, F.I., Orozco-Mosqueda, M. del C., et al. (2023). *Trichoderma* species: our best fungal allies in the biocontrol of plant diseases- A Review. Plants, 12(3): 1–35. https://doi.org/10.3390/plants12030432

Inayat, R., Khurshid, A., Boamah, S., Zhang, S. and Xu, B. (2022). Mortality, Enzymatic antioxidant activity and gene expression of cabbage aphid (*Brevicoryne brassicae* L.) in response to *Trichoderma longibrachiatum* T6. Front. Physiol., 13: 1–11. https://doi.org/10.3389/fphys.2022.901115

Jafarbeigi, F., Samih, M. A., Alaei, H. and Shirani, H. (2020). Induced tomato resistance against bemisia tabaci triggered by salicylic acid, β-Aminobutyric Acid, and *Trichoderma*. Neotrop. Entomol., 49(3): 456–467. https://doi.org/10.1007/s13744-020-00771-0

Kaushik, N., Díaz, C. E., Chhipa, H., Fernando Julio, L., Fe Andrés, M. et al. (2020). Chemical composition of an aphid antifeedant extract from an endophytic fungus, *Trichoderma* sp. Efi671. Microorganisms, 8(3): 1–14. https://doi.org/10.3390/microorganisms8030420

Khan, R.A.A., Najeeb, S., Hussain, S., Xie, B. and Li, Y. (2020). Bioactive secondary metabolites from *Trichoderma* spp. against phytopathogenic fungi. Microorganisms, 8(6). https://doi.org/10.3390/microorganisms8060817

Kubicek, C.P., Steindorff, A.S., Chenthamara, K., Manganiello, G., Henrissat, B., et al. (2019). Evolution and comparative genomics of the most common *Trichoderma* species. BMC Genomics, 20(1): 1–24. https://doi.org/10.1186/s12864-019-5680-7

Kushiyev, R., Tuncer, C., Erper, I. and Özer, G. (2021). The utility of Trichoderma spp. isolates to control of *Xylosandrus germanus* Blandford (Coleoptera: Curculionidae: Scolytinae). J. Plant Dis. Prot., 128(1): 153–160. https://doi.org/10.1007/s41348-020-00375-1

Laib, D.E., Benzara, A., Akkal, S. and Bensouici, C. (2020). The anti-acetylcholinesterase, insecticidal and antifungal activities of the entophytic fungus *Trichoderma* sp. isolated from *Ricinus communis* L. against *Locusta migratoria* L. and *Botrytis cinerea* Pers.: Fr. ANS., 7(1): 112–125. https://doi.org/10.2478/asn-2020-0011

Leppik, E. and Frérot, B. (2014). Maize field odorscape during the oviposition flight of the European corn borer. Chemoecology, 24(6): 221–228. https://doi.org/10.1007/s00049-014-0165-2

Li, J., Zou, C., Xu, J., Ji, X., Niu, X., et al. (2015). Molecular mechanisms of nematode-nematophagous microbe interactions: basis for biological control of plant-parasitic nematodes. Annu. Rev. Phytopathol., 53: 67–95. https://doi.org/10.1146/annurev-phyto-080614-120336

Manzar, N., Kashyap, A.S., Goutam, R.S., Rajawat, M.V.S., Sharma, P.K., et al. (2022). *Trichoderma*: advent of versatile biocontrol agent, its secrets and insights into mechanism of biocontrol potential. Sustainability, 14(19): 1–32. https://doi.org/10.3390/su141912786

Martinez, Y., Ribera, J., Schwarze, F.W.M.R. and De France, K. (2023). Biotechnological development of *Trichoderma*-based formulations for biological control. Appl. Microbiol. Biotechnol., 107(18): 5595–5612. https://doi.org/10.1007/s00253-023-12687-x

Mendoza-Mendoza, A., Zaid, R., Lawry, R., Hermosa, R., Monte, E., et al. (2018). Molecular dialogues between *Trichoderma* and roots: Role of the fungal secretome. Fungal Biol. Rev., 32(2): 62–85. https://doi.org/10.1016/j.fbr.2017.12.001

Mohamed, G. and Taha, E. (2017). Potency of entomopathogenic fungi, *Trichoderma album* preuss in controlling, *Rhzopertha dominica* F. (Coleoptera: Bostrichidae) under laboratory conditions. J. Plant Prot. Res., 8(11): 571–576. https://doi.org/10.21608/jppp.2017.46855

Morán-Diez, M.E., Martínez de Alba, Á.E., Rubio, M.B., Hermosa, R. and Monte, E. (2021). *Trichoderma* and the plant heritable priming responses. J. Fungi, 7(4): 318. https://doi.org/10.3390/jof7040318

Muvea, A.M., Meyhöfer, R., Subramanian, S., Poehling, H.M., Ekesi, S. et al. (2014). Colonization of onions by endophytic fungi and their impacts on the biology of thrips tabaci. *PLoS ONE*, 9(9): 1–7. https://doi.org/10.1371/journal.pone.0108242

Narayanan, P., Parthasarathy, S., Rajalakshmi, J., Arunkumar, K. and Vanitha, S. (2016). Systemic elicitation of defense related enzymes suppressing *Fusarium* wilt in mulberry (*Morus* spp.). Afr. J. Microbiol. Res., 10(22): 813–819. https://doi.org/10.5897/ajmr2015.7900

Nasution, L., Corah, R., Nuraida, N. and Siregar, A. Z. (2018). Effectiveness *Trichoderma* and *Beauveria bassiana* on larvae of *Oryctes rhinoceros* on palm oil plant (*Elaeis Guineensis* Jacq.) *in vitro*. Intern. J. Env. Agr. Biotech., 3(1): 158–169. https://doi.org/10.22161/ijeab/3.1.20

Nawaz, A., Ali, H., Sufyan, M., Gogi, M.D., Arif, M.J., Ali, A., et al. (2020). In-vitro assessment of food consumption, utilization indices and losses promises of leafworm, Spodoptera litura (Fab.), on okra crop. J. Asia-Pac. Entomol., 23(1): 60–66. https://doi.org/10.1016/j.aspen.2019.10.015

Omar, A. and Abdul-Wahid. (2012). Evaluation of the insecticidal activity of *Fusarium solani* and *Trichoderma harzianum* against cockroaches; *Periplaneta Americana*. Afr. J. Microbiol. Res., 6(5): 1024–1032. https://doi.org/10.5897/ajmr-11-1300

Parrilli, M., Sommaggio, D., Tassini, C., Di Marco, S., Osti, F., et al. (2019). The role of *Trichoderma* spp. and silica gel in plant defence mechanisms and insect response in vineyard. Bull. Entomol. Res., 109(6): 771–780. https://doi.org/10.1017/S0007485319000075

Poveda, J. (2021). *Trichoderma* as biocontrol agent against pests: New uses for a mycoparasite. Biol. Cont., 159: 104634. https://doi.org/10.1016/j.biocontrol.2021.104634

Rahim, S. and Iqbal, M. (2019). Exploring enhanced insecticidal activity of mycelial extract of *Trichoderma harzianum* against *Diuraphis noxia* and *Tribolium castaneum*. Sarhad J. Agric., 35(3): 757–762. https://doi.org/10.17582/journal.sja/2019/35.3.757.762

Razinger, J., Lutz, M., Schroers, H. J., Palmisano, M., Wohler, C., et al. (2014). Direct plantlet inoculation with soil or insect-associated fungi may control cabbage root fly maggots. J. Invertebr. Pathol., 120: 59–66. https://doi.org/10.1016/j.jip.2014.05.006

Rodríguez-González, Á., Campelo, M.P., Lorenzana, A., Mayo-Prieto, S., González-López, Ó., et al. (2020). Spores of *Trichoderma* strains sprayed over *Acanthoscelides obtectus* and *Phaseolus vulgaris* L. beans: Effects in the biology of the bean weevil. J. Stored Prod. Res., 88. https://doi.org/10.1016/j.jspr.2020.101666

Rodríguez-González, Á., Casquero, P.A., Suárez-Villanueva, V., Carro-Huerga, G., Álvarez-García, S., et al. (2018). Effect of trichodiene production by *Trichoderma harzianum* on *Acanthoscelides obtectus*. J. Stored Prod. Res., 77: 231–239. https://doi.org/10.1016/j.jspr.2018.05.001

Rodríguez-González, Á., Mayo, S., González-López, Ó., Reinoso, B., Gutierrez, S. and Casquero, P.A. (2017). Inhibitory activity of *Beauveria bassiana* and *Trichoderma* spp. on the insect pests *Xylotrechus arvicola* (Coleoptera: Cerambycidae) and *Acanthoscelides obtectus* (Coleoptera: Chrisomelidae: Bruchinae). Environ. Monit. Assess., 189(1). https://doi.org/10.1007/s10661-016-5719-z

Saharan, R., Patil, J.A., Yadav, S., Kumar, A. and Goyal, V. (2023). The nematicidal potential of novel fungus, *Trichoderma asperellum* FbMi6 against *Meloidogyne incognita*. Sci. Rep., 13(1): 1–9. https://doi.org/10.1038/s41598-023-33669-z

Salwa, S. and Kottb, M.R. (2017). Bioactivity of *Trichoderma* (6-Pentyl α -pyrone) against *Tetranychus urticae* Koch (Acari: Tetranychidae). Egypt. Acad. J. Biol., 10(3): 29–34.

Salwan, R., Sharma, A., Kaur, R., Sharma, R. and Sharma, V. (2022). The riddles of *Trichoderma* induced plant immunity. *Biological Control*, 174: 105037. https://doi.org/10.1016/j.biocontrol.2022.105037

Savary, S., Willocquet, L., Pethybridge, S.J., Esker, P., McRoberts, N. et al. (2019). The global burden of pathogens and pests on major food crops. Nat. Ecol. Evol., 3(3): 430–439. https://doi.org/10.1038/s41559-018-0793-y

Schmoll, M., Dattenböck, C., Carreras-Villaseñor, N., Mendoza-Mendoza, A., Tisch, D., et al. (2016). The genomes of three uneven siblings: footprints of the lifestyles of three Trichoderma species. Microbiol. Mol. Biol., 80(1), 205–327. https://doi.org/10.1128/mmbr.00040-15

Schmoll, M. and Schuster, A. (2010). Biology and biotechnology of *Trichoderma*. Appl. Microbiol. Biotechnol., 87: 787–799. https://doi.org/10.1007/s00253-010-2632-1

Shakeri, J. and Foster, H.A. (2007). Proteolytic activity and antibiotic production by *Trichoderma harzianum* in relation to pathogenicity to insects. Enzyme Microb. Technol., 40(4): 961–968. https://doi.org/10.1016/j.enzmictec.2006.07.041

Sharma, A., Gupta, B., Verma, S., Pal, J., Mukesh, Akanksha, et al. (2023). Unveiling the biocontrol potential of *Trichoderma*. Eur. J. Plant Pathol., 167(4): 569–591. https://doi.org/10.1007/s10658-023-02745-5

Silva, B.B., Banaay, C.G. and Salamanez, K. (2019). *Trichoderma*-induced systemic resistance against the scale insect (*Unaspis mabilis* Lit and Barbecho) in lanzones (*Lansium domesticum* corr.). Agric. For., 65(2): 59–78. https://doi.org/10.17707/AgricultForest.65.2.05

Sood, M., Kapoor, D., Kumar, V. and Sheteiwy, M.S. (2020). *Trichoderma*: The "Secrets" of a Multitalented. Plants, 9: 762.

Sundaravadivelan, C. and Padmanabhan, M.N. (2014). Effect of mycosynthesized silver nanoparticles from filtrate of *Trichoderma harzianum* against larvae and pupa of dengue vector *Aedes aegypti* L. Environ. Sci. Pollut. Res., 21(6): 4624–4633. https://doi.org/10.1007/s11356-013-2358-6

TariqJaveed, M., Farooq, T., Al-Hazmi, A.S., Hussain, M.D. and Rehman, A.U. (2021). Role of *Trichoderma* as a biocontrol agent (BCA) of phytoparasitic nematodes and plant growth inducer. J. Invertebr. Pathol., 183: 107626. https://doi.org/10.1016/j.jip.2021.107626

Vijayakumar, N. and Alagar, S. (2017). Consequence of chitinase from *Trichoderma viride* integrated feed on digestive enzymes in *Corcyra cephalonica* (Stainton) and antimicrobial potential. Biosci., Biotech. Res. Asia., 14(2): 513–519. https://doi.org/10.13005/bbra/2473

Woo, S. L., Hermosa, R., Lorito, M. and Monte, E. (2023). *Trichoderma*: a multipurpose, plant-beneficial microorganism for eco-sustainable agriculture. Nat. Rev. Microbiol., 21(5): 312–326. https://doi.org/10.1038/s41579-022-00819-5

Zahran, Z., Mohamed Nor, N.M.I., Dieng, H., Satho, T. and Ab Majid, A.H. (2017). Laboratory efficacy of mycoparasitic fungi (*Aspergillus tubingensis* and *Trichoderma harzianum*) against tropical bed bugs (*Cimex hemipterus*) (Hemiptera: Cimicidae). Asian Pac. J. Trop. Biomed., 7(4): 288–293. https://doi.org/10.1016/j.apjtb.2016.12.021

Zhou, D., Huang, X.F., Guo, J., dos-Santos, M.L. and Vivanco, J.M. (2018). Trichoderma gamsii affected herbivore feeding behaviour on Arabidopsis thaliana by modifying the leaf metabolome and phytohormones. Microb. Biotechnol. 11(6): 1195–1206. https://doi.org/10.1111/1751-7915.13310

10 Entomopathogenic Fungi for the Control of *Aedes* Mosquitoes

Anderson Ribeiro, Adriano Rodrigues de Paula, Ricardo de Oliveira Barbosa Bitencourt, Gerson Adriano Silva, Josiane Pessanha Ribeiro, Leila Eid Imad Silva and Richard Ian Samuels*

1. Introduction

Incidences of mosquito-vectored pathogens are increasing worldwide; a major concern for societies and governments (Tolle, 2009; WHO, 2023). There are over 3,500 species of mosquitoes, of which, from an epidemiological point of view, one of the most important genera is *Aedes* (Becker et al., 2010), responsible for transmitting viruses (arboviruses) of wide-ranging significance, such as dengue (DENV), Zika (ZIKV), urban yellow fever (UYF), chikungunya (CHICKV), and Mayaro virus (Smith and Francy, 1991; Benedict et al., 2007; Proestos et al., 2015; Dickens et al., 2018; Kantor et al., 2019; Dieme et al., 2020; Valencia-Marín et al., 2020). These arboviruses are responsible for outbreaks of diseases affecting millions of people annually. Unfortunately, there is no vaccine to combat Zika or chikungunya, and there are no antiviral drug treatments available. On the other hand, a vaccine against UYF is available, but not all at-risk populations have been vaccinated (Barret, 2017). Currently, there are vaccines approved for dengue, however, they are expensive and have restrictions on their use. Furthermore, they do not provide complete protection, especially against reinfection with different serotypes (Ferguson et al., 2016; Wilder-Smith, 2020).

Originating in Africa, *Aedes aegypti* can now be found in 167 countries, distributed in the Americas, Africa, Asia, the Indian Ocean islands, and northern Australia (Kraemer et al., 2015; Dickens et al., 2018). *Aedes albopictus* is native to southeast Asia and has now become established in 126 tropical and sub-tropical countries (Kraemer et al., 2015; Wilkerson et al., 2020; Laporta et al., 2023; Ainsworth, 2023). Both of these anthropophilic species deposit their eggs in improperly discarded human waste, such as plastic bags, bottles, and tires, as well as in water storage containers, soil located in areas prone to water accumulation, and inside certain plants, such as bromeliads (Matthews, 2019; Benelli et al., 2020; McGregor and Connelly, 2021). The global distribution of these mosquitoes has increased rapidly due to several factors,

Universidade Estadual do Norte Fluminense Darcy Ribeiro, Department of Entomology and Plant Pathology, Av. Alberto Lamego 2000, Parque California, Campos dos Goytacazes, RJ 28013-602 Brazil.

* Corresponding author: richardiansamuels@gmail.com

including increasing urban populations, freight transport, air travel, uncontrolled mosquito populations, and climate changes such as increased precipitation, favorable temperatures and humidity, among others (Jones et al., 2008; Weaver and Reisen, 2010; WHO, 2021; Piovezan-Borges et al., 2022).

The most common approaches for reducing incidences of these mosquitoes are prevention and control through the elimination of breeding sites, and the application of synthetic insecticides (Zara et al., 2016). However, due to the development of insecticide-resistance and environmental persistence, one of the main focuses of the scientific community is to find novel ways to control mosquito populations (Yi et al., 2014). Currently, a bioinsecticide based on the entomopathogenic bacteria *Bacillus thuringiensis,* that specifically targets *Ae. aegypti* larvae, is available. Several researchers have studied the host-pathogen interactions with other agents, such as filamentous fungi, against all life stages of *Aedes* sp. and how these bioagents may be incorporated as part of an integrated vector management strategy (Alkhaibari et al., 2018a; Carolino et al., 2019; Paula et al., 2021; de Sousa et al., 2021; Reyes-Villanueva et al., 2021; Rodrigues et al., 2021; Bitencourt et al., 2023; Ramirez et al., 2023).

Reducing the population of *Aedes* mosquitoes is the most effective method to reduce incidences of related disease (Vannavong et al., 2017). Thus, biological control using entomopathogenic fungi (EPF) has emerged as an innovative approach with the potential to control mosquito populations (Paula et al., 2008; Silva et al., 2017; Paula et al., 2018; Bitencourt et al., 2021a) and assist in limiting the transmission of vector-borne diseases. In the following, we summarize current methods to control *Aedes* species and present new, related studies. This chapter aims to condense and synthesize current literature on the biocontrol of *Ae. aegypti* and *Ae. albopictus* using EPF, highlighting the potential of these microrganisms for mosquito control.

2. Mosquito Control Methods and Strategies

Several methods, encompassing chemical, biological, and mechanical approaches, are employed in the indirect control of mosquito-borne diseases, resulting in a reduction in their incidence (Achee et al., 2015; Roberts et al., 2016). However, due to a lack of data, it is not known to what extent these methods have prevented epidemics and the spread of arboviral diseases (Horstick et al., 2010). The use of natural enemies such as predators, parasites, or pathogens is a biological method to manage populations of an organism considered harmful. A variety of biological control agents have been studied, and some have been employed against mosquitoes (Lima et al., 2016; McGregor and Connelly, 2021), as described below.

2.1 Predators

There are over 250 species of invertebrate and vertebrate predators of mosquito larvae, especially larviparous fish (*Oreochromise* and *Poeciliareticulata*) and nematodes of the Mermithidae family (Consoli and Oliveira, 1998; McGregor and Connelly, 2021). A group of larvivorous arthropods of the genus *Toxorhynchites* (Diptera: Culicidae) have been studied for the control of *Ae. aegypti*. *Toxorhynchites* larvae typically attack other mosquito larvae to obtain the proteins necessary for their development and egg production, since these mosquitoes do not feed on blood during the adult stage (Vinogradov et al., 2022). Uejio et al. (2014) found in field experiments that although *Toxorhynchites mectezuma* was effective in controlling pupae, their efficiency decreased when prey populations increased beyond their predatory capacity.

Alkhaibari et al. (2018b) verified the interaction between *Toxorhynchites bevipalpis* larvae and *Metarhizium brunneum* propagules for the control of *Ae. aegypti* larvae. In laboratory bioassays, it was observed that blastospores at low concentrations complemented the predatory performance of *T. brevipalpis* resulting in greater control than if either agent was used alone. Certain challenges limit the use of this group of larvivorous arthropods. Maintaining colonies of *Toxorhynchites* can be labor-intensive and requires rearing additional mosquitoes to act as food sources for the larvae. In addition, distributing *Toxorhynchites* to treatment areas can be cumbersome, especially if the released individuals do not breed naturally to increase their population (Schreiber and Jones, 1994; Collins and Blackwell, 2000).

2.2 Viruses

The capacity of certain entomopathogenic viruses to infect *Aedes* sp. has been explored, but their application in biological control has been constrained due to low infectivity or production methods that are unsuitable for field testing (Perrin et al., 2020). One such entomopathogenic virus belongs to the genus *Brevidensovirus* (densovirus). These viruses are icosahedral non-enveloped DNA viruses with two types: densovirus 1 and 2, and they specifically target *Ae. aegypti.* For *Ae. albopictus,* densovirus 1 (AalDV-125), densovirus 2 and 3 (AalDV-228 and AalDV-329, respectively) have been studied as potential biocontrol agents (Afanasiev et al., 1991; O'Neil et al., 1995; Sivaram et al., 2009; Chen et al., 2004). Li et al. (2019) tested the pathogenicity of *Ae. albopictus* densovirus-7 (AalDV-7) against *Ae. albopictus*, *Ae. aegypti*, and *Culex quinquefasciatus* larvae and observed positive results for the control of *Ae. albopictus* using AalDV-7. Perrin et al. (2020) examined the susceptibility of both *Aedes* species to various densoviruses under laboratory conditions. The findings showed that *Aedes albopictus* densovirus 2 (AalDV-2) demonstrated higher aggressiveness against *Ae. aegypti.* Batool et al. (2022) conducted a study on the formulation of a larvicide using a combination of *Bacillus* var. israelensis (Bti) and *Ae. albopictus* densovirus (AalDV-5). Subsequently, this larvicide was tested against *Ae. albopictus* larvae and evaluated under experimental semi-field and open-field conditions. The results affirmed virulence enhancement using this combination.

The first mosquito baculovirus was isolated from *Aedes sollicitans* in Louisiana in 1969 (Clark, et al., 1969). These viruses can be used both as biological insecticides for pest control in forests or crops (Moscardi, 1999), and as expression vectors in biotechnological applications (Kost et al., 2005). In addition, studies have shown that baculoviruses exhibit high infectivity for other mosquito species such as *Culex nigripalpus* and *Culex quinquefasciatus* larvae, leading to the death of these insects (Becnel et al., 2001; Williams et al., 2017). However, most of them have a limited ability to overcome the larval midgut barrier (Moser et al., 2001). Additionally, the difficulty of isolating and cultivating these viruses in the laboratory, combined with the scarcity of significant changes in pathology and cellular morphology post-infection, has resulted in limited progress in investigating *baculovirus* infections in mosquitoes (Afonso et al., 2001).

2.3 Nematodes

Entomopathogenic nematodes (EPNs) are obligate insect parasites and are recognized biocontrol agents for crop pests and as promising *Aedes* larvicides (Suwannaroj et al., 2020; Subkrasae et al., 2022). They form associations with symbiotic bacteria such as the genera *Xenorhabdus* or *Photorhabdus*, and their life cycle encompasses eggs, juveniles, and adults. Liu et al. (2020) reported the efficacy of *Steinernema abbasi* in the infection of *Ae. albopictus* larval hemocoel, which led to death. Suwannaroj et al. (2020) employed a baiting technique and White traps were used to isolate 35 strains of *Steinernema surkhetense* and 14 strains of *Heterorhabditis indica* from soil samples. Subsequently, *Xenorhabdus stockiae* was also isolated. The larvicidal activity of this bacteria against *Ae. aegypti* was then assessed under laboratory conditions. The study reported a significant mortality rate of *Ae. aegypti*, reaching 99%, 96 hours after exposure to *X. stockiae*. Although several researchers have used EPNs to control *Ae. aegypti,* their efforts have concentrated mainly on the larval stages (Peschiutta et al., 2014; Cardoso et al., 2015; Ulvedal et al., 2017; da Silva et al., 2017; Shah et al., 2021; Thanwisai et al., 2022). In addition, as far as we know, there is no information regarding commercializing EPNs for controlling *Aedes* larvae.

2.4 Bacterial Pathogens

Bacillus thuringiensis var. *israelensis* (*Bti*) and *Bacillus sphaericus* are mosquito larval pathogens (Consoli and Oliveira, 1998; Lutinski et al., 2017), and their effectiveness has now been demonstrated in vector control programs worldwide (Flacio et al., 2015; Brühl et al., 2020). The main active ingredients of these microbial larvicides are crystals containing protoxins that are highly toxic to dipteran larvae such as *Aedes*, *Culex*, *Anopheles*, and *Simulium* (Lacey, 2007). Following ingestion of *Bti* crystals by the larvae, they are solubilized in the midgut, releasing protoxins that are proteolytically converted into toxins. These activated toxins interact with receptors in the midgut, where they insert themselves into the membrane, forming pores that lead to cell permeability and osmotic lysis, causing fatal damage to the larval epithelium (Vachon et al., 2012).

There has been an increase in the use of *Bti* to control *Aedes* in tropical countries, not only for the conventional treatment of breeding sites but also for its use in association with approaches such as in lethal ovitraps (Regis et al., 2013; Johnson et al., 2017). Another genus of bacteria that has been gaining ground in research on reducing disease transmission, is *Wolbachia*. These bacteria are intracellular endosymbionts that modify host reproduction (Hugo et al., 2022). They were first observed in the reproductive tissues of *Culex pipiens* by Hertig and Wolbach in 1924 (Werren, 1997). These bacteria were believed to be absent in *Ae. aegypti* and *Anopheles* mosquitoes (Kittayapong et al., 2000). However, recent studies have reported the presence of *Wolbachia* in wild *Ae. aegypti* (Bennett et al., 2019) and *Ae. albopictus* (Huang et al., 2020).

The first releases of *Ae. aegypti* mosquitoes infected with *Wolbachia* occurred in Australia, and subsequently the "World Mosquito Program" was set up to expand collaborations with municipalities and communities in different countries to reduce the incidence of arboviruses. Mosquitoes are being released in communities with high rates of arboviruses (dengue, Zika, and chikungunya), and then possible reductions of these diseases are being assessed. According to data from this program, the project has been implemented in 14 countries and has protected nearly 11 million people (www.worldmosquitoprogram.org).

2.5 *Entomopathogenic Fungi*

Entomopathogenic fungi (EPF) are extensively used as biocontrol agents to manage crop pests and may also be used in an innovative approach to controlling insect species of public health concern (Silva et al., 2017; Paula et al., 2018; Mascarin et al., 2019; Bitencourt et al., 2021a). EPF are generally multicellular eukaryotic organisms that can infect arthropods through ingestion or external contact of conidia with the cuticle (Alves, 1998; Alkhaibari et al., 2016; Samuels et al., 2016; Bitencourt et al., 2023), demonstrating advantages over viruses and bacterial toxins since they do not need to be ingested by insects for infection to occur (Costa et al., 2010; Bilgo et al., 2018). Another advantage of using EPF is the extensive genetic diversity, which allows scientists to choose isolates that are more virulent for a specific insect target (Schrank and Vainstein, 2010, Wang and Wang, 2017). Fungi are ubiquitous organisms, easily obtained from soil samples or insect cadavers (Niu et al., 2019; Domingues et al., 2023). Biological control methods using EPF are safer, less harmful to the environment, and more cost-effective than chemical or physical control methods (Huang et al., 2017). As EPF produces various toxins during infection, the selection pressure for resistance is likely to be less intense when compared to fast-acting insecticides, making the evolution of fungal resistance much slower compared to insecticide resistance development (Ffrench-Constant, 2005).

3. Mechanisms of Infection

Infection begins with conidial adhesion (Fig. 1) onto the insect cuticle through nonspecific hydrophobic interactions and the secretion of enzymes, such as *Metarhizium/Beauveria*-type adhesin *Mad1*, and hydrophobins *Hyd1* and *Hyd2* (Pedrini, 2018; Liu et al., 2021; Zhou et al., 2021). Filamentous fungi can secrete a complex mixture of enzymes, including subtilisin-like Pr1, trypsin-like Pr2, and lipases, which assist the fungus in disrupting lipids on the cuticle surface (Wang and Wang, 2017). After adhesion, the spore germinates, forming a structure called appressorium with a penetration peg. These structures mechanically overcome the physical barrier, aided by proteases, chitinases, and lipases (Ryder and Talbot, 2015; Butt et al., 2016; Wang et al., 2016; Wang and Wang, 2017). On reaching the hemocoel, the fungus multiplies rapidly, as hyphal bodies or blastospores, disseminating throughout the host. Death is caused by the destruction of host tissues and, in certain cases, the production of cyclic hexadepsipeptide toxins such as destruxin, beauvericin, and oosporein, capable of immunosuppressing some insect species (Hu et al., 2016;

Wang et al., 2019). These toxins also inhibit V-type ATPase hydrolytic activity and open calcium channels, causing flaccid paralysis and accelerating the death of insects (Samuels, 1998; Ravindran et al., 2016; Shakeel et al., 2017; Chen et al., 2022). The fungus emerges from the body of the insect to produce spores on the surface of the cuticle, which are dispersed by wind, rain, or contact with other insects (Alves, 1998). EPF might also be ingested by immature stages of mosquitoes, disrupting the peritrophic membrane and enterocytes (Mannino et al., 2019; Noskov et al., 2019; Alkhaibari et al., 2016; Bitencourt et al., 2023).

After infection by EPF, insects activate their protective mechanisms, involving melanization, and cellular and humoral immune responses. The outcome depends on various factors, including the fungal strain, insect species, life stage, environmental conditions, and route of infection. In addition to immune responses, insects have developed several mechanisms to fend off pathogens, including production of cuticular antimicrobial lipids and behavioral-environmental adaptations such as self-grooming (Jacquet et al., 2012; Ortiz-Urquiza and Keyhani, 2013). This interplay may either lead to insect death, or the host successfully eliminating the fungus. Studies have shown that EPF can also affect insect physiology, such as altering blood-feeding behavior (Scholte et al., 2006). Changes in feeding behavior are one of the first symptoms of infected hosts (Hajek and St. Leger, 1994).

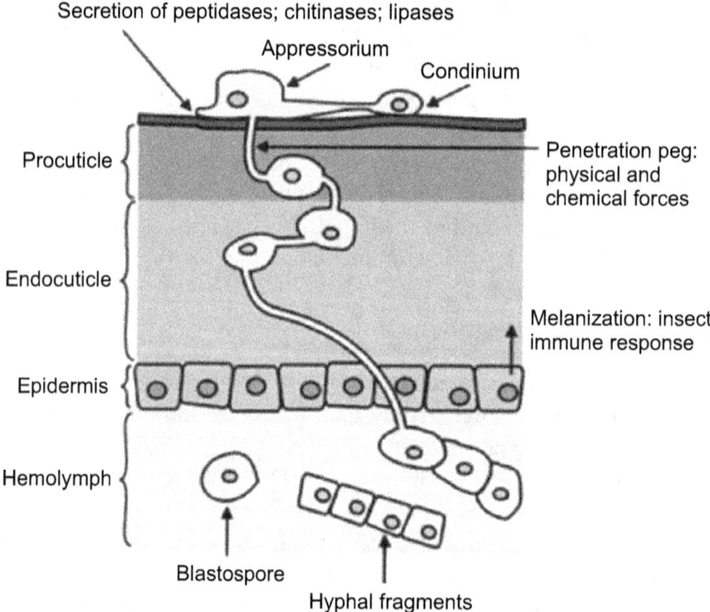

Fig. 1 Summary of initial events involved in the infection cycle of an entomopathogenic fungus attacking an insect host. From adhesion of the conidia on the cuticle surface to colonization of the hemocoel. (This figure was created by Richard Ian Samuels.)

Applications of filamentous fungi of the genera *Metarhizium* and *Beauveria* are highly studied due to their effectiveness in controlling insects (Alves, 1998). These fungi belong to the class Sordariomycetes (Ascomycota) and are found naturally in soils worldwide, where they survive for long periods (Alves, 1998; Franceschini et al., 2001). While both genera of fungi share a similar mechanism of action, as described previously, they exhibit distinct features such as conidial morphology and colony color. Furthermore, these genera encompass a wide range of fungal isolates, each varying in selectivity and virulence towards specific species of pests or vectors (Araújo and Hughes, 2016; Wang and Wang, 2017; Barrera-López et al., 2020).

The entomopathogenic activity of *M. anisopliae* (Metch.) Sorok (Hypocreales: Clavicipitaceae) and *Beauveria bassiana* (Bals.) Vuill. (Hypocreales: Cordycipitaceae) against *Aedes* sp. has been demonstrated in several studies as they can infect all stages of the mosquito life cycle, from egg to adult (Scholte et al., 2007; Albernaz et al., 2009; Darbro et al., 2012; Butt et al., 2013; Gomes et al., 2015; Ravindran et al., 2016; Ramirez et al., 2018; Carolino et al., 2019; Noskov et al., 2019; Bitencourt et al., 2023; Paula et al., 2021; Rodrigues et al., 2021). Scholte et al. (2004) reviewed the importance of a variety of fungal genera, such as *Lagenidium*, *Coelomomyces*, *Entomophthora*, and *Culicinomyces*, as potential biological control agents for mosquitoes. Evans et al. (2018) reviewed the use of a range of natural *Ae. aegypti* pathogens as biological control agents and suggested that a search of the natural habitats of this species (African forests) could yield interesting biocontrol candidates for use in South America and other regions.

3.1 Fungal Infection of Mosquito Eggs and Control Strategies in the Field

Although EPF can cause infection in both immature and mature life-stages of mosquitoes, few studies have reported their ovicidal activity. This is likely because, beyond being resistant to desiccation, the mosquito eggshell is highly resistant to infection, given that it comprises three layers: exochorion, endochorion, and serosal cuticle, all of which are rich in chitin (Farnesi et al., 2015; Flor-Weiler et al., 2017). In one of the first studies on egg infection, Clark et al. (1968) evaluated the ability of the fungus *B. bassiana* to infect *Aedes* and *Culex* eggs. *Beauveria bassiana* did not infect the eggs of these two species; sprayed with dry conidia. The results showed that 100% of the larvae hatched after 14 days of exposure to the fungus. However, Russell et al. (2001) obtained positive results in field experiments using the fungus *Penicillium citrinum* against *Ae. aegypti* eggs.

Luz et al. (2007) evaluated 21 isolates against *Ae. aegypti* eggs and found that only nine isolates, comprising *Paecilomyces* sp., *Isaria* sp., *Penicillium* sp., and *Metarhizium* sp., exhibited ovicidal activity after 25 days of fungal exposure, kept at relative humidity (RH) above 70%. Albernaz et al. (2009) also observed that *M. anisopliae*, when formulated in vegetable oil, was highly virulent against *Ae. aegypti* eggs at RH $\geq$ 98%. In another study, conidia of *M. anisopliae* formulated in water required long periods ($\geq$ 15 days) to show high ovicidal activity (Santos et al., 2009). Leles et al. (2012) demonstrated laboratory infection of *Ae. aegypti* eggs in soils inoculated with *M. anisopliae* isolate IP 46 when maintained at RH $\geq$ 98%. The results of this work showed that 53% of the eggs had external development of mycelia and conidia after incubation for 15 days. The larvae that hatched from these eggs died, and the fungus was observed on the cadavers.

In contrast, Flor-Weiler et al. (2017) reported the effect of *Tolypocladium cylindrosporum* IBT 41712 against eggs from *Ae. aegypti* and *Ae. albopictus* at different temperatures (4°C, 12°C, 15°C, 21°C, 28°C, 33°C, 37°C, and 40°C), considering that this fungus is found in cold climates. The authors observed a high infection rate of *T. cylindrosporum* on both types of mosquito eggs at 15°C. However, despite observing a successful infection, they also noted a premature eclosion of the larvae. The authors speculated that the alteration of amino acids, proteins, and phosphate salts, along with a decrease in oxygen levels, could stimulate egg hatching.

Recently, Sousa et al. (2021) demonstrated that *Metarhizium humberi* infected *Ae. aegypti* eggs, forming appressoria that ruptured the chorion, followed by hyphal colonization of the interior of the eggs. High numbers of conidia were produced on the egg surface, with the capacity to infect other hosts. Furthermore, Sousa et al. (2023) reported the susceptibility of newly deposited eggs to *M. humberi* IP 46 infection. Unfortunately, advances in the biological control of *Aedes* eggs have not reached the stage of developing EFP based traps or bioinsecticides. Moreover, some strategies require in-depth research, such as evaluating and comparing the oviposition preferences of both mosquito species in containers with water and EFP. Developing an effective application strategy against mosquito eggs presents a challenge, particularly when considering reliable and efficient delivery systems for use in the field.

3.2 Fungal Infection of Mosquito Larvae and Pupae

Many researchers have focused on EPF efficacy against *Aedes* larvae. Mosquito larvae present intriguing options for the integrated management of this disease vector. Applying bioinsecticides to containers that serve as mosquito breeding sites, can be an effective solution for mosquito control.

Clark et al. (1968) demonstrated that *B. bassiana* dry conidia sprayed on water surfaces was efficient at killing *Ae. aegypti*, *Aedes albimanus*, *Aedes sierrensis*, *Aedes nigromaculis,* and other species of mosquito larvae. They also found that *Anopheles* and *Culex* larvae were more susceptible to the fungus than *Aedes* larvae. Daoust and Roberts (1982) infected *Ae. aegypti*, *Cx. pipiens*, and *An. stephensi* larvae using different *M. anisopliae* isolates. They observed that the larvae died after coming into contact with the fungus through the respiratory siphon.

Pereira et al. (2009) evaluated, in the laboratory, the entomopathogenic activity of eight *M. anisopliae* isolates and two *B. bassiana* isolates against *Ae. aegypti* larvae. It was observed that two isolates of *M. anisopliae* (ESALQ 818 and CG 144) and one isolate of *B. bassiana* (CG 24) were the most virulent, resulting in high mortality rates. They also observed that 20% of the larvae exposed to the fungi, which subsequently developed into pupae, did not complete their development cycle to become adults, indicating that even following metamorphosis, the infection continued to develop inside the insects' bodies.

Metarhizium anisopliae conidia attached to rice were also used in virulence experiments against *Ae. aegypti* larvae under laboratory conditions. Paula et al. (2013a) observed that the application of 20 grains of rice with *M. anisopliae* conidia to water containers with larvae reduced the survival rate to 5%, whereas the application of 10 grains of rice with fungus resulted in 52% survival.

A study indicated that *M. anisopliae* did not adhere to the mosquito larval cuticle and that ingested conidia germinate within the insect gut, being expelled in fecal pellets (Butt et al., 2013). The results of this study by Butt et al. (2013) indicated that *M. anisopliae* conidia caused stress-induced mortality, a very different mode of action to that normally seen in terrestrial hosts. Interestingly, Alkhaibari et al. (2016) reported blastospores penetrating the midgut peritrophic membrane. Furthermore, Noskov et al. (2019) reported similar results when exposing *Ae. aegypti* larvae to *Metarhizium robertsii* conidia combined with avermectins. Recently, disrupted enterocytes were observed in larvae exposed to both *M. anisopliae* and *B. bassiana* propagules (Bitencourt et al., 2022, 2023), and these authors suggested that toxins could have caused this response, but that in-depth studies need to be conducted.

Recent work has shown that the yeast-like forms of EPF, analogs of hyphal bodies, named blastospores, are highly virulent to mosquito larvae (Alkhaibari et al., 2016, 2017). Blastospores are thin-walled, pleomorphic hydrophilic spores, which can be produced using liquid fermentation. Blastospores were found to be more virulent than conidia to mosquito larvae. The fact that blastospores have multiple routes of entry (cuticle and gut) may explain why this form of the inoculum killed *Ae. aegypti* larvae in a relatively short time (12-24 hrs), significantly quicker than when larvae were exposed to conidia. However, considerable discussion has ensued regarding the challenges associated with introducing blastospores as bioinsecticides in the field. These difficulties stem from their susceptibility to and limited persistence in adverse environmental conditions, such as elevated temperatures, UV radiation, or desiccation (Mascarin et al., 2018; Corval et al., 2021). Corval et al. (2021) investigated the varying UV-B tolerance among EFP isolates and found that *M. anisopliae* LCM S05 exhibited UV-B tolerance as blastospores and microsclerotia, but not in the form of conidia. In contrast, *M. anisopliae* LCM S10 and LCM S08 demonstrated heightened UV-B tolerance of microsclerotia or conidia, though not in the form of blastospores.

Bernardo et al. (2020) conducted a study comparing the susceptibility to heat and UV-B radiation of blastospores and conidia of *M. robertsii* IP 146, *M. anisopliae* s.l. IP 363, *Metarhizium acridum* ARSEF 324, and *B. bassiana* s.l. IP 361 and CG 307. The results showed that variations in susceptibility depended on the isolate. For example, IP 146 and IP 363 conidia and blastospores were similarly tolerant to UV-B radiation, and that CG 307 blastospores manifested superior UV-B tolerance.

Field and semi-field studies on the control of *Aedes* larvae have been carried out (Gomes et al., 2023; Mendonça et al., 2023). Gomes et al. (2023) showed the virulence and persistence of *M. anisopliae* blastospores against *Ae. aegypti* larvae under field conditions. This represents a significant advance in terms of bioprospecting. Entomopathogenic fungi naturally inhabit the soil and rhizosphere of plants, yet they do not occur naturally in aquatic environments. While studies have reported the larvicidal activity of these fungi, it is imperative for researchers to increasingly focus on developing formulations that enhance the persistence and virulence of these fungi in aquatic settings. Some studies have explored associations of blastospores and conidia with additional agents such as mineral oil (Bitencourt et al., 2021b), vegetable oils (Gomes et al., 2015; Paula et al., 2019), essential oils (Bitencourt et al., 2022), and synthetic insecticides (Noskov et al., 2019). Furthermore, technologies such as the application of microsclerotia have been pursued to improve fungal virulence, increase resistance, and ensure prolonged persistence, particularly in aquatic environments, for the effective control of mosquito larvae (Paixão et al., 2024). Therefore, selecting the appropriate form of inoculum is important for the efficacious control of disease vectors such as *Ae. aegypti*.

When considering mass production, Mascarin et al. (2018) demonstrated new approaches to increase yields. This enhancement occurred following the addition of various nitrogen sources, such as soy flour, autolyzed yeast, and cottonseed flour, which have a crucial role in industrial fermentation. Hence, this study could prove instrumental in the development of EPF-based products targeting mosquito larvae or other life stages.

Pupae represent a challenge to control because, unlike larvae, this immature stage is incapable of feeding due to the lack of mouthparts. Furthermore, pupae are normally more resistant to biological control measures due to the short duration of this phase of the life cycle; therefore, the infection process needs to be very quick and efficient. Only one study reported the pupicidal effect of EPF. Carolino et al. (2019) demonstrated the infection of *Ae. aegypti* pupae by propagules of the fungus *M. anisopliae*. They observed that the application of blastospores to water resulted in 100% mortality following a 24 h exposure. In contrast, conidia of the same fungus were significantly less efficient at killing the pupae, and many of them developed into larvae. By killing the immature forms of this vector (eggs, larvae, and pupae), it is possible to reduce the population of the adults, lowering the vectorial capacity of this insect.

3.3 Fungal Infection of Adult Mosquitoes

The utilization of EPF against adult mosquitoes has been explored as an eco-friendly alternative to insecticide application (Scholte et al., 2004; Blanford et al., 2005; Farenhorst et al., 2008). As mentioned earlier, EPF naturally infect terrestrial insects through the cuticle, presenting an advantageous approach for controlling adult mosquitoes. This natural infection process might allow the fungus to express its full infectious potential in a way that aligns with the mosquito's natural environment.

In-vitro and field tests have already been conducted against adult mosquitoes. Clark et al. (1968), in laboratory tests, verified that dry conidia of *B. bassiana* caused 100% mortality in adult *Culex tarsalis*, *Cx. pipiens*, *Ae. aegypti*, *Aedes sierrensis*, *Aedes nigromaculis*, and *Aedes albimanusem*, five days after fungal exposure. In 1992, Cuebas-Incle reported, for the first time, the virulence of the entomopathogenic fungus *Erynia conica* against adults of *Ae. aegypti*. Recent articles reported the adulticidal activity of *Culicinomyces* spp. (Rodrigues et al., 2018), *Trichoderma* sp., and *Aspergillus* sp. (Cisneros-Vázquez et al., 2023). Interestingly, in 2022, Rodrigues and coworkers reported the natural occurrence of the entomopathogenic fungus *Clonostachys* spp. (Hypocreales: Bionectriaceae) in *Ae. aegypti*. Following re-isolation, the authors conducted bioassays, confirming not only the fungus's infective capacity against adults but also its effectiveness in targeting *Ae. aegypti* eggs and larvae. Although these researchers investigated the virulence of other fungal species against mosquitoes, the majority of studies have focused on *Ae. aegypti* and *Ae. albopictus* adults using *Metarhizium* sp. and *Beauveria* sp. (Leles et al., 2010; Lee et al., 2019; Paula et al., 2011, 2018, 2021; Mehmood et al., 2023). Scholte et al. (2007) evaluated that *M. anisopliae* was highly virulent against male and female *Ae. aegypti* and *Ae. albopictus*. The results showed that, on average, 87% of *Ae. aegypti* and 89% of *Ae. albopictus* became infected with the fungus and died. The virulence of different *B. bassiana* and *M. anisopliae* isolates was evaluated in the laboratory by Paula et al. (2008) against female *Ae. aegypti*.

Carolino et al. (2014) tested the persistence of *M. anisopliae* conidia, impregnated on black cloths, when formulated with vegetable and synthetic oil (isoparaffin) against female *Ae. aegypti*. They found that the conidia remained viable, causing 28 to 60% reductions in mosquito survival even when the cloths were kept on a covered veranda for up to 18 days under natural conditions, but without being exposed to direct sunlight, thus reducing the need for frequent changes of black cloths when deployed in residences. Lobo et al. (2015) showed that *M. anisopliae* (s.l. IP 46) suspended in low concentrations of vegetable oil did not repel adult *Ae. aegypti* when tested in oviposition devices, and that larvae hatching from eggs laid on fungus-impregnated filter paper subsequently died.

Experiments involving the combination of EPF and other agents have also been conducted. According to Paula et al. (2011, 2013b), the combination of *M. anisopliae* with a low concentration of the insecticide imidacloprid (IMI) significantly reduced the survival of *Ae. aegypti* females fed on sucrose or blood. The use of insecticides generally leads to an increase in the frequency of resistant mosquitoes (Munywoki et al., 2021; Juarez et al., 2021), as mosquito resistance is typically associated with the induction of detoxification enzymes (Samal et al., 2022).

Rodrigues et al. (2021) tested a granular formulation of *M. humberi* microsclerotia *in-vitro* and achieved promising results when infecting adult *Ae. aegypti*. EPF can also be combined with chemical insecticides, showing a synergistic effect against certain insects (Ismail et al., 2020; Salem et al., 2023).

Blastospores are not only virulent to mosquito larvae but were recently shown to kill adult *Ae. aegypti* (Paula et al., 2021). Results showed that blastospores were more virulent to adult female *Ae. aegypti* than conidia when sprayed onto the insects or applied to black cloths. However, the persistence of blastopores was much shorter than that of conidia. When using blastospores in vegetable oil emulsions, it was possible to maintain virulence for longer periods.

Experiments were conducted to verify behavioral changes following exposure to EPF. Darbro et al. (2012) observed an 80% reduction in blood feeding in *Ae. aegypti* exposed to fungus in semi-field tests. In experiments using an acrylic observation chamber, Paula et al. (2013c) found that black cloths impregnated with the fungus *M. anisopliae* were not repellant to *Ae. aegypti* mosquitoes.

4. Strategies for Deploying Entomopathogenic Fungi for Mosquito Control

The use of fungal pathogens to infect agricultural pests in the field requires either direct spraying or treating surfaces (Alves, 1998). A distinct advantage of EPF is that it can be combined with other control strategies, such as synthetic insecticides (Paula et al., 2011).

When considering the use of EPFs as vector control agents, different approaches are needed to those commonly used for agricultural pest control; it is probably not viable to carry out direct spraying techniques of adult vector insects. However, applying fungi to surfaces to which the target vector will contact is a highly promising strategy (Paula et al., 2008; Silva et al., 2017; Paula et al., 2018). Using visually attractive objects or odors, could be an effective approach .

Studies have shown that certain materials or containers impregnated with fungus were attractive and efficient in reducing mosquito survival rates; black cotton cloths, curtains, clay pots, amongst others (Scholte et al., 2006; Paula et al., 2008; Lwetoijera et al., 2010; Mmbando et al., 2015; Silva et al., 2017). Black cloths impregnated with different *M. anisopliae* and *B. bassiana* isolates, formulated with sunflower oil and hung in cages, were efficient in attracting mosquitoes and decreasing the survival rate of *Ae. aegypti* females. In this case, high levels of mosquito infection were obtained, with a 70% reduction in survival rates (Paula et al., 2008).

Silva et al. (2017) developed a trap made from a transparent 2 L PET bottle containing a black cotton cloth impregnated with conidia, that was tested against *Ae. aegypti* adults. It was found that exposure of mosquitoes to black cloths impregnated with *M. anisopliae* or *B. bassiana* significantly reduced the survival rate of *Ae. aegypti* females under laboratory conditions. Combining a synthetic attractant (Atr*Aedes*®), placed in the PET traps together with the fungus-impregnated black cloths, increased the effectiveness of the trap, with a further reduction in the survival of *Ae. aegypti* females observed when tested under semi-field conditions (Paula et al., 2018). Martinez and co-workers (2021) recently published a study showing the potential of an oviposition device impregnated with *M. humberi* conidia for the control of adult *Ae. aegypti*. This device attracted both male and female *Ae. aegypti* in laboratory, semi-field, and field settings.

A trap that combines an EPF, insecticide, and attractive bait is currently being marketed in more than 25 countries by "In2Care" (http://www.in2care.org/). Mosquitoes are attracted to the trap by a bait (yeast tablets releasing CO_2), become infected with *B. bassiana*, and they are also contaminated with a synthetic larvicide (pyriproxyfen), which is highly efficient in reducing the survival of larvae and adult *Ae. aegypti* (Snetselaar et al., 2014). Buckner et al. (2017) concluded that the presence of an oviposition attractant in the In2Care trap increased its efficiency for *Ae. aegypti* and *Cx. quinquefasciatus* gravid females and any larvae that hatched from eggs laid in the trap were killed by pyriproxyfen. However, as stated earlier, the continued use of insecticides results in the development of resistant mosquitoes (Schaefer and Mulligan, 1991).

Fungus-impregnated ovitraps could be deployed to control more than one life stage of mosquitoes. These ovitraps could be used to infect eggs laid in the traps, larvae that subsequently hatch, and adults that land to rest or lay their eggs.

Recently, in Brazil, a trap named "Mata*Aedes*" has been developed utilizing *M. anisopliae* for the control of adult *Ae. aegypti* and *C. quinquefasciatus* in indoor environments. The trap is currently in the registration phase and undergoing field validation before being commercialized (www.mataaedes.com.br).

Conclusion and Prospects

Integrated vector management strategies, which include elimination of breeding sites and the use of synthetic and biological control methods, are crucial for efficiently reducing mosquito populations. The use of chemical insecticides saves thousands of lives every year, however, increased levels of insecticide resistance and the toxic effects non-target organisms demonstrate the need for alternative methods. Biocontrol approaches employing EPFs are viable and promising alternatives that can be incorporated into mosquito control strategies. The ability of EPF to kill insects by exploiting several routes of infection, such as the cuticle, and digestive and respiratory tracts, offers a substantial advantage when targeting all life stages of mosquitoes. To advance the future development of biological control agents based on entomopathogenic fungi, it is essential to focus on the selection of specific fungi or isolates with high efficacy for mosquito control. Furthermore, it is important to understand the pathogenic mechanisms and identify factors that influence EPF effectiveness at each stage of infection. We propose conducting in-depth studies on the utilization of entomopathogenic fungi in combination with attractants within traps for the control of adult mosquitoes. Additionally, technologies for formulation development, including encapsulation and combination with other agents, should be further investigated to create larvicidal, pupicidal, and ovicidal products, further contributing to the reduction of adult mosquito populations in the environment. Finally, the scientific community and governments need to consider improving financial resources for developing in-depth studies on EPF biopesticides to control mosquitoes; a serious public health concern. Governments also need to facilitate the registration and commercialization of innovative, technology-based mosquito control products that use entomopathogenic fungi. This would enable the adoption of these technologies for widespread use.

References

Achee, N.L., Gould, F., Perkins, T., Reiner, R.C., Morrison, A.C. et al. (2015). A critical assessment of vector control for dengue prevention. PLoS Negl. Trop. Dis., 5: e0003655.

Afanasiev, B.N., Galyov, E.E., Buchatsky, L.P. and Kozlov, Y.V. (1991). Nucleotide sequence and genomic organization of *Aedes* densonucleosis virus. Virology, 185: 323–36.

Afonso, C.L., Tulman, E.R., Lu, Z., Balinsky, C.A., Moser, B.A. et al. (2001). Genome sequence of a baculovirus pathogenic for *Culex nigripalpus*. J. Virol., 75: 11157–65.

Ainsworth, C. (2023). Tropical diseases move north. Nature. doi: 10.1038/d41586-023-03476-7. Online ahead of print.

Albernaz, D.A.S., Tai, M.H.H. and Luz, C. (2009). Enhanced ovicidal activity of an oil formulation of the fungus *Metarhizium anisopliae* on the mosquito *Aedes aegypti.* Med. Vet. Entomol., 23: 141–147.

Alkhaibari, A.M., Carolino, A.T., Bull, J.C., Samuels, R.I. and Butt, T.M. (2017). Differential pathogenicity of *Metarhizium blastospores* and conidia against larvae of three mosquito species. J. Med. Entomol., 54: 696–704.

Alkhaibari, A.M., Carolino, A.T., Yavasoglu, S.I., Maffeis, T., Mattoso, T.C. et al. (2016). *Metarhizium brunneum* blastospore pathogenesis in *Aedes aegypti* larvae: attack on several fronts accelerates mortality. PLoS Pathog., 12: e1005715.

Alkhaibari, A.M., Lord, A.M., Maffeis, T., Bull, J.C., Olivares, F.L. et al. (2018a). Highly specific host-pathogen interactions influence *Metarhizium brunneum* blastospore virulence against *Culex quinquefasciatus* larvae. Virulence, 9: 1449–1467.

Alkhaibari, A.M., Maffeis, T., Bull, J.C. and Butt, T.M. (2018b). Combined use of the entomopathogenic fungus, *Metarhizium brunneum*, and the mosquito predator, *Toxorhynchites brevipalpis*, for control of mosquito larvae: Is this a risky biocontrol strategy. J. Invertebr. Pathol., 153: 38–50.

Alves, S.B. (1998). Controle microbiano de insetos. Fundação de Estudos Agrários Luiz de Queiroz. 1163.

Araújo, J.P. and Hughes, D.P. (2016). Diversity of entomopathogenic fungi: which groups conquered the insect body? Adv. Genet., 94: 1–39.

Barrera-López, A.A., Guzmán-Franco, A.W., Santillán-Galicia, M., Tamayo-Mejía, F., Bujanos-Muñiz, R. et al. (2020). Differential susceptibility of *Bagrada hilaris* (Hemiptera: Pentatomidae) to different species of fungal pathogens. J. Econ. Entomol., 113: 50–54.

Barrett, A.D. (2017). Yellow fever live attenuated vaccine: a very successful live attenuated vaccine but still we have problems controlling the disease. Vaccine, 35: 5951–5955.

Batool, K., Xiao, J., Xu, Y., Yang, T., Tao, P. et al. (2022). Densovirus oil suspension significantly improves the efficacy and duration of larvicidal activity against *Aedes albopictus*. Viruses, 14: 475.

Becker, N., Petric, D., Zgomba, M., Boase, C., Madon, M. et al. (2010). Mosquitoes and their control. Springer Science & Business Media, pp 577.

Becnel, J.J., White, S.E., Moser, B.A., Fukuda, T., Rotstein, M.J. et al. (2001). Epizootiology and transmission of a newly discovered baculovirus from the mosquitoes *Culex nigripalpus* and *C. quinquefasciatus*. J. Gen. Virol., 82: 275–282.

Benedict, M.Q., Levine, R.S., Hawley, W.A. and Lounibos, L.P. (2007). Spread of the tiger: global risk of invasion by the mosquito *Aedes albopictus*. Vector Borne Zoonotic Dis., 7: 76–85.

Benelli, G., Wilke, A.B.B. and Beier, J.C. (2020). *Aedes albopictus* (Asian Tiger Mosquito). Trends Parasitol. 36: 942–943.

Bennett, K.L., Gómez-Martínez, C., Chin, Y., Saltonstall, K., McMillan, W.O. et al. (2019). Dynamics and diversity of bacteria associated with the disease vectors *Aedes aegypti* and *Aedes albopictus*. Sci. Rep., 9: 1–12.

Bernardo, C.D.C., Pereira-Junior, R.A., Luz, C., Mascarin, G.M. and Kamp Fernandes, É.K. (2020). Differential susceptibility of blastospores and aerial conidia of entomopathogenic fungi to heat and UV-B stresses. Fungal Biol, 124: 714–722.

Bilgo, E., Lovett, B., Leger, R.J.S., Sanon, A., Dabiré, R.K. et al. (2018). Native entomopathogenic *Metarhizium spp.* from Burkina Faso and their virulence against the malaria vector *Anopheles coluzzii* and non-target insects. Parasit. Vectors, 11: 209.

Bitencourt, R.O.B., de Souza Faria, F., Marchesini, P., Reis Dos Santos-Mallet, J., Guedes Camargo, M. at al. (2022). Entomopathogenic fungi and *Schinus molle* essential oil: The combination of two eco-friendly agents against *Aedes aegypti* larvae. J. Invertebr. Pathol., 194: 107827.

Bitencourt, R.O.B., Reis dos Santos Mallet, J., Mesquita, E., Silva Gôlo, P., Fiorotti, J. et al. (2021b). Larvicidal activity, route of interaction and ultrastructural changes in *Aedes aegypti* exposed to entomopathogenic fungi. Acta Trop., 213: 105732.

Bitencourt, R.O.B., Salcedo-Porras, N., Umaña-Diaz, C., da Costa Angelo, I. and Lowenberger, C. (2021a). Antifungal immune responses in mosquitoes (Diptera: Culicidae): A review. J. Invertebr. Pathol., 178: 107505.

Bitencourt, R.O.B., Santos-Mallet, J.R.D., Lowenberger, C., Ventura, A., Gôlo, P.S. et al. (2023). A novel model of pathogenesis of *Metarhizium anisopliae* propagules through the midguts of *Aedes aegypti* larvae. Insects, 14: 328.

Blanford, S., Chan, B.H., Jenkins, N., Sim, D., Turner, R.J. et al. (2005). Fungal pathogen reduces potential for malaria transmission. Science, 308: 1638–1641.

Brühl, C.A., Després, L., Frör, O., Patil, C.D., Poulin, B. et al. (2020). Environmental and socioeconomic effects of mosquito control in Europe using the biocide *Bacillus thuringiensis* subsp. *israelensis* (Bti). Sci. Total Environ., 724: 137800.

Buckner, E.A., Williams, K.F., Marsicano, A.L., Latham, M.D. and Lesser, C.R. (2017). Evaluating the vector control potential of the In2Care® mosquito trap against *Aedes aegypti* and *Aedes albopictus* under semifield conditions in Manatee County, Florida. J. Am. Mosq. Control Assoc., 33: 193–199.

Butt, T.M., Coates, C.J., Dubovskiy, I.M. and Ratcliffe, N.A. (2016). Entomopathogenic fungi: new insights into host-pathogen interactions. Adv. Genet., 94: 307–364.

Butt, T.M., Greenfield, B.J., Greig, C., Maffeis, T.G.G. and Taylor, J.W.D. (2013). *Metarhizium anisopliae* pathogenesis of mosquito larvae: a verdict of accidental death. PLoS One, 8: e81686.

Cardoso, D.O., Gomes, V.M., Dolinski, C. and Souza, R.M. (2015). Potential of entomopathogenic nematodes as biocontrol agents of immature stages of *Aedes aegypti*. Nematoda, Vol. 2.

Carolino, A.T., Gomes, S.A., Pontes Teodoro, T.B., Mattoso, T.C. and Samuels, R.I. (2019). *Aedes aegypti* pupae are highly susceptible to infection by *Metarhizium anisopliae* blastospores. J. Pure Appl. Microbiol., 13: 1629–1634.

Carolino, A.T., Paula, A.R., Silva, C.P., Butt, T.M. and Samuels, R.I. (2014). Monitoring persistence of the entomopathogenic fungus *Metarhizium anisopliae* under simulated field conditions with the aim of controlling adult *Aedes aegypti* (Diptera: Culicidae). Parasit. Vectors, 7: 198.

Chen, S., Cheng, L., Zhang, Q., Lin, W., Lu, X. et al. (2004). Genetic, biochemical, and structural characterization of a new densovirus isolated from a chronically infected *Aedes albopictus* C6/36 cell line. Virology, 318: 123–33.

Chen, X., Zhang, W., Wang, J., Zhu, S., Shen, X. et al. (2022). Transcription factors BbPacC and Bbmsn2 jointly regulate oosporein production in *Beauveria bassiana*. Microbiol-Spectr., 10: e0311822.

Cisneros-Vázquez, L.A., Penilla-Navarro, R.P., Rodríguez, A.D., Ordóñez-González, J.G., Valdez-Delgado, K.M. et al. (2023). Entomopathogenic fungi for the control of larvae and adults of *Aedes aegypti* (Diptera: Culicidae) vector of dengue, chikungunya and Zika viruses in Mexico. Salud Publica Mex., 65: 144–150.

Clark, T.B., Chapman, H.C. and Fukuda, T. (1969). Nuclear-polyhedrosis and cytoplasmic-polyhedrosis virus infections in Louisiana mosquitoes. J. Invertebr. Pathol., 14: 284–286.

Clark, T.B., Kellen, W.R., Fukuda, T. and Lindegren, J.E. (1968). Field and laboratory studies on the pathogenicity of the fungus *Beauveria bassiana* to three genera of mosquitoes. J. Invertebr. Pathol., 11: 1–7.

Collins, L.E. and Blackwell, A. (2000). The biology of *Toxorhynchites* mosquitoes and their potential as biocontrol agents. Biocontrol News and Information, 21: 105N–116N.

Consoli, R.A.G.B. and Oliveira, R.L. (1998). Principais mosquitos de importância sanitária no Brasil, Publisher: Editora Fiocruz, Brazil 1–228.

Corval, A.R.C., Mesquita, E., Corrêa, T.A., Silva, C.S.R., Bitencourt, R.O.B. et al. (2021). UV-B tolerances of conidia, blastospores, and microsclerotia of *Metarhizium* spp. entomopathogenic fungi. J. Basic. Microbiol., 61: 15–26.

Costa, J.R., Rossi, J.R., Marucci, S.C., Alves, E.C.D.C., Volpe, H.X. et al. (2010). Atividade tóxica de isolados de *Bacillus thuringiensis* a larvas de *Aedes aegypti* (L.) (Diptera: Culicidae). Neotrop. Entomol., 39: 757–766.

Cuebas-Incle, E.L. (1992) Infection of adult mosquitoes by the entomopathogenic fungus *Erynia conica* (Entomophthorales: Entomophthoraceae). J. Am. Mosq. Control. Assoc., 8: 367–71.

da Silva, R., Luiz, J., Schwalm, F.U., Silva, C.E., da Costa, M. et al. (2017). Larvicidal and growth Inhibitory activity of entomopathogenic bacteria culture fluids against *Aedes aegypti*. J. Econ. Entomol., 110.2: 378–85.

Daoust, R.A., and Roberts, D.W. (1982). Virulence of natural and insect-passaged strains of *Metarhizium anisopliae* to mosquito larvae. J. Invertebr. Pathol., 40: 107–117.

Darbro, J.M., Johnson, P.H., Thomas, M.B., Ritchie, S.A., Kay, B.H. et al. (2012). Effects of *Beauveria bassiana* on survival, blood-feeding success, and fecundity of *Aedes aegypti* in laboratory and semi-field conditions. Am. J. Trop. Med. Hyg., 86(4): 656–64.

de Sousa, N.A., Rodrigues, J., Arruda, W., Humber, R.A. and Luz, C. (2021). Development of *Metarhizium humberi* in *Aedes aegypti* eggs. J. Invertebr. Pathol., 184: 107648.

Dickens, B.L., Sun, H., Jit, M., Cook, A.R. and Carrasco, L.R. (2018). Determining environmental and anthropogenic factors which explain the global distribution of *Aedes aegypti* and *Ae. albopictus*. BMJ Global Health, 3: e000801.

Dieme, C., Ciota, A.T. and Kramer, L.D. (2020). Transmission potential of Mayaro virus by *Aedes albopictus*, and *Anopheles quadrimaculatus* from the USA. Parasit. Vectors, 13: 613.

Domingues, M.M., Santos, P.L.D., Gêa, B.C.C., Carvalho, V.R., Zanuncio, J.C. et al. (2023). Diversity of entomopathogenic fungi from soils of eucalyptus and soybean crops and natural forest areas. Braz. J. Biol., 6: 82:e263240.

Evans, H.C., Elliot, S.L. and Barreto, R.W. (2018). Entomopathogenic fungi and their potential for the management of *Aedes aegypti* (Diptera: Culicidae) in the Americas. Mem. Inst. Oswaldo Cruz, 3: 113.

Farenhorst, M., Farina, D., Scholte, E.J., Takken, W., Hunt, R.H. et al. (2008). African water storage pots for the delivery of the entomopathogenic fungus *Metarhizium anisopliae* to the malaria vectors *Anopheles gambiae ss* and *Anopheles funestus*. Am. J. Trop. Med. Hyg., 78: 910–916.

Farnesi, L.C., Menna-Barreto, R.F., Martins, A.J., Valle, D. and Rezende, G.L. (2015). Physical features and chitin content of eggs from the mosquito vectors *Aedes aegypti*, *Anopheles aquasalis* and *Culex quinquefasciatus*: Connection with distinct levels of resistance to desiccation. J. Insect Physiol., 83: 43–52.

Ferguson, N.M., Rodríguez-Barraquer, I., Dorigatti, I., Mier-y-Teran-Romero, L., Laydon, D.J. (2016). Benefits and risks of the Sanofi-Pasteur dengue vaccine: Modeling optimal deployment. Science, 353: 1033–1036.

Ffrench-Constant, R.H. (2005). Something old, something transgenic, or something fungal for mosquito control? Trends Ecol. Evol., 20: 577–579.

Flacio, E., Engeler, L., Tonolla, M., Lüthy, P. and Patocchi, N. (2015). Strategies of a thirteen year surveillance program on *Aedes albopictus* (*Stegomyia albopicta*) in southern Switzerland. Parasit. Vectors, 8: 208.

Flor-Weiler, L.B., Rooney, A.P., Behle, R.W. and Muturi, E.J. (2017). Characterization of *Tolypocladium cylindrosporum* (Hypocreales: *Ophiocordycipitaceae*) and its impact against *Aedes aegypti* and *Aedes albopictus* eggs at low temperature. J. Am. Mosq. Control. Assoc., 33(3): 184–192.

Franceschini, M., Guimarães, A.P., Camassola, M., Frazzon, A.P., Baratto, C.M. et al. (2001). Biotecnologia aplicada ao controle biológico. Biotecnol. Ciênc. Desenvolv., 31: 32–37.

Gomes, S.A., Carolino, A.T., Teodoro, T.B.P., Silva, G.A., Bitencourt, R.O.B. et al. (2023). The potential of *Metarhizium anisopliae* blastospores to control *Aedes aegypti* larvae in the field. J. Fungi, 9: 759.

Gomes, S.A., Paula, A.R., Ribeiro, A., Moraes, C.O., Santos, J.W. (2015). Neem oil increases the efficiency of the entomopathogenic fungus *Metarhizium anisopliae* for the control of *Aedes aegypti (*Diptera: Culicidae) larvae. Parasit. Vectors, 8: 669.

Hajek, A.E. and St. Leger, R.J. (1994). Interactions between fungal pathogens and insect hosts. Annu. Rev. Entomol., 39: 293–322.

Horstick, O., Runge-Ranzinger, S., Nathan, M.B. and Kroeger, A. (2010). Dengue vector-control services: How Do They Work? A Systematic Literature Review and Country Case Studies. Trans. R. Soc. Trop. Med. Hyg. 104: 379–386.

Hu, W., He, G., Wang, J. and Hu, Q. (2016). The effects of destruxin A on relish and rel rene regulation to the suspected immune-related genes of Silkworm. Molecules, 22: 41.

Huang, E.Y., Wong, A.Y., Lee, I.H., Qu, Z., Yip, H.Y. et al. (2020). Infection patterns of dengue, Zika and endosymbiont *Wolbachia* in the mosquito *Aedes albopictus* in Hong Kong. Parasit. Vectors, 13: 361.

Huang, Y.J.S., Higgs, S. and Vanlandingham, D.L. (2017). Biological control strategies for mosquito vectors of arboviruses. Insects, 8: 21.

Hugo, L.E., Rašić, G., Maynard, A.J., Ambrose, L., Liddington, C. et al. (2022). *Wolbachia* wAlbB inhibit dengue and Zika infection in the mosquito *Aedes aegypti* with an Australian background. PLoS Negl. Trop. Dis., 16: e0010786.

Ismail, H.M., Freed, S., Naeem, A., Malik, S. and Ali, N. (2020). The effect of entomopathogenic fungi on enzymatic activity in chlorpyrifos-resistant mosquitoes, *Culex quinquefasciatus* (Diptera: Culicidae). J. Med. Entomol, 57: 204–213.

Jacquet, M., Lebon, C., Lemperiere, G. and Boyer, S. (2012). Behavioural functions of grooming in male *Aedes albopictus* (Diptera: Culicidae), the Asian tiger mosquito. Appl. Entomol. Zool., 47: 359–363.

Johnson, B.J., Ritchie, S.A. and Fonseca, D.M. (2017). The state of the art of lethal oviposition trap-based mass interventions for arboviral control. Insects, 8: 5.

Jones, K.E., Patel, N.G., Levy, M.A., Storeygard, A., Balk, D. et al. (2008). Global trends in emerging infectious diseases. Nature, 451: 990–993.

Juarez, J.G., Garcia-Luna, S.M., Roundy, C.M., Branca, A., Banfield, M.G. et al. (2021). Susceptibility of south Texas *Aedes aegypti* to pyriproxyfen. Insects, 12: 460.

Kantor, A.M., Lin, J., Wang, A., Thompson, D.C. and Franz, A.W.E. (2019). Infection pattern of Mayaro virus in *Aedes aegypti* (Diptera: Culicidae) and transmission potential of the virus in mixed infections with Chikungunya virus. J. Med. Entomol., 56: 832–843.

Kirsch, J.M. and Tay, J.W. (2022). Larval mortality and ovipositional preference in Aedes albopictus (Diptera: *Culicidae*) induced by the entomopathogenic fungus *Beauveria bassiana* (Hypocreales: *Cordycipitaceae*). J. Med. Entomol., 59: 1687–1693.

Kittayapong, P., Baisley, K.J., Baimai, V. and O'Neill, S.L. (2000). Distribution and diversity of *Wolbachia* infections in Southeast Asian mosquitoes (Diptera: Culicidae). J. Med. Entomol., 37: 340–345.

Kost, T.A., Condreay, J.P. and Jarvis, D.L. (2005). Baculovirus as versatile vectors for protein expression in insect and mammalian cells. Nat. Biotechnol., 23: 567–575.

Kraemer, M.U., Sinka, M.E., Duda, K.A., Mylne, A.Q.N., Shearer, F.M. et al. (2015). The global distribution of the arbovirus vectors *Aedes aegypti* and *Ae. albopictus*. eLife, 4: e08347.

Lacey, L.A. (2007). *Bacillus thuringiensis* serovariety *israelensis* and *Bacillus sphaericus* for mosquito control. J. Am. Mosq. Control. Assoc., 23: 133–163.

Laporta, G.Z., Potter, A.M., Oliveira, J.F.A., Bourke, B.P., Pecor, D.B. et al. (2023). Global distribution of *Aedes aegypti* and *Aedes albopictus* in a climate change scenario of regional rivalry. Insects, 14: 49.

Lee, J.Y., Woo, R.M., Choi, C.J., Shin, T.Y., Gwak, W.S. et al. (2019). *Beauveria bassiana* for the simultaneous control of *Aedes albopictus* and *Culex pipiens* mosquito adults shows high conidia persistence and productivity. AMB Express, 9: 206.

Leles, R.N., D'Alessandro, W.B. and Luz, C. (2012). Effects of *Metarhizium anisopliae* conidia mixed with soil against the eggs of *Aedes aegypti*. Parasitol. Res., 110: 1579–1582.

Leles, R.N., Sousa, N.A., Rocha, L.F., Santos, A.H., Silva, H.H. and (2010). Pathogenicity of some hypocrealean fungi to adult *Aedes aegypti* (Diptera: *Culicidae*). Parasitol. Res., 107: 1271–4.

Li, J., Dong, Y., Sun, Y., Lai, Z., Zhao, Y. et al. (2019). A novel densovirus isolated from the Asian tiger mosquito displays varied pathogenicity depending on its host species. Front. Microbiol., 10: 1549.

Lima, A., Lovin, D.D., Hickner, P.V. and Severson, D.W. (2016). Evidence for an overwintering population of *Aedes aegypti* in Capitol Hill neighborhood, Washington, DC. Am. J. Trop. Med. Hyg., 94: 231.

Liu, J., Ling, Z., Wang, J., Xiang, T., Xu, L. et al. (2021). In vitro transcriptomes analysis identifies some special genes involved in pathogenicity difference of the *Beauveria bassiana* against different insect hosts. Microb. Pathog., 154: 104824.

Liu, W.T., Chen, T.L., Hou, R.F., Chen, C.C. and Tu, W.C. (2020). The invasion and encapsulation of the entomopathogenic nematode, *Steinernema abbasi*, in *Aedes albopictus* (Diptera: Culicidae) larvae. Insects, 11: 832.

Lobo, L.S., Rodrigues, J. and Luz, C. (2015). Effectiveness of *Metarhizium anisopliae* formulations against dengue vectors under laboratory and field conditions. Biocontrol Sci. Technol., 26: 386–401.

Lutinski, J.A., Kuczmainski, A.G., de Quadros, S., Busato, M.A., Weirich, C.M.M. and (2017) *Bacillus thuringiensis* var. *israelensis* como alternativa para o controle populacional de *Aedes aegypti* (Linnaeus, 1762) (Diptera: Culicidae). Ciência e Natura, 39: 211–220.

Luz, C., Tai, M.H.H., Santos, A.H., Rocha, L.F.N., Albernaz, D.A.S. et al. (2007). Ovicidal activity of entomopathogenic hyphomycetes on *Aedes aegypti* (Diptera: Culicidae) under laboratory conditions. J. Med. Entomol., 44: 799–804.

Lwetoijera, D.W., Sumaye, R.D., Madumla, E.P., Kavishe, D.R., Mnyone, L.L. et al. (2010). An extra-domiciliary method of delivering entomopathogenic fungus, *Metarhizium anisopliae* IP 46 for controlling adult populations of the malaria vector, *Anopheles arabiensis*. Parasit. Vectors 3: 18.

Mannino, M.C., Huarte-Bonnet, C., Davyt-Colo, B. and Pedrini, N. (2019). Is the insect cuticle the only entry gate for fungal infection? Insights into alternative modes of action of entomopathogenic fungi. J. Fungi, 5: 33.

Martinez, J.M., Rodrigues, J., Marreto, R.N., Mascarin, G.M., Fernandes, E.K.K. et al. (2021). Efficacy of focal applications of a mycoinsecticide to control *Aedes aegypti* in Central Brazil. Appl. Microbiol. Biotechnol., 105: 8703–8714.

Mascarin, G.M., Kobori, N.N., Jackson, M.A., Dunlap, C.A. and Delalibera, Í.Jr. (2018). Nitrogen sources affect productivity, desiccation tolerance and storage stability of *Beauveria bassiana* blastospores. J. Appl. Microbiol., 124: 810–820.

Mascarin, G.M., Lopes, R.B., Delalibera, Í.Jr., Fernandes, É.K.K., Luz, C. et al. (2019). Current status and perspectives of fungal entomopathogens used for microbial control of arthropod pests in Brazil. J. Invertebr. Pathol., 165: 46–53.

Matthews, B.J. (2019). *Aedes aegypti*. Trends Genet., 35: 470–471.

McGregor, B.L. and Connelly, C.R. (2021). A review of the control of *Aedes aegypti* (Diptera: Culicidae) in the continental United States. J. Med. Entomol., 58: 10–25.

Mehmood, N., Hassan, A., Zhong, X., Zhu, Y., Ouyang, G. et al. (2023). Entomopathogenic fungal infection following immune gene silencing decreased behavioral and physiological fitness in *Aedes aegypti* mosquitoes. Pestic. Biochem. Physiol., 195: 105535.

Mendonça, G.R.Q., Peters, L.P., Lopes, L.M., Sousa, A.H. and Carvalho, C.M. (2023). Native fungi from Amazon with potential for control of *Aedes aegypti* L. (Diptera: Culicidae). Braz. J. Biol., 83: e274954.

Mmbando, A.S., Okumu, F.O., Mgando, J.P., Sumaye, R.D., Matowo, N.S. et al. (2015). Effects of a new outdoor mosquito control device, the mosquito landing box, on densities and survival of the malaria vector, *Anopheles arabiensis*, inside controlled semi-field settings. Malar. J., 14: 1–13.

Moscardi, F. (1999). Assessment of the application of baculoviruses for control of Lepidoptera. Annu. Rev. Entomol., 44: 257–289.

Moser, B.A., Becnel, J.J., White, S.E., Afonso, C., Kutish, G. et al. (2001). Morphological and molecular evidence that *Culex nigripalpus* baculovirus is an unusual member of the family *Baculoviridae*. J. Gen. Virol., 82: 283–297.

Munywoki, D.N., Kokwaro, E.D., Mwangangi, J.M., Muturi, E.J. and Mbogo, C.M. (2021). Insecticide resistance status in *Anopheles gambiae* (*s.l.*) in coastal Kenya. Parasit. Vectors, 14: 207.

Niu, X., Xie, W., Zhang, J. and Hu, Q. (2019). Biodiversity of entomopathogenic fungi in the soils of south China. Microorganisms, 7: 311.

Noskov, Y.A., Polenogova, O.V., Yaroslavtseva, O.N., Belevich, O.E., Yurchenko, Y.A. et al. (2019). Combined effect of the entomopathogenic fungus *Metarhizium robertsii* and avermectins on the survival and immune response of *Aedes aegypti* larvae. Peer J, 7: e7931.

O'Neill, S.L., Kittayapong, P., Braig, H.R., Andreadis, T.G., Gonzalez, J.P. et al. (1995). Insect densoviruses may be widespread in mosquito cell lines. J. Gen. Virol., 76: 2067–74.

Ortiz-Urquiza, A. and Keyhani, N.O. (2013). Action on the surface: entomopathogenic fungi versus the insect cuticle. Insects, 4: 357–74.

Paixão, F.R.S., Falvo, M.L., Huarte-Bonnet, C., Santana, M., García, J.J. et al. (2024). Pathogenicity of microsclerotia from *Metarhizium robertsii* against *Aedes aegypti* larvae and antimicrobial peptides expression by mosquitoes during fungal-host interaction. Acta Trop., 249: 107061.

Paula, A.R. Brito, E.S., Pereira, C.R., Carrera, M.P. and Samuels, R.I. (2008). Susceptibility of adult *Aedes aegypti* (Diptera: Culicidae) to infection by *Metarhizium anisopliae* and *Beauveria bassiana*: prospects for dengue vector control. Biocontrol Sci. Technol., 18: 1017–1025.

Paula, A.R., Carolino, A.T., Paula, C.O. and Samuels, R.I. (2011). The combination of the entomopathogenic fungus *Metarhizium anisopliae* with the insecticide Imidacloprid increases virulence against the dengue vector *Aedes aegypti* (Diptera: Culicidae). Parasit. Vectors, 4: 8.

Paula, A.R., Carolino, A.T., Silva, C.P., Pereira, C.R. and Samuels, R.I. (2013c). Testing fungus impregnated cloths for the control of adult *Aedes aegypti* under natural conditions. Parasit. Vectors. 6: 256.

Paula, A.R., Carolino, A.T., Silva, C.P. and Samuels, R.I. (2013b). Efficiency of fungus-impregnated black cloths combined with Imidacloprid for the control of adult *Aedes aegypti* (Diptera: Culicidae). Lett. Appl. Microbiol., 57: 157–163.

Paula, A.R., Ribeiro, A., Morais, C.O.P., Dias, R.S.B., Gomes, C.R.P. et al. (2013a). Utilização de grãos de arroz com *Metarhizium anisopliae* contra larvas de *Aedes aegypti* em condição de semicampo. Encontro Latino Americano de Iniciação Científica – INIC.

Paula, A.R., Silva, L.E., Ribeiro, A., Butt, T.M., Silva, C.P. et al. (2018). Improving the delivery and efficiency of fungus-impregnated cloths for control of adult *Aedes aegypti* using a synthetic attractive lure. Parasit. Vectors, 11: 285.

Paula, A.R., Ribeiro, A., Lemos, F.J.A., Silva, C.P. and Samuels, R.I. (2019). Neem oil increases the persistence of the entomopathogenic fungus *Metarhizium anisopliae* for the control of *Aedes aegypti* (Diptera: Culicidae) larvae. Parasit. Vectors, 12: 163.

Paula, A.R., Silva, L.E.I., Ribeiro, A., Silva, G.A. and Samuels, R.I. (2021). *Metarhizium anisopliae* blastospores are highly virulent to adult *Aedes aegypti*, an important arbovirus vector. Parasit. Vectors, 14: 555.

Pedrini, N. (2018). Molecular interactions between entomopathogenic fungi (*Hypocreales*) and their insect host: Perspectives from stressful cuticle and hemolymph battlefields and the potential of dual RNA sequencing for future studies. Fungal Biol., 122: 538–545.

Pereira, C.R., Paula, A.R., Gomes, S.A., Pedra, Jr. P.C.O. and Samuels, R.I. (2009). The potential of *Metarhizium anisopliae* and *Beauveria bassiana* isolates for the control of *Aedes aegypti* (Diptera: Culicidae) larvae. Biocontrol Sci. Technol., 19: 881–886.

Perrin, A., Gosselin-Grenet, A.S., Rossignol, M., Ginibre, C., Scheid, B., Lagneau, C. et al. (2020). Variation in the susceptibility of urban *Aedes* mosquitoes infected with a densovirus. Sci. Rep., 10: 18654.

Peschiutta, M.L., Cagnolo, S.R. and Almirón, W.R. (2014). Susceptibility of larvae of *Aedes Aegypti* (Linnaeus) (Diptera: Culicidae) to entomopathogenic nematode *Heterorhabditis bacteriophora* (Poinar) (Rhabditida: Heterorhabditidae). Rev. Soc. Entomol. Argent., 73: 3–4.

Piovezan-Borges, A.C., Valente-Neto, F., Urbieta, G.L., Laurence, S.G.W. and de Oliveira Roque, F. (2022). Global trends in research on the effects of climate change on *Aedes aegypti*: international collaboration has increased, but some critical countries lag behind. Parasit. Vectors, 15: 346.

Proestos, Y., Christophides, G.K., Ergüler, K., Tanarhte, M., Waldock, J. et al (2015). Present and future projections of habitat suitability of the Asian tiger mosquito, a vector of viral pathogens, from global climate simulation. Philos. Trans. R. Soc. B, Biol. Sci., 370: 20130554.

Ramirez, J.L., Dunlap, C.A., Muturi, E.J., Barletta, A.B.F. and Rooney, A.P. (2018). Entomopathogenic fungal infection leads to temporospatial modulation of the mosquito immune system. PLoS. Negl. Trop. Dis., 12: e0006433.

Ramirez, J.L., Hampton, K.J., Rosales, A.M. and Muturi, E.J. (2023). Multiple mosquito AMPs are needed to potentiate their antifungal effect against entomopathogenic fungi. Front. Microbiol., 13: 1062383.

Ravindran, K., Akutse, K.S., Sivaramakrishnan, S. and Wang, L. (2016). Determination and characterization of destruxin production in *Metarhizium anisopliae* Tk6 and formulations for *Aedes aegypti* mosquitoes control at the field level. Toxicon, 120: 89–96.

Regis, L.N., Acioli, R.V., Silveira, Jr.J.C., Melo-Santos, M.A.V., Souza, W.V. et al. (2013). Sustained reduction of the dengue vector population resulting from an integrated control strategy applied in two Brazilian cities. PLoS One, 8: e67682.

Reyes-Villanueva, F., Russell, T.L. and Rodríguez-Pérez, M.A. (2021). Estimating contact rates between *Metarhizium anisopliae*-exposed males with female *Aedes aegypti*. Front. Cell Infect. Microbiol., 11: 616679.

Roberts, D., Tren, R., Bate, R. and Zambone, J. (2016). The Excellent Powder: DDT's Political and Scientific History. Outlooks on Pest Management, Dog Ear Publishing, LLC.

Rodrigues, J., Campos, V.C., Humber, R.A. and Luz, C. (2018). Efficacy of *Culicinomyces* spp. against *Aedes aegypti* eggs, larvae and adults. J. Invertebr. Pathol., 57: 104–111.

Rodrigues, J., Catão, A.M.L., Dos Santos, A.S., Paixão, F.R.S., Santos, T.R. et al. (2021). Relative humidity impacts development and activity against *Aedes aegypti* adults by granular formulations of *Metarhizium humberi* microsclerotia. Appl. Microbiol. Biotechnol., 105(20): 2725–2736.

Rodrigues, J., Rocha, L.F.N., Martinez, J.M., Montalva, C., Humber, R.A., et al. (2022). *Clonostachys* spp., natural mosquito antagonists, and their prospects for biological control of *Aedes aegypti*. Parasitol. Res., 121: 2979–2984.

Russell, B.M., Kay, B.H., Shipton, W. (2001). Survival of *Aedes aegypti* (Diptera: Culicidae) eggs in surface and subterranean breeding sites during the northern Queensland dry season. J. Med. Entomol., 38: 441–445.

Ryder, L.S. and Talbot, N.J. (2015). Regulation of appressorium development in pathogenic fungi. Curr. Opin. Plant Biol., 26: 8–13.

Salem, H.H.A., Mohammed, S.H., Eltaly, R.I., Moustafa, M.A.M., Fónagy, A. et al. (2023). Co-application of entomopathogenic fungi with chemical insecticides against *Culex pipiens*. J. Invertebr. Pathol., 198: 107916.

Samal, R.R., Panmei, K., Lanbiliu, P. and Kumar, S. (2022). Reversion of CYP450 monooxygenase-mediated acetamiprid larval resistance in dengue fever mosquito, *Aedes aegypti* L. Bull. Entomol. Res., 112: 557–566.

Samuels, R.I. (1998). A sensitive bioassay for destruxins, cyclodepsipeptides from the culture filtrates of the entomopathogenic fungus *Metarhizium anisopliae* (Metsch.) Sorok. An. Soc. Entomol. Bras., 27: 229–235.

Samuels, R.I., Paula, A.R., Carolino, A.T., Gomes, S.A., Paula, C.O. et al. (2016). Entomopathogenic organisms: conceptual advances and real-world applications for mosquito biological control. Open Access Insect Physiol., 6: 25–31.

Santos, A.H., Tai, M.H.H., Rocha, L.F.N., Silva, H.H.G. and Luz, C. (2009). Dependence of *Metarhizium anisopliae* on high humidity for ovicidal activity on *Aedes aegypti*. Biol. Control, 50: 37–42.

Schaefer, C.H. and Mulligan III, F.S. (1991). Potential for resistance to pyriproxyfen: a promising new mosquito larvicide. J. Am. Mosq. Control. Assoc., 7: 409–411.

Scholte, E.J., Knols, B.G., Samson, R.A. and Takken, W. (2004). Entomopathogenic fungi for mosquito control: a review. J. Insect Sci., 4: 19.

Scholte, E.J., Knols, B.G. and Takken, W. (2006). Infection of the malaria mosquito *Anopheles gambiae* with the entomopathogenic fungus *Metarhizium anisopliae* reduces blood feeding and fecundity. J. Invertebr. Pathol., 91: 43–9.

Scholte, E.J., Takken, W. and Knols, B.G. (2007). Infection of adult *Aedes aegypti* and *Ae. albopictus* mosquitoes with the entomopathogenic fungus *Metarhizium anisopliae*. Acta Trop., 102: 151–8.

Schrank, A. and Vainstein, M.H. (2010). *Metarhizium anisopliae* enzymes and toxins. Toxicon, 56: 1267–1274.

Schreiber, E.T and Jones, C.J. (1994). Evaluation of inoculative releases of *Toxorhynchites splendens* (Diptera: Culicidae) in urban environments in Florida. Environ. Entomol., 23: 770–777.

Shah, F.A., Abdoarrahem, M.M., Berry, C., Touray, M., Hazir, S. et al. (2021). Indiscriminate ingestion of entomopathogenic nematodes and their symbiotic bacteria by *Aedes aegypti* larvae: a novel strategy to control the vector of chikungunya, dengue and yellow fever Turk. J. Zool., 45.SI-1: 372–83.

Shakeel, M., Xu, X., Xu, J., Zhu, X., Li, S. et al. (2017). Identification of immunity-related genes in *Plutella xylostella* in response to fungal peptide destruxin A: RNA-Seq and DGE analysis. Sci. Rep., 7: 10966.

Silva, L.E.I., Paula, A.R., Ribeiro, A., Butt, T.M., Silva, C.P. et al. (2017). A new method of deploying entomopathogenic fungi to control adult *Aedes aegypti* mosquitoes. J. Appl. Entomol., 142: 59–66.

Sivaram, A., Barde, P.V., Kumar, S.R., Yadav, P., Gokhale, M.D. et al. (2009). Isolation and characterization of densonucleosis virus from *Aedes aegypti* mosquitoes and its distribution in India. Intervirology, 52: 1–7.

Smith, G.C. and Francy, D.B. (1991). Laboratory studies of a Brazilian strain of *Aedes albopictus* as a potential vector of Mayaro and Oropouche viruses. J. Am. Mosq. Control. Assoc., 7: 89–93.

Snetselaar, J., Andriessen, R., Suer, R.A., Osinga, A.J., Knols, B.G. et al. (2014). Development and evaluation of a novel contamination device that targets multiple life-stages of *Aedes aegypti*. Parasit. Vectors, 7: 200.

Sousa, N.A., Rodrigues, J., Arruda, W., Humber, R.A. and Luz, C. (2021). Development of *Metarhizium humberi* in *Aedes aegypti* eggs. J. Invertebr. Pathol., 184: 107648.

Sousa, N.A., Rodrigues, J., Luz, C. and Humber, R.A. (2023). Exposure of newly deposited *Aedes aegypti* eggs to *Metarhizium humberi* and fungal development on the eggs. J. Invertebr. Pathol., 197:107898.

Subkrasae, C., Ardpairin, J., Dumidae, A., Janthu, P., Meesil, W. et al. (2022). Molecular identification and phylogeny of Steinernema and *Heterorhabditis* nematodes and their efficacy in controlling the larvae of *Aedes aegypti*, a major vector of the dengue virus. Acta Trop., 228: 106318.

Suwannaroj, M., Yimthin, T., Fukruksa, C., Muangpat, P., Yooyangket, T. et al. (2020). Survey of entomopathogenic nematodes and associate bacteria in Thailand and their potential to control *Aedes aegypti*. J. Appl. Entomol., 144: 212–223.

Thanwisai, A., Muangpat, P., Meesil, W., Janthu, P., Dumidae, A. et al. (2022). Entomopathogenic nematodes and their symbiotic bacteria from the national parks of Thailand and larvicidal property of symbiotic bacteria against *Aedes aegypti* and *Culex quinquefasciatus*. Biology, 11: 1658.

Tolle, M.A. (2009). Mosquito-borne diseases. Curr. Probl. Pediatr. Adolesc. Health Care, 39: 97–140.

Uejio, C.K., Hayden, M.H., Zielinski-Gutierrez, E., Lopez, J.L.R., Barrera, R. et al. (2014). Biological control of mosquitoes in scrap tires in Brownsville, Texas, USA and Matamoros, Tamaulipas, Mexico. J. Am. Mosq. Control. Assoc., 30: 130–135.

Ulvedal, C., Bertolotti, M.A., Cagnolo, S.R. and Almirón, W.R. (2017). Laboratory susceptibility tests of *Aedes aegypti* and *Culex quinquefasciatus* larvae to the entomopathogenic nematode *Heterorhabditis bacteriophora*. Biomédica, 37: 67–76.

Vachon, V., Laprade, R. and Schwartz, J.L. (2012). Current models of the mode of action of *Bacillus thuringiensis* insecticidal crystal proteins: a critical review. J. Invertebr. Pathol., 111: 1–12.

Valencia-Marín, B.S., Gandica, I.D. and Aguirre-Obando, A.O. (2020). The Mayaro virus and its potential epidemiological consequences in Colombia: an exploratory biomathematics analysis. Parasit. Vectors, 13: 508.

Vannavong, N., Seidu, R., Stenström, T.A., Dada, N., and Overgaard, H.J. (2017). Effects of socio-demographic characteristics and household water management on *Aedes aegypti* production in suburban and rural villages in Laos and Thailand. Parasit. Vectors. 10: 170.

Vinogradov, D.D., Sinev, A.Y. and Tiunov, A.V. (2022). Predators as Control Agents of Mosquito Larvae in Micro-Reservoirs (Review). Inland Water Biol., 15: 39–53.

Wang, C. and Wang, S. (2017). Insect pathogenic fungi: genomics, molecular interactions, and genetic improvements. Annu. Rev. Entomol., 62: 73–90.

Wang, J.B., St Leger, R.J. and Wang, C. (2016). Advances in genomics of entomopathogenic fungi. Adv. Genet., 94: 67–105.

Wang, J., Weng, Q. and Hu, Q. (2019). Effects of destruxin A on silkworm's immunophilins. Toxins, 11: 349.

Weaver, S.C. and Reisen, W.K. (2010). Present and future arboviral threats. Antivir. Res., 85: 328–345.

Werren, J.H. (1997). Biology of *Wolbachia*. Annu. Rev. Entomol., 42: 587–609.

Wilder-Smith, A. (2020). Dengue vaccine development: status and future. Bundesgesundheitsblatt - Gesundheitsforsch. Gesundheitsschutz., 63: 40–44.

Wilkerson, R.C., Linton, Y-M. and Strickman, D. (2020). Mosquitoes of the World; Johns Hopkins University Press: Baltimore, MD, USA.

Williams, T., Virto, C., Murillo, R. and Caballero, P. (2017). Covert Infection of Insects by Baculoviruses. Front. Microbiol., 8: 1337.

World Health Organization (2021). Available online: http://www.who.int/mediacentre/factsheets/fs117/en (accessed on December 01, 2023).

World Health Organization (2023). Available online: https://www.who.int/news-room/fact-sheets/detail/dengue-and-severe-dengue (accessed on December 18, 2023).

Yi, H., Devkota, B.R., Yu, J.S., Oh, K., Kim, J. et al. (2014). Effects of global warming on mosquitoes & mosquito-borne diseases and the new strategies for mosquito control. Entomol. Res., 44: 215–235.

Zara, A.L.S.A., Santos, S.M., Fernandes-Oliveira, E.S., Carvalho, R.G. and Coelho, G.E. (2016) Estratégias de controle do *Aedes aegypti*: uma revisão. Epidemiol. Serv. Saúde, Brasília, DF, 25: 391–404.

Zhou, S., Liu, X., Sun, W., Zhang, M., Yin, Y. et al. (2021). The COMPASS-like complex modulates fungal development and pathogenesis by regulating H3K4me3-mediated targeted gene expression in *Magnaporthe oryzae*. Mol. Plant Pathol., 22: 422–439.

11 Entomopathogenic Fungi and Groundnut Pests: Applications, Challenges and Prospects

S. Karthick Raja Namasivayam,[1*] R.S. Arvind Bharani,[2] Arun John[3] and G.S. Amrish Varshan[4]

1. Introduction

There are numerous remarkable alternatives to chemical insect control available. Some have already been implemented with great success; others are currently undergoing laboratory testing; and some are still just ideas in the minds of creative scientists, awaiting the chance to be tested. What they all share in common is that they are biological solutions.

Agriculture is a cornerstone of India's economy and a major source of livelihood for a substantial portion of the population. It makes a significant contribution to India's Gross Domestic Product (GDP). Despite its decreasing share of overall GDP due to the growth of other sectors, agriculture remains a critical component of the economy. India's socioeconomic development heavily relies on agriculture, which employs a large percentage of the workforce. The sector faces challenges such as small landholdings and low productivity, which can create difficulties for farmers. Several factors have driven the growth of the Indian agricultural sector in recent years, including increased consumption, rising incomes, the expansion of the food processing sector, and improvements in the transport of agricultural products. The private sector has played a key role in this growth, especially through initiatives in organic farming and the application of information technology, which have revolutionized processing and irrigation methods.

Recognising the importance of the agricultural sector, the government of India has taken ample steps towards several vital developments. These steps include corporate credit for farmers and promoting and facilitating scientific warehousing infrastructure, such as cold chains and storage, to increase agricultural product shelf life. The Indian agriculture sector will gain momentum and flourish in the next few

[1] Centre for Applied Research, Saveetha School of Engineering, Saveetha Institute of Medical and Technical Sciences (SIMATS), Chennai-602105, Tamil Nadu, India.

[2] Institute of Obstetrics and Gynaecology, Madras Medical College, Egmore, Chennai-600008, Tamil Nadu, India.

[3] Departmnet of Computational Biology, Saveetha School of Engineering, Saveetha Institute of Medical and Technical Sciences (SIMATS), Chennai-602105, Tamil Nadu, India

[4] Centre for Applied Research, Saveetha School of Engineering, Saveetha Institute of Medical and Technical Sciences (SIMATS), Chennai-602105, Tamil Nadu, India.

* Corresponding author: biologiask@gmail.com

years due to these investments in infrastructure. In contrast, factors like budgetary incentives, reduced transaction costs, and time will drive this upward momentum. Moreover, the increased adoption of genetically modified crops has resulted in higher yields for Indian farmers. The Indian government projects that storage capacity will expand to 35 million metric tons by the end of the 12th Five-Year Plan (2012–2017). Additionally, it aims for a comprehensive growth rate of 4%, which will significantly transform the Indian agriculture sector in the coming years.

The aim and scope of this chapter on entomopathogenic fungi is to present comprehensive information on fungi that are pathogenic to insects and pests associated with groundnut ecosystems. Key aspects covered include synthetic chemical pesticides, groundnut pests, mainly *Spodoptera litura*, biological control principles, including biocontrol agents, the most employed entomopathogenic fungal strains and their bioactive metabolites, and challenges and future prospects. By addressing these aspects, the chapter aims to provide a comprehensive understanding of the applications of entomopathogenic fungi, making it a valuable resource for researchers, students, and practitioners in the field.

2. Synthetic Chemical Pesticides

Among the many challenges in agriculture, insect pests feature significantly (Aidroos and Roberts, 1978; Guliy et al., 2003). Chemical pesticides are substances used to control or eliminate pests, including insects, weeds, fungi, and rodents, that can damage crops, harm livestock, or negatively impact human health (Amakiri, 1982). These pesticides are a part of modern agricultural practices and are employed to increase crop yields and protect agricultural investments. However, the use of chemical pesticides raises environmental and health concerns. Shah and Devakota (2009) and Pimental et al. (1990) outlined key concerns about chemical pesticides in their studies.

Pesticides can be categorized based on their intended purpose, mode or duration of action, or chemical function. Examples include insecticides (for insects), fungicides (for fungi), rodenticides (for rodents), molluscicides (for snails), defoliants (for leaf removal), and desiccants (for drying foliage). They are widely used as pesticides, nematicides, and herbicides, and are classified into groups such as organochlorides, organophosphates, and pyrethroids. Many types of pesticides have been in use for decades, largely in agricultural applications (Cardeal et al., 2011). In recent years, organophosphorus and its pesticidal compounds have been the largest cluster of pesticides used worldwide (Singh and Walker, 2006; Eissa et al., 2014).

3. Groundnut

Legumes are an important and diverse family of flowering plants, representing a significant group among higher plants, with more than 650 genera and 18,000 species. Groundnut is a chief economic crop and is extensively grown in the USA, Africa, and other countries like China and India. Groundnuts are a significant protein and vegetable-oil source for human nutrition. Almost 70% of overall production is used for oil, with the remaining 30% used as food.

Groundnut cultivation is also crucial for sustainable fertilization, increased agricultural productivity as a rotational crop, global food security, fuel, nutrition, and energy. Groundnut have numerous health advantages and are essential as a major food allergen. However, farmers face excessive production costs, primarily due to disease control using chemical pesticides. Aflatoxin contamination of groundnuts is another significant food safety issue that poses a threat to human health (Feng et al., 2012).

Poor yields of groundnut are primarily due to biotic and abiotic factors a like intermittent rainfall, low and inconsistent moisture levels, lack of high-yield reformed cultivars, deprived agronomic practices, and, crucially, mutilation by pests and diseases (Verma et al., 2011).

3.1 Groundnut Pests

Groundnut production comprises measures like shielding crops from insects. Most harvests can endure some insect harm without major yield loss. The groundnut crops' ability to endure insect mutilation is well known. The plant's resilience to losses enables growers to decrease pesticide use and crop production costs. Cultivators of cash crops such as groundnuts have complained about the economic loss when damage caused by pests has exceeded the tolerable level. All parts of the groundnut plant, foliage, roots, pegs, and pods are vulnerable to one or more insect pests. Initially, insects primarily damage the foliage and impair pods, pegs, and roots. Some insect pests can be challenging to detect, despite the clear damage they cause. Foliage mutilation is caused by several pests, such as thrips, leafhoppers, two-spotted spider mites, the silver leaf whitefly, granulate cutworms, green cloverworms, corn earworms, fall armyworms, beet armyworms, southern armyworms, yellow-striped armyworms, tropical armyworms, loopers, and rednecked peanutworms. Peg-and-pod feeding pests also come in many forms: lesser cornstalk borer, southern corn rootworm, wireworms, white fringed beetles, and burrowing stink bugs. Chemical treatment is usually necessary when the average quantity of foliage-feeding pests or peg-pod-root-feeding pests in a field exceeds acceptable thresholds.

3.2 Spodoptera litura

Spodoptera litura, commonly known as the tobacco cutworm or cluster caterpillar, is a polyphagous insect pest that belongs to the Noctuidae family. This species is widely distributed in the tropical and subtropical regions of Asia, Africa, and Australia. It is known for its voracious feeding habits and can cause significant damage to a variety of crops. Roa et al. (1993) and Armes et al. (1997) outlined some key features and information about Spodoptera litura in their studies. Spodoptera litura is highly polyphagous, meaning it feeds on a wide range of host plants. It can infest various crops, including cotton, tobacco, groundnut, soybean, maize, vegetables, and ornamental plants (Jiang and Ma, 2000; Kaur et al., 2014).

3.3 The Life Cycle of Spodoptera litura

Approximately 1000-2000 eggs, or egg clusters, are deposited on the underside of leaves on the host plant. It is worth noting that the female moth lays eggs within 2-5 days after entering this stage. These clusters are covered with hair-like scales originating from the tip of the insect's abdomen (Miyahara et al., 1971). Generally, approximately 960 eggs are laid at a temperature of 30°C and a relative humidity (RA) of around 90%. The remaining 145 eggs are laid at a slightly higher temperature, around 35°C, and a relative humidity of around 30%. The larval stage undergoes around six instar levels over a period of 15-23 days at around 25°C to 26°C.

Young larvae in clusters are inactive on the epidermis of the leaf. Meanwhile, the mature larvae disband and conceal themselves beneath the leaf during the day and feed at night and early in the morning. The pupal period is habitually spent in encapsulation and persists for 11 to 13 days at 25°C. The average life span of an adult is between 4 and 10 days, which is based on temperature and low relative humidity. The average life cycle of a generation will be completed, mostly within 5 weeks (Carasi et al., 2014).

4. Biological Control

It is important to strike a balance between the benefits of using chemical pesticides for agricultural productivity and the potential risks to the environment and human health. Sustainable and responsible pesticide use, along with the exploration of alternative methods, is crucial for the long-term health of ecosystems and human populations. Biological control is the use of one or more favourable organisms, typically called natural enemies, to decrease the quantity of another kind of organism, the pest. Also to be considered are the actions of living organisms with respect to the target pests and surroundings. It follows that biological control practices are more complex than outdated, specific pest control procedures, such as the use of chemical pesticides.

Three comprehensive tactics deal with the biological control of pests; importation, augmentation, and conservation. Importing biological enemies addresses the issue of the most operative natural opponents not naturally occurring in areas where a pest is causing damage. Augmentation of natural enemies entails increasing the populations of natural enemies through human intervention. There are two overall approaches to the enhanced delivery of natural enemies: inundation and inoculation. Inundation is the mass release of great numbers of natural enemies to swiftly curb pest populations. This strategy is effective only when the natural predator can rapidly eradicate the pest. If the natural predator's actions are delayed, the pest population may potentially cause significant damage before being brought under control. Inoculation dispenses a limited number of natural enemies, even when the pest population is at a lower level and stable. Inoculation relies on released natural enemies reproducing rapidly enough to counter any buildup of the pest population. It is a natural approach, as it relies on the increase in the number of favourable natural enemies occurring through natural processes and benefits from their biological characteristics to seek distant prey. Conservation of natural enemies involves preserving predators, pathogens, and parasites that occur naturally or that have been introduced through importation or augmentation plans (Sanda and Sunusi, 2014).

4.1 Biocontrol Agents

Biocontrol agents, otherwise termed natural enemies, are vital in controlling pests. These biocontrol agents are predators, parasites, and pathogens that attack pests and control their population.

4.2 Predators

Predatory insects play a vital role as they consume insect pests like aphids, making them advantageous in pest control. Common examples of predatory insects include ladybugs, lacewings, and mantids. Interestingly, cultures of certain predatory insects, such as ladybugs and beetles with predatory behavior, have been documented for centuries. The larvae and adults of these predatory insects consume a considerable number of small, soft-bodied pests, like aphids. However, it's worth mentioning that these same pests also prey on small, delicate-bodied larvae, eggs, and mites.

4.3 Parasitoids

A parasite is an organism smaller than its host that derives benefits from the host organism. Typically, parasites do not kill the host and have a shorter life cycle than the host organism. They spend a significant part of their lives within or on a single host organism. In contrast, a parasitoid, while similar to a true parasite, eventually kills and sometimes consumes the host. Once the parasitoid completes its life cycle, it becomes a free-living insect no longer dependent on the host.

4.4 Pathogens

Entomopathogens are microorganisms that infect and kill insects. They play a crucial role in natural pest control and are considered an environmentally friendly alternative to chemical pesticides (Lacey et al., 2001). Entomopathogens are integral components of integrated pest management (IPM) programs, where multiple strategies are combined to manage pest populations effectively. Biologically occurring entomopathogens have vital monitoring factors in insect populations.

4.5 Entomopathogenic Nematodes

Nematodes are simple, colourless roundworms that lack appendages and segments. They may be free-living, predaceous, or parasitic. Many parasitic species cause significant diseases in plants, animals, and humans (Stuart et al., 2006). Entomopathogenic nematodes from the families Steinernematidae and Heterorhabditidae have been used to control the populations of insect pests in a variety of agro-environments, and, in several circumstances, their beneficial influence on crop yield has been evident (Mracek et al., 1993). Entomopathogenic nematodes comprising *Steinernema carpocapsae* and *Heterorhabditis indica* have been used to infect and destroy soil-dwelling larval populations (Yan et al., 2013). They may feasibly be applied to the pupal stages of the striped flea beetle, *Phyllotreta striolata*, and serve to curb the occurrence of future generations of adult flea beetles from pupae.

Entomopathogenic nematodes are produced in quantities for use in biopesticides by in vivo or in vitro methods (Shapiro-Ilan et al., 2010). Entomopathogenic nematodes are deadly to numerous prominent insect pests. However, they are safe for plants and animals.

4.6 Entomopathogenic Bacteria

Entomopathogenic bacteria include two unique insect pathogenic bacteria, *Photorhabdus* (Proteobacteria: Enterobacteriaceae) and *Bacillus* (Firmicutes: Bacillaceae). *Bacillus thurigiensis* (Bt) is a well-known bacterium used as a biological insecticide. It produces toxins that are lethal to certain insect pests, including caterpillars like those of Spodoptera litura (Watkins et al., 2012; Gatehouse, 2008). Bt is a Gram-positive, soil-dwelling bacterium that has gained importance in agriculture and pest control. Discovered in 1901, Bt is known for producing insecticidal crystal proteins during sporulation, which are toxic to various insect pests. Bt is effective against various pests, including lepidopteran larvae (caterpillars), coleopteran larvae (beetle larvae), and dipteran larvae.

5. Entomopathogenic Fungi

Entomopathogenic fungi are a group of fungi that naturally infect and kill insects. They play a crucial role in natural pest control and are used as biopesticides in agriculture because they cause natural epizootics on target pests and mites (Smith et al., 1981; Carruthers and Soper, 1987; Burges, 1981; McCoy et al., 1988). Entomopathogenic fungi are specific to insects, minimizing harm to non-target organisms and reducing environmental impact. (Roberts and Humber, 1981). Entomopathogenic fungi contribute to biological control by maintaining a balance in insect populations and preventing outbreaks. Species belonging to *Beauveria*, *Metarhizium*, *Nomuraea*, and *Lecanicillium* are commonly employed in EPF to control diverse groups of insect pests associated with different crops (DeFaria, 2007).

Their environmentally friendly nature makes entomopathogenic fungi a valuable tool in sustainable agriculture, aligning with the principles of organic farming and reducing chemical inputs. These fungi are often integrated into integrated pest management programs, alongside other control methods, to enhance the overall effectiveness of pest management strategies (St. Leger et al., 1996; Butt et al., 2001; Wang and St. Leger, 2007; St. Leger and Wang, 2009; Butt, 2002). The use of entomopathogenic fungi can help manage the development of resistance in pest populations since they often have different modes of action compared to chemical pesticides (Hajek and St. Leger, 1994; Qazi and Khachatourians, 2005; Fan et al., 2007; Thomas and Read, 2007).

5.1 Biological Life Cycle of Entomopathogenic Fungi

The life cycles of entomopathogenic fungi involve several stages, from spore germination to the infection and eventual death of the host insect. The following is a general overview of the life cycle of entomopathogenic fungi, with specific variations depending on the fungal species (Khan et al., 2012; Shah and Pell, 2003; Sierotzki et al., 2000; Pell et al., 2001; Shaw et al., 2002; Roberts and Humber, 1981).

1. *Spore germination:* The life cycle begins with the germination of fungal spores. These spores are often produced on the surface of the infected insect or are dispersed in the environment (Pell et al., 2001).

2. *Attachment and adhesion:* The germinated spores attach to the cuticle (outer surface) of the host insect. The attachment is facilitated by specialized structures such as appressoria or adhesive substances produced by the fungus.
3. *Penetration:* The fungus penetrates the insect cuticle using enzymes and mechanical pressure. The penetration may occur through natural openings, joints, or weak points in the cuticle.
4. *Colonization:* Once inside the insect, the fungus starts to grow and proliferate, forming hyphae (thread-like structures). The hyphae spread throughout the insect body, competing with the host tissues for nutrients.
5. *Nutrient absorption:* The fungus absorbs nutrients from the insect's body, leading to the weakening and eventual death of the host. Some entomopathogenic fungi release toxins that contribute to the mortality of the host.
6. *Coniophores with conidia formation:* Conidiophores are specialized structures that produce asexual spores called conidia. These spores are essential for the fungus's reproductive cycle and are responsible for spreading the infection to new hosts. Conidia are formed on the conidiophores and released into the environment. These airborne conidia can land on other susceptible insects, plants, or surfaces.
7. *Dissemination:* Conidia serve as the primary means of spreading the fungal infection. They can be carried by wind, water, or other environmental factors, facilitating the search for new host insects.
8. *Secondary infection:* Conidia that land on new host insects initiate the infection process, completing the cycle. Secondary infections can contribute to the persistence of the entomopathogenic fungus in the environment.
9. *Resting structure:* Some entomopathogenic fungi produce resting structures, such as thick-walled survival structures or sclerotia, which allow them to withstand unfavorable environmental conditions and persist between host availability.

5.2 Common Examples of Entomopathogenic Fungi

5.2.1 Nomuraea rileyi

Nomuraea rileyi, formerly known as Nomuraea atypicola, is an entomopathogenic fungus that belongs to the order Hypocreales. This fungus is known for its effectiveness in controlling lepidopteran pests, particularly caterpillars. Nomuraea rileyi plays a significant role in biological control strategies, providing an environmentally friendly alternative to chemical pesticides for managing insect pests in agriculture (Singh and Gangrade, 1975; Patil and Abhilash, 2014).

Nomuraea rileyi primarily infects and targets insects belonging to the order Lepidoptera, which includes various moth and butterfly species. Caterpillars of different agricultural pests are particularly susceptible to infection (Ignoffo, 1981; Ignoffo et al., 1976; Pinango et al., 2002). The application of Nomuraea rileyi in pest management contributes to the overall goal of reducing the environmental impact of agriculture while effectively controlling lepidopteran pests that can cause significant damage to crops (Pavone et al., 2008). The fungus infects the host insect by penetrating through the cuticle, which is the outer protective layer of the insect, followed by spore germination and proliferation, which feed on the insect's tissues, and ultimately the death of the host (Boucias, and Pendland., 1984). Infected caterpillars often display characteristic symptoms, including a loss of body color and "chalky" appearance due to surface fungal growth, (Schwarz, 1995; Shah et al., 1999; Connick et al., 1998; Stirling and Smith, 1998).

Fungi produce asexual spores known as conidia as part of their reproductive cycle. These conidia are essential for spreading the fungus to new hosts (Lewis and Larkin, 1998). Conidia are released into the environment when the infected host cadaver deteriorates. These airborne conidia can be carried by wind and can land on other susceptible caterpillars. *Nomuraea rileyi* thrives under specific environmental conditions, including high humidity. This preference for humid conditions is typical of many entomopathogenic fungi. *Nomuraea rileyi* is often used in integrated pest management programs, where multiple strategies are combined to control pest populations sustainably.

5.2.2 *Metarhizium anisopliae*

Metarhizium anisopliae is another well-known entomopathogenic fungus used to control various groups of insect pests. This fungus was popularly known as green muscardine fungi, as infected larvae turned green (Roddam and Rath, 1997). This fungus was the first to be commercially produced for pest control (Krassilstschik, 1888; Steinhaus, 1975). The fungal strain can be cultured in a wide range of synthetic and non-synthetic mediums. This fungal species was found to be compatible with the host plant and various agro-active compounds. This fungal strain produces potential bioactive metabolites, mainly destruxin. Destruxin is a potential cyclodepsi peptide known to exhibit insecticidal activity. Both in-vitro and in-silico analyses have clearly revealed the uniqueness of destruxin in insect pest control (Morel et al., 1983; Chen et al., 1995; Chen et al., 1997; Ast et al., 2001; Togashi et al., 1997; Pedras and Biesenthal, 2000; Pedras et al., 2001; Dumas et al., 1996). Destruxin A and Destruxin B acquire identical amino acid sequences but vary in their hydroxyl acid residues. These Dtxs are antiviral and immuno-depressant agents against insect cells (Quiot et al., 1985; Huxham et al., 1989).

The docking study below clearly reveals the insecticidal activity of destruxin (Fig. 1). Insect cuticle proteins play crucial roles in the formation, maintenance, and regulation of the insect exoskeleton (cuticle). The cuticle is not only a physical barrier protecting insects from environmental stresses, pathogens, and physical damage but is also crucial for processes such as molting (ecdysis), which is necessary for growth and development (Neville et al., 1976; Muthukrishnan et al., 2020). The docking of destruxin (CID_11421831) against the cuticular protein 47Eg (AF-Q7JZV0) represents a compelling frontier in insect pest control research. Destruxin, a cyclic hexadepsipeptide produced by entomopathogenic fungi, has shown remarkable insecticidal properties, influencing various physiological processes (Muthukrishnan et al., 2020). In this context, the computational docking of destruxin with the cuticular protein 47Eg aims to elucidate potential interactions that could disrupt the structural integrity of insect cuticles.

Cuticular protein 47Eg plays a pivotal role in cuticle formation, a process critical for insect development and survival. Computational docking studies employ molecular modeling techniques to predict and analyze the binding affinity and modes between destruxin and 47Eg. This approach offers insights into the specific molecular interactions at the binding site, guiding the design of insecticides with enhanced selectivity and efficacy.

In this study, the AutoDock software was used to understand the binding modes between a protein receptor and a small-molecule ligand. The initial step involves preparing both the protein receptor and ligand structures, incorporating actions such as hydrogen atom addition, charge assignment, and the definition of the ligand's binding site on the protein receptor. Following this, a three-dimensional grid is constructed around the receptor's binding site, outlining the spatial parameters for ligand docking. Ultimately, the ligand undergoes docking into the protein receptor's binding site, exploring diverse conformations and assessing the energy of the interaction between the ligand and the protein receptor (Morris et al., 2009). The post-docking analysis reveals strong hydrogen bonding with amino acids SER64 and SER74 with a docking score of –7.56 kcal/mol, vdW energy of –7.25 kcal/mol, and electrical energy of 18.76 kcal/mol. Serine residues in the cuticular protein 47 Eg play a significant role in mediating protein-ligand interactions, contributing to the overall functionality and structural integrity of the insect cuticle. Therefore, serine, with its hydroxyl group, can form hydrogen bonds with ligands or other amino acids. This capability is pivotal in stabilizing the binding interface between the cuticular protein 47 Eg and potential ligands, which may interrupt its activity by inhibiting the protein. Understanding the detailed molecular interplay between destruxin and 47Eg opens avenues for the development of targeted insecticides. Such tailored approaches hold the promise of minimizing environmental impact and off-target effects by specifically disrupting the cuticular structure of pest insects. This knowledge contributes to the broader field of biopesticides and aligns with the growing emphasis on sustainable pest management practices.

5.2.3 Beauveria bassiana

Beauveria bassiana, commonly known as white muscardian fungi, is the most important deuteromycete entomopathogenic fungus, (Kirkland et al., 2004; Leathers et al., 1993; McCoy, 1990) that is active against a wide range of insect pests, including mosquitoes, sandflies, and ticks (Scholte et al., 2005; Samish et al., 2004; Reithinger et al., 1997). This fungal strain is able to colonize and lead an endophytic relationship with plant tissues (Bing and Lewis, 1991; Bing and Lewis, 1992). B. bassiana's wide range of metabolites, mainly enzymes like chitinases, proteases, and other hydrolytic enzymes, help in the degradation of insect structural integrity (El-Sayed et al., 1993; Fuguet et al., 2004; Kirkland et al., 2005). This fungal strain can be cultivated on a wide range of natural, semi-natural, and synthetic mediums and is commercially available on the market under different names. It produces various bioactive metabolites, primarily including beauvericin, bassianolide, and the diketomorpholine bassiatin (Eley et al., 2007). Among the metabolites, beauvericin is a well-known cyclodepsipeptide known

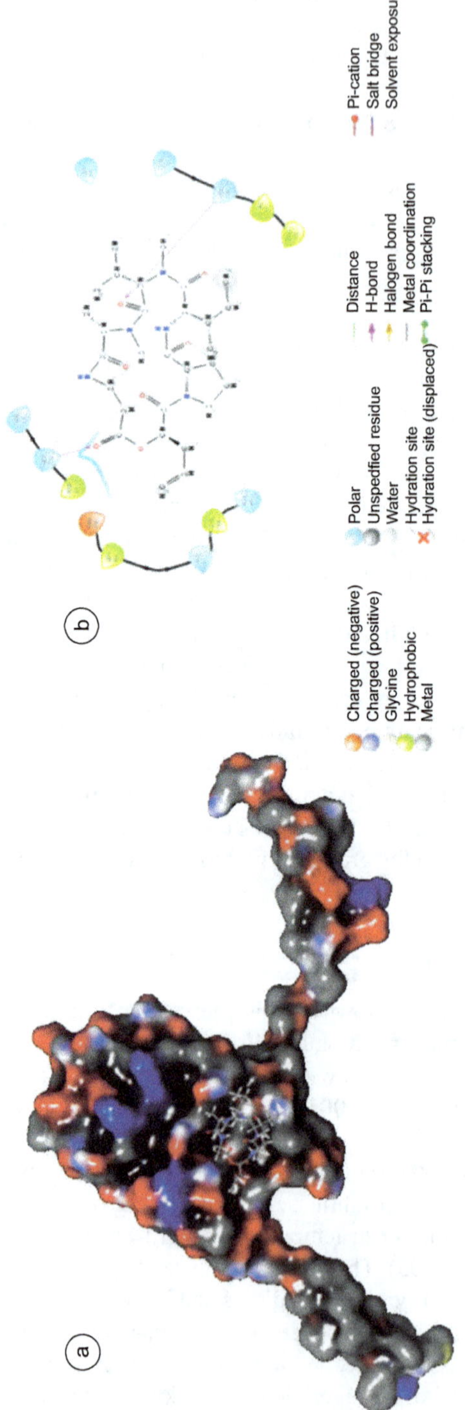

Fig. 1 (a) The docked complex of cuticular protein 47Eg with destruxin showed in surface charge; (b) 2D interaction diagram of cuticular protein 47Eg with destruxin.

to exhibit diverse biological activities (Steinrauf, 1985; Hamill et al., 1969; Zhang et al., 2007; Jow et al., 2004; Carmeliet, 2003).

The docking study below shows the molecular mechanism of insecticidal activity (Fig. 2). Beauvericin, known for its ionophoric properties and multifaceted biological activities, has emerged as a compelling candidate for pest control (Hasuda and Bracarense, 2023). In this context, the Calcium-Binding Domain 2 (CBD2) of the *Drosophila* Na(+)/Ca(2+) exchanger presents an intriguing target due to its essential role in ion homeostasis and cellular signaling (Hryshko and Zheng., 2009). The CBD2 domain of the Na(+)/Ca(2+) exchanger is instrumental in sensing and regulating calcium levels within cells. Calcium signaling plays a pivotal role in various physiological processes, including muscle contraction, neurotransmission, and developmental pathways (Schwarz and Benzer, 1997; Dyck et al., 1998). Disrupting the finely tuned calcium homeostasis mediated by CBD2 could have profound consequences for insect physiology, making it an attractive target for pest control strategies. Beauvericin's ionophoric activity, specifically its ability to facilitate the transport of ions across cellular membranes, aligns with the role of the Na(+)/Ca(2+) exchanger in maintaining ion gradients. This shared ionophoric characteristic prompts the investigation into whether beauvericin could interfere with the calcium-binding and regulatory functions of CBD2, potentially disrupting the insect's essential physiological processes.

The docking studies involving beauvericin (CID_105014) and CBD2 (PDB ID:3E9U) employ sophisticated computational tools such as AutoDock to simulate and predict binding interactions. Initial phases involve the preparation of both the ligand and receptor structures, followed by the generation of three-dimensional grids defining the spatial parameters for docking. Subsequent analyses explore various conformations and evaluate the energy of the interaction between beauvericin and CBD2 (Morris et al., 2009).

The post-docking analysis revealed strong hydrogen bonding with amino acids MET1 and ARG134 with a docking score of –8.07 kcal/mol, vdW energy of –22.7 kcal/mol, and electrical energy of 5.87 kcal/mol (Fig. 2a,b). Methionine residues can play essential roles in protein structure due to their unique chemical properties. They are often found at the N-terminus of proteins, contributing to the stability and folding of the protein. Further, arginine is a positively charged amino acid and is known for its ability to interact with negatively charged molecules, including ions. In the context of the CBD2 domain, ARG134 may contribute to calcium binding or interact with other negatively charged residues, influencing the domain's calcium-sensing capabilities. Therefore, the binding of Beauvericin may disrupt the domain structure stability and also affect calcium binding.

Insights gained from computational docking studies can shed light on the potential binding sites, affinity, and modes of interaction between beauvericin and CBD2. Understanding these molecular interactions is crucial for assessing the feasibility of utilizing beauvericin as an insecticide or as a lead compound for the design of more targeted pest control agents. While computational studies provide a valuable theoretical framework, translating these findings into practical applications requires rigorous experimental validation.

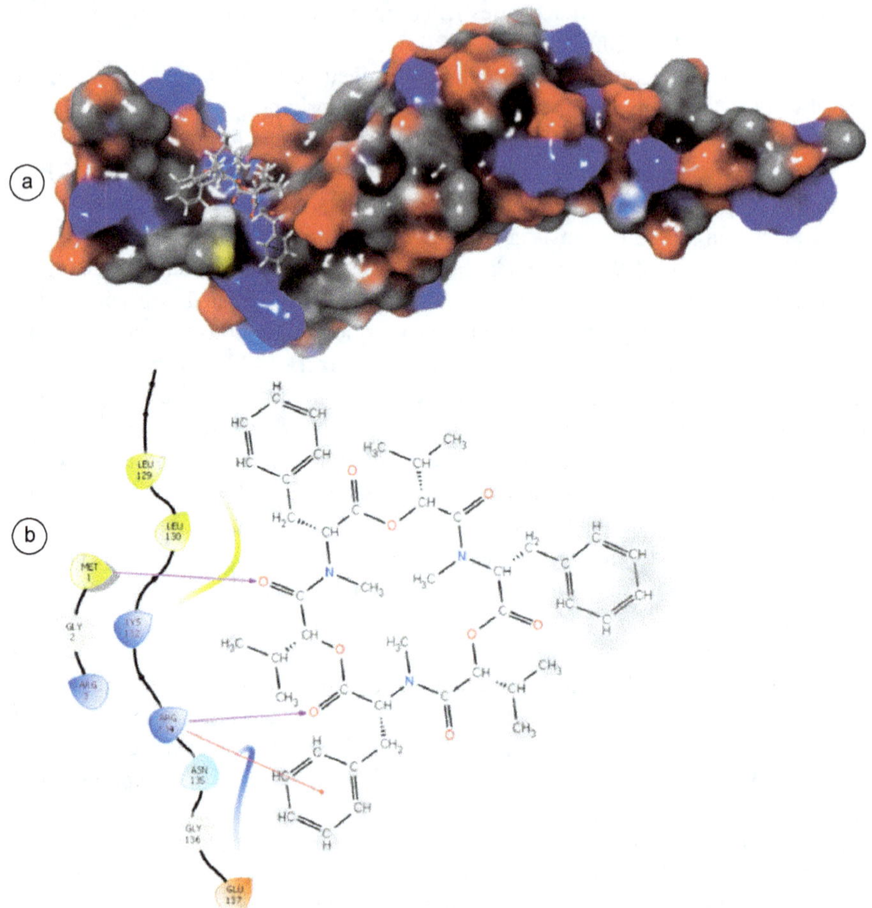

Fig. 2 (a) The docked complex of CBD2 with Beauvericin shown in surface charge; (b) 2D interaction diagram of CBD2 with Beauvericin.

6. Challenges and Prospects

6.1 Challenges

Fungal biocontrol agents, which involve the use of fungi to control pests and diseases in agriculture, face various challenges and offer promising future prospects:

1. *Efficiency and biocontrol potential:* The effectiveness of fungal biocontrol agents can vary depending on environmental conditions, making it challenging to achieve consistent results across different regions and seasons.
2. *Mass production and commercialization:* Large-scale production of fungal biocontrol agents can be costly and technically challenging. Maintaining product quality and shelf life during production and storage is crucial.
3. *Host range and specificity:* Some fungal biocontrol agents may have a narrow host range, limiting their effectiveness against a broader spectrum of pests

or pathogens. Expanding the host range without causing harm to beneficial organisms is a challenge.

4. *Compatibility with agrochemicals:* Chemical pesticides often provide quick and reliable pest control. Convincing farmers to adopt biocontrol agents may be challenging, especially if they perceive chemical alternatives as more effective.
5. *Regulatory approval:* Stringent regulatory requirements for biocontrol agents can hinder their commercialization. Demonstrating safety, efficacy, and consistency according to regulatory standards is a time-consuming and costly process.
6. *Public perception:* Overcoming scepticism and building public trust in the effectiveness and safety of fungal biocontrol agents may be challenging, especially when compared to more established chemical alternatives.
7. *Bioactive metabolite production:* Most of the fungal strains are known to produce bioactive metabolites. However, the yield of the final product, including stability, is very minimal.

6.2 Future Prospects

1. Biotechnology principles - Continued advancements in biotechnology, including genetic engineering, can lead to the development of improved fungal strains with enhanced biocontrol capabilities, broader host ranges, and increased resistance to environmental stressors.
2. Microbiome research - Understanding the interactions between fungi, plants, and other microorganisms in the soil can provide insights into optimizing the performance of biocontrol agents and developing tailored solutions for specific agricultural systems.
3. Awareness and public concern - Creating awareness and educating farmers, stakeholders, and the public about the benefits of fungal biocontrol agents, as well as addressing misconceptions, can contribute to wider acceptance and adoption.
4. Collaboration and partnerships - Collaboration between researchers, industry, and regulatory bodies can expedite the development, testing, and commercialization of fungal biocontrol agents, addressing challenges more effectively.
5. Sustainable developmental goals - Growing interest in sustainable agriculture and environmentally friendly practices may drive the demand for biocontrol agents, creating a favorable market for fungal-based solutions.
6. Scale-up and optimization for metabolite production - Designing mediums using suitable optimization studies will be helpful to support the growth, multiplication, and production of bioactive metabolites.

Conclusion

Entomopathogenic fungi (EPF) are a fascinating group of microorganisms that play a crucial role in biological pest control. These fungi have evolved unique strategies to infect and kill various insect pests, making them valuable tools in integrated pest management (IPM) programs. As we conclude our exploration of entomopathogenic fungi in groundnut pest management, several key points emerge which include the general principles of EPFs. The most commonly used EPF are *Beauveria*, *Nomuraea*, and *Metarhizium,* including their bioactive metabolites. This study has explored various challenges and future prospects of EPF in IPM. In conclusion, entomopathogenic fungi hold great promise as eco-friendly alternatives for pest control. Continued research, technological advancements, and effective integration into agricultural practices will further enhance their role in sustainable pest management, contributing to the overall goal of environmentally responsible and economically viable agriculture.

References

Aidroos, A.K. and Roberts, D.W. (1978). Mutants of *Metarhizium anisopliae* with increased virulence toward mosquito larvae, Can. J. Genet, Cytol., 20(2): 211–219. https://doi.org/10.1139/g78-024.

Amakiri, M.A. (1982). Microbial degradation of soil applied herbicides, Niger. J. Microbiol., 2: 17–21.

Armes, N.J., Wightman, JA, Jadhav, D.R. and Rao, G.V.R. (1997). Status of insecticide resistance in *Spodoptera litura* in Andhra Pradesh, India. Pestic. Sci., 50(3): 240–248.

Ast, T., Barron, E., Kinne, L., Schmidt. M., Germeroth, L., Simmons, K. and Wenschuh, H. (2001). Synthesis and biological evaluation of destruxin A and related analogs. J. Pept. Res., 58: 1–11.

Bing, L.A. and Lewis, L.C. (1991). Suppression of *Ostrinia nubilalis* (Hubner) (Lepidoptera: Pyralidae) by endophytic *Beauveria bassiana* (Balsamo) Vuillemin. Environ. Entomol., 20(4): 1207–1211.

Bing, L.A. and Lewis, L.C. (1992). Endophytic *Beauveria bassiana* (Balsamo) Vuillemin in corn: the influence of the plant growth stage and *Ostrinia nubilalis* (Hubner), Biocontrol. Sci. Techn., 2(1): 39–47.

Boucias, D.G. and Pendland, J.C. (1984). Nutritional requirements for spore germination of several host range pathotypes of *Nomuraea rileyi*, J. Invertebr. Pathol., 43(2): 288–292. https://doi.org/10.1016/0022-2011(84)90153-8.

Burges, H.D. (1981). Safety, safety testing and quality control of microbial pesticides. In: Burges, H. D. (Ed.) Microbial control of pests and plant diseases. London, Academic Press, 737–767.

Butt, T.M., Jackson, C. and Magan, N. (2001). Introduction Fungal biological control agents: progress, problems and potential. In: Butt, T.M., Jackson, C. and Magan, N.(Eds.), Fungi as biocontrol agents: progress, problems and potential, Wallingford, CAB International, 1–8.

Butt, T.M. (2002). Use of entomogenous fungi for the control of insect pests, In: Esser, K. and Bennett, J.W. (Eds.), Mycota, Springer, Berlin, 111–134.

Carasi, R.C., Telan, I.F. and Pera, B.V. (2014). Bioecology of common cutworm (*S. Litura*) of mulberry, Int. J. Sci. Res Publ., 4(4): 1–8.

Cardeal, Z.L., Souza, A.G. and Amorim, L.C.A. (2011). Analytical methods for performing pesticide degradation studies in environmental samples. In: Stoytcheva, M. (Ed.) Pesticides-Formulations, Effects, Fate, InTech, Croatia, Europe, 595–617.

Carmeliet, P. (2003). Blood vessels and nerves: common signals, pathways and diseases, Nat. Rev. Genet., 4(9): 710–720.

Carruthers, R.I. and Soper, R.S. (1987). Fungal diseases. In: Fuxa J.R. and Tanada, Y. (Eds.) Epizootiology of Insect Diseases, New York, John Wiley and Sons. Pp. 357–416.

Chen, H.C., Yeh, S.F., Ong, G.T., Wu, S.H., Sun, C.M. and Chou, C.K. (1995). The novel desmethyldestruxin B_2, from *Metarhizium Anisopliae* that suppresses Hepatitis B virus surface antigen production in human hepatoma cells, J. Nat. Prod., 58(4): 527–531.

Chen, H.C., Chou, C.K., Sun, C.M. and Yeh, S.F. (1997). Suppressive effects of destruxin B on hepatitis B virus surface antigen gene expression in human hepatoma cells. Antivir. Res., 34(3): 137–144.

Connick, W., Daigle, D., Pepperman, A., Hebbar, K., Lumsden, R., Anderson, T. and Sands, D. (1998). Preparation of stable, granular formulations containing *Fusarium oxysporum* pathogenicity narcotic plants, Biol. Control, 13: 79–84.

DeFaria, M. R. and Wraight, S. P. (2007). Mycoinsecticides and Mycoacaricides: a comprehensive list with worldwide coverage and international classification of formulation types, Biol. Control, 43: 237–256.

Dumas, C., Matha, V., Quiot, J.M. and Vey, A. (1996). Effects of destruxins, cyclic depsipeptide mycotoxins, on calcium balance and phosphorylation of intracellular proteins in Lepidopteran cell lines. Comp. Biochem. Physiol. C Pharmacol. Toxicol. Endocrinol., 114(3): 213–219. https://doi.org/10.1016/0742-8413(96)00041-2.

Dyck, C., Maxwell, K., Buchko, J., Trac, M., Omelchenko, A., Hnatowich, M., & Hryshko, L. V. (1998). Structure-Function Analysis of CALX1. 1, a Na^{+}-$Ca2^{+}$ Exchanger from Drosophila: mutagenesis of ionic regulatory sites. J. Biol. Chem., 273(21): 12981–12987.

Eissa, F.I., Mahmoud, H.A., Massoud, O.N., Ghanem, K.M. and Gomaa, I.M. (2014). Biodegradation of chlorpyrifos by microbial strains isolated from agricultural wastewater, J Am. Sci., 10(3): 98–108.

Eley.K.L., Halo.L.M., Song.Z., Powles.H., Cox.R.J., Bailey.A.M., Lazarus.C.M., Simpson.T.J (2007). Biosynthesis of the 2-pyridone tenellin in the insect pathogenic fungus *Beauveria bassiana*. Chembiochem, 8(3): 289–297.

El-Sayed, G.N., Ignoffo, C.M., Leathers, T.D. and Gupta, S.C. (1993). Effects of cuticle source and concentration on the expression of hydrolytic enzymes by an entomopathogenic fungus, *Nomuraea rileyi*, Mycopathologia, 122(3): 149-152.

Fan, Y., Fang, W., Guo, S., Pei, X., Zhang, Y., Xiao, Y., Li, D., Jin, K., Bidochka, M.J. and Pei, Y. (2007). Increased insect virulence in *Beauveria bassiana* strains overexpressing an engineered chitinase, Appl. Environ. Microbiol., 73(1): 295–302.

Feng, S., Wang, X., Zhang, X., Dang, P.M., Holbrook, C.C., Albert, K., Culbreath, A.K., Wu,Y. and Guo, B. (2012). Peanut (*Arachis hypogaea*) expressed sequence tag project: Progress and application. Comp. Funct. Genomics., 2012: 1–9. https://doi.org/10.1021/acs.jnatprod.1c00021.

Fuguet, R., Theraud, M. and Vey, A. (2004). Production *In vitro* of toxic macromolecules by strains of *Beauveria bassiana*, and purification of a chitosanase-like protein secreted by a melanising isolate, Comp. Biochem. Physiol. Part - C: Toxicol. Pharmacol., 138(2): 149–161.

Gatehouse, J.A. (2008). Biotechnological prospects for engineering insect-resistant plants, Plant Physiol., 146(3): 881–887.

Guliy, O.I., Ignatov, O.V., Makarov, O.E. and Ignatov, V.V. (2003). Determination of organophosphorus aromatic nitro insecticides and p-nitrophenol by microbial-cell respiratory activity. Biosens. Bioelectron, 18(8): 1005–1013. https://doi.org/10.1016/s0956-5663(02)00224-5

Hajek, A.E. and St.Leger, R.J. (1994). Interactions between fungal pathogens and insect hosts, Annu. Rev. Entomol., 39: 293–322.
https://doi.org/10.1146/annurev.en.39.010194.001453.

Hamill, Hamill R.L., Higgens, C.E., Boaz, M.E. and Gorman, M. (1969). The structure of beauvericin, a new depsipeptide antibiotic toxic to *Artemia salina*. Tetrahedron Lett., 10(49): 4255–4258. https://doi.org/10.1016/S0040-4039(01)88668-8.

Hasuda, A.L. and Bracarense, A.P.F. (2023). Toxicity of the emerging mycotoxins beauvericin and enniatins: A mini-review. *Toxicon*, 107534.

Hryshko, L.V. and Zheng, L. (2009). Crystal Structure of CBD2 from the Drosophila Na^{+}/$Ca2^{+}$ Exchanger: Diversity of $Ca2^{+}$ Regulation and Its Alternative Splicing Modification. *J. Mol. Biol*, *387*, 104–112.

Huxham, I.M., Lackie, A.M. and McCorkindale, N.J. (1989). Inhibitory effects of cyclodepsipeptides, destruxins, from the fungus *Metarhizium anisopliae*, on cellular immunity in insects, J. Insect Physiol., 35(2): 97–105. https://doi.org/10.1016/0022-1910(89)90042-5.

Ignoffo, C., Marston, N., Hostetter, D., Puttler, B. and Bell, J. (1976). Natural and induced epizootics of *Nomuraea rileyi* in soybean caterpillars, J. Invertebr. Pathol., 27(2): 191–198.

Ignoffo, C.M. (1981). The fungus *Nomuraea rileyi* as microbial insecticides. In: Burges, H.D. (Ed), Microbial control of pests and plant diseases, Academic Press, 513–538.

Jiang, L. and Ma, C.S. (2000). Progress of researches on biopesticides, Pesticides, 16: 73–77.

Jow, G.M., Chou, C.J., Chen, B.F. and Tsai, J.H. (2004). Beauvericin induces cytotoxic effects in human acute lymphoblastic leukemia cells through cytochrome c release, caspase 3 activation: the causative role of calcium, Cancer Lett., 216(2): 165–173. https://doi.org/10.1016/j.canlet.2004.06.005.

Kaur, T., Vasudev, A., Sohal, S.K. and Manhas, R.K. (2014). Insecticidal and growth inhibitory potential of *Streptomyces hydrogenans* DH16 on major pest of India, *Spodoptera litura* (Fab.) (Lepidoptera: Noctuidae), BMC Microbiol., 14: 227. https://doi.org/10.1186/s12866-014-0227-1.

Khan, S., Guo, L., Maimaiti, Y., Mijit, M. and Qiu, D. (2012). Entomopathogenic fungi as microbial biocontrol agent, Mol. Plant Breed., 3(7): 63–79.

Kirkland, B.H., Eisa, A. and Keyhani, N.O. (2005). Oxalic acid as a fungal acaracidal virulence factor, J. Med. Entomol., 42(3): 346-351. https://doi.org/10.1093/jmedent/42.3.346.

Kirkland, B.H., Westwood, G.S. and Keyhani, N.O. (2004). Pathogenicity of entomopathogenic fungi *Beauveria bassiana* and *Metarhizium anisopliae* to Ixodidae tick species *Dermacentor variabilis*, *Rhipicephalus sanguineus*, and *Ixodes scapularis*, J. Med. Entomol., 41(4): 705–711. https://doi.org/10.1603/0022-2585-41.4.705.

Kookana, R.S., Baskaran, S. and Naidu, R. (1998). Pesticide fate and behavior in Australian soils in relation to contamination and management of soil and water: A review, Aust. J. Soil Res., 36(5): 715–764.

Krassilstschik, I.M. (1888). La production industrielle des parasites vegetaux pour la destruction des insects nuisibles. Bull. Soc. chim. Fr., 19: 461–472.

Lacey, L.A., Frutos, R., Kaya, H.K. and Vail, P. (2001). Insect pathogens as biological control agents: Do they have a future?, Biol. Control, 21(3): 230–248. https://doi.org/10.1006/bcon.2001.0938.

Leathers, T.D., Gupta, S.C. and Alexander, N.J (1993). Mycopesticides: status, challenges, and potential, J. Ind. Microbiol., 12(2): 69–75. https://doi.org/10.1007/BF01569904.

Lewis, J. and Larkin, R. (1998). Formulation of the biocontrol fungus *Cladorrhinum foecundissimum* to reduce damping-off diseases caused by *Rhizoctonia solani* and *Pythium ultimum*, Biol. Control, 12(3): 182–190.

McCoy, C.W. (1990). Entomogenous fungi as microbial pesticides. In New Directions in Biological Control, Baker R.R. and Dunn P.E. (Eds.) pp. 139–159. New York: A.R. Liss.

McCoy, C.W., Samson, R.A. and Boucias, D.G. (1988). Entomogenous fungi. In: Handbook of Natural Pesticides, Vol. 5, Microbial Insecticides, Part A: Entomogenous Protozoa and Fungi, Ignoffo C.M. and N.B. Mandava, N.B. (Eds.) Boca, Raton, Fla: Mric Press. 2:157–72.

Miyahara,Y., Wakikado,T. and Tanaka, A. (1971). Seasonal changes in the number and size of the egg-masses of Prodenia litura, Jpn. J. Appl. Entomol. Zool., 15(3): 139–143. https://doi.org/10.1303/jjaez.15.139.

Morel, E., Pais, M., Turpin, M. and Guyour, M. (1983). Cytotoxicity of cyclodepsipeptides on murine lymphocytes and on L 1210 leukemia cells, Biomed. Pharmacother., 37(4): 184–185.

Morris, G.M., Huey, R., Lindstrom, W., Sanner, M.F., Belew, R.K., Goodsell, D.S. and Olson, A.J. (2009) Autodock4 and AutoDockTools4: automated docking with selective receptor flexibility. J. Comput. Chem.. 16: 2785–91.

Mracek, Z., Jiskra, K. and Kahounova, L. (1993). Efficiency of steinernematid nematodes (Nematoda: Steinernematidae) in controlling larvae of the black vine weevil, *Otiorrhynchus sulcatus* (Coleoptera: Curculionidae), in laboratory and field experiments, Eur. J. Entomol., 90(1): 71-76. https://doi.org/10.1016/j.mib.2012.04.006.

Muthukrishnan, S., Mun, S., Noh, M.Y., Geisbrecht, E.R. and Arakane, Y. (2020). Insect cuticular chitin contributes to form and function. Curr. Pharm. Des. *26*(29): 3530–3545.

Neville, A. C., Parry, D. A. D. and Woodhead-Galloway, J. (1976). The chitin crystallite in arthropod cuticle. *Journal of cell science*, *21*(1), 73–82.

Nielsen-LeRoux, C., Gaudriault, S. and Ramarao, N. (2012). How the insect pathogen bacteria *Bacillus thuringiensis* and *Xenorhabdus / Photorhabdus* occupy their hosts, Curr Opin Microbiol., 15(3): 220–231.

Ochman, H., Lawrence, J G. and Groisman, E.A. (2000). Lateral gene transfer and the nature of bacterial innovation, Nature, 405: 299–304. https://doi.org/10.1038/35012500.

Patil, R.H. and Abhilash, C. (2014). *Nomuraea Rileyi* (Farlow) Samson: a bio-pesticide ipm component for the management of leaf eating caterpillars in soybean ecosystem, International Conference on Biological, Civil and Environmental Engineering (BCEE-2014) March 17-18, 2014 Dubai (UAE), 131–133.

Pavone, D., Diaz, M., Trujillo, L. and Dorta, B. (2008). A granular formulation of *Nomuraea rileyi* Farlow (Samson) for the control of *Spodoptera frugiperda* (Lepidoptera: Noctuidae), Interciencia, 34(2): 130–134.

Pedras, M.S.C. and Biesenthal, C.J. (2000). Vital staining of plant cell suspension cultures: evaluation of the phytotoxic activity of the phytotoxins phomalide and destruxin B, Plant Cell Rep., 19(11): 1135–1138.

Pedras, M.S.C., Zaharia, I.L., Gai, Y., Zhou,Y. and Ward, D.E. (2001). In planta sequential hydroxylation and glycosylation of a fungal phytotoxin: Avoiding cell death and overcoming the fungal invader. Proc Natl Acad Sci USA, 98(2): 747–752. https://doi.org/10.1073/pnas.98.2.747.

Pell, J.K., Eilenberg, J., Hajek, A.E. and Steinkraus, D.C. (2001). Biology, ecology and pest management potential of entomophthorales. In: Butt, T.M. Jackson, C. and Magan, N. (Eds) Fungi as biocontrol agents: progress, problems and potential. CAB International, Wallingford, 1–153.

Pimentel, D., McLaughlin, L., Zepp, A., Lakitan, A., Lakitan, B., Kraus, T., Kleinman, P., Vancini, F., Roach, W.J., Graap, E., Keeton, W.S. and Selig, G. (1990). Environmental and economic impacts of reducing U.S. agricultural pesticide use. In: Pimentel, D. and Hanson, A.A. (Eds.) CRC Handbook of Pest Management in Agriculture. I: 679-718.

Pinango, L., Arnal, E., Rodriguez, B., Guerrero, I., Perez, H. and Ramos, F. (2002). Influenciade di ferentespracticas demanejosobre la f luctuacionpoblacional del gusanocogollero (Spodoptera frugiperda) en el cult ivo del ma iz. I nMem. VI Fest ivalNacionaldelMaízy VI Jornada CientíficaNacional del Maíz, (11/20-23/2002). Facultad deAgronomía, Universidad Centralde Venezuela. Maracay, Venezuela.

Qazi, S.S. and Khachatourians, G.G. (2005). Insect pests of Pakistan and their management practices: prospects for the use of entomopathogenic fungi, Biopestic. Int. 1(1-2): 13–24.

Quiot, J.M., Vey, A. and Vago, C. (1985). Effects of mycotoxins on invertebrate cells In vitro, Adv Cell Culture, 4: 199–212.

Rao, G.V.R., Wightman, J.A. and Rao, D.V.R. (1993). World review of the natural enemies and diseases of *Spodoptera litura* (F.) (Lepidoptera: Noctuidae), Insect Sci. Appl., 14(3): 273–284.

Reithinger, R., Davies, C.R., Cadena, H. and Alexander, B. (1997). Evaluation of the fungus *Beauveria bassiana* as a potential biological control agent against phlebotomine sand flies in Colombian coffee plantations. J. Invertebr. Pathol., 70(2): 131–135. https://doi.org/10.1006/jipa.1997.4671.

Roberts, D.W. and Humber, R.A. (1981). Entomogenous fungi. In: Biology of Conidial Fungi, Cole G.T. and Kendrick B.(Eds.) Academic Press, New York, 1–236.

Roddam, L.F. and Rath, A.C. (1997). Isolation and characterization of *Metarhizium anisopliae* and *Beauveria bassiana* from sub Antarctic Macquarie Island, J. Invertebr. Pathol., 69(3): 285–288. https://doi.org/10.1006/jipa.1996.4644.

Samish, M., Ginsberg, H. and Glazer, I. (2004). Biological control of ticks, Parasitology 129: S389–S403.

Samson, R.A., Evans, H.C. and Latg, J.P. (1988). Atlas of entomopathogenic fungi. Springer, Berlin Heidelberg, NewYork.

Sanda, N.B. and Sunusi, M. (2014). Fundamentals of biological control of pests, Int. J. Chem. Biol. Sci., 1(6): 1–11.

Sarkar, S., Satheshkumar, A. and Premkumar, R. (2009). Biodegradation of dicofol by Pseudomonas strains isolated from tea rhizosphere microflora, Int. J. Integr. Biol., 5(3): 164–166.

Scholte, E.J., Nghabi, K., Kihonda, J., Takken, W., Paaijmans, K., Abdulla, S., Killeen, G.F. and Knols, B.G. (2005). An entomopathogenic fungus for control of adult African malaria mosquitoes, Science, 308(5728): 1641–1642.

Schwarz, M. (1995). *Metarhizium anisopliae* for soil pest control. In: Biorational Pest Control Agents: Formulation and delivery, Hall. F. R and Barry. J. W (Eds.), American Chemical Society, Washington DC, USA. 183-196. https://doi.org/10.1021/bk-1995-0595.ch012.

Schwarz, E. M. and Benzer, S. (1997). Calx, a Na-Ca exchanger gene of Drosophila melanogaster. Proc Natl Acad Sci USA, 94(19): 10249–10254

Shah, B.P. and Devkota, B. (2009). Obsolete pesticides: Their environmental and human health hazards, J. Agric. Environ., 10: 51–56.

Shah, P., Aebi, M. and Tuor, U. (1999). Production factors involved in the formulation of *Erynianeo aphidis* as alginate granules, Biocontrol Sci. Technol., 9: 19-28. https://doi.org/10.1080/09583159929875.

Shah, P.A. and Pell, J.K. (2003). Entomopathogenic fungi as biological control agents, Appl. Microbiol. Biotechnol., 61 (5-6): 413–423.

Shapiro-Ilan, D.I., Cottrell, T.E., Mizell, R.F., Horton, D.L., Behle, B. and Dunlap, C. (2010). Efficacy of *Steinernema carpocapsae for* control of the lesser peachtree borer, *Synanthedon pictipes*: Improved aboveground suppression with a novel gel application, Biol. Control, 54(1): 23–28. https://doi.org/10.1016/j.biocontrol.2009.11.009.

Shaw, K.E., Davidson, G., Clark, S.J., Ball, B.V., Pell, J.K., Chandler, D. and Sunderland, K.D. (2002). Laboratory bioassays to assess the pathogenicity of mitosporic fungi to *Varroa destructor* (Acari: Mesostigmata), an ectoparasitic mite of the honeybee, *Apis mellifera*, Biol. Control, 24(3): 266–276.

Sierotzki, H., Camastral, F., Shah, P.A., Aebi, M. and Tuor, U. (2000). Biological characteristics of selected *Erynia neoaphidis* isolates, Mycol. Res., 104: 213–219. https://doi.org/10.1017/S0953756299001392.

Singh, B.K. and Walker, A. (2006). Microbial degradation of organophosphorus compounds, FEMS Microbiol. Rev., 30(3): 428–471.

Singh, O.P. and Gangrade, G.A. (1975). Parasites, predators and diseases of larvae of *Diaerisia obliqua* Walker (Lepi: Arctidiae) on soybean, Curr. Sci., 44: 481–482.

Smith, R.J., Pekrul, S. and Grula, E.A. (1981). Requirement for sequential enzymatic activities for penetration of the integument of the corn earworm, J. Invertebr. Pathol., 38(3): 335–344. https://doi.org/10.1016/0022-2011(81)90099-9.

St.Leger, R.J., Joshi, L., Bidochka, M.J. and Roberts, D.W. (1996). Construction of an improved mycoinsecticide over expressing a toxic protease, Proc. Natl. Acad. Sci. U.S.A., 93: 6349–6354. https://doi.org/10.1073/pnas.93.13.6349.

St.Leger, R.J. and Wang, C. (2009). Entomopathogenic fungi and the genomic era. In: Insect Pathogens: Molecular Approaches and Techniques, Stock, S.P., J. Vandenberg, J., Glazer, I. and Boemare, N. (Eds.), CABI, Wallingford, UK, 365–400.

Steinhaus, E.A. (1975). Disease in a minor chord. Ohio State University Press, Columbus, Ohio.

Steinrauf, L.K. (1985). Beauvericin and the other enniatins, Metal Ions Biol. Syst, 19: 139–171.

Stirling, G. and Smith, L. (1998). Field tests of formulated products containing either *Verticillium lecanii* or *Arthrobotrys dactyloides* for biological control of root-knot nematodes, Biol. Control, 11: 231–239.

Stuart, R.J., Barbercheck, M.E., Grewal, P.S., Taylor, R.A.J. and Hoy, C.W. (2006). Population biology of entomopathogenic nematodes: Concepts, issues, and models, Biol. Control, 38(1): 80–102. https://doi.org/10.1016/j.biocontrol.2005.09.019.

Thomas, M.B. and Read, A.F. (2007). Can fungal biopesticides control malaria?, Nat. Rev. Microbiol, 5: 377–383.

Togashi, K.I., Kataoka,T. and Nagai, K. (1997). Characterization of a series of vacuolar type H+-ATPase inhibitors on CTL-mediated cytotoxicity, Immunol. Lett., 55(3): 139–144. https://doi.org/10.1016/s0165-2478(97)02698-9.

Verma, A., Malik, C.P., Gupta, V.K. and Bajaj, B.K. (2011). Effects of In vitro triacontanol on growth, antioxidant enzymes, and photosynthetic characteristics in Arachis hypogaea L, Braz. J. Plant Physiol., 23(4): 271–277.

Wang, C. and St.Leger, R.J. (2007). The *Metarhizium anisopliae* perilipin homolog MPL1 regulates lipid metabolism, appressorial turgor pressure, and virulence, J. Biol. Chem., 282(29): 21110–21115. https://doi.org/10.1074/jbc.M609592200.

Watkins, P.R., Huesing, J.E. and Margam,V. (2012). Insects, nematodes, and other pests. In: Plant Biotechnology and Agriculture Prospects for the 21st Century, Altmanand, A. and Hasegawa, P.M. (Eds.), Elsevier, London, UK, : 353–370.

Yan, X., Han, R.C., Moens, M., Chen, S.L. and DeClercq, P. (2013). Field evaluation of entomopathogenic nematodes for biological control of striped flea beetle, Phyllotreta striolata (Coleoptera: Chrysomelidae), BioControl, 58(2): 247–256. https://doi.org/10.1007/s10526-012-9482-y.

Zhang, L., Yan, K., Zhang, Y., Huang, R., Bian, J., Zheng, C., Sun, H., Chen, Z., Sun, N., An, R. and Min, F. (2007). High-throughput synergy screening identifies microbial metabolites as combination agents for the treatment of fungal infections, Proc. Nat. Acad. Sci., 104(11): 4606–4611. https://doi.org/10.1073/pnas.0609370104

12 Metabolic Profile of Entomopathogenic Fungi

M. Leyte-Lugo,[1] S. Páez-León,[2,3] L. Aguilar-Marcelino,[2]
J.F. Palacios-Espinosa[4] and G.S. Castañeda-Ramírez[3*]

1. Introduction

The kingdom Fungi comprises a significant group of heterotrophic eukaryotic microorganisms in terrestrial and aquatic ecosystems, thriving across diverse substrates and climatic conditions (Araújo and Hughes, 2016; Aguilar-Marcelino et al., 2020). This group is characterized by chitin and glucan cell walls. They usually consist of filaments and multiple cells, displaying straightforward structures composed of tube-shaped hyphae that form modular structures with unlimited growth potential. However, certain species may exist in a single-celled vegetative state. They obtain nutrition by releasing degradative enzymes and absorbing the resulting broken-down products (Chandler, 2017).

Fungi commonly coexist near each other within their natural environments, engaging in positive or negative interactions with various microbes and organisms. These fungi assume different symbiotic associations, including saprophytic, pathogenic, or endophytic. Beneficial associations assist hosts by stimulating their growth and secondary metabolite production and improving resistance to biotic and abiotic stresses. Conversely, pathogenic fungi can pose a significant threat, causing diseases that can adversely impact microorganisms, organisms, and environmental ecosystems (Dai et al., 2023). Fungi encompass five distinct phyla, namely Chytridiomycota, Zygomycota, Ascomycota, Basidiomycota, and Glomeromycota (Naranjo-Ortiz and Gabaldón, 2019; Aguilar-Marcelino et al., 2020).

As a parasite class, entomopathogenic fungi (EPF) represent a distinctive category of soil-dwelling microorganisms that serve as crucial natural regulators by infecting and eliminating insects and other arthropods. In essence, EPF play a vital role as biocontrol agents in regulating insect populations (Goettel et al., 2005; Sinha et al., 2016; Chandler, 2017; Aguilar-Marcelino et al., 2020; Mantzoukas et al., 2022). These

[1] Investigadora por México CONAHCYT Comisionado al Departamento de Sistemas Biológicos, División de Ciencias Biológicas y de la Salud, Universidad Autónoma Metropolitana–Xochimilco (UAM–X), C.P. 04960. México City, México.

[2] Centro Nacional de Investigación Disciplinaria en Sanidad e Inocuidad Animal, INIFAP, Km 11 Carretera Federal Cuernavaca-Cuautla, No. 8534, C.P. 62550. Jiutepec, Estado de Morelos, México.

[3] Centro de Investigación en Biotecnología (CEIB), Universidad Autónoma del Estado de Morelos (UAEM), C.P. 62209. Cuernavaca, Morelos, México.

[4] Departamento de Sistemas Biológicos, División de Ciencias Biológicas y de la Salud, Universidad Autónoma Metropolitana–Xochimilco (UAM–X), C.P. 04960. México City, México.
Email: sarahi.castaneda@uaem.mx

* Corresponding author: gs.castanedaramirez@gmail.com

fungi infect various types of insects at different life stages, including eggs, larvae, pupae, nymphs, and adults. They are found in 20 of the 31 insect orders (Araújo and Hughes, 2016; Kaczmarek and Boguś, 2021a; Mantzoukas et al., 2022).

Unlike conventional insecticides, EPF do not necessarily need to be ingested to impede or kill target pests; mere physical contact is sufficient (Fig. 1). These fungi generate conidia that adhere to the insect's outer covering (the insect cuticle), triggering an enzymatic process involving proteases, peptidases, chitinases, and lipases. This enzymatic activity, combined with the formation of appressoria, mechanically breaches the insect's protective layer, facilitating entry into the insect's body for invasion and extraction of nutrients from its tissues (Brunner-Mendoza et al., 2019; Aguilar-Marcelino et al., 2020; Sharma and Sharma, 2021; Shaurub, 2022).

EPF are any fungal species causing insect infection, remaining within four major classes: Oomycetes, Chytridiomycetes, Ascomycetes, and Zygomycetes (Goettel et al., 2005). Although some insect-killing fungi belong to other fungal groups, most are found within the Ascomycete and Zygomycete classes (Sinha et al., 2016; Chandler, 2017; Esparza Mora et al., 2017). It is worth noting that the limited presence of insect-killing lineages in Basidiomycota indicates that their evolutionary diversification has likely been influenced by ecological roles distinct from parasitizing insects (Esparza Mora et al., 2017).

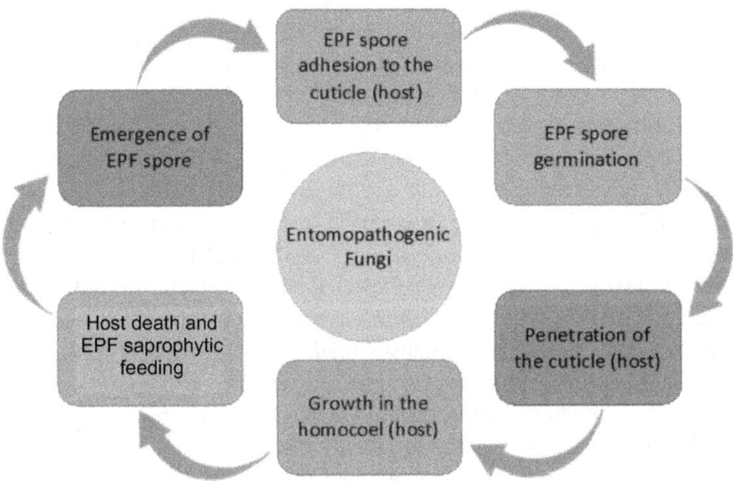

Fig. 1 EPF mode of infection against host.

EPF as biocontrol agents offer several advantages over traditional insecticides, including high reproductive capabilities, target-specific activity, short generation times, and the ability to enter a resting or saprobic phase, ensuring prolonged survival without hosts. They are cost-effective, efficient, safe for beneficial organisms, reduce environmental residues, and help enhance biodiversity in ecosystems managed by humans (Sinha et al., 2016). Hence, utilizing entomopathogenic fungi for myco-biocontrol holds great importance, addressing environmental and food safety considerations (Sinha et al., 2016; Shaurub, 2022).

Despite approximately 800 identified fungal species pathogenic to insects, most commercial mycoinsecticides are developed based on only three EPF species: *Beauveria bassiana*, *Metarhizum anisopliae*, and *Isaria fumosoroseus* (Maina et al., 2018; Aguilar-Marcelino et al., 2020).

2. Entomopathogens of Different Phyla

The classification of fungal organisms has relied heavily on their ultrastructure and morphology as fundamental criteria, a proposition substantiated by subsequent DNA sequence analyses (Alexopoulos et al., 1996; Sharma and Sharma, 2021). A more precise categorization of EPF into well-defined scientific taxa has been achieved through detailed examinations of their digestive systems and the application of genomic analysis (Sharma and Sharma, 2021).

Fungi, classified as true fungi, are currently organized into four legitimate phyla: Basidiomycota, Ascomycota, Zygomycota, and Chytridiomycota (Naranjo-Ortiz and Gabaldón, 2019; Sharma and Sharma, 2021). Conversely, oomycetes have found their place within the kingdom Stramenopila, specifically in the phylum Oomycota (Barr, 1992; Sharma and Sharma, 2021).

2.1 Ascomycota

The Ascomycota phylum, the largest group within the kingdom Fungi, boasts approximately 64,000 described species (Lutzoni et al., 2004; Schoch et al., 2009; McLaughlin and Spatafora, 2015; Esparza Mora et al., 2017); most of these fungi are filamentous, forming regularly segmented hyphae, with a notable feature being the production of sexual spores (ascospores) within sac-like structures called asci. Mating triggers the development of dikaryotic hyphae, eventually leading to the formation of the ascus. Unlike Basidiomycota, the ascus in Ascomycota typically has a very short life (Goettel et al., 2005; Naranjo-Ortiz and Gabaldón, 2019).

Given its status as the most species-rich fungal group, Ascomycota exhibits diverse ecological roles, serving as decomposers, plant pathogens, human and animal pathogens, and participating in mutualistic relationships such as lichens (Alexopoulos et al., 1996). Most entomopathogenic species within Ascomycota exhibit a well-established parasitic phase that targets the host's body. After killing the insect, this group shifts to a saprophytic mode of nutrition (hemibiotrophic), sustaining hyphal growth even after the host's demise (Joop and Vilcinskas, 2016; Araújo and Hughes, 2016). This phylum encompasses many hosts and includes insect pathogens such as Pleosporales, Myriangiales, and Ascosphaerales, albeit with relatively few species. The hyper-diverse Hypocreales stands out as the largest group of entomopathogens (Araújo and Hughes, 2016). The entomopathogenic species are from over 40 genera, with representative species in *Aschersonia*, *Aspergillus*, *Beauveria*, *Culicinomyces*, *Gibellula*, *Hirsutella*, *Hymenostilbe*, *Lecanicillium* (formerly *Verticillium*), *Metarhizium*, *Nomuraea*, *Paecilomyces*, *Sorosporella*, and *Tolypocladium* (Goettel et al., 2005). Some species of Ascomycota are shown in Table 1.

Table 1 Examples of EPF and their hosts. Main lineages within Ascomycota.

Fungi	*Host*	*Reference*
Beauveria bassiana	Aphids, whiteflies, plant bugs, silkworms, and mites	Basnet et al., 2022 Butt et al., 2016 de Freitas Soares et al., 2020 Goettel et al., 2005
Cordyceps spp.	Silkworm and caterpillars	
Metarhizium brunneum (*M. anisopliae*)	Thrips, whiteflies, mites, weevils, and ticks	
Isaria fumosorosea (*Paecilomyces fumosoroseus*)	Whiteflies, aphids, thrips, plant bugs, mites, and some soil pests	Basnet et al., 2022
Purpureocillium lilacinum	Plant parasitic nematodes	Basnet et al., 2022
Verticillium lecanii (*Lecanicillium lecanii*)	Aphids, whiteflies, and scales	

2.2 Basidiomycota

This group, alongside Ascomycota, constitutes the subkingdom Dikarya, characterized by a dikaryotic phase (Hibbett et al., 2007). Dikarya is distinguished by a sexual reproduction process where hyphae fuse without undergoing meiosis, forming hyphae containing two distinct nuclei (dikaryotic hyphae). Additionally, numerous species have septate hyphae, utilizing ergosterol as a membrane sterol, and specific lineages possess the ability to develop multicellular reproductive or vegetative structures (Naranjo-Ortiz and Gabaldón, 2019; Sinha et al., 2016).

Basidiomycota are among the most complex fungi considering their cell cycles and multicellular structures. The group is distinguished by its production of sexual spores called basidiospores, which are formed externally in specialized reproductive cells called basidia. The mating process usually involves the fusion (anastomosis) of two hyphae and the creation of a dikaryon, where hyphae contain two stable nuclei. An essential and unique feature within the group is the presence of clamp connections, structures formed during nuclear division at the tips of growing septate hyphae, facilitating the maintenance of the dikaryotic condition (Alexopoulos et al., 1996; Naranjo-Ortiz and Gabaldón, 2019; Sinha et al., 2016).

Basidiomycetes exhibit significant ecological characteristics, such as colonizing dead wood and serving as decomposers of leaf litter on forest floors. Also, these species cause substantial agricultural losses and impact forest environments by affecting trees and conifers (Kendrick, 2000; Braga-Neto et al., 2008; Araújo and Hughes, 2016; Sinha et al., 2016). Some species are animal pathogens and attack nematodes. Specific genera are recognized as pathogens of insects, infecting scale insects (e.g., *Septobasidium* and *Uredinella*, order Septobasidiales) and termite eggs of the genus *Reticulitermes* (e.g., *Fibularhizoctonia*, order Atheliales) (Goettel et al., 2005; Sharma et al., 2020; Sinha et al., 2016).

2.3 Chytridiomycota

This category encompasses zoosporic fungi, commonly called chytrids, characterized by coenocytic hyphae and chitin cell walls (Gleason et al., 2012). Within this group, single-flagellated zoospores and gametes settle and develop into a thallus, eventually transforming into resting spores or coenocytic hyphae. Chytridiomycetes are the largest class of zoosporic fungi, boasting approximately 1000 described species (Naranjo-Ortiz and Gabaldón, 2019). These organisms exist in unicellular, colonial, or filamentous forms, utilizing absorptive nutrition and reproducing through the generation of motile zoospores (Barr, 2001). Remarkably, *Script letter* is the only phylum within the Fungi kingdom to incorporate motile cells at least once during the life cycle (Araújo and Hughes, 2016).

Most chytrids are prevalent as saprophytic organisms, particularly in freshwater and damp soils, with some species even thriving in marine environments (Gleason et al., 2012; Araújo and Hughes, 2016). Nonetheless, a noteworthy portion of the chytrid species is recognized for their parasitic relationships with plants, algae, amphibians, animals, rotifers, tardigrades, and protists, as well as various invertebrates, including nematodes and insects (Betancourt-Román et al., 2016; Sinha et al., 2016; Kaczmarek and Boguś, 2021a).

The genus *Coelomomyces* within the Blastocladiales houses the most entomopathogenic Chytridiomycetes (Barr, 2001; Goettel et al., 2005). These species, primarily mosquitoes and Heteroptera, have been identified in Diptera. More entomopathogenic species have been identified within *Coelomycidium* (Blastocladiales) and *Myriophagus* (Chytridiales). *Coelomycidium* is linked to blackflies and mosquitoes, whereas *Myriophagus* has been observed as a pathogen affecting dipterous pupae (Goettel et al., 2005).

Recent research has reclassified the nephridiophagids within *Chytridiomycota* (Strassert et al., 2021; Kaczmarek and Boguś, 2021b). Nephridiophagids are unicellular, spore-forming parasites that target the Malpighian tubules of insects, with a particular affinity for cockroaches (*Dictyoptera*) and beetles (*Coleoptera*); their primary localization is within the lumen of these tubules (Radek et al., 2011). Some examples of entomopathogenic Chytridiomycota are shown in Table 2.

Table 2 Examples of EPF and their hosts. Main lineages within Chytridiomycota.

Fungi	*Host*	*Reference*
Coelomomyces africanus	*Anopheles nigerrimus*	Scholte et al., 2004
Coelomycidium simulii	Larval stage of black flies (Diptera): *Simulium asakoae*, *S. chiangmaiense*	Jitklang et al., 2012; Sinha et al., 2016; Sharma et al., 2020; Kaczmarek and Boguś, 2021b
Myriophagus spp.	Hemipterans and dipterans, mosquitoes and flies	Sinha et al., 2016; Sharma et al., 2020
Nephridiophaga archimandrite	*Archimandrita tessellate* (Giant Peppered Roach)	Radek et al., 2011; Kaczmarek and Boguś, 2021a
N. maderae	*Leucophaea maderae* (Madeira cockroach)	Radek et al., 2011
N. lucihormetica	*Lucihormetica verrucosa* (warty glowspot cockroach)	Radek et al., 2011; Kaczmarek and Boguś, 2021a

2.4 *Mucoromycota*

Mucoromycota encompasses the most extensive and extensively studied category of zygomycete fungi. Many of these fungi function as saprobes, although some also serve as non-haustorial parasites of plants and other fungi or engage in ectomycorrhizal associations (Zhang and Hyde, 2014; Araújo and Hughes, 2016; Naranjo-Ortiz and Gabaldón, 2019).

Mucoromycotina is the largest and most morphologically diverse order among zygomycetes. Only one entomopathogenic fungus—*Sporodiniella umbellate*—is identified within the subphylum. This species is known to thrive on various insects, particularly membracids, members of the *Umbonia* genus, which are plant-feeding insects belonging to the order Hemiptera (Evans and Samson, 1977; Araújo and Hughes, 2016; Esparza Mora et al., 2017).

2.5 *Oomycota*

The Oomycota comprise a genetically and morphologically diverse group, able to form hyphae or exist as uncomplicated holocarpic thalli. This phylum encompasses a minimum of 1500 fungal species distributed across 100 genera (Beakes and Thines, 2017; Kaczmarek and Boguś, 2021a). Oomycetes exhibit a range of ecological roles, serving as saprotrophs by thriving on decaying matter or acting as parasites on higher plants. Additionally, they can adapt to various habitats, including aquatic, amphibious, or terrestrial environments.

The oomycetes, classified within the kingdom Straminipila and alternatively referred to as Heterokonta, Heterokontobionta, or Chromista (Piepenbring, 2015; Kaczmarek and Boguś, 2021a), are filamentous eukaryotic microorganisms. Distinguishing them from true fungi, the cell walls of oomycetes consist of cellulose derivatives, serving as crucial structural elements, rather than the chitin found in true fungi (Hassett et al., 2019; Klinter et al., 2019; Kaczmarek and Boguś, 2021a). The coenocytic hyphae of these fungi contain cellulose without chitin, and they produce biflagellate zoospores (Goettel et al., 2005). The process of sexual reproduction occurs either on the same hyphae or different hyphae, involving gametangia such as antheridia and oogonia, with an exception being *Periplasma isogametum* (Sinha et al., 2016; Martin and Warren, 2020; Kaczmarek and Boguś, 2021a).

Pathogenic oomycetes function as parasites, targeting various organisms, including algae, plants, protists, fungi, arthropods, and even vertebrates, including humans. Among them, twelve species of entomopathogenic Oomycota have been identified, distributed across six genera: *Lagenidium*, *Leptolegnia*, *Pythium*, *Crypticola*, *Couchia*, and *Aphanomyces* (Miao and Nauwerck, 1999; Araújo and Hughes, 2016; Mendoza et al., 2018; Kaczmarek and Boguś, 2021b; Sinha et al., 2016). It is important to highlight that *Lagenidium giganteum*, a parasite of mosquito larvae, is likely the most extensively studied and well-known entomopathogenic Oomycota. Certain *Lagenidium* species are pathogenic to crabs and other aquatic crustaceans (Goettel et al., 2005; Hatai et al., 2000). Examples of Oomycota species are shown in Table 3.

Table 3 Examples of EPF and their hosts. Main lineages within Oomycota.

Fungus	*Host*	*Reference*
Aphanomyces laevis	*Anopheles* sp. (mosquito larvae)	Kaczmarek and Boguś, 2021a
Lagenidium giganteum		Goettel et al., 2005; Mumo Sila et al., 2023
Leptolegnia chapmanii		Sinha et al., 2016 Muniz et al., 2018
L. caudate	*Leptodora kindti*	Miao and Nauwerck, 1999; Gutierrez et al., 2017
Lagenidium ajelloi	*Anopheles* sp. (mosquito)	Mumo Sila et al., 2023
Pythium carolinianum		Su et al., 2001
P. guiyangense		Shen et al., 2019
Crypticola clavulifera		Mendoza et al., 2018
Couchia amphora		Wallace Martin, 2000
C. linnophila		

2.6 *Zoopagomycota*

This phylum represents the earliest diverging group among non-flagellated fungi, encompassing three primary lineages: Zoopagomycotina, Entomophthoromycotina, and Kickxellomycotina (Araújo and Hughes, 2016; Esparza Mora et al., 2017). The ability to form true mycelia is a distinctive feature within these lineages. While Kickxellomycotina and Zoopagomycotina primarily consist of saprobes, certain families within Zoopagomycotina are recognized for their predation on nematodes (Zhang and Hyde, 2014; Araújo and Hughes, 2016).

Entomophthoromycotina is divided into three main classes: Basidiobolomycetes, Neozygitomycetes, and Entomophthoromycetes (Humber, 2012; Naranjo-Ortiz and Gabaldón, 2019). Basidiobolomycetes, the earliest-diverging group, includes the genus Basidiobolus. This genus is a saprotrophic organism found in the digestive tracts of amphibians and reptiles, and it occasionally causes infections in humans. The second class, Neozygitomycetes, comprises several genera that parasitize mites and aphids. However, due to limited sequencing data, there are uncertainties regarding its taxonomic classification, placement in the phylogenetic tree, and whether it truly belongs to this subphylum (White et al., 2006; Gryganskyi et al., 2013; Naranjo-Ortiz and Gabaldón, 2019).

Entomophthoromycetes, the most species-rich and well-characterized class in the subphylum, predominantly feature the genus *Conidiobolus* (family Ancylistaceae). *Conidiobolus* is described as a saprophyte, a facultative insect parasite, and an occasional human pathogen. This class further includes various families primarily specializing in insect parasitism, forming a distinct monophyletic group, along with small genera exhibiting diverse ecological strategies (Naranjo-Ortiz and Gabaldón, 2019; Kaczmarek and Boguś, 2021b).

2.7 *Zygomycota*

Zygomycete fungi were initially categorized under a single phylum, Zygomycota, characterized by sexual reproduction through zygospores, filamentous and non-flagellated in nature, a lack of multicellular sporocarps, and the generation of coenocytic hyphae, with a few exceptions (Araújo and Hughes, 2016; Naranjo-Ortiz and Gabaldón, 2019; Sinha et al., 2016).

A more refined phylogenetic classification is proposed in this study, introducing two phyla, six subphyla, four classes, and 16 orders (NCBI, 2024). Based on these findings, the phyla Mucoromycota and Zoopagomycota are outlined (Spatafora et al., 2016). Zoopagomycota includes Entomophthoromycotina, Kickxellomycotina, and Zoopagomycotina, constituting the most anciently diverged lineage (paraphyletic phylum) of zygomycetes. This group comprises species predominantly functioning as parasites and pathogens of small animals (such as amoeba and insects) and other fungi. Mucoromycota, on the other hand, encompasses Glomeromycotina, Mortierellomycotina, and Mucoromycotina (Naranjo-Ortiz and Gabaldón, 2019). Table 4 presents several instances of entomopathogenic Zygomycota.

Table 4 Examples of EPF and their hosts. Main lineages within Zygomycota.

Fungi	*Host*	*Reference*
Batkoa apiculate	*Anopheles* sp. (mosquito)	Sánchez et al., 2010; Sharma et al., 2020
Conidiobolus coronatus	*Sarcophaga argyrostoma* (flesh fly)	Kaczmarek and Boguś, 2021b
Entomophaga grylli	*Praxibulus* sp. (Australian grasshopper)	Bidochka et al., 1995
Entomophthora planchoniana and *Pandora neoaphidis*	*Sitobion avenae Rhopalosiphum padi* (cereal aphids)	Ben Fekih et al., 2019
Pandora delphacis	Brown plant hopper and green hopper	Basnet et al., 2022
Neozygites parvispora	*Frankliniella occidentalis*	Montserrat et al., 1998; Sharma et al., 2020
N. fresenii	*Aphis gossypii*	Steinkraus and Boys, 2005; Sharma et al., 2020

3. Secondary Metabolites of Entomopathogenic Fungi

There are many reviews concerning production systems, biosynthesis of secondary metabolites (SM), and the biological activities of EPF. In this section, we describe the most important SM biosynthesized by EPF from the species of the genus *Beauveria*, *Penicillium*, *Cordyceps*, *Metharizium*, and others. Additionally, we present information gathered about the most recent discoveries of new SM from this important group of fungi from 2020 to 2023. We include the most recent reviews from the previous five years.

In the reviews performed by Zhang et al. (2020 a, b), the first reports recent advances in the studies of biosynthetic gene clusters (BCG) that may be linked to SM product biosynthesis, concentrating on BGCs with polyketide synthase (PKS), nonribosomal peptide synthase (NRPS), or PKS-NRPS hybrids. The review details that between 2014 and 2019, 31 new genome sequences were reported, representing a vast amount of material to analyze in the relationship between BCGs and SMs in fungi. The second review reports the structure and biological activities of SMs in Hypocrealan Entomopathogenic Fungi (HEF), covering the period from 2014 to 2019, with special attention to several genera, and describes many groups of SMs, such as polyketides, peptides (cyclic and acyclic), pyridones, alkaloids, terpenoids, and others.

We take this opportunity to demonstrate some representative SMs, but we encourage the readers to consult these two documents. Another interesting review, by Niu et al. (2020), was conducted concerning the diversity of linear nonribosomal peptides from biocontrol fungi. These kinds of compounds are biosynthesized through NRPS. As examples, we include select SMs from this document, like the cicadapeptins from *Cordyceps heteropoda* ARSEF1880, leucinostatins from *Paecilomyces lilacinus* CG-189, and Linear nonribosomal peptides (L-NRP) d-(L-a-aminoadipyl)-L-cysteinyl-D-valine (ACV).

One of the most significant EPF species is *Beauveria bassiana.* It has a broad spectrum of action against arthropods and insects. Its pathogenic mechanism depends not only on its ability to evade the insect's immune response but also on the production of SMs with antimicrobial properties. Some of these SMs include the cyclic peptides beauvericin, bassianolide, and beauverolides, the quinone oosporein, and the pigments tenellin and bassianin (Keswani et al., 2019; Ávila-Hernández et al., 2020). These compounds are included in this section.

A complete review of extracts, mycotoxins, and SMs obtained from *Penicillium* species was published by Nicoletti et al. (2023). This document includes an extensive list of *Penicillium* species isolated from insects. The review includes information about the evaluation of culture extracts. However, it does not report the characterization of bioactive compounds, description of the principal mycotoxins isolated from *Penicillium* species (for example, citrinin, ochratoxin A, or brevianamide A), or other metabolites.

3.1 Recent findings of SMs from Entomopathogenic Fungi

In a study performed by Rosas-García et al. (2020), 11 strains of *B. bassiana* were analyzed to identify two genes (*bbBeas* and *bbBsls*) related to the production of beauvericin and bassianolide. The researchers found the presence of both genes in almost all the strains analyzed. Also, the researchers quantitated both compounds in mycelial extracts by mass spectrometry. However, they identified only the presence of bassianolide in all strains. Additionally, they evaluated the mortality provoked by bassianolide-rich extracts on *Spodoptera exigua* and *S. frugiperda*. They observed different behaviors between strains, with a maximal mortality of 44% in *S. frugiperda* neonate larvae.

In the *Cordyceps* genus, there are many species of medicinal and industrial importance; one of them is *Cordyceps militaris*. As active metabolites, nucleosides are the main compounds biosynthesized by this fungus group. 3'-deoxyadenosine, also named Cordycepin, is one of the nucleosides isolated from *C. militaris*. Woolley et al. (2020) described the evaluation of this nucleoside on immune-related gene expression in *Drosophila melanogaster* cells and the insect *Galleria mellonella.* In the coadministration of 25 mg/mL of Cordycepin and the immunostimulant curdlan (20 mg/mL) in *Drosophila* S2r+ cells, they found that the increase in the expression of the immune effector gene *metchnikowin* was not significant. Besides, Cordycepin injection into *G. mellonella* larvae provoked mortality with an LC_{50} of 2.1 mg/insect.

According to the review by Zhang et al. (2020a), there are reports of 31 new genome sequences for BCGs that could be implicated in the production of SMs in fungi, and this represents valuable information to analyze. Most of the time, related SMs are biosynthesized by intimately related fungal species. Yin et al. (2020) reported the production of beauveriolides (BVDs) by fungi that infect insects. *Beauveria bassiana* and *B. brongniartii* produced, from a structural point of view, distinct BVD compounds. Meanwhile, *B. brongniartii* and *Cordyceps militaris* biosynthesized structural-related SMs under the same growth conditions. The authors confirmed that BCG possessed four genes involved in the production of BVD, which were present in the three fungi.

The discovery of new compounds in EPF has continued worldwide. Elshamy et al. (2020) reported the isolation, structural determination, and antioxidant effect of Ophiocordylongiiside A, a new cerebroside-type SM produced by *Ophiocordyceps longiissima* (NBRC 106965). Furthermore, a known holestane-derivative was also isolated, and *n*-butanol extract from the fungus *O. longiissima* NBRC 106965 afforded Ophiocordylongiiside A, in addition to 22*E*,24*R*-24-methylcholest-5,22-diene-3-*O*-b-D-*O*-glucopyranoside. The IC_{50} of antioxidant activity on the DPPH-free radical scavenging method for Ophiocordylongiiside A was 55.72 mg/mL.

From an ethyl acetate extract of culture broth of *Isaria* sp., a new compound named 7-*O*-methylisariotin C and two known isariotin-type compounds (TK-57-164A and B) were isolated. In their paper, Yahagi et al. (2020) report the antiproliferative and antimigration activities of the new compound. It showed inhibition of PANC-1 cell growth with an IC_{50} of 4.8 mM, and it provoked and inhibited migration by upregulating the expression of the *E-cadherin* and lowering the *N-cadherin* and *Snail* genes in the range of 0.1-1 mM.

Alkaloids are a relevant group of secondary metabolites produced by EPF. Anggiani et al. (2020) reported the identification of quinidine in chloroform extracts of *Beauveria* sp. and *Aspergillus sclerotiorum* cultures. Each fungus grew in a PDA medium in static conditions, pH of 6.2, at room temperature for a 7, 14, and 21-day incubation period. The researchers concluded that the production of alkaloids depends on fungus species and the incubation period. *Beauveria* sp. produced a higher concentration of quinidine.

Other studies report quantifying select secondary metabolite groups, such as phenols and flavonoids. For example, from the methanolic extract of the *Isaria tenuipes* mycelium, Chhetri et al. (2020) reported the quantification of phenols and flavonoids.

Furthermore, they evaluated the antioxidant effects in DPPH, ABTS, hydroxyl radical scavenging, and metal-chelating assays, as well as the antiproliferative evaluation on three cancer cells. The results showed a content of phenols of 140.09 ± 3.51 mg gallic acid equivalent/mg and 9.02 ± 0.95 mg quercetin/mg of flavonoids. Concerning the antioxidant activity, the extract showed an IC_{50} of 5.04 mg/mL in the DPPH test and a percentage inhibition of an iron-chelating assay of 86.76% (IC_{50} of 4.43 mg/mL). On the other hand, the extract exhibited percentages of radical scavenging of 44.42% and 49.82% at 10 mg/mL, using the ABTS and hydroxyl assays. Finally, in the antiproliferative assays using HeLa, PC3, and HepG2 cells, the IC_{50} calculated for the extract was 43.45 mg/mL, 119.33 mg/mL, and 125.55 mg/mL, respectively.

Certain studies document the evaluation and identification of main constituents in extracts employing instrumental techniques like GC-MS, IR, or NMR. In this sense, Vivekanandhan et al. (2020) report the larvicidal effect of the ethyl acetate-soluble fraction obtained from the biomass of *Metarhizum anisopliae*. The researchers used *Aedes aegypti*, *Anopheles stephensi*, and *Culex quinquefasciatus* larvae in the study. They identified five main constituents from this organic extract: hexatriacontane, 9-octyleicosane, heptacosane, nanocosane, and (*Z,Z*)-3,6-cis-9,10-epoxyheptadecadie, and found that these SMs showed strong larvicidal activity.

Arunthirumeni et al. (2023) identified 20 distinct compounds, predominantly esters, and long-chain fatty acids, from the bioactive ethyl acetate soluble fraction of *Penicillium* sp. This extract exhibited efficacy against *Spodoptera litura* and *Culex quinquefasciatus* insects. These compounds include acetic acid, propanoic acid-ethyl ester (propyl ester, isopentyl acetate, and 2-methylpropyl ester), organic acids (*n*-hexadecanoic acid and octadecanoic acid), alcohols (behenic alcohol, 1-hexacosanol, and 1-tetradecanol), long-chain hydrocarbons (1-hexadecene, 1-octadecene, and 1-dodecene), and nitrogen compounds (tetrydamine, phenylmethyl N-1-[(3-hydroxyphenyl)amino carbonyl]).

As we can see, EPF is a source of SMs; with increasing numbers of investigations conducted. We can see that the number of reports describing new compounds is growing and reviewing "state of the art" understanding on this subject is a frequent necessity. Some of the secondary metabolites that have been obtained from EPF are illustrated below (Table 5).

Microscopic EPF produces SM derived from the interaction between the fungus and insect. Some of these compounds have been observed to exhibit insecticidal activity (Vey et al., 2001). Some of the mechanisms of action of SM that have been reported involve the opening of Ca^{2+} channels, changes in morphology, reduced peptide expression, and oxidative stress (Samuels et al., 2002). The metabolites reported in EPF are presented in Table 6.

However, insecticidal activity has also been reported in the basidiomata and mycelium of macroscopic fungi (De Silva et al., 2012). In a study, 175 fungi were evaluated against *Drosophila melanogaster* and *Spodoptera littoralis*, reporting insecticidal activity in different fungi (Mier et al., 1996). Some studies identified molecules that could be responsible for insecticidal activity, as shown in Table 7.

Table 5 Secondary metabolites from entomopathogenic fungi (classification based on chemical structure).

Non-ribosomal cyclooligomeric depsipeptides		
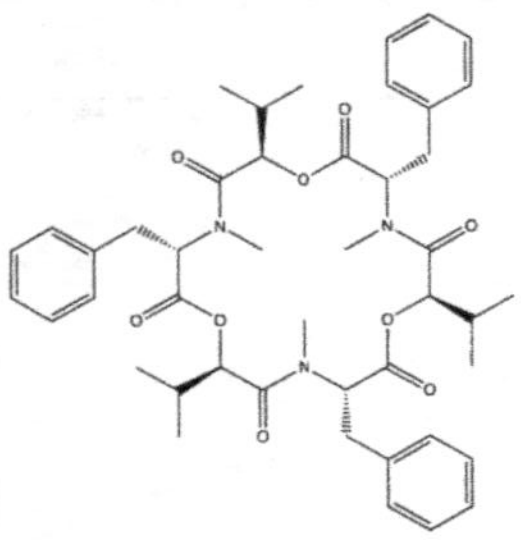	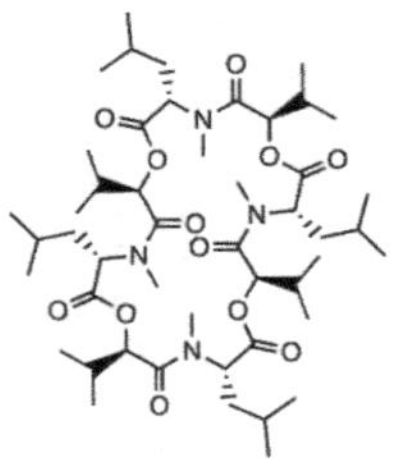	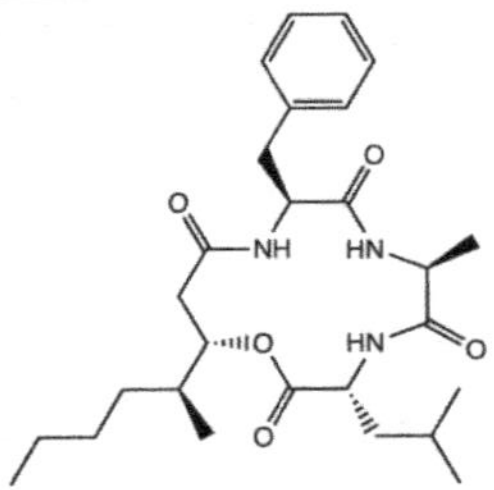
Beauvericin *Beauveria bassiana* *Isaria* sp. Insecticidal, antitumor, antimicrobial, antiviral Keswani et al., 2019; Amobonye et al., 2020; Ávila-Hernández et al., 2020; Wang et al., 2021; Yahagi et al., 2020	**Bassianolide** *Beauveria bassiana* Toxic for *Spodoptera exigua* and *S. frugiperda* Keswani et al., 2019; Amobonye et al., 2020; Rosas García et al., 2020; Wang et al., 2021	**Beauverolide I** *Beauveria* sp. FO-6979 *Cordyceps militaris* Inhibitor of acyl-CoA: cholesterol acyltransferase (ACAT) Keswani et al., 2019; Wang et al., 2020
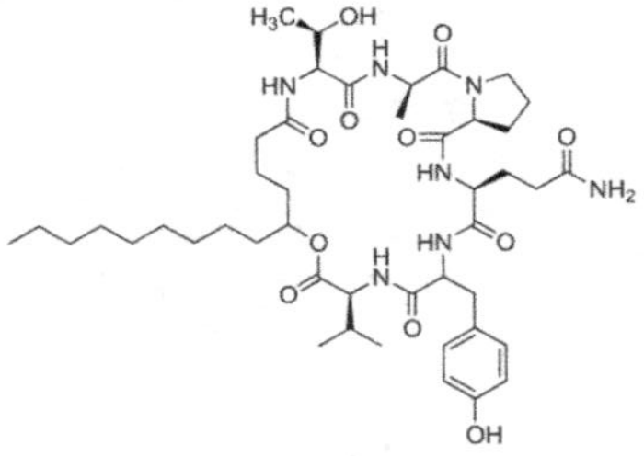	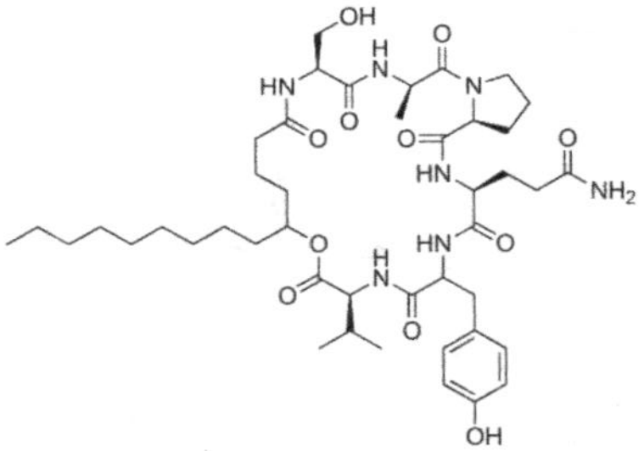	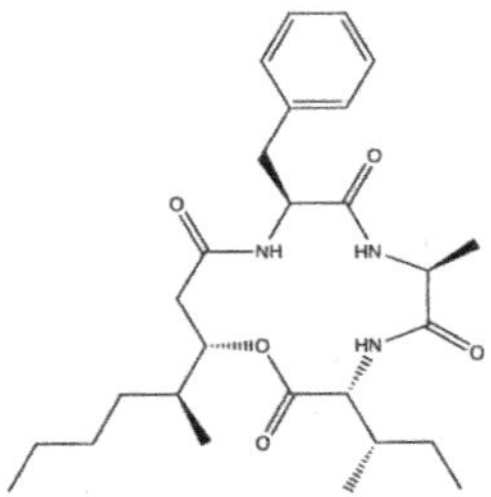
Verlamelin A *Lecanicillium* sp. HF627 Zhang et al., 2020a	**Verlamelin B** *Lecanicillium* sp. HF627 Zhang et al., 2020a	**Beauveriolide III** *Beauveria* sp. FO-6979 Namatame et al., 1999; Keswani et al., 2019

Contd.

Table 5 *Contd.*

Non-ribosomal cyclooligomeric depsipeptides		
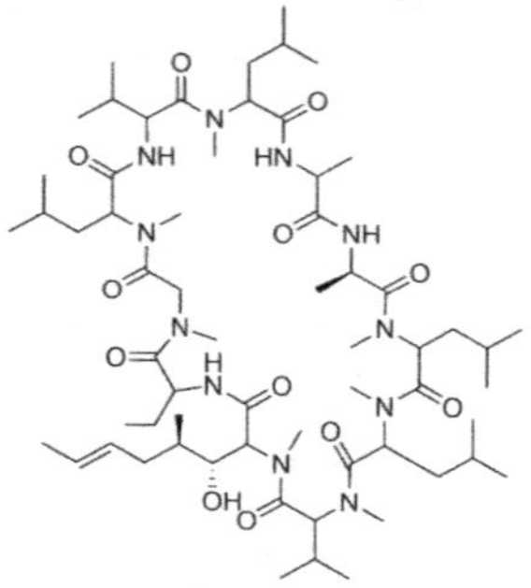	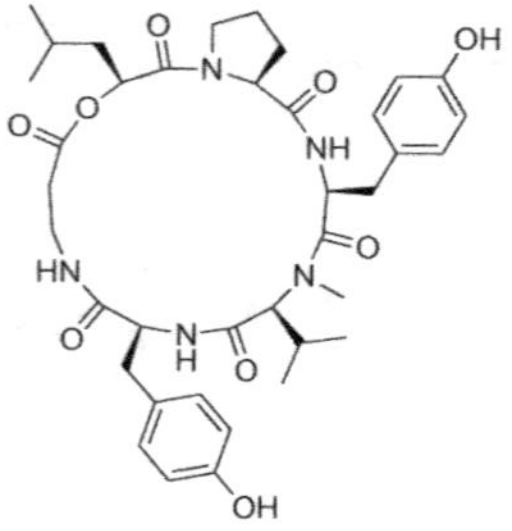	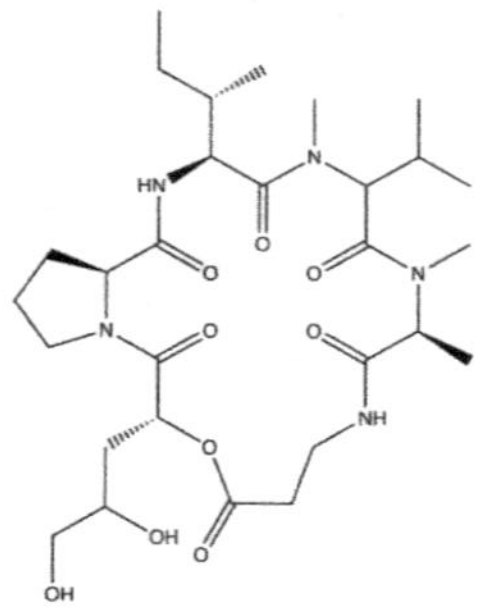
Cyclosporine A *Tolypocladium inflatum* Zhang et al., 2020b	**Isaridin H** *Beauveria feline* Moderate nematicidal effect against *Meloidogyne incognita* Antagonist effect against soil-pathogens and weeds Du et al., 2020	**Destruxin *E*-diol** *Metarhizium robertsii* Insecticidal against *Anastrepha obliqua* Lozano-Tovar et al., 2023
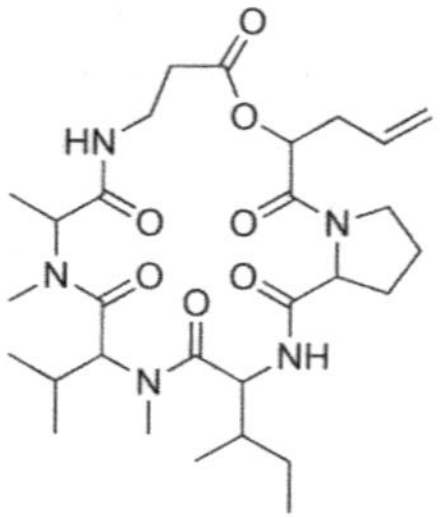		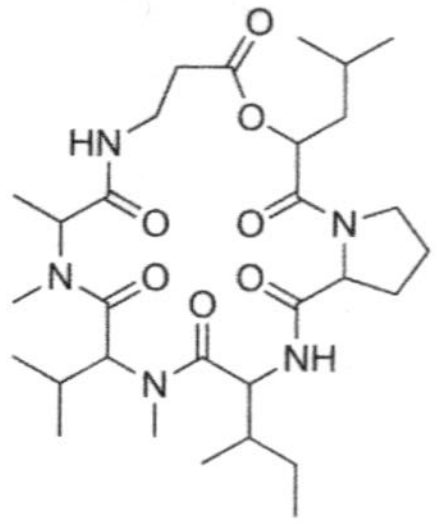
Destruxin A		**Destruxin B**
Metarhizium pinghaense 15R Insecticidal against cotton aphid in Korea Heo et al., 2023		

Contd.

Table 5 *Contd.*

Linear non-ribosomal peptides	
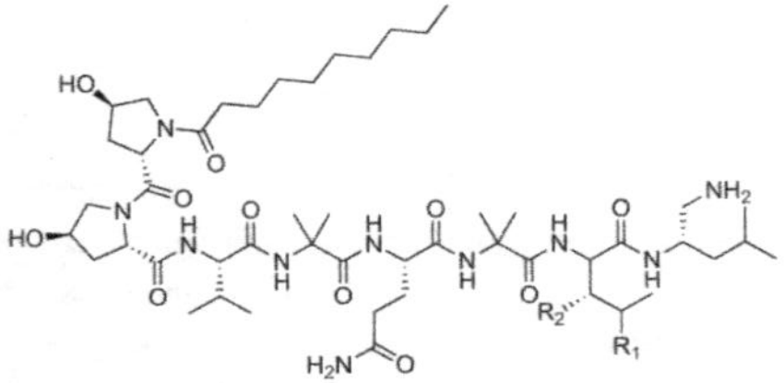 Cicadapeptin I = R_1 = CH_3, R_2 = H Cicadapeptin II = R_1 = H, R_2 = CH_3	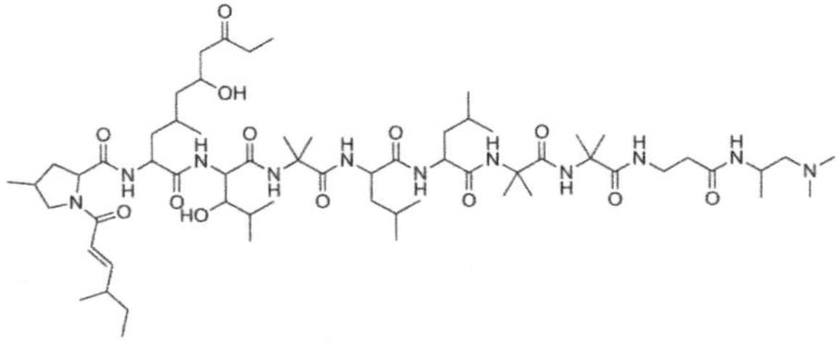
Cicadapeptin I and II *Cordyceps heteropoda* ARSEF1880 Antibacterial (Gram-positive and -negative) Niu et al., 2020	**Leucinostatin A** *Paecilomyces lilacinus CG-189* Niu et al., 2020
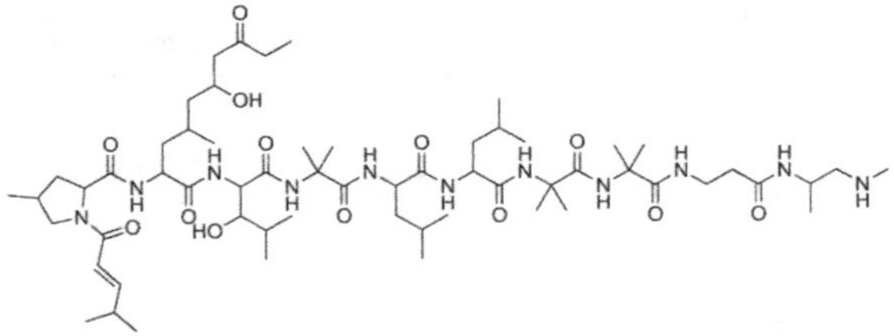	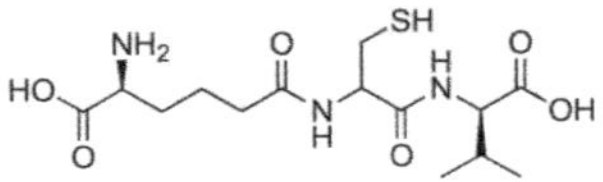
Leucinostatin B *Paecilomyces lilacinus CG-189* Niu et al., 2020	**d-(L-a-aminoadipyl)-L-cysteinyl-D-valine (ACV)** *Penicillium chrysogenum, Cephalosporins acremonium* and *Aspergillus nidulans* Niu et al., 2020

Contd.

Table 5 *Contd.*

Terpenes		
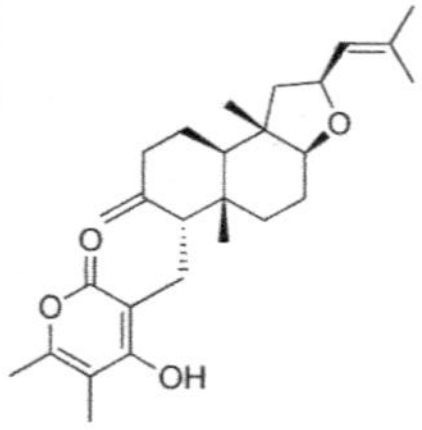	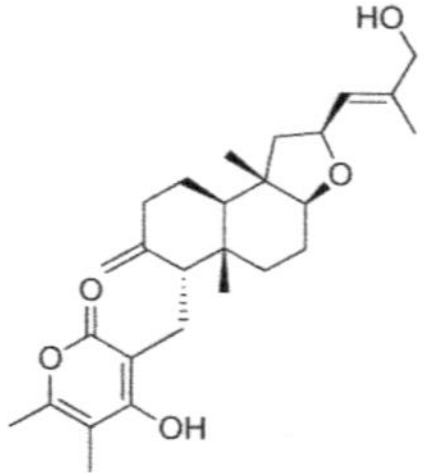	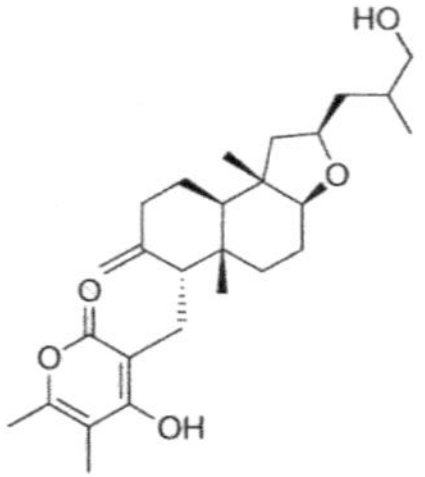
Subglutinol A	**Subglutinol C**	**Subglutinol D**
Metarhizium robertsii Zhang et al., 2020b		
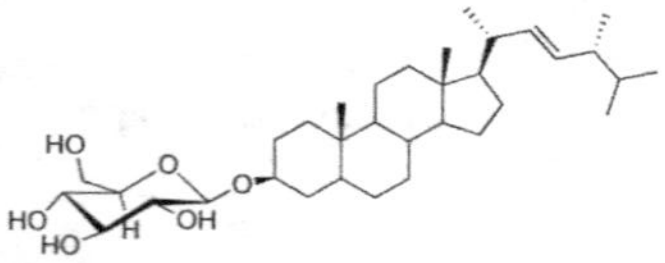	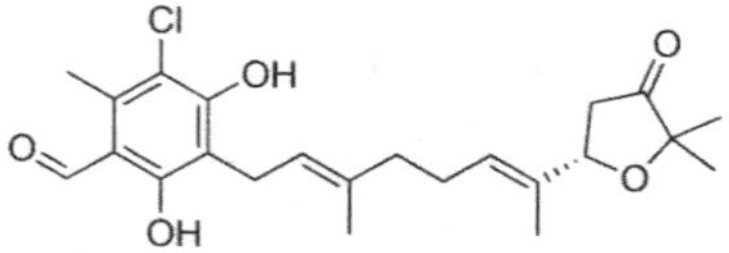	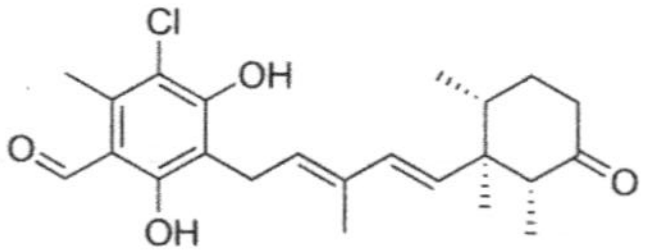
22*E*,24*R*-24-methylcholest-5,22-diene-3-*O*-b-D-*O*-glucopyranoside *Ophiocordyceps longiissima* NBRC 106965 Elshamy et al., 2020	**Ascofuranone**	**Ascochlorin**
	Acremonium egyptiacum Zhang et al., 2020b	

Contd.

Table 5 *Contd.*

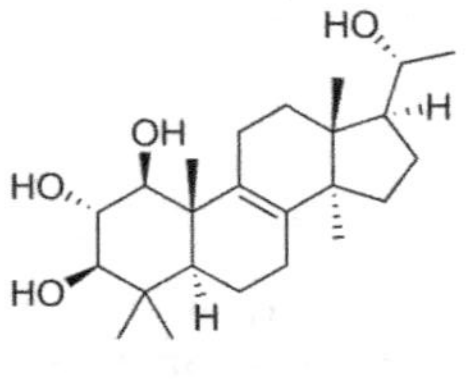	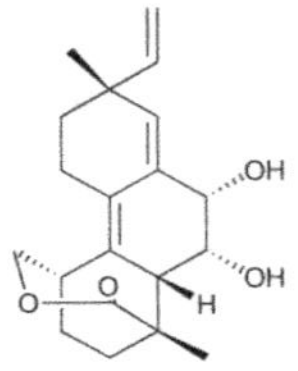	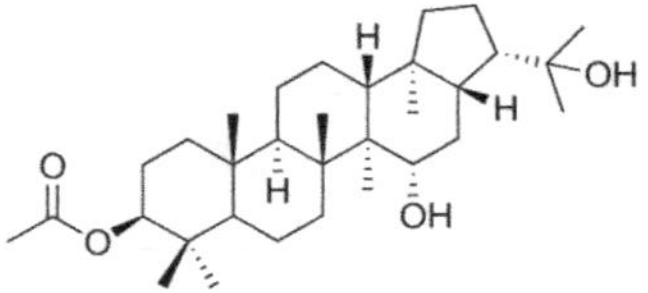
Felinane A	**Felinane B**	**3β-acetoxy-15α,22-dihydroxyhopane**
Beauveria feline Antifungal effect against carbendazim-resistant strains of *Botrytis cinerea* Antagonist effect against soil-pathogens and weeds Du et al., 2020		*Moelleriella raciborskii* Toxicity against *Tetranychus runcates* Suradet et al., 2022
Amides and esters from fatty acids long-chain		
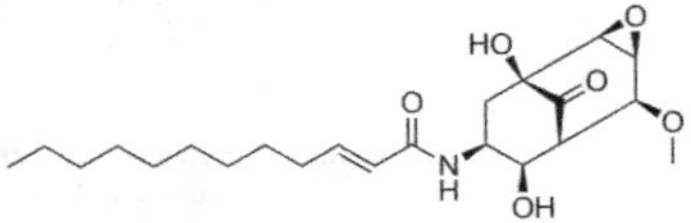	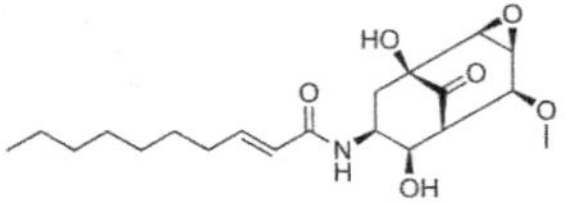	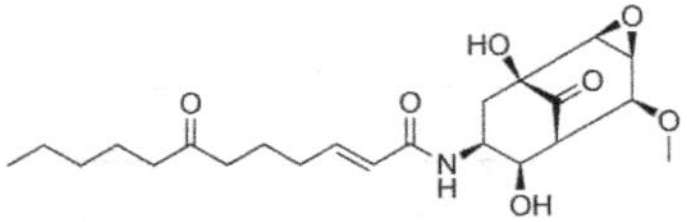
TK-57-164[a]	**TK-57-164B**	**7-*O*-methylisariotin C**
Isaria sp. Yahagi et al., 2020		
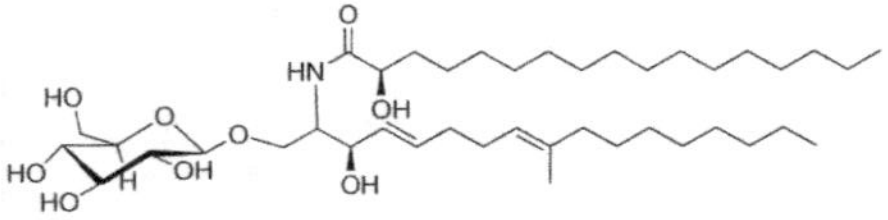	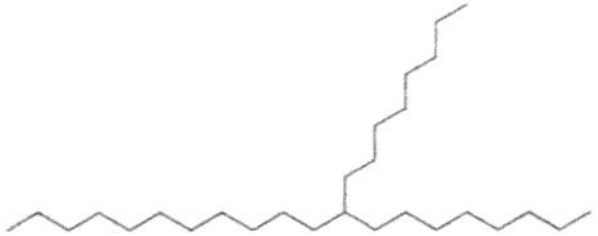	
Ophiocordylongiiside A *Ophiocordyceps longiissima* NBRC 106965 antioxidant Elshamy et al., 2020	**9-octylicosane**	**(6*Z*,9*Z*)-cis-3,4-epoxyheptadecadiene**
	Metarhizium anisopliae Vivekanandhan et al., 2020	

Contd.

Table 5 *Contd.*

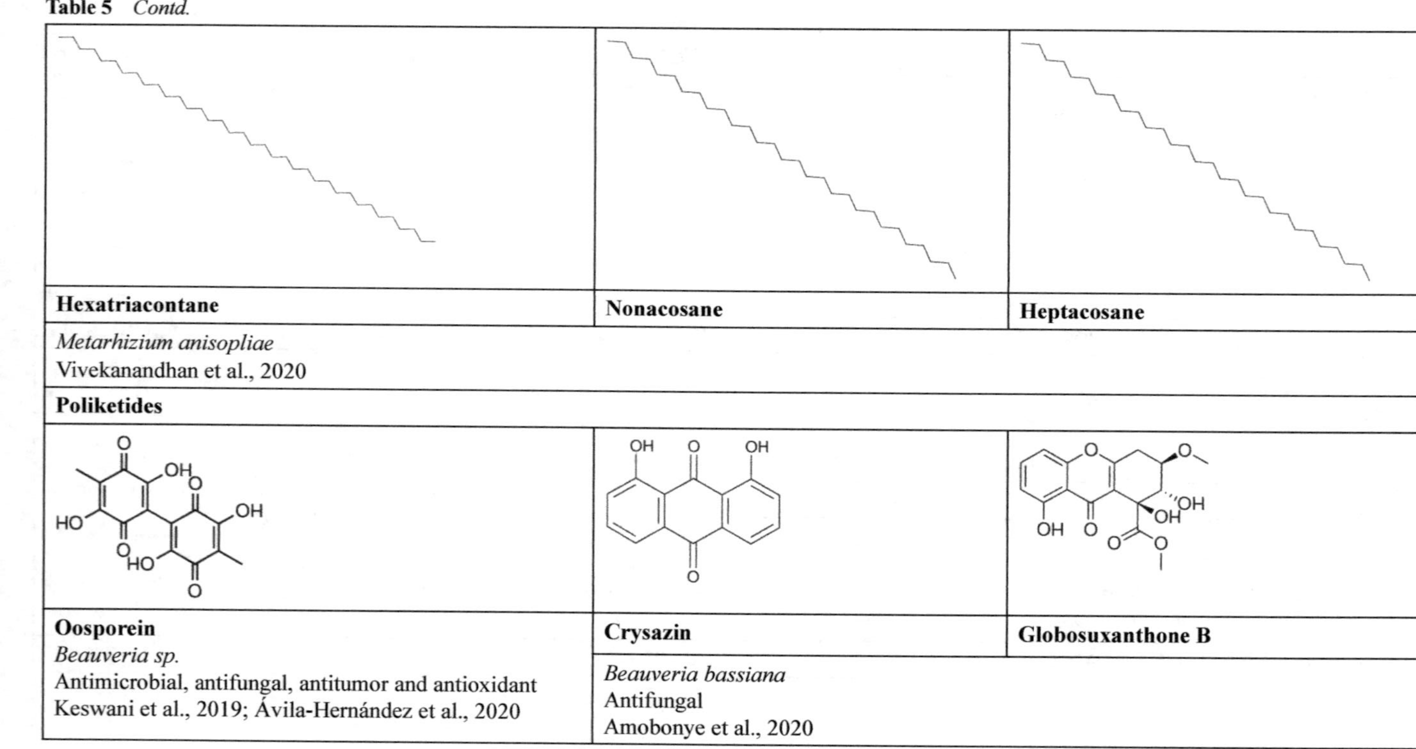

Hexatriacontane	**Nonacosane**	**Heptacosane**
Metarhizium anisopliae Vivekanandhan et al., 2020		
Poliketides		
Oosporein *Beauveria sp.* Antimicrobial, antifungal, antitumor and antioxidant Keswani et al., 2019; Ávila-Hernández et al., 2020	**Crysazin**	**Globosuxanthone B**
	Beauveria bassiana Antifungal Amobonye et al., 2020	

Contd.

Table 5 *Contd.*

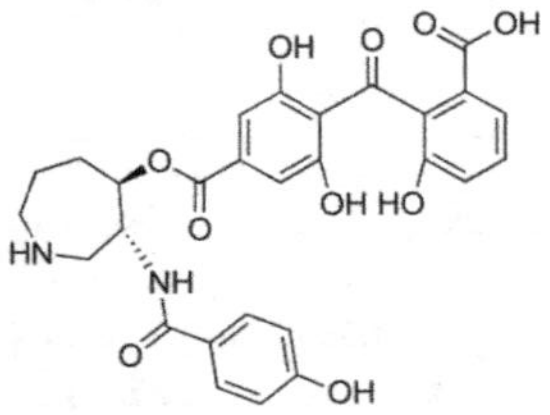 **Balanol** *Tolypocladium ophioglossoides* Zhang et al., 2020a	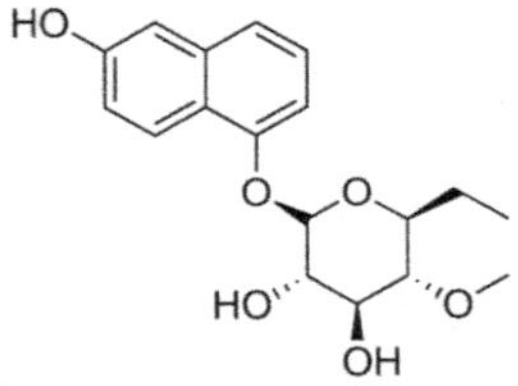 **Akanthol** *Hevansia novoguineensis* BCC47849 (Formely *Akanthomyces)* Zhang et al., 2020b	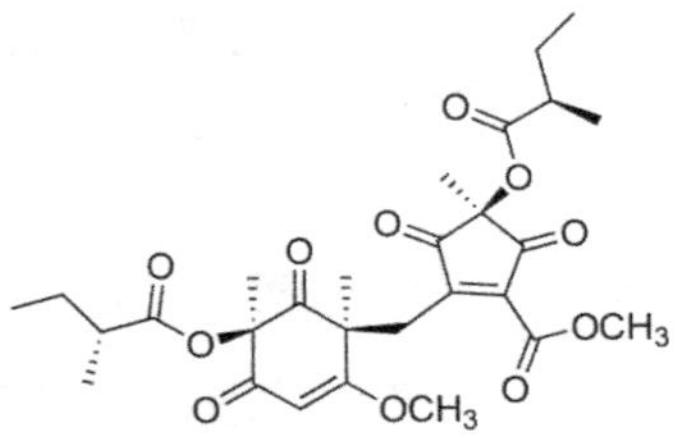 **Beauvetetraones A** *Beauveria bassiana* JMRC ST000047 Zhang et al., 2020b
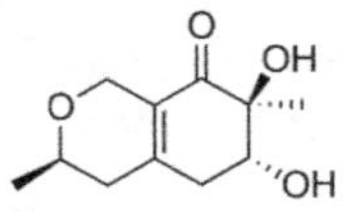 **Felinone A** *B. felina* EN-135 Zhang et al., 2020b	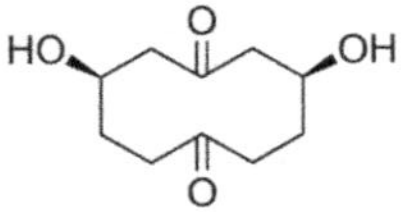 **Cephalosporolide-J** *Cordyceps tenuipes* Yoneyama et al., 2022	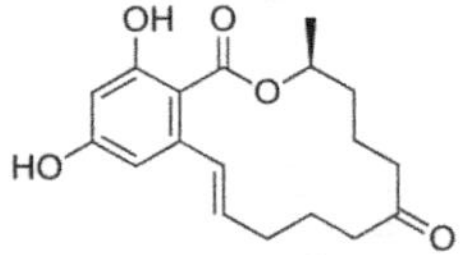 **Zearalenone** *Fusarium pernambucanum* *Fusarium caatingaense* de H.C. Maciel et al., 2021

Contd.

Table 5 *Contd.*

<table>
<tr><td colspan="3">Poliketides</td></tr>
<tr><td colspan="2">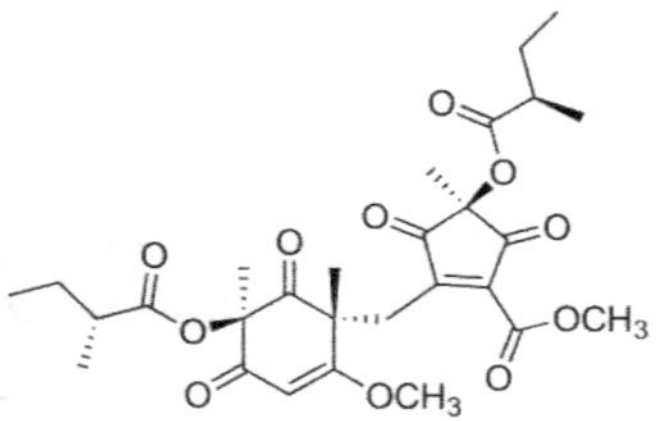</td><td>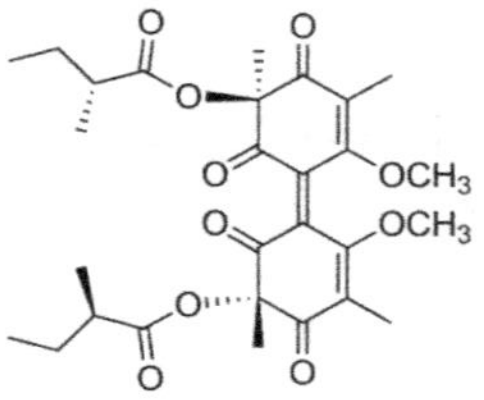</td></tr>
<tr><td colspan="2">Beauvetetraones B</td><td>Beauvetetraones C</td></tr>
<tr><td colspan="3">B. bassiana JMRC ST000047
Zhang et al., 2020b</td></tr>
<tr><td>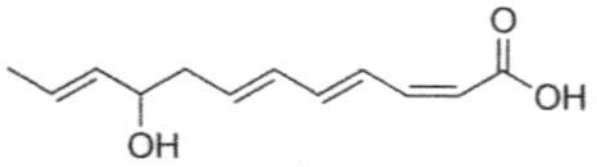</td><td>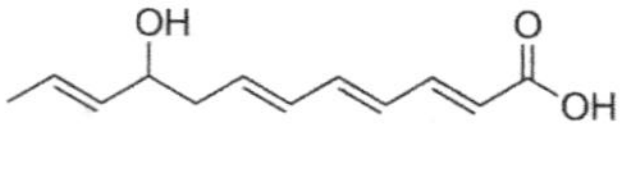</td><td>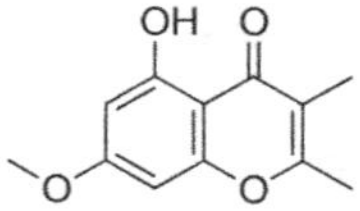</td></tr>
<tr><td>(2Z,4E,6E,10E)-9-hydroxydodeca -2,4,6,10-tetraenoic acid
Metarhizium anisopliae
Weak antifungal activity against Saccharomyces cerevisiae
Increased Candida albicans growth
Sbaraini et al., 2021</td><td>(2E,4E,6E,10E)-9-hydroxydodeca-2,4,6,10-tetraenoic acid
Metarhizium anisopliae
Increased Candida albicans growth
Sbaraini et al., 2021</td><td>5-hydroxy-2,3-dimethyl-7-methoxychromone
Trichoderma harzianum M10
Antimicrobial and antitumor activities
Staropoli et al., 2023</td></tr>
<tr><td>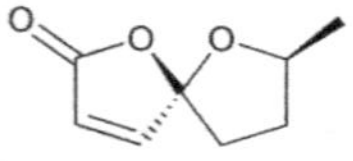</td><td>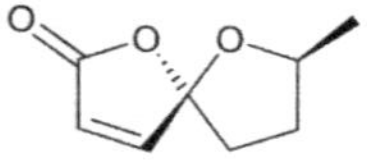</td><td>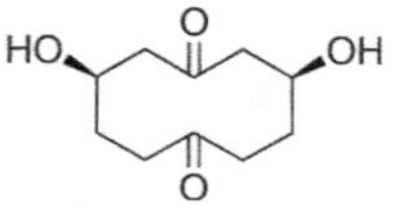</td></tr>
<tr><td>Tenuipesone A</td><td>Tenuipesone B</td><td>Cephalosporolide A</td></tr>
<tr><td colspan="3">Cordyceps tenuipes
Yoneyama et al., 2022</td></tr>
</table>

Contd.

Table 5 *Contd.*

Poliketides		
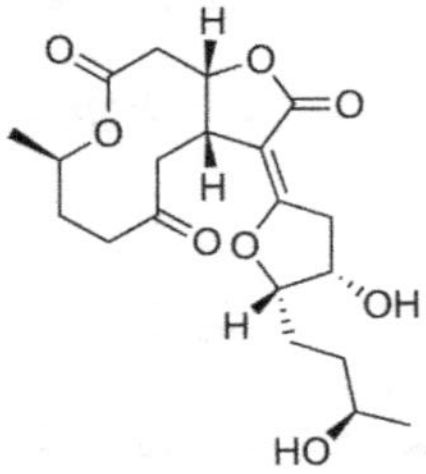	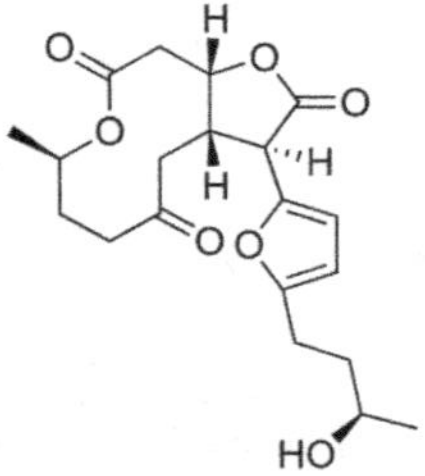	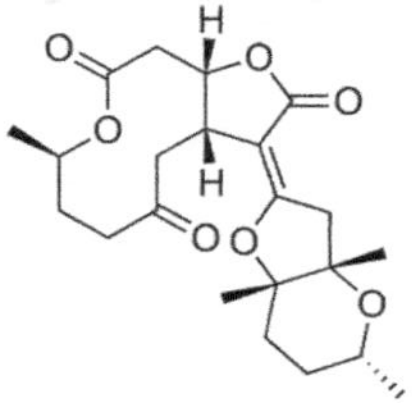
Cordycicadin A *Cordyceps cicadae* JXCH1 Antifeedant activity against silkworm larvae *Bombyx mori* Li et al., 2022	**Cordycicadin B** *Cordyceps cicadae* JXCH1 Antifeedant activity against silkworm larvae *Bombyx mori* Li et al., 2022	**Cordycicadin C** *Cordyceps cicadae* JXCH1 Li et al., 2022
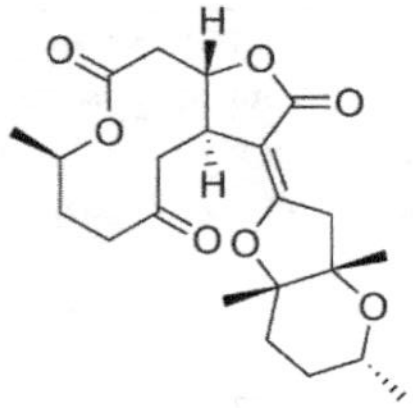	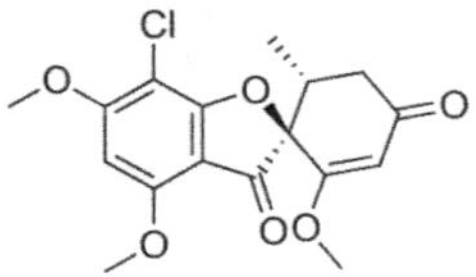 Cl	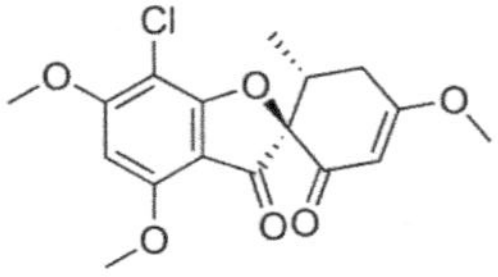 Cl
Cordycicadin D *Cordyceps cicadae* JXCH1 Li et al., 2022	**Griseofulvin** *Aspergillus tamarii* NL3 Antifungal activity Ton That Huu et al., 2022	**Isogriseofulvin**

Contd.

Table 5 *Contd.*

Adenosin related derivatives		
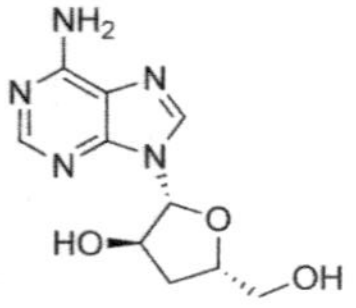 **Cordycepin** *Cordyceps militaris* strain ARSEF 11703 Reduce the immune-related gene expression in *Drosophila melanogaster* cells and toxic in *Galleria mellonella*. Woolley et al., 2020		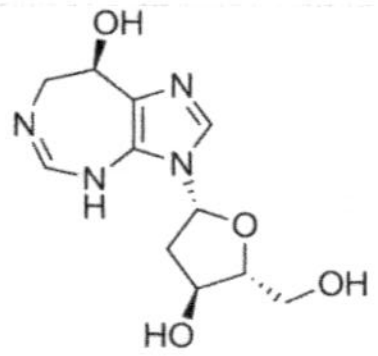 **Pentostatin** *Cordyceps militaris* Zhang et al., 2020a
Alkaloids		
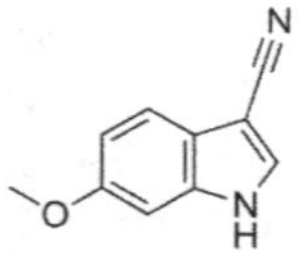 **6-methoxy-1*H*-indole-3-carbonitrile** *Pseudogibellula formicarum* Mongkolsamrit et al., 2021	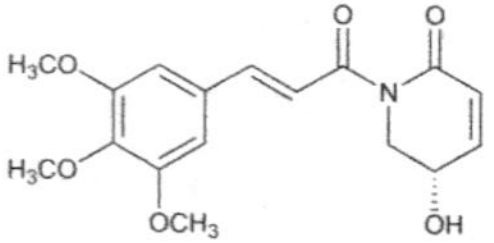 **5-hydroxylpiperlongumine** *Beauveria bassiana* ATCC 7159 Amobonye et al., 2020	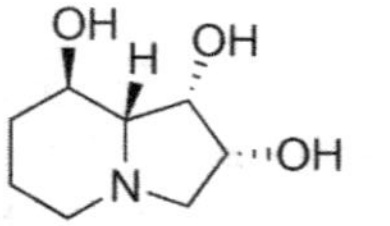 **Swainsonine** *Metarhizium robertsii* Zhang et al., 2020a
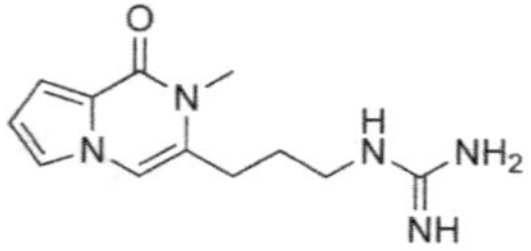 **Peramine** *Metarhizium rileyi* Zhang et al., 2020a	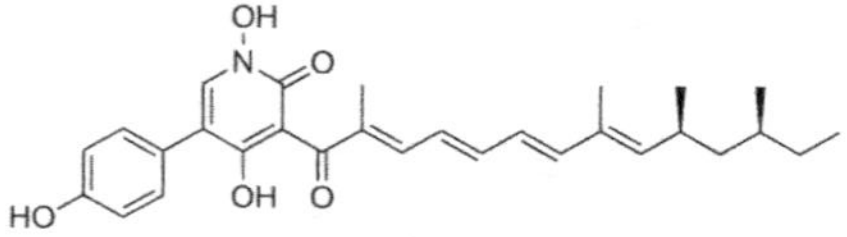 **Fumosorinone** *Cordyceps fumosorosea* Noncompetitive inhibitor of protein tyrosine phosphatase 1B Deshmukh et al., 2023	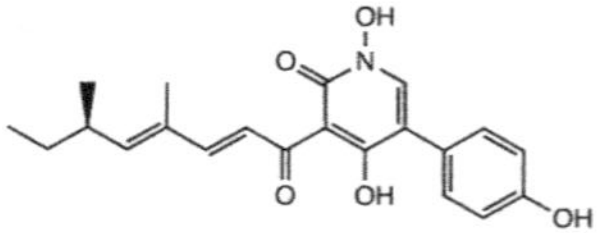 **Tenellin** *Beauveria bassiana* Pigment Zhang et al., 2020b

Contd.

Table 5 *Contd.*

Alkaloids		
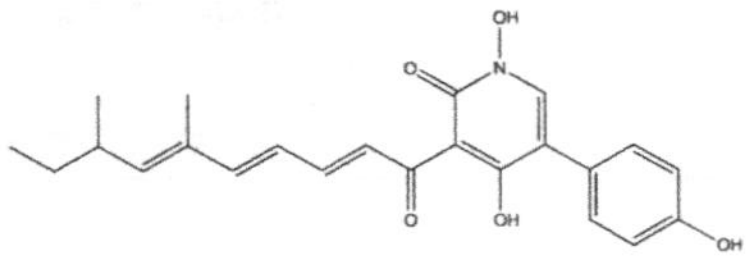	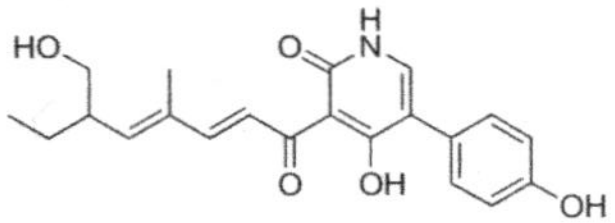	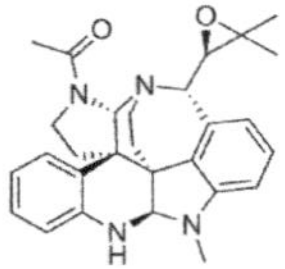
Bassianin *Beauveria bassiana* Pigment Zhang et al., 2020b	**Pyridovericin** *Beauveria bassiana* Anti-allergic Zhang et al., 2020b	**Communesin A** *Penicillium expansum* Insecticidal against *Bombyx mori* Nicoletti et al., 2023
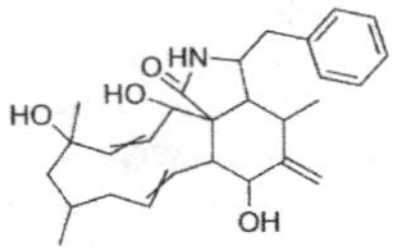	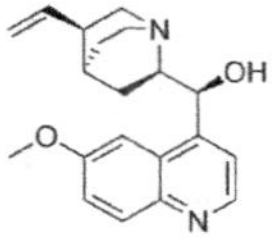	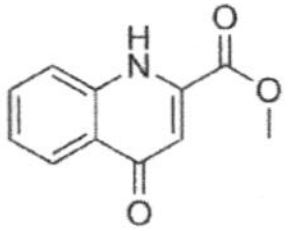
Cytochalasin J *Aspergillus tamarii* NL3 Weak antimicrobial activity Ton That Huu et al., 2022	**Quinidine** *Beauveria* sp. IPBCC.19.1499 *Aspergillus sclerotiorum* IPBCC.19.1500 Antimalarial, antimicrobial, pesticide, anticancer, antifungal, antihelmintic, anticonvulsivant, antiinflammatory, analgesic. Anggiani et al., 2020	**Methyl 4-oxo-1,4-dihydroquinoline-2-carboxylate** *Cordyceps tenuipes* Yoneyama et al., 2022
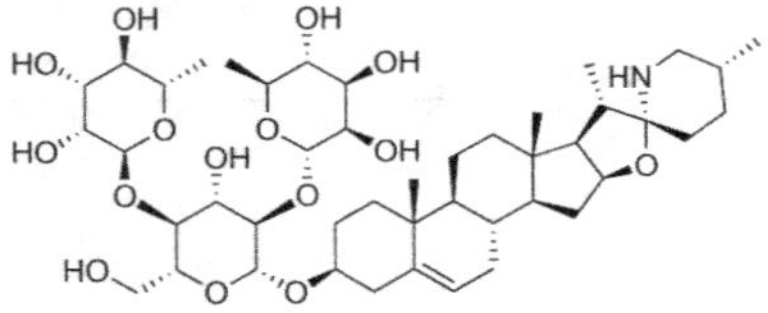		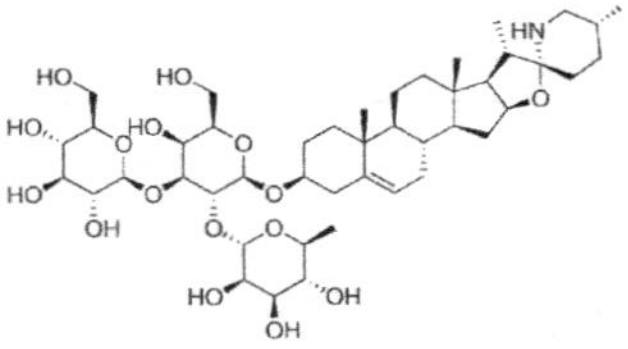
Solamargine		**Solasonine**
Aspergillus tamarii NL3 Stronger antifungal than antibacterial activity Ton That Huu et al., 2022		

Contd.

Table 5 *Contd.*

Mycotoxins		
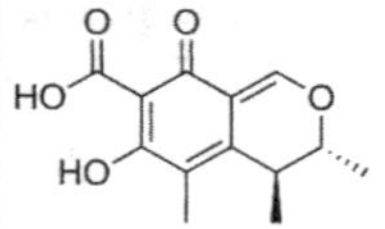	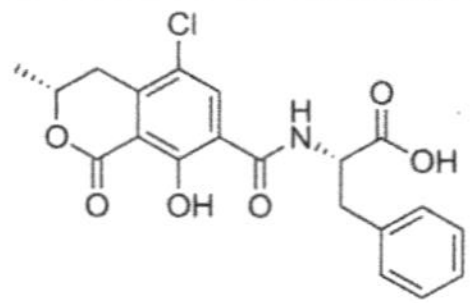	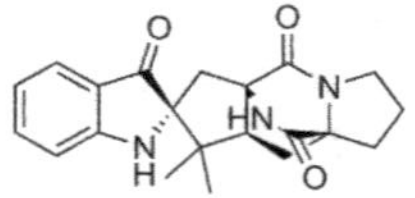
Citrinin *Penicillium* sp. Repelent of Tribolium confusum Nicoletti et al., 2023	**Ochratoxin A** *Penicillium* sp. Inhibited larval growth of *Attagenus megatoma* Nicoletti et al., 2023	**Brevianamide A** *P. brevicompactum* Toxicity against *Spodoptera littoralis* Nicoletti et al., 2023
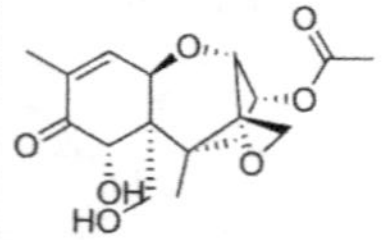	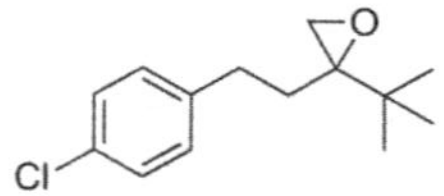	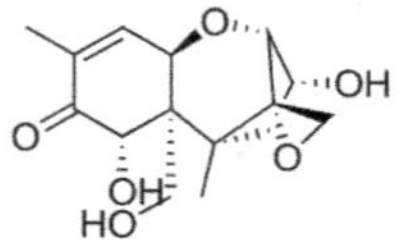
3-Acetyldeoxynivalenol *Fusarium pernambucanum* *Fusarium caatingaense* de H. C. Maciel et al., 2021	**2-[2-(4-chlorophenyl)ethyl]-2-(1,1-dimethylethyl)-oxirane** *Ophiocordyceps flavida (new EPF)* Mongkolsamrit et al., 2021	**Deoxynivalenol** *Fusarium pernambucanum* *Fusarium caatingaense* de H. C. Maciel et al., 2021
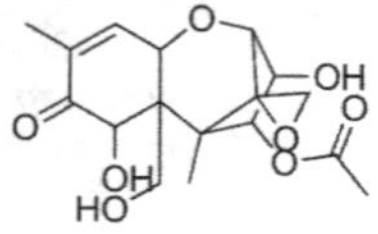	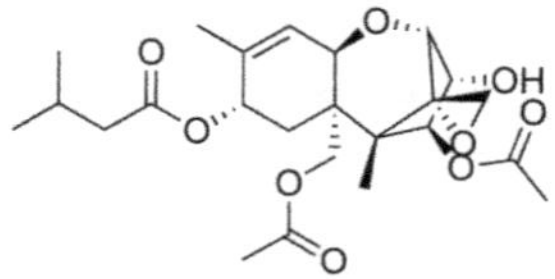	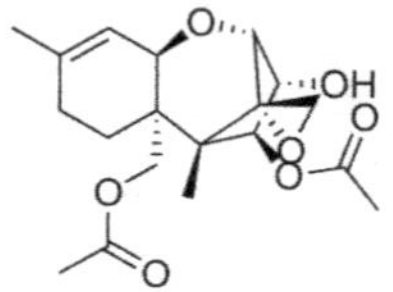
Fusarenone X	**T-2 Toxin**	**Diacetoxyscirpenol**
Fusarium pernambucanum *Fusarium caatingaense* de H. C. Maciel et al., 2021		

Contd.

Table 6 EPF molecules reported with insecticidal activity.

Secondary metabolite	*Synthesized by species of EPF*	*Reference*
Bassianolide	*Beauveria bassiana* *Lecanicillium* sp.	Xu et al., 2009
Bassiacridin	*Beauveria bassiana*	Quesada-Moraga and Vey, 2004
Beauvericin	*Beauveria bassiana* *Paecilomycestenuipes* *Isaria* sp.	Wang and Lijian, 2012 Yahagi et al., 2020 Wang et al., 2021
Cephalosporolides	*Cephalosporium aphidicola*	Ackland et al., 1985
Destruxin A	*Metarhizium anisopliae*	Heo et al., 2023
Destruxin B	*Metarhizium anisopliae*	Molnár et al., 2010 Wang et al., 2012 Heo et al., 2023
Dipicolinic acid	*Beauveria bassiana* *Isaria fumosorosea* *Cordyceps militaris* *Lecanicilium* sp. *Verticillium lecanii*	Paterson, 2008
Efrapeptins	*Tolypocladium cylindrosporum*	Gledhill and Walker, 2006
Oosporein	*Chaetomium* sp. *Beauveria* sp.	Pegram and Wyatt, 1981 Keswani et al., 2019
(+)-Phomalactone	*Nigrospora* sp. *Phoma* sp.	Molnár et al., 2010
Vertilecanin A	*Verticillium* sp.	Soman et al., 2001

Table 7 Molecules of macroscopic EPF reported with insecticidal activity.

Secondary metabolite	*Synthesized by specie of EPF*	*Reference*
Ibotenic acid	*Amanita muscaria* *A. pantherina*	Wieland, 1968
Amatoxins	*Amanita* sp.	Jaenike et al., 1983
2-amino-3-(1,2-dicarboxyethylthio) propanoic acid	*A. pantherina*	Fushiya et al., 1993
Ergosterol	*Pleurotus salmoneostramineus*	Alexandre et al., 2017
Aegerolysin Ostreolysin A6, Pleurotolysin A2 and Erysin	*Pleurotus* sp.	Anastasija et al., 2020

4. *In-vitro* Tests of Entomopathogenic Fungi

Fungi are microorganisms that need certain care in the laboratory, like any other organism. The climatic conditions for fungal growth must be controlled, and sterile material must be kept clean to avoid contamination (Cañedo and Ames, 2004). In addition, the most favorable culture medium for the fungus should be identified after all the conditions for maintaining the entomopathogenic fungus and its morphological or molecular identification have been established.

Initially, *in-vitro* tests are carried out for EPF to determine pathogenicity, virulence, and lethal time. The organisms to be tested against are recommended to be laboratory-reared insects. The tests are performed in Petri dishes, transparent flasks, or other types of containers. Placing the fungi using a desired concentration of conidia is recommended. The different tests that are performed are spray, immersion, contact, and feeding.

In the immersion test, to evaluate the fungus's *in-vitro* effect, the insects are immersed for seconds in the fungal suspension at different concentrations and then placed in containers to observe the effect. Observations are carried out over several days. Care must be taken with the number of individuals placed and the necessary repetitions. The toxicity assay was carried out through exposure: 300 μL of fungal suspension was applied in Petri dishes, and ten *Rhyssomatus nigerrimus* adult insects were placed in each Petri dish. Treated insects were observed daily, and the number of insects killed each day was recorded to calculate mortality. Sterile water was used as a negative control, and fipronil insecticide solution was used as a positive control.

5. Metabolomic Analysis of Entomopathogenic Fungi

Entomopathogenic fungi are microorganisms that infect and typically kill their insect hosts as a crucial part of their development; these insect-fungi interactions are intricate. The success of a fungal infection relies on various events working together, with a key factor being the production of numerous toxic secondary metabolites. The metabolites produced by these fungi serve dual roles, either aiding in fungal invasion or functioning as immunosuppressive agents to thwart host defenses (Lu and St. Leger, 2016; Zhang et al., 2020a; Altimira et al., 2022). A significant aspect of their pathogenicity lies in the secretion of secondary metabolites. To gain insights into the infection life cycle of these environmentally impactful fungi, an exploration into metabolomics becomes imperative, offering a valuable avenue for revealing such critical information (de Bekker et al., 2013; Pedrini, 2022).

Metabolomics offers a comprehensive approach to assessing and analyzing the metabolites generated by organisms when subjected to pathophysiological stimuli. This approach involves a quantitative measurement of multi-target analyses across diverse organisms. Leveraging its holistic characteristics, metabolomics is a novel method for unraveling the pathogenic mechanisms of diseases.

Moreover, contemporary metabolomics techniques, including liquid chromatography coupled to mass spectrometry (LC–MS), gas chromatography-mass spectrometry (GC–MS), and nuclear magnetic resonance spectroscopy (NMR), are employed for the detection, identification, and characterization of mycotoxins, as well as for determining metabolic profiles (Chen et al., 2018; Hu et al., 2018; Qasim et al., 2020).

Among the extensively researched microorganisms are species belonging to the genera *Metarhizium* and *Beauveria*, which are commonly employed as biological pesticides (de Bekker et al., 2013; Xu et al., 2015; Zhang et al., 2016; Mani et al., 2023; Zhang et al., 2023). Researchers have focused on metabolomic analyses, exploring aspects such as the metabolites present in fungal extracts, the impact of fungal infection on insect metabolism, and the fungal metabolic profiles under specific conditions.

One species within the *Beauveria* genus, namely *Beauveria bassiana*, stands out as a well-established biological control agent with a broad spectrum of effectiveness. Moreover, *B. bassiana* exhibits the ability to thrive either as a saprophyte in the soil or as an endophyte within plants (Boomsma et al., 2014; Harith-Fadzilah et al., 2021; Pedrini, 2022; Mani et al., 2023). Despite grasping the general mechanisms underlying *B. bassiana* infection, the intricate molecular processes at each stage of infection still need to be explored. Unraveling the complexity of pathogenicity holds the key to enhancing pest control efficacy. This involves synergizing multiple pesticides or biopesticides with distinct modes of action, avoiding overlap, and uncovering novel toxic molecules to expand the biopesticide arsenal.

Xu et al. (2015) utilized metabolomic techniques, integrating GC-MS and LC-MS, to analyze the metabolic shifts in silkworm (*Bombyx mori*) larvae upon infection with the pathogenic fungus *B. bassiana*. The metabolomic analysis revealed increased carbohydrates, amino acids, fatty acids, and lipids, while eicosanoids and amines were downregulated (Xu et al., 2015). Additionally, insecticidal and cytotoxic mycotoxins such as oosporein and beauveriolides were identified in insects during the later stages of infection (Pedrini, 2022). Microscopic observations from other assays involving the infection of silkworm larvae with fungal spores showed a correlation between the depletion of insect hemocytes and the proliferation of the fungus in the insect body cavity.

Hemolymph samples were also collected for metabolomic analyses. The results of the immune responses showed a significant impact of the fungal infection on the insect's defense mechanisms. Notably, immune reactions such as hemocyte encapsulation and melanization of fungal cells were observed in response to *B. bassiana* infection, mirroring similar immune responses seen during infection with *Metarhizium robertsii* (Wang et al., 2012, Xu et al., 2015). Furthermore, the analysis indicated substantial alterations in insect energy and nutrient metabolisms due to fungal infection. The insect antifeedant effect was apparent as the levels of maclurin, a component found in mulberry leaves, decreased in infected insects but increased in control insects. The integration of metabolomics data suggests that insect immune responses entail energy-intensive reactions. Additionally, the fungus employs a coordinated approach, including nutrient deprivation, inhibition of host immune responses, and toxin production, to combat and eliminate insects effectively.

Zhang et al. (2016) conducted a study on the metabolites produced by *B. bassiana* (RCEF0383) in response to oxidative stress induced by hydrogen peroxide (H_2O_2), utilizing an LC–MS approach. When insects face *B. bassiana* infestation, reactive oxygen species (ROS) are generated as part of their defense mechanism against the fungus (Azevedo et al., 2014; Lavine and Strand, 2002). Paradoxically, the fungus can counteract ROS-induced damage by hydrolyzing glycerophospholipids, enabling fungal hyphae to invade and lethally affect the insects. Consequently, the insects' defensive strategy of producing H_2O_2 or ROS turns counterproductive as it inadvertently enhances the virulence of the fungi.

Results indicated that H_2O_2 treatment could amplify mycotoxin production, suggesting that oxidative stress may boost the fungus's virulence. Compared to the control group, citric acid and UDP-N-acetylglucosamine were downregulated, implying a metabolic shift towards the tricarboxylic acid (TCA) cycle and an enhancement in carbohydrate metabolism (Zhang et al., 2016). Concurrently, as glycerophospholipids decreased during the antioxidation of *B. bassiana*, there was a substantial increase in both saturated and unsaturated fatty acids. The fungus can convert unsaturated fatty acids into oxylipins through autoxidation, enabling *B. bassiana* to evade reactive oxygen generated by H_2O_2, thus protecting vital molecules like DNA and enzymes.

The results of the analysis demonstrated that an increase in oxylipins correlated with elevated mycotoxins such as beauvericin and beauvericin A in H_2O_2-treated mycelia. Since oxylipins function as signaling molecules, they may circumvent oxidative stress by engaging in nonribosomal peptide toxin synthesis, consequently augmenting the production of beauvericins (Christensen and Kolomiets, 2011), and the heightened levels of beauvericins may potentially enhance the virulence of *B. bassiana*, as these compounds constitute the principal insecticidal metabolites of the fungus (Grove and Pople, 1980).

The fungus mitigated oxidative stress through enhanced lipid catabolism and glycometabolism. Moreover, the observed decrease in glycerophospholipids and the concurrent rise in fatty acids and choline suggest that oxidative stress can promote fungal glycerophospholipid hydrolysis. This hydrolysis releases glycerol, which drives fungal hyphae to invade the host insect (Zhang et al., 2016).

Considering the same species, *B. bassiana*, Nithya, et al. (2019) analyzed the metabolome profiles in the culture filtrates of three isolates of the species utilizing GC-MS: MH590235 (TM), MK918495 (BR), and KX263275 (BbI8). The study revealed notable distinctions in the metabolite profiles among the isolates. A total of 63 metabolites were identified across all isolates: the TM isolate displayed 29 different metabolites (alkanes, carboxylic acid derivatives, glucopyranose, galactofuranose derivatives, unsaturated fatty acids, and hexadecanoic acid derivatives); BR and BbI8 isolates exhibited 29 and 26 metabolites, respectively, with similarities in the level of metabolite production between them.

Particularly, an anhydrase was detected, the 1,6-anhydro-α-D-Glucopyranose (levoglucosan), present in all three isolates, and a γ-lactone, 5-Oxotetrahydrofuran-2-carboxylic acid, found in TM and BbI8 isolates (Nithya et al., 2019). The latter refers to a derivative of bassialone, an antimicrobial secondary metabolite synthesized by *B. bassiana*. However, it was absent in the BR isolate, indicating a clear variation in the metabolite profile, potentially suggesting reduced virulence.

Correlation analysis further supported these findings, revealing a positive significant correlation between the BR and BbI8 isolates of *B. bassiana*. The metabolome of the TM isolate exhibited notable distinctions from the other two isolates, featuring a variety of compounds with diverse biological roles. These included insecticidal and antimicrobial activities, lipid and fatty acid metabolisms, and factors that enhance virulence. These insights contribute to a better understanding of the heterogeneity in metabolome profiles among *B. bassiana* isolates and have implications for the selection of strains with enhanced pest control potential (Nithya et al., 2019).

In a previous study, Chaithra et al. (unpublished data mentioned in Chaithra et al., 2022) documented a primary assessment of the virulence exhibited by various strains of *B. bassiana*, sourced from different locations and environments, against *Tetranychus truncates* mites. Three were identified as virulent from a pool of 30 strains (Bb6, Bb12, and Bb15), while three were non-virulent (Bb5, Bb8, and Bb21). These selected strains were then utilized to establish a dose-response relationship, examining the correlation between the concentration of *B. bassiana* strains and the mortality rate of *T. truncatus*.

The six B. *bassiana* strains underwent GC-MS analysis using ethyl acetate extracts. *In-vitro* leaf disc bioassays indicated that the potential *B. bassiana* strains (Bb6, Bb12, and Bb15) exhibited significantly higher *T. truncatus* mortality rates than the commercial strain Beveroz (67.30%). Conversely, Bb5, Bb8, and Bb21 demonstrated lower mortality rates against *T. truncatus*, recording 6.49%, 5.02%, and 8.71%, respectively.

Potted bioassays showed that Bb6, Bb12, and Bb15 strains reduced mite numbers 48 to 72 hours after *B. bassiana* application. Conversely, potted bean plants sprayed with Bb5, Bb8, and Bb21 strains exhibited an increase in the *T. truncatus* population during the same timeframe.

Regarding metabolite analysis, the highly promising strains, namely Bb6, Bb12, and Bb15, were identified to contain 13, 12, and 12 metabolites, respectively. These metabolites encompass alkanes, alkenes, carboxylic acid derivatives, alcohol, unsaturated fatty acids, and hexadecanoic acid derivatives. Conversely, in the non-potential strains (Bb5, Bb8, and Bb21), 15, 15, and 12 metabolites were detected, representing similar chemical categories.

Interestingly, although the number of metabolites was greater in the non-potential strains, the potentially acaricidal metabolites were found in higher abundance in the more promising strains. Notably, there was a higher concentration of bis (dimethylethyl)-phenol—5.79%, 6.15%, and 4.69%—in the potential strains Bb6, Bb12, and Bb15, respectively. This compound not only exhibits acaricidal activity but also possesses repellent and oviposition deterrent properties (Chen and Dai, 2015).

Besides their acaricidal effects, Bb6 and Bb12 potential strains showed elevated nonadecene levels of this defense secretion compound (19.51% and 4.72%, respectively). This component may contribute to increased pathogenicity by overcoming the insect's innate immunity. Additionally, insect-derived plant regulators, such as docosene, were detected in Bb8 (3.54%), Bb12 (8.81%), Bb15 (4.30%), and Bb21 (7.90%) strains (Doss et al., 2000). Among these strains, Bb12 exhibited a higher production of long-chain hydrocarbons, potentially aiding in penetrating the outer coverage of mites (Chaithra et al., 2022).

In a separate study, Mani et al. (2023) investigated the impact of commercial formulations of *B. bassiana* on the growth characteristics and hemolymph metabolomics of black soldier fly (BSF; *Hermetia illucens*) larvae. They employ an LC–MS/MS approach; the research aimed to elucidate the biological mechanisms governing the interaction between *B. bassiana* and BSF, specifically concerning BSF physiology and immunity.

The experiments involved feeding (mixing spores with the diet; BF) and contact treatments (dipping larvae in the spore solution; BD), both of which were compared to a control group treated with water. The analysis revealed a total of 1072 metabolites. Significantly, metabolites such as 5-guanidino-2-oxopentanoic acid, imidazolelactic acid, and 3,4-dihydroxyphenylpropionic acid showed increased levels in both the BD treatment and the control groups. Adenosine, however, displayed higher accumulation in the control group. Palmitic acid and 1-vinyl-2-pyrrolidone were over-accumulated in the BD treatment group. Conversely, other metabolites, such as purine (xanthine, hypoxanthine, and guanosine monophosphate), exhibited increased accumulation in the BF treatment group.

This analysis reinforces that the BF treatment over-accumulates several metabolites, significantly impacting BSF larval metabolism. Between the two tested methods (BD and BF), the BF method had a pronounced negative effect on larval growth, resulting in decreased larval weight, adult emergence, and adult weight compared to the control (Mani et al., 2023). The metabolomics analysis uncovered a significant over-accumulation of metabolites spanning diverse classes, including purine and purine derivatives, organooxygen compounds, organonitrogen compounds, carboxylic acids and their derivatives (encompassing peptides and oligopeptides), and fatty acyls and their derivatives. Fatty acyls, known for their pivotal roles in insects, including providing energy for growth and reproduction, were notably elevated in the hemolymph following BF treatment, indicating energy-intensive immunoreactions in BSF larvae (van Meer et al., 2008; Mani et al., 2023). Noteworthy among the observed alterations is that Mani et al. (2023) reported a significant impact of BF treatment on purine metabolism, potentially facilitating the bonding of fungal spores with host tissue for hyphal development.

Furthermore, the upregulation of taurine metabolism in both BF- and BD-treated larval hemolymphs directly points to the stress experienced by BSF larvae due to *B. bassiana* infection (Mani et al., 2023). To summarize, the formulation of *B. bassiana* proved to be a potent bio-control agent, significantly impacting crucial growth variables of BSF, including larval growth, adult emergence, and adult body weight. These effects are likely attributed to the secretion of metabolites by *B. bassiana*. However, it is noteworthy that the study did not detect conventional insecticidal metabolites commonly associated with *B. bassiana*, such as beauvercin, bassianin, or beauverolide.

In a distinct investigation, da Costa Stuart et al. (2023) directed their efforts toward comprehending the biocontrol processes involved in the interaction between a fungal consortium of *B. bassiana* and *Duponchelia fovealis* caterpillars, a prominent pest affecting strawberries in Brazil (Zawadneak et al., 2016). Both deceased and surviving caterpillars underwent analysis using GC–MS and LC–MS to assess the effects.

The GC–MS analysis identified 137 molecules in deceased caterpillars. Among these, da Costa Stuart et al. (2023) pinpointed various metabolites within the glycerophospholipids (GPL) class, including lysophosphatidic acid (LPA), phosphatidic acid (PA), phosphatidylserine (PS), lysophosphatidylserine (LPS), lysophosphatidylethanolamine (LPE), and phosphatidylethanolamine (PE). This set of metabolites encompassed L-α-amino acids like ornithine and the non-proteinogenic amino acid L-norleucine, monosaccharides such as glucopyranose phosphate, pentose D-xylose, and 2,5-dimethoxymandelic acid from the benzenes and derivatives class.

Notably, some of these metabolites exhibited primarily antioxidant properties, while others displayed potential insecticidal characteristics, including diltiazem-like and tamsulosin-like compounds, along with 2,5-dimethoxymandelic acid.

Conversely, in samples from surviving caterpillars, 141 molecules were detected. Among these, four markers showed significantly higher abundance in caterpillars treated with fungi compared to the control: N-acyl amine molecule 2-palmitoylglycerol, N-acyl-α-amino acid N-α-acetyl-L-Lysine, pyrimidine nucleoside uridine, and secondary alcohol 1,2,3-butanetriol. The primary mechanisms involved pro-inflammatory effects from the 2-palmitoylglycerol metabolite and the antifungal action of the Aegle marmelos alkaloid-C metabolite (da Costa Stuart et al., 2023).

Having identified the metabolites implicated in the demise and survival of *D. fovealis*, the observations revealed that the heightened mortality induced by the fungal consortium stemmed from its antioxidant mechanism. This mechanism suppresses the caterpillars' immune systems and exerts insecticidal action. Simultaneously, metabolites linked to virulence enhance fungal virulence, leading to caterpillar mortality through direct and indirect pathways.

In surviving caterpillars, primary resistance mechanisms seem to revolve around immune stimulation and a defense mechanism characterized by the activation of inflammatory processes, the production of oxidative molecules, and the presence of metabolites with potent fungicidal properties. Furthermore, the analysis revealed molecules exhibiting delayed insecticidal effects. Given that these compounds induce a slower demise, surviving insects in a seven-day study may die due to the delayed insecticidal compounds (da Costa Stuart et al., 2023).

Recent additional research focused on investigating the impact of *B. bassiana* invasion on the activities of three major antioxidative enzymes in *S. frugiperda* (fall armyworm). Additionally, the study employed UPLC-MS with multivariate statistical analyses to discern the variations in the metabolomes of *B. bassiana*-infected and uninfected larvae of *S. frugiperda* (Zhang et al., 2023). The findings unveiled a distinct metabolic profile in *B. bassiana*-infected samples compared to their uninfected counterparts. These contrasting metabolites encompass amino acids and their derivatives, carbohydrates, lipids, and organic heterocyclic compounds. Their close association with amino acid, lipid, nucleotide, and carbohydrate metabolism was evident, indicating a significant impact of *B. bassiana* on the metabolic pathways of the fall armyworm larvae.

For instance, in *B. bassiana*-infected larvae, the concentrations of N-acetyl-glutamate, ornithine, and citrulline significantly increased by 1.33, 1.56, and 1.34 times, respectively, compared to uninfected larvae. Conversely, arginine experienced a significant decrease, possibly attributed to fungal infection downregulating the activity of arginine succinate lyase in larvae. Another noteworthy change in the metabolic profile was the notable increase in acetylcholine levels, reaching 3.72 times that of uninfected larvae. Acetylcholine, an insect neurotransmitter, binds to various acetylcholine receptors to ensure normal nerve signal transmission (Zibaee et al., 2009). The elevated acetylcholine levels in *B. bassiana*-infected larvae suggested that fungal infection inhibits acetylcholinesterase activity in *S. frugiperda* larvae, leading to the production of abnormally high levels of acetylcholine to compensate for the inhibition and maintain nerve signal transmission.

Zhang et al. (2023) demonstrated that *B. bassiana* infection in larvae induces heightened cellular oxidative stress and significantly alters the activities of larval antioxidant enzymes and metabolite levels. These metabolites play crucial roles in various immune activities, encompassing energy, nucleotide, and amino acid metabolism. Consequently, the infection of *S. frugiperda* larvae by *B. bassiana* may be linked to oxidative stress, potentially impairing insect immune defenses.

De Bekker et al.'s (2013) investigations employed a comprehensive metabolomics approach with a newly developed *ex-vivo* insect culturing assay. They utilized LC–MS following extended contact between two generalist isolates of the genus *Metarhizium* and *Beauveria* and tissues (brains and muscles extracted from heads) of carpenter ants (*Camponotus pennsylvanicus*), to analyze the metabolomic profile. The results unveiled that both fungi employed tissue-specific strategies, as evidenced by distinct metabolite secretion patterns on the tested insect tissues. Metabolically, these fungi exhibited varying behaviors on different insect tissues. The metabolomics analyses in these experimental setups illustrated that entomopathogenic fungi can discern between dead and live host material, leading to distinct metabolic responses. Notably, this study identified several novel destruxins and beauverolides that were previously undocumented, likely because earlier surveys did not utilize insect tissues as a culturing system. While *Beauveria* secreted these cyclic depsipeptides upon encountering live insect tissues, *Metarhizium* predominantly utilized them on deceased tissue. This suggests that, although these fungi employ comparable strategies in terms of entomopathogenesis, there are undoubtedly significant molecular-level differences that warrant further investigation (de Bekker et al., 2013).

M. anisopliae, which is extensively utilized for controlling agricultural and forestry pests, undergoes degeneration after multiple successive cultures, resulting in a significant virulence reduction, a decrease in conidia production, and a loss of application value. To determine this, Yang et al. (2023) compared the metabolic profiles based on the metabolomics approach of GC–MS. The study involved analyses of secondary metabolite profiles and sporulation degeneration in different strains.

Seventy-four metabolites with normal sporulation were detected in both strains, and those displaying impaired sporulation showed 40 differential metabolites that contributed significantly to the model. The analyses identified 23 downregulated metabolites and 17 upregulated metabolites. The findings suggest a noteworthy involvement of metabolites, particularly amino acids, in the processes of sporulation and degeneration in *M. anisopliae*. Specifically, amino acid metabolism, focusing on glutamate, aspartate, serine, glycine, arginine, and leucine, appears to play a crucial role in the sporulation mechanism of *M. anisopliae* (Yang et al., 2023).

6. *In-silico* Studies of Insecticidal Compounds

In-silico studies constitute a branch of computational biology that investigates and explores biological processes through computer simulations (Fig. 2). This digital approach enables the visualization of three-dimensional structures of various peptides, the observation of the function and role of diverse molecules and proteins, and even the assessment of an entire cell. Additionally, it allows the anticipation of responses to stimuli through modeling and conducting *in-silico* experiments to confirm hypotheses posed in *in-vitro* experiments (Vargas-Aguilar and Fuentes-Condori, 2021). Once predictions of protein structures are completed, they become routine so that molecular and cellular researchers can receive automated predictions from protein servers and vice versa before experimental studies are implemented.

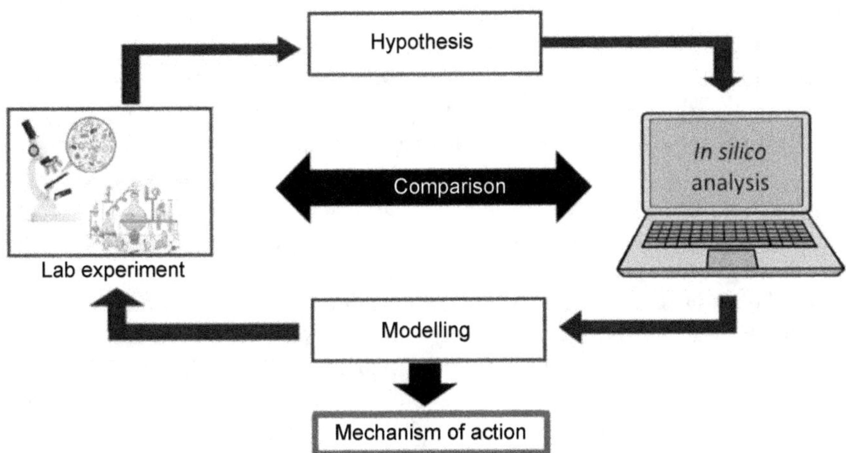

Fig. 2 Diagram showing the relationship between *in-vitro* experiments and computational design (*in-silico*) for hypothesis testing (created by Páez-León, S.Y.).

In *in-silico* studies, attention has been primarily directed toward proteins and enzymes, as it has been reported that entomopathogenic fungi cause infections by invading their insect hosts through the external skeleton (cuticle). There is a possibility that enzymes such as chitinase, protease, and lipase, responsible for degrading the cuticle, can penetrate the refractory cuticle of the insect. Proteases that break peptide bonds in proteins and peptides are used in agriculture as biocontrol agents against phytopathogenic insects and nematodes (Khan et al., 2016).

Keppanan et al. (2017) conducted molecular docking of proteases from three isolates of *M. anisopliae* (Tk6, Tk29, and Tk37) that exhibit cuticle-degrading and insecticidal properties against *G. mellonella*. The study analyzed the toxic impact of the isolates on *G. mellonella* larvae and was conducted with extracellular protease enzymes from the isolates. Among the three isolates, Tk6 had the maximum capacity to bind and naturally degrade the host protein.

B. bassiana is the focus of intense studies as a promising biopesticide against insects and pests, effectively contributing to food production systems as a natural biological control agent (Mondal and Singh, 2022). Further, it has been highlighted that the virulence of *B. bassiana* is associated with the cell surface's hydrophobicity, especially the conidia's hydrophobic layer.

The study in question used refined protein models generated from sequences of hydrophobins H1 and H2 of *B. bassiana*, simulating and docking them with simulated protein models derived from protein sequences of the surface cuticles of *Nilaparvata lugens* and *S. exigua*. The selected protein models for simulation exhibited more amino acid residues that participate in protein-protein interactions. A significant increase in hydrophobic bonds was observed in protein complexes formed by *B. bassiana* hydrophobins linked with cuticular proteins of its pathogens (*N. lugens* and *S. exigua*).

Identification of molecular actors in *B. bassiana* surface protein interactions has revealed the contribution of rodlet layers and their hydrophobin constituents to the hydrophobicity and adhesion of fungal cells to their pathogenic insects. Understanding the forces governing interactions between *B. bassiana* and its hosts can be utilized to design formulation conditions specifically targeted to objectives, and future experiments promise valuable information on host-pathogen interactions.

In another study, the molecular docking of compounds present in the mycelial extract of *Aspergillus* sp., isolated from almonds of *Bertholletia excelsa* in the Brazilian Amazon, was conducted due to its larvicidal activity against *Aedes aegypti*. The extract exhibited larvicidal activity with an LC_{50} of 26.86 µg/mL at 24 hours and 18.75 µg/mL at 48 hours. Molecular docking studies revealed that the compound Aspergillol B acts as a potent larvicide by inhibiting acetylcholinesterase. This enzyme catalyzes the hydrolysis of acetylcholine to yield choline and acetate in the central nervous system. These findings suggest that SM from *Aspergillus* sp., obtained from almonds of *B. excelsa*, possesses significant biological potential in combating *A. aegypti* vectors (Araujo et al., 2022).

While there are few studies involving computational aspects, researchers are expected to utilize *in-silico* studies as a powerful and complementary tool to traditional research methods, enabling the analysis and prediction of a variety of biological, chemical, and physical processes more efficiently and cost-effectively.

Conclusion

In conclusion, this review has endeavored to shed light on the multifaceted world of entomopathogenic fungi, their key role in pest management, the wide variety of secondary metabolites they biosynthesize, and their potential effects. Furthermore, a diverse range of entomopathogenic fungi, including *B. bassiana*, *M. anisopliae*, and others, exhibit remarkable potential as biocontrol agents against various insect pests. The extensive research presented in this review highlights the efficacy of these fungi in pest control in diverse agricultural settings. This revision allows us to visualize entomopathogenic fungi as a sustainable alternative to chemical pesticides. Their specificity in targeting insects and minimal impact on non-target organisms position them as valuable tools in integrated pest management strategies. Moreover, the interaction between some entomopathogenic fungi and their hosts has been intricately investigated. Insight into the molecular mechanisms underlying fungal pathogenicity provides a foundation for developing tailored strategies to improve biopesticide formulations. Therefore, it is important to continue studying fungi, their mechanisms of pathogenicity, and their secondary metabolites.

References

Ackland, M.J., Hanson, J.R., Hitchcock, P.B. and Ratcliffe, A.H. (1985). Structures of the cephalosporolides B–F, a group of C 10 lactones from *Cephalosporium aphidicola*. J. Chemical Soc. Perkin Trans 1: 843–847. https://doi.org/10.1039/P19850000843

Aguilar-Marcelino, L., Mendoza-de-Gives, P., Tawfeeq Al-Ani, L.K., López-Arellano, M.E., Gómez-Rodríguez, O., Villar-Luna, E. and Reyes-Guerrero, D.E. (2020). Using molecular techniques applied to beneficial microorganisms as biotechnological tools for controlling agricultural plant pathogens and pest. In: Molecular Aspects of Plant Beneficial Microbes in Agriculture, Sharma, V., Salwan, R., Tawfeeq Al-Ani, L.K., (Eds.). Academic Press 333–349. https://doi.org/10.1016/B978-0-12-818469-1.00027-4

Alexandre, T.R., Lima, M.L., Galuppo, M.K., Mesquita, J.T., do Nascimento, M.A., dos Santos, A.L., Sartorelli, P., Pimenta, D. and Tempone, A. (2017). Ergosterol isolated from the basidiomycete *Pleurotus salmoneostramineus* affects *Trypanosoma cruzi* plasma membrane and mitochondria. J. Venom. Anim. Toxins incl. Trop. Dis. 23:30. https://doi.org/10.1186/s40409-017-0120-0

Alexopoulos, C.J., Mims, C.W. and Blackwell, M. (1996). Introductory mycology, 4th ed. Wiley, New York, p 880.

Altimira, F., Arias-Aravena, M., Jian, L., Real, N., Correa, P., González, C., Godoy, S., Castro, J.F., Zamora, O., Vergara, C., Vitta, N. and Tapia, E. (2022). Genomic and experimental analysis of the insecticidal factors secreted by the entomopathogenic fungus *Beauveria pseudobassiana* RGM 2184. J. Fungi, 8: 253. https://doi.org/10.3390/jof8030253

Amobonye, A., Bhagwat, P., Pandey, A., Singh, S. and Pillai, S. (2020). Biotechnological potential of *Beauveria bassiana* as a source of novel biocatalysts and metabolites. Critical Rev. Biotechnol. 40: 1019–1034. https://doi.org/10.1080/07388551.2020.1805403

Anastasija, P., Matej, S., Špela, M., Jaka, R. and Kristina, S. (2020). Aegerolysins from the fungal genus *Pleurotus* - Bioinsecticidal proteins with multiple potential applications. J. Invertebr. Pathol. 186: 107474. https://doi.org/10.1016/j.jip.2020.107474

Anggiani, J. P., Listiyowati, S. and Rahayu, G. (2020). Entomopathogenic fungi *Beauveria* sp. and *Aspergillus sclerotiorum* can produce secondary metabolite quinidine. IOP Conference Series: Earth and Environmental Science, 457:012032. https://doi.org/10.1088/1755-1315/457/1/012032

Araújo, I.F., Marinho, V.H.S., Sena, I.S., Curti, J.M., Ramos, R.D.S., Ferreira, R.M.A., Souto, R.N.P. and Ferreira, I.M. (2022). Larvicidal activity against *Aedes aegypti* and molecular docking studies of compounds extracted from the endophytic fungus *Aspergillus* sp. isolated from *Bertholletia excelsa* Humn. & Bonpl. Biotechnol. Lett. 44: 439–459. https://doi.org/10.1007/s10529-022-03220-7

Araújo, J.P.M. and Hughes, D.P. (2016). Diversity of entomopathogenic fungi: which groups conquered the insect body? In: Genetics and Molecular Biology of Entomopathogenic Fungi. Brian Lovett, B. and St. Leger, R.J. (Eds.) Book series: Advances in Genetics. Academic Press, 1–39. https://doi.org/10.1016/bs.adgen.2016.01.001

Arunthirumeni, M., Vinitha, G. and Shivakumar, M.S. (2023). Antifeedant and larvicidal activity of bioactive compounds isolated from entomopathogenic fungi *Penicillium* sp. for the control of agricultural and medically important insect pest (*Spodoptera litura* and *Culex quinquefasciatus*). Parasitol. Int. 92:102688. https://doi.org/10.1016/j.parint.2022.102688

Ávila-Hernández, J.G., Carrillo-Inungaray, M.L., De la Cruz-Quiroz, R., Wong-Paz, J.E., Muñiz-Márquez, D.B., Parra, R., Aguilar, C.N. and Aguilar-Zarate, P. (2020). *Beauveria bassiana* secondary metabolites: A review inside their production systems, biosynthesis, and bioactivities. Mex. J. Biotechnol. 5: 1–33. https://doi.org/10.29267/mxjb.2020.5.4.1

Azevedo, R.F.F., Souza, R.K.F., Braga, G.U.L. and Rangel, D.E.N. (2014). Responsiveness of entomopathogenic fungi to menadione-induced oxidative stress. Fungal Biol., 118: 990–995. https://doi.org/10.1016/j.funbio.2014.09.003

Barr, D.J.S. (1992). Evolution and kingdoms of organisms from the perspective of a mycologist. Mycologia 84: 1–11. https://doi.org/10.2307/3760397

Barr, D.J.S. (2001). Chytridiomycota. In: Systematics and Evolution. The Mycota. McLaughlin, D.J., McLaughlin, E.G. and Lemke, P.A. (Eds.) Springer Berlin, Heidelberg, 93–112. https://doi.org/10.1007/978-3-662-10376-0_5

Basnet, P., Dhital, R. and Rakshit, A. (2022). Biopesticides: a genetics, genomics, and molecular biology perspective. In Advances in Bio-inoculant Science, Biopesticides.Rakshit, A., Singh Meena, V., Abhilash, P.C., Sarma, B.K., Singh, H.B., Fraceto, L., Parihar, M., Kumar Singh, A., (Eds.) Woodhead Publishing, pp. 107–116. https://doi.org/10.1016/B978-0-12-823355-9.00019-5

Beakes, G.W. and Thines, M. (2017). Hyphochytriomycota and Oomycota. In: Handbook of the Protists. 2nd ed., Archibald, J.M., Simpson, A.G.B. and Slamovits, C.H., (Eds.) Springer International Publishing, Cham., pp. 435–505. https://doi.org/10.1007/978-3-319-32669-6_26-1

Ben Fekih, I., Jensen, A.B., Boukhris-Bouhachem, S., Pozsgai, G., Rezgui, S., Rensing, C. and Eilenberg, J. (2019) Virulence of two entomophthoralean fungi, *Pandora neoaphidis* and *Entomophthora planchoniana*, to their conspecific (*Sitobion avenae*) and heterospecific (*Rhopalosiphum padi*) aphid hosts. Insects 10: https://doi.org/10.3390/insects10020054

Betancourt-Román, C.M., O'Neil, C.C. and James, T.Y. (2016). Rethinking the role of invertebrate hosts in the life cycle of the amphibian chytridiomycosis pathogen. Parasitol., 143: 1723–1729. https://doi.org/10.1017/S0031182016001360

Bidochka, M.J., Walsh, S.R.A., Ramos, M.E., St. Leger, R.J., Silver, J.C. and Roberts, D.W. (1995). Pathotypes in the *Entomophaga grylli* species complex of grasshopper pathogens differentiated with random amplification of polymorphic DNA and cloned-DNA probes. Appl. Environ. Microbiol., 61: 556–560. https://doi.org/10.1128/aem.61.2.556-560.1995

Boomsma, J.J., Jensen, A.B., Meyling, N.V. and Eilenberg, J. (2014). Evolutionary interaction networks of insect pathogenic fungi. Annu. Rev. Entomol. 59: 467–485. https://doi.org/10.1146/annurev-ento-011613-162054

Braga-Neto, R., Luizão, R.C.C., Magnusson, W.E., Zuquim, G. and Castilho, V.C. (2008). Leaf litter fungi in a central Amazonian Forest: the influence of rainfall, soil and topography on the distribution of fruiting bodies. Biodivers. Conserv. 17: 2701–2712. https://doi.org/10.1007/s10531-007-9247-6

Brunner-Mendoza, C., Reyes-Montes, M.R., Moonjely, S., Bidochka, M.J. and Toriello, C., (2019). A review on the genus *Metarhizium* as an entomopathogenic microbial biocontrol agent with emphasis on its use and utility in Mexico. Biocontrol. Sci. Technol. 29: 83–102. https://doi.org/10.1080/09583157.2018.1531111

Butt, T.M., Coates, C.J., Dubovskiy, I.M. and Ratcliffe, N.A. (2016). Entomopathogenic Fungi: New Insights into Host–Pathogen Interactions. Adv. Genet. 94: 307–364. https://doi.org/10.1016/bs.adgen.2016.01.006

Cañedo, V. and Ames, T. (2004). Manual de laboratorio para el manejo de hongos entomopatógenos. Lima, Peru, 62 p. ISBN 92-9060-238-4

Chaithra, M., Prameeladevi, T., Prasad, L., Kundu, A., Bhagyasree, S.N., Subramanian, S. and Kamil, D. (2022). Metabolomic diversity of local strains of *Beauveria bassiana* (Balsamo) Vuillemin and their efficacy against the cassava mite, *Tetranychus truncatus* Ehara (Acari: Tetranychidae). PloS One 17: e0277124. https://doi.org/10.1371/journal.pone.0277124

Chandler, D. (2017) Basic and applied research on entomopathogenic fungi. In: Microbial Control of Insect and Mite Pests. From Theory to Practice, Lacey, L.A., (Ed.). Academic Press, pp. 69–89. https://doi.org/10.1016/B978-0-12-803527-6.00005-6

Chen, S., Li, M., Zheng, G., Wang, T., Lin, J., Wang, S., Wang, X., Chao, Q., Cao, S., Yang, Z. and Yu, X. (2018). Metabolite profiling of 14 wuyi rock tea cultivars using UPLC-QTOF MS and

UPLC-QqQ MS combined with chemometrics. Molecules 23: 104. https://doi.org/10.3390/molecules23020104

Chen, Y. and Dai, G. (2015). Acaricidal, repellent, and oviposition-deterrent activities of 2,4-ditert-butylphenol and ethyloleate against the carmine spider mite *Tetranychus cinnabarinus*. J. Pest. Sci., 88: 645–655. https://doi.org/10.1007/s10340-015-0646-2

Chhetri, D.R., Chhetri, A., Shahi, N., Tiwari, S., Karna, S.K.L., Lama, D. and Pokharel, Y.R. (2020). *Isaria tenuipes* Peck, an entomopathogenic fungus from Darjeeling Himalaya: evaluation of *in-vitro* antiproliferative and antioxidant potential of its mycelium extract. BMC Complement. Med. Ther. 20: 185. https://doi.org/10.1186/s12906-020-02973-w

Christensen, S.A. and Kolomiets, M.V. (2011). The lipid language of plant-fungal interactions. Fungal Genet. Biol., 48: 4–14. https://doi.org/10.1016/j.fgb.2010.05.005

da Costa Stuart A.K., Lee Furuie, J., Regiani Cataldi, T., Makowiecky Stuart, R., Cassilha Zawadneak, M.A., Labate, C.A. and Chapaval Pimentel, I. (2023). Metabolomics of the interaction between a consortium of entomopathogenic fungi and their target insect: Mechanisms of attack and survival. Pestic. Biochem. Physiol. 191: 105369, https://doi.org/10.1016/j.pestbp.2023.105369

Dai, D.Q., Suwannarach, N., Chathuranga Bamunuarachchige, T. and Chandranath Karunarathna, S. (2023). Plant-fungal interactions. Front Microbiol 14: 1236394 https://doi.org/10.3389/fmicb.2023.1236394

de Bekker, C., Smith, P.B., Patterson, A.D. and Hughes, D.P. (2013). Metabolomics reveals the heterogeneous secretome of two entomopathogenic fungi to ex vivo cultured insect tissues. PLoS One 5:70609. https://doi.org/10.1371/journal.pone.0070609

de Freitas Soares, F.E., Silva Gôlo, P. and Fernandes, E.K.K. (2020). Nematophagous and entomopathogenic fungi: new insights into the beneficial fungus-plant interaction. In: Molecular Aspects of Plant Beneficial Microbes in Agriculture, Sharma, V., Salwan, R. and Khalil Tawfeeq Al-Ani, L. (Eds.) Academic Press 295–304. https://doi.org/10.1016/B978-0-12-818469-1.00024-9

de H.C. Maciel, M., do Amaral, A.C.T., da Silva, T.D., Bezerra, J.D.P., de Souza-Motta, C.M., da Cota, A.F., Vieira Tiago, P. and de Oliveira, N.T. (2021). Evaluation of mycotoxin production and phytopathogenicity of the entomopathogenic fungi *Fusarium caatingaense* and *F. pernambucanum* from Brazil. Current. Microbiol., 78: 1218–1226. https://doi.org/10.1007/s00284-021-02387-y

De Silva, D.D., Rapior, S., Hyde, K.D. and Bahkali, A.H. (2012). Medicinal mushrooms in prevention and control of *diabetes mellitus*. Fungal Divers. 56:1–29. https://doi.org/10.1007/s13225-012-0187-4

Deshmukh, S.K., Agrawal, S. and Gupta, K.M. (2023). Fungal metabolites: a potential source of antidiabetic agents with particular reference to PTP1B Inhibitors. Curr. Pharm. Biotechnol., 24: 927–945. http://dx.doi.org/10.2174/1389201023666220506104219

Doss, R.P., Oliver, J.E., Proebsting, W.M., Potter, S.W., Kuy, S., Clement, S.L., Williamson, R.T., Carney, J.R. and DeVilbiss, E.D. (2000). Bruchins: insectderived plant regulators that stimulate neoplasm formation. Proc. Natl. Acad. Sci. 97: 6218–6223. https://doi.org/10.1073/pnas.110054697

Du, F.Y., Li, X.M., Sun, Z.C., Meng, L.H. and Wang, B.G. (2020). Secondary metabolites with agricultural antagonistic potentials from *Beauveria felina*, a marine-derived entomopathogenic fungus. J. Agric. Food. Chem. 68: 14824–14831. https://doi.org/10.1021/acs.jafc.0c05696

Elshamy, A.I., Yoneyama, T., Trang, N.V., Son, N.T., Okamoto, Y., Ban, S., Noji, M. and Umeyama, A. (2020). A new cerebroside from the entomopathogenic fungus *Ophiocordyceps longiissima*: Structural-electronic and antioxidant relations. Experimental and DFT calculated studies. J. Mole. Struc. 1200: 127061. https://doi.org/https://doi.org/10.1016/j.molstruc.2019.127061

Esparza Mora, M.A., Conteiro Castilho, A.M. and Fraga, M.E. (2017). Classification and infection mechanism of entomopathogenic fungi. Agricul. Microbiol. 84: 1–10, e0552015. https://doi.org/10.1590/1808-1657000552015

Evans, H.C. and Samson, R.A. (1977). *Sporodiniella umbellate*, an entomogenous fungus of the Mucorales from cocoa farms in Ecuador. Can. J Bot., 55: 2981–2984. https://doi.org/10.1139/b77-334

Fushiya, S., Gu, Q.Q., Ishikawa, K., Funayama, S. and Nozoe, S. (1993). (2R), (1'R) and (2R), (1'S)-2-Amino-3-(1,2-dicarboxyethylthio)propanoic acids from *Amanita pantherina*. Antagonists of N-methyl-D-aspartic acid (NMDA) receptors. Chem. Pharm. Bull. 41: 484–486. https://doi.org/10.1248/cpb.41.484

Gleason, F.H., Carney, L.T., Lilje, O. and Glockling, S.L. (2012). Ecological potentials of species of *Rozella* (Cryptomycota). Fungal Ecol., 5: 651–656. https://doi.org/10.1016/j.funeco.2012.05.003

Gledhill, J.R. and Walker, J.E. (2006). Inhibitors of the catalytic domain of mitochondrial ATP synthase. Biochem. Soc. Trans., 34(Pt5), 989–992. https://doi.org/10.1042/BST0340989

Goettel, M.S., Eilenberg, J. and Glare, T. (2005). Entomopathogenic fungi and their role in regulation of insect populations. In: I*nsect Control*. Gilbert, L.I. and Gill, S.S. (Eds.), Academic Press, 69–89.

Grove, J.F. and Pople, M. (1980). The insecticidal activity of beauvericin and the enniatin complex. Mycopathologia 70: 103–105. https://doi.org/10.1007/BF00443075

Gryganskyi, A.P., Humber, R.A., Smith, M.E., Hodge, K., Huang, B., Voigt, K. and Vilgalys, R. (2013). Phylogenetic lineages in Entomophthoromycota. Persoonia 30: 94–105. https://doi.org/10.3767/003158513X666330

Gutierrez, A.C., Rueda Páramo, M.E., Falvo, M.L., López Lastra, C.C. and García, J.J. (2017). *Leptolegnia chapmanii* (Straminipila: Peronosporomycetes) as a futurebiorational tool for the control of *Aedes aegypti* (L.). Acta Trop 169: 112–118. http://dx.doi.org/10.1016/j.actatropica.2017.01.021

Harith-Fadzilah, N., Ghani, IA. and Hassan, M. (2021). Omics-based approach in characterising mechanisms of entomopathogenic fungi pathogenicity: A case example of *Beauveria bassiana*. J. King Saud Univ. Sci., 33: 101332. https://doi.org/10.1016/j.jksus.2020.101332

Hassett, B.T., Thines, M., Buaya, A., Ploch, S. and Gradinger, R. (2019). A glimpse into the biogeography, seasonality, and ecological functions of arctic marine Oomycota. IMA Fungus. 10: https://doi.org/10.1186/s43008-019-0006-6

Hatai, K., Roza, D. and Nakayama, T. (2000). Identification of lower fungi isolated from larvae of mangrove crab, *Scylla serrata*, in Indonesia. Mycoscience, 41: 565–572. https://doi.org/10.1007/BF02460922

Heo, I., Kim, S., Han, G.H., Im, S., Kim, J.W., Hwang, D.Y., Jang, J.W., Lee, J.Y., Woo, S.D. and Shin, T.Y. (2023). Characteristics of insecticidal substances from the entomopathogenic fungus *Metarhizium pinghaense* 15R against cotton aphid in Korea. J. Asia-Pacific Entomol., 26: 102013. https://doi.org/10.1016/j.aspen.2022.102013

Hibbett, D.S., Binder, M., Bischoff, J.F., Blackwell, M., Cannon, P.F., Eriksson,O.E., Huhndorf, S., James,T., Kirk, P.M., Lücking,R., Thorsten Lumbsch, H., Lutzoni, F., Matheny, P.B., Mclaughlin, D.J., Powell, M.J., Redhead,S., Schoch, C.L., Spatafora,J.W., Stalpers, J.A., Vilgalys, R. and Zhang, N. (2007). A higher-level phylogenetic classification of the Fungi. Mycol. Res. 111: 509–47. https://doi.org/10.1016/j.mycres.2007.03.004

Hu, W., Pan, X., Li, F. and Dong, W. (2018). UPLC-QTOF-MS metabolomics analysis revealed the contributions of metabolites to the pathogenesis of *Rhizoctonia solani* strain AG-1-IA. PloS One 13: e0192486. https://doi.org/10.1371/journal.pone.0192486

Humber, R.A. (2012) Entomophthoromycota: a new phylum and reclassification for entomophthoroid fungi. Mycotaxon 120: 477–492. https://doi.org/10.5248/120.477

Jaenike, J., Grimaldi, D.A., Sluder, A.E., and Greenleaf, A.L. (1983). agr-Amanitin tolerance in mycophagous *Drosophila*. Science 221: 165–167. https://doi.org/10.1126/science.221.4606.165

Jitklang, S., Ahantarig, A., Kuvangkadilok, C., Baimai, V. and Adler, P.H. (2012). Parasites of larval black flies (Diptera: Simuliidae) in Thailand. Songk J Sci Technol 34: 597–599.

Joop, G. and Vilcinskas, A. (2016). Coevolution of parasitic fungi and insect hosts. Zoology (Jena), 119: 350–358. https://doi.org/10.1016/j.zool.2016.06.005

Kaczmarek, A. and Boguś, M.I. (2021a). Fungi of entomopathogenic potential in Chytridiomycota and Blastocladiomycota, and in fungal allies of the Oomycota and Microsporidia. IMA Fungus, 12: 1–13. https://doi.org/10.1186/s43008-021-00074-y

Kaczmarek, A. and Boguś, M.I. (2021b). The impact of the entomopathogenic fungus *Conidiobolus coronatus* on the free fatty acid profile of the flesh fly *Sarcophaga argyrostoma*. Insects. 12: 970. https://doi.org/10.3390/insects12110970

Kendrick, B (2000) The fifth kingdom. 3rd ed. Newburyport: Focus Publishing.

Keppanan, R., Sivaperumal, S., Chadra Kanta, D., Akutse, K.S. and Wang, L. (2017). Molecular docking of protease from *Metarhizium anisopliae* and their toxic effect against model insect *Galleria mellonella*. Pestic. Biochem. Phys. 138: 8–14. https://doi.org/10.1016/j.pestbp.2017.01.013

Keswani, C., Singh, H.B., Hermosa, R., García-Estrada, C., Caradus, J., He, Y.-W., Mezaache-Aichour, A., Glare, T.R., Borriss, R., Vinale, F. and Sansinenea, E. (2019). Antimicrobial secondary metabolites from agriculturally important fungi as next biocontrol agents. Appl. Microbiol. Biotechnol. 103: 9287–9303. https://doi.org/10.1007/s00253-019-10209-2

Khan, S., Nadir, S., Wang, X., Khan, A., Xu, J., Li, M., Tao, L., Khan, S., Karunarathna, S.C. (2016). Using *in silico* techniques: Isolation and characterization of an insect cuticle-degrading-protease gene from *Beauveria bassiana*. Microb. Pathog. 97: 189–197. https://doi.org/10.1016/j.micpath.2016.05.024

Klinter, S., Bulone, V. and Arvestad, L. (2019). Diversity and evolution of chitin synthases in oomycetes (Straminipila: Oomycota). Mol. Phylogenetics Evol. 139: 106558. https://doi.org/10.1016/j.ympev.2019.106558

Lavine, M.D. and Strand, M.R. (2002) Insect hemocytes and their role in immunity. Insect. Biochem. Mol. Biol. 32: 1295–1309. https://doi.org/10.1016/s0965-1748(02)00092-9

Li, X., Chen, H.-P., Zhou, L., Fan, J., Awakawa, T., Mori, T., Ushimaru, R., Abe, I., Liu, J.-K. (2022). Cordycicadins A–D, antifeedant polyketides from the entomopathogenic fungus *Cordyceps cicadae* JXCH1. Org. Lett. 24: 8627–8632. https://doi.org/10.1021/acs.orglett.2c03432

Lozano-Tovar, M.D., Ballestas Álvarez, K.L., Sandoval-Lozano, L.A., Palma Mendez, G. M. and Barrera-Cubillos, G.P. (2023). Study on the insecticidal activity of entomopathogenic fungi for the control of the fruit fly (*Anastrepha obliqua*), the main pest in mango crop in Colombia. Arch. Microbiol. 205: 83. https://doi.org/10.1007/s00203-023-03405-2

Lu, H.L. and St. Leger, R.J. (2016). Insect immunity to entomopathogenic fungi. Adv. Genet. 94: 251–285. https://doi.org/10.1016/bs.adgen.2015.11.002

Lutzoni, F., Kauff, F., Cox, C.J., McLaughlin, D., Celio, G., Dentinger, B., Padamsee, M., Hibbett, D., James, T.Y., Baloch, E., Grube, M., Reeb, V., Hofstetter, V., Schoch, C., Arnold, A.E., Miadlikowska, J., Spatafora, J., Johnson, D., Hambleton, S., Crockett, M., Shoemaker, R., Sung, G.H., Lücking, R., Lumbsch, T., O'Donnell, K., Binder, M., Diederich, P., Ertz, D., Gueidan, C., Hansen, K., Harris, R.C., Hosaka, K., Lim, Y.W., Matheny, B., Nishida, H., Pfister, D., Rogers, J., Rossman, A., Schmitt, I., Sipman, H., Stone, J., Sugiyama, J., Yahr, R. and Vilgalys, R. (2004). Assembling the fungal tree of life: progress, classification, and evolution of subcellular traits. Am. J. Bot. 91: 1446–1480. https://doi.org/10.3732/ajb.91.10.1446

Maina, U.M., Galadima, I.B., Gambo, F.M. and Zakaria, D. (2018). A review on the use of entomopathogenic fungi in the management of insect pests of field crops. J. Entomol. Zool. Stud. 6: 27–32.

Mani, K., Vitenberg, T., Khatib, S., and Opatovsky, I. (2023). Effect of entomopathogenic fungus *Beauveria bassiana* on the growth characteristics and metabolism of black soldier fly larvae. Pestic. Biochem. Phys. 197: 105684. https://doi.org/10.1016/j.pestbp.2023.105684

Mantzoukas, S., Kitsiou, F., Natsiopoulos, D. and Eliopoulos, P.A. (2022). Entomopathogenic Fungi: Interactions and Applications. Encyclopedia 2: 646–656. https://doi.org/10.3390/encyclopedia2020044

Martin, W.W. and Warren, A. (2020). *Periplasma*, gen. Nov., a new oomycete lineage with isogamous sexual reproduction. Mycologia 5: 1–14. https://doi.org/10.1080/00275514.2020.1797385

McLaughlin, D.J. and Spatafora, J.W. (2015). The Mycota: A Comprehensive Treatise on Fungi as Experimental Systems for Basic and Applied Research. VII Systematics and Evolution. Part B, 2nd ed.

Mendoza, L., Vilela, R. and Humber, R.A. (2018). Taxonomic and phylogenetic analysis of the Oomycota mosquito larvae pathogen *Crypticola clavulifera*. Fungal Biol., 122: 847–855. https://doi.org/10.1016/j.funbio.2018.04.010

Miao, S. and Nauwerck, A. (1999). Fungal Infection of *Eudiaptomus gracilis* (Copepoda, Crustacea) in Lake Mondsee. Limnologica, 29: 168–173. https://doi.org/0075-9511/99/29/02-168

Mier, N., Canete, S., Klaebe, A., Chavant, L., and Fourniertf, D. (1996). Insecticidal properties of mushroom and toadstool carpophores. Phytochemistry 41: 1293–1299. https://doi.org/10.1016/0031-9422(95)00773-3

Molnár, I., Gibson, D.M. and Krasnoff, S.B. (2010). Secondary metabolites from entomopathogenic Hypocrealean fungi. Nat Prod Rep 27: 1241–1275. https://doi.org/10.1039/c001459c

Mondal, S. and Singh, P. (2022). In silico host-pathogen surface interaction of entomopathogeic fungi, *Beauveria bassiana*. Cambridge Open Engage. https://doi.org/10.33774/coe-2022-b5hvg

Mongkolsamrit, S., Noisripoom, W., Pumiputikul, S., Boonlarppradab, C., Samson, R.A., Stadler, M., Becker, K. and Luangsa-ard, J.J. (2021). *Ophiocordyceps flavida* sp. nov. (Ophiocordycipitaceae), a new species from Thailand associated with *Pseudogibellula formicarum* (Cordycipitaceae), and their bioactive secondary metabolites. Mycol. Prog. 20: 477–492. https://doi.org/10.1007/s11557-021-01683-y

Montserrat, M., Castañé, C., Santamaria, S. (1998). *Neozygites parvispora* (Zygomycotina: Entomophthorales) causing an epizootic in *Frankliniella occidentalis* (Thysanoptera: Thripidae) on cucumber in Spain. J. Invertebr. Pathol. 71: 165–16. https://doi.org/10.1006/jipa.1997.4730

Mumo Sila, M., Mutie Musila, F., Wafula Wekesa, V. and Sangilu, I.S. (2023). Evaluation of pathogenicity of entomopathogenic oomycetes *Lagenidium giganteum* and *L. ajelloi* against anopheles mosquito larvae. Psyche, (Camb. Mass.) 2023: 2806034. https://doi.org/10.1155/2023/2806034

Muniz, E.R., Catão, A.M.L. Rueda-Páramo, M.E., Rodrigues, J., López Lastra, C.C., García, J.J., Fernandes, E.K.K., Luz, C. (2018). Impact of short-term temperature challenges on the larvicidal activities of the entomopathogenic watermold *Leptolegnia chapmanii* against *Aedes aegypti*, and development on infected dead larvae. Fungal Biol. 6: 430–435. https://doi.org/10.1016/j.funbio.2017.10.002

Namatame, I., Tomoda, H., Tabata, N., Si, S. and Omura, S. (1999). Structure elucidation of fungal Beauveriolide III, a novel inhibitor of lipid droplet formation in mouse macrophages. J. Antibiot., 52: 7–12. https://doi.org/10.7164/antibiotics.52.7

Naranjo-Ortiz, M.A. and Gabaldón, T. (2019). Fungal evolution: diversity, taxonomy and phylogeny of the Fungi. Biol. Rev. 94:387–432. https://doi.org/10.1111/brv.12550

NCBI (2024). Taxonomy Browser. Research support [online]. Available from: https://www.ncbi.nlm.nih.gov/Taxonomy/Browser/wwwtax.cgi?id=4751 NCBI [accessed January 2024]

Nicoletti, R., Andolfi, A., Becchimanzi, A. and Salvatore, M.M. (2023). Anti-insect properties of *Penicillium* secondary metabolites. Microorganisms 11: 1302. https://doi.org/10.3390/microorganisms11051302

Nithya, P.R., Manimegalai, S., Nakkeeran, S., Mohankumar, S., Nelson, S.J. (2019). Metabolome heterogeneity in the isolates of entomopathogenic fungus, *Beauveria bassiana* (Balsamo) Vuillemin. J. Biol, Control, 33: 326–335. https://doi.org/10.18311/jbc/2019/24302

Niu, X., Thaochan, N., Hu, Q. (2020). Diversity of linear non-ribosomal peptide in biocontrol fungi. J Fungi. 6(2): 61. https://doi.org/10.3390/jof6020061

Paterson, R.R.M. (2008). Cordyceps –a traditional Chinese medicine and another fungal therapeutic biofactory? Phytochemistry 69: 1469–1495. https://doi.org/10.1016/j.phytochem.2008.01.027

Pedrini, N. (2022). The entomopathogenic fungus *Beauveria bassiana* shows its toxic side within insects: expression of genes encoding secondary metabolites during pathogenesis. J Fungi. 8: 488. https://doi.org/10.3390/jof8050488

Pegram, R.A. and Wyatt, R.D. (1981). Avian gout caused by oosporein, a mycotoxin produced by *Chaetomium trilaterale*. Poult. Sci., 60: 2429–2440. https://doi.org/10.3382/ps.0602429

Piepenbring, M. (2015). Straminipila (Heterokonta) - Fungus-Like Organisms. In: Introduction to Mycology in the Tropics. Piepenbring, M., (Ed.) The American Phytopathological Society (Mycology), 301–312. https://doi.org/10.1094/9780890546130.008

Qasim, M., Islam, S.U., Islam, W., Noman, A., Khan, K.A., Hafeez, M., Hussain, D., Dash, C.K., Bamisile, B.S., Akutse, K.S., Rizwan, M., Nisar, M.S., Jan, S. and Wang, L. (2020). Characterization of mycotoxins from entomopathogenic fungi (*Cordyceps fumosorosea*) and their toxic effects to the development of asian citrus psyllid reared on healthy and diseased citrus plants. Toxicon 188: 39–47. https://doi.org/10.1016/j.toxicon.2020.10.012

Quesada-Moraga, E. and Vey, A. (2004). Bassiacridin, a protein toxic for locusts secreted by the entomopathogenic fungus *Beauveria bassiana*. Mycol. Res., 108: 441–452. https://doi.org/10.1017/s0953756204009724

Radek, R., Wellmanns, D. and Wolf, A. (2011). Two new species of *Nephridiophaga* (Zygomycota) in the *Malpighian tubules* of cockroaches. Parasitol. Res., 109: 473–482. https://doi.org/10.1007/s00436-011-2278-7

Rosas García, N.M., Mireles Martínez, M. and Villegas-Mendoza, J.M. (2020). Detección de bassianolida y beauvericina en cepas de *Beauveria bassiana* y su participación en la actividad patogénica hacia *Spodoptera* sp. Biotecnia, 22:93–99. https://doi.org/10.18633/biotecnia.v22i3.1060

Samuels, R.I., Coracini, D.L.A., Martins Dos Santos, C.A. and Gava, C.A.T. (2002). Infection of *Blissus antillus* (Hemiptera:Lygaeidae) eggs by the entomopathogenic fungi *Metarhizium anisopliae* and *Beaveria bassiana*. Biol. Control, 23: 269–273 http://dx.doi.org/10.1006/bcon.2001.1009

Sánchez, S.E.M., Humber, R., Freitas, A. and Pinheiro, A.M.C. (2010). *Batkoa apiculata* (Thaxter) Humber affecting Anopheles (Diptera: Culicidae) in the municipality of Una, Southern Bahia, Brazil. Entomotro 25: 63–68.

Sbaraini, N., Hu, J., Roux, I., Phan, C.-S., Motta, H., Rezaee, H., Schrank, A., Chooi, Y.-H., Staats, C.C. (2021). Polyketides produced by the entomopathogenic fungus *Metarhizium anisopliae* induce *Candida albicans* growth. Fungal Genet Biol. 152: 103568. https://doi.org/10.1016/j.fgb.2021.103568

Schoch, C.L., Sung, G.-H., L´opez-Gir´aldez, F., Townsend, J.P., Miadlikowska, J., Hofstetter, V., Robbertse, B., Matheny, P.B., Kauff, F., Wang, Z., Gueidan, C., Andrie, R.M., Trippe, K., Ciufetti, L.M., Wynns, A., Fraker, E., Hodkinson, B.P., Bonito, G., Groenewald, J.Z., Arzanlou, M., de Hoog, G.S., Crous, P.W., Hewitt, D., Pfister, D.H., Peterson, K., Gryzenhout, M., Wingfield, M.J., Aptroot, A., Suh, S.O., Blackwell, M., Hillis, D.M., Griffith, G.W., Castlebury, L.A., Rossman, A.Y., Lumbsch, H.T., Lücking, R., Büdel, B., Rauhut, A., Diederich, P., Ertz, D., Geiser, D.M., Hosaka, K., Inderbitzin, P., Kohlmeyer, J., Volkmann-Kohlmeyer, B., Mostert, L., O'Donnell, K., Sipman, H., Rogers, J.D., Shoemaker, R.A., Sugiyama, J., Summerbell, R.C., Untereiner, W., Johnston, P.R., Stenroos, S., Zuccaro, A., Dyer, P.S., Crittenden, P.D., Cole, M.S., Hansen, K., Trappe, J.M., Yahr, R., Lutzoni, F. and Spatafora, J.W. (2009). The Ascomycota tree of life: a phylum-wide phylogeny clarifies the origin and evolution of fundamental reproductive and ecological traits. Syst. Biol., 58: 224–239. https://doi.org/224–239. 10.1093/sysbio/syp020

Scholte, E.J., Knols, B.G., Samson, R.A. and Takken, W. (2004). Entomopathogenic fungi for mosquito control: a review. J. Insect Sci. 4: 19. https://doi.org/10.1093/jis/4.1.19

Sharma, A., Srivastava, A., Shukla, A.K., Srivastava, K., Srivastava, A.K. and Saxena, A.K. (2020). Entomopathogenic fungi: a potential source for biological control of insect pests. In: *Phytobiomes:* Current Ins Fut Vistas, Solanki, M., Kashyap, P. and Kumari, B. (Eds.) Springer, Singapore. https://doi.org/10.1007/978-981-15-3151-4_9

Sharma, R. and Sharma, P. (2021). Fungal entomopathogens: a systematic review. Egypt. J. Biol. Pest. Control, 31: 57. https://doi.org/10.1186/s41938-021-00404-7

Shaurub, E.S.H. (2022). Review of entomopathogenic fungi and nematodes as biological control agents of tephritid fruit flies: current status and a future vision. Entomol. Exp. Appl., 171: 17–34. https://doi.org/10.1111/eea.13244

Shen, D., Tang, Z., Wang, C., Wang, J., Dong, Y., Chen, Y., Wei, Y., Cheng, B., Zhang, M., Grenville-Briggs, L.J., Tyler, B.M., Dou, D. and Xia, A. (2019). Infection mechanisms and putative effector repertoire of the mosquito pathogenic oomycete *Pythium guiyangense* uncovered by genomic analysis. PLoS Genetics 15: e1008116. https://doi.org/10.1371/journal.pgen.1008116

Sinha, K.K., Choudhary, A.Kr. and Kumari, P. (2016). Entomopathogenic Fungi. In: Ecofriendly Pest Management for Food Security, Omkar (Ed.) Academic Press, 475–505. https://doi.org/10.1016/B978-0-12-803265-7.00015-4

Soman, A.G., Gloer, J.B., Angawi, R.F., Wicklow, D.T. and Dowd, P.F. (2001). Vertilecanins: New phenopicolinic acid analogues from *Verticillium lecanii.* J. Nat. Prod. 64: 189–192. https://doi.org/10.1021/np000094q

Spatafora, J.W., Chang, Y., Benny, G.L., Lazarus, K., Smith, M.E., Berbee, M.L., Bonito, G., Corradi, N., Grigoriev, I., Gryganskyi, A., James, T.Y., O'Donnell, K., Roberson, R.W., Taylor, T.N., Uehling, J., Vilgalys, R., White, M.M. and Stajich, J.E. (2016). A phylum-level phylogenetic classification of zygomycete fungi based on genome-scale data. Mycologia, 108: 1028–1046. https://doi.org/10.3852/16-042

Staropoli, A., Iacomino, G., De Cicco, P., Woo, S.L., Di Costanzo, L. and Vinale, F. (2023). Induced secondary metabolites of the beneficial fungus *Trichoderma harzianum* M10 through OSMAC approach. Chem. Biol. Technol. Agric. 10: 28. https://doi.org/10.1186/s40538-023-00383-x

Steinkraus, D.C. and Boys, G.O. (2005). Mass harvesting of the entomopathogenic fungus, Neozygites fresenii, from natural field epizootics in the cotton aphid, *Aphis gossypii.* J. Invertebr. Pathol. 88: 212–217. https://doi.org/10.1016/j.jip.2005.01.008

Strassert, J.F.H., Wurzbacher, C., Hervé, V. Antany, T., Brune, A. and Radek, R. (2021). Long rDNA amplicon sequencing of insect-infecting nephridiophagids reveals their affiliation to the Chytridiomycota and a potential to switch between hosts. Sci. Rep., 396 https://doi.org/10.1038/s41598-020-79842-6

Su, X., Zou, F., Guo, Q., Huang, J. and Chen, T. (2001). A report on a mosquito-killing fungus, *Pythium carolinianum.* Fungal Divers, 7: 129–133.

Suradet, B., War War May, Z., Winanda, H., Pabhop, S. and Anake, K. (2022). Laboratory-based toxicity of scale insect pathogen *Moelleriella raciborskii* (Zimm.) (Hypocreales: Clavicipitaceae) crude extracts and isolated compounds against *Tetranychus truncatus* Ehara (Acari: Tetranychidae). Syst. App. Acarol. 27: 1152–1165. https://doi.org/10.11158/saa.27.6.13

Ton That Huu, D., Phuong, H.T., Diem Tran, P.T., Souvannalath, B., Trung, H.L., Ho, D.V., and Canh Viet, C.L. (2022). Secondary metabolites from the grasshopper-derived entomopathogenic fungus *Aspergillus tamarii* NL3 and their biological activities. Nat. Prod. Comm. 17: https://doi.org/10.1177/1934578X221141548

van Meer, G., Voelker, D.R. and Feigenson, G.W. (2008). Membrane lipids: where they are and how they behave. Nat. Rev. Mol. Cell Biol. 9: 112–124. https://doi.org/10.1038/nrm2330

Vargas-Aguilar A.A. and Fuentes-Condori, R. (2021). Estudios *in Silico*, simulando vida en un entorno virtual. Gac. Méd. Boliv. 44: 278–279. https://doi.org/10.47993/gmb.v44i2.263

Vey, A., Hoagland, R.E. and Butt, T.M. (2001). Toxic metabolities of fungal biocontrol agents. In: Fungi as Biocontrol Agents: Progress, Problems and Potential, Butt, T.M., Jackson, C.W. and Magan, N. (Eds.). CABI Publishing, Wallingford, UK, pp. 311–346.

Vivekanandhan, P., Swathy, K., Kalaimurugan, D., Ramachandran, M., Yuvaraj, A., Kumar, A.N., Manikanda, A.T., Poovarasan, N., Shivakumar, M.S. and Kweka, E.J. (2020). Larvicidal toxicity of *Metarhizium anisopliae* metabolites against three mosquito species and non-targeting organisms. PLoS One 15: e0232172. https://doi.org/10.1371/journal.pone.0232172

Wallace Martin, W. (2000). Two new species of *Couchia* parasitic in midge eggs. Mycologia, 92: 1149–1154. https://doi.org/10.1080/00275514.2000.12061262

Wang, B., Kang, Q., Lu, Y., Bai, L. and Wang, C. (2012). Unveiling the biosynthetic puzzle of destruxins in *Metarhizium* species. Proc. Natl. Acad. Sci. USA 109: 1287–1292. https://doi.org/10.1073/pnas.1115983109

Wang, Q. and Lijian, X. (2012) Beauvericin, a Bioactive Compound Produced by Fungi: A Short Review. Molecules 17(3): 2367–2377. https://doi.org/10.3390/molecules17032367

Wang, H., Peng, H., Li, W., Cheng, P. and Gong, M. (2021). The toxins of *Beauveria bassiana* and the strategies to improve their virulence to insects. Front Microbiol., 12:705343. https://doi.org/10.3389/fmicb.2021.705343

Wang, X., Gao, Y.L., Zhang, M.L., Zhang, H.-D., Huang, J.Z. and Li, L. (2020). Genome mining and biosynthesis of the Acyl-CoA:cholesterol acyltransferase inhibitor beauveriolide I and III in *Cordyceps militaris*. J. Biotechnol., 309: 85–91. https://doi.org/10.1016/j.jbiotec.2020.01.002

White, M.M., James, T.Y., O'Donnell, K., Cafaro, M.J., Tanabe, Y. and Sugiyama, J. (2006). Phylogeny of the Zygomycota based on nuclear ribosomal sequence data. Mycologia, 98: 872–884. https://doi.org/10.3852/mycologia.98.6.872

Wieland, T. (1968) Poisonous principles of mushrooms of the genus *Amanita*. Four-carbon amines acting on the central nervous system and cell-destroying cyclic peptides are produced. Science, 59(3818): 946–952. https://doi.org/10.1126/science.159.3818.946

Woolley, V.C., Teakle, G.R., Prince, G., de Moor, C.H. and Chandler, D. (2020). Cordycepin, a metabolite of *Cordyceps militaris*, reduces immune-related gene expression in insects. J. Inver. Pathol., 177: 107480. https://doi.org/10.1016/j.jip.2020.107480

Xu, Y., Orozco, R., Kithsiri Wijeratne, E.M., Espinosa-Artiles, P., Gunatilaka, A.A.L., Stock, S.P. and Molnár, I. (2009). Biosynthesis of the cyclooligomer depsipeptide bassianolide, an insecticidal virulence factor of *Beauveria bassiana*. Fungal Genet Biol. 46: 353–364. https://doi.org/10.1016/j.fgb.2009.03.001

Xu, Y.J., Luo, F., Gao, Q., Shang, Y. and Wang, C. (2015). Metabolomics reveals insect metabolic responses associated with fungal infection. Anal. Bioanal. Chem. 407: 4815–21. https://doi.org/10.1007/s00216-015-8648-8

Yahagi, H., Yahagi, T., Furukawa, M. and Matsuzaki, K. (2020). Antiproliferative and antimigration activities of beauvericin isolated from *Isaria* sp. on pancreatic cancer cells. Molecules, 25: https://doi.org/10.3390/molecules25194586

Yang, H., Tian, L., Qiu, H., Qin, C., Ling, S. and Xu, J. (2023). Metabolomics analysis of sporulation-associated metabolites of *Metarhizium anisopliae* based on Gas Chromatography–Mass Spectrometry. J. Fungi, 9: 1011. https://doi.org/10.3390/jof9101011

Yin, Y., Chen, B., Song, S., Li, B., Yang, X. and Wang, C. (2020). Production of diverse beauveriolide analogs in closely related fungi: a rare case of fungal chemodiversity. mSphere, 5. https://doi.org/10.1128/msphere.00667-20

Yoneyama, T., Elshamy, A.I., Yamada, J., El-Kashak, W.A., Kasai, Y., Imagawa, H., Ban, S., Noji, M. and Umeyama, A. (2022). Antimicrobial metabolite of *Cordyceps tenuipes* targeting MurE ligase and histidine kinase via in silico study. App Microbiol. Biotechnol. 106: 6483–6491. https://doi.org/10.1007/s00253-022-12176-7

Zawadneak, M.A.C., Gonçalves, R.B., Pimentel, I.C., Schuber, J.M., Santos, B., Poltronieri, A.S. and Solis, M.A. (2016). First record of Duponchelia fovealis (Lepidoptera: Crambidae) in South America. Idesia 34: 91–95. https://doi.org/10.4067/S0718-34292016000300011

Zhang, C., Teng, B., Liu, H., Wu, C., Wang, L. and Jin, S. (2023). Impact of *Beauveria bassiana* on antioxidant enzyme activities and metabolomic profiles of *Spodoptera frugiperda*. J. Invertebr. Pathol. 198: 107929. https://doi.org/10.1016/j.jip.2023.107929

Zhang, C., Wanga, W., Lu, R., Jin, S., Chen, Y., Fan, M., Huang, B., Li, Z., Hu, F. (2016). Metabolic responses of *Beauveria bassiana* to hydrogen peroxide-induced oxidative stress using an LC-MS-based metabolomics approach. J. Invertebr. Pathol. 137: 1–9. http://dx.doi.org/10.1016/j.jip.2016.04.005

Zhang, K.Q. and Hyde, K.D. (2014). Nematode-trapping fungi. In: Fungal Diversity Research Series. Zhang, K.Q. and Hyde, K.D. (Ed.) Springer Science & Business.

Zhang, L., Fasoyin, O.E., Molnár, I., Xu, Y. (2020b). Secondary metabolites from hypocrealean entomopathogenic fungi: novel bioactive compounds. Nat. Prod. Rep. 37: 1181–1206. https://doi.org/10.1039/C9NP00065H

Zhang, L., Yue, Q., Wang, C., Xu, Y., Molnár, I. (2020a). Secondary metabolites from hypocrealean entomopathogenic fungi: genomics as a tool to elucidate the encoded parvome. Nat. Prod. Rep., 37(9): 1164–1180. https://doi.org/10.1039/D0NP00007H

Zibaee, A., Bandani, A.R., Tork, M. (2009). Effect of the entomopathogenic fungus, *Beauveria bas*siana, and its secondary metabolite on detoxifying enzyme activities and acetylcholinesterase (AChE) of the Sunn pest, *Eurygaster integriceps* (Heteroptera: Scutellaridae). Bio. Sci. Technol., 19: 485–498. https://doi.org/10.1080/09583150902847127

13 Pest Control Strategies Using Secondary Metabolites of Entomopathogenic Fungi

Carlos A. Granados-Echegoyen,[1*] Nancy Calderón-Cortés,[2] Alfonso Vásquez-López,[3] Pedro Mendoza de Gives,[4] Manuela Reyes-Estebanez,[5] Luis G. Sarmiento-López,[6] Ileana Vera-Reyes,[7] Nadia S. Gómez-Domínguez,[8] Nadia Landero-Valenzuela,[9] Esperanza Loera-Alvarado,[10] Florinda García-Pérez,[11] Beatriz Quiroz-González[3] and César Sánchez-Hernández[11]

1. Introduction

Several microorganisms in nature play a crucial role in regulating insect populations and are considered pests. They can be used as biological control agents to safeguard agricultural and forest crops and regulate vector insects (Shamim et al., 2024). Comparing microbial insecticides and chemical products requires a focused cost-benefit analysis. Microbial insecticides offer significant benefits, including ensuring human safety and reducing pesticide residues in food. This, in turn, helps to conserve natural enemies and promote biodiversity in ecosystems (Saroop and Tamchos, 2024). Due to their adaptability, fungi can thrive in both invertebrate and vertebrate hosts. Some species target insects and nematodes to control their populations. Entomopathogenic

[1] CONAHCYT-Instituto Politécnico Nacional (IPN), CIIDIR-Oaxaca, Santa Cruz Xoxocotlán 71230, Oaxaca, Mexico.
[2] National School of Higher Studies Morelia Unit, National Autonomous University of Mexico (UNAM), Morelia 58190, Michoacán, Mexico.
[3] Instituto Politécnico Nacional (IPN), CIIDIR-Oaxaca, Santa Cruz Xoxocotlán 71230, Oaxaca, Mexico.
[4] Laboratory of Helminthology, National Centre for Disciplinary Research in Animal Health and Innocuity (INIFAP-Mexico), Jiutepec-62550, Morelos, Mexico.
[5] Research Center in Environmental Microbiology and Biotechnology (CIMAB), Autonomous University of Campeche, San Francisco de Campeche 24039, Campeche, Mexico.
[6] Departamento Académico de Ciencias Naturales y Exactas, Universidad Autónoma de Occidente Unidad Regional Los Mochis. Sinaloa, 81223, Mexico.
[7] Department of Biosciences and Agrotechnology, Applied Chemistry Research Center, Saltillo 25294, Coahuila, Mexico.
[8] Instituto Politécnico Nacional (IPN), Biotic Product Development Center (CEPROBI), Yautepec 62739, Morelos, Mexico.
[9] Department of Horticulture, Universidad Autónoma Agraria Antonio Narro, Calzada Antonio Narro, 25294, Coahuila, Mexico.
[10] CONAHCYT-Universidad Autónoma Chapingo, Centro Regional Universitario Centro Occidente (CRUCO), Morelia, Michoacán, 58170, Mexico.
[11] Universidad NovaUniversitas, Oaxaca-Puerto Ángel, Ocotlán de Morelos 71513, Oaxaca, Mexico.
* Corresponding author: granados.cgranadose@ipn.mx

fungi, which belong to phyla such as Ascomycota, Zygomycota, and Chytridiomycota (Hibbett et al., 2007), include both generalist species such as *Beauveria bassiana* and *Metarhizium anisopliae* (Hypocreales: Clavicipitaceae) and specific species such as *Aschersonia aleyrodis, Hirsutella thompsonii*, and *Entomophthora* spp. The former infects multiple orders of insects, while the latter targets specific species, families, or groups (Vidhate et al., 2023).

Because of the wide range of hosts, entomopathogenic fungi show remarkable variation. They display pathotypes with similar morphologies but distinct host ranges, variations in virulence, and specific developmental requirements (Bidochka et al., 2002). For example, *B. bassiana* is a diverse group characterized by its ability to exchange cellular material, leading to a wide range of isolates with significant genetic polymorphism and strong adaptability to different environments (Meyling et al., 2009). Furthermore, it has been suggested that host insects exert selective pressure, favoring certain genotypes and leading to specialization within *Beauveria* species (Shapiro-Ilan et al., 2005). Levels of infection and mortality depend on various factors, such as host and pathogen characteristics, the pathogen's capability to invade the host, and its capacity to persist in the environment. Environmental factors such as temperature, humidity, wind speed, and vegetation structure play a critical role in this dynamic process (Páez and Fleming-Davies, 2020).

Application methods, mycoinsecticide formulations, and various biotic and abiotic environmental factors play a critical role in determining the persistence and spatial distribution of pathogen propagules (Shahid et al., 2012). Transmission can occur horizontally from infected individuals as well as through residues in vegetation, which is an additional route of infection for adults (Thomas, 1997). The ability to disperse is crucial for entomopathogenic fungi to spread within the host environment (Bidochka et al., 2001). Understanding the complex interactions between host and pathogen within the environmental conditions in which these events take place is essential. Disease transmission between individuals depends on the virulence of the isolate or pathotype, which is evaluated using a sensitive bioassay (Hajek and St Leger, 1994). Entomopathogenic microorganisms are assessed using three different bioassays: the pathogenicity test, the biological response interval, and bioassays designed to determine the concentration and duration required to induce 50% mortality in treated insects. These assays help determine the suitable concentration or dosage for field application (CORPOICA, 2011, Maistrou et al., 2020).

Entomopathogenic fungi, integrated into pest management strategies, are considered environmentally friendly and are widely utilized in organic, extensive, protected, and horticultural agriculture (Jackson and Jaronski, 2012). These products must be standardized, highly effective, long-lasting, and cost-effective. However, biopesticides differ from agrochemicals in that they are composed of living organisms, which presents challenges in their production and storage (Jackson et al., 2010). Entomopathogenic fungi have shown effectiveness in controlling agricultural arthropod pests and vector-borne diseases, such as aphids, caterpillars, thrips, and whiteflies (Tapfuma et al., 2022). Several strategies, such as classical biological control, augmenting through periodic applications, self-propagating through the dispersal of conidia by contaminated insects, and conserving through environmental manipulation, can be used to effectively manage pests with these fungi (Alatorre-Rosas, 2007).

Entomopathogenic fungi, traditionally known as pathogens of arthropods, play several ecological roles in nature. Recent research suggests that they act as phytopathogenic fungi adversaries, colonizers of the rhizosphere, and promoters of plant growth (Jaber and Salem, 2014). While they are commonly found in soils, their natural distribution seems to be linked to specific habitats, and their soil populations are affected by agricultural practices (Meyling et al., 2006, Jaronski, 2010). Approximately 750 species of entomopathogenic fungi in over 100 genera have been documented (Ramanujam et al., 2014). These include *Akanthomyces, Aschersonia, Beauveria, Entomophthora, Erynia, Eryniopsis, Fusarium, Hirsutella, Hymenostilbe, Isaria, Lecanicillium, Metarhizium, Paecilomyces, Verticillium*, and *Zoophthora* (Maina et al., 2018; Sharma and Sharma, 2021).

Beauveria and *Metarhizium* are globally recognized and extensively studied genera because of their effectiveness and ease of propagation (Rodríguez et al., 2006). While microbial insecticides are considered effective and environmentally friendly alternatives for pest control (Escamilla et al., 2022), their limited use is attributed to their slow action and lack of rapid control. The prevalence of specific insects, influenced by factors such as extensive monocultures, justifies the use of chemical pesticides, albeit with consideration of the associated environmental risks (Khan et al., 2023).

Research has been conducted to demonstrate the synergistic efficacy of combining strategies that individually aim to reduce pesticide doses. This synergy not only effectively addresses the pest problem, but also facilitates the successful introduction of entomopathogens into a given ecosystem. Once established, these entomopathogens have the potential to effectively control pest populations in an ecologically sustainable manner. Over time, this could lead to a significant decrease or complete elimination of pesticide use. Entomopathogens enter the insect through the cuticle, and the initial interaction between the infectious unit and the host is crucial for initiating infection (Gómez et al., 2014). The disease process involves nine sequential steps: (1) spore attachment to the cuticle; (2) germination; (3) penetration; (4) multiplication within the hemocoel; (5) synthesis of toxic metabolites; (6) host mortality; (7) mycelial invasion of organs; (8) mycelial growth outside the insect; and (9) generation of new conidia (Abdel-Raheem, 2020).

In the process of mycosis development by entomopathogenic fungi, several critical stages describe the infection: (1) Adhesion to the cuticle: Spores adhere to the insect cuticle through specific and non-specific interactions (Gómez et al., 2014; Butt et al., 2016); (2) Conidial germination: Triggered by compounds on the insect's epicuticle, germination is influenced by environmental and biological factors and can lead to appressorium formation (Valero-Jiménez et al., 2016); (3) Penetration: The fungus uses enzymes such as CYP, lipases, proteases, and chitinases to penetrate the cuticle (Wang and Wang, 2017); (4) Multiplication: Fungi such as *B. bassiana* utilize proteolytic enzymes, while insects activate immune responses, including cellular and systemic responses (Zimmermann, 2007; Pedrini, 2018); (5) Fungal development: Within the insect, the fungus forms hyphal bodies that multiply and facilitate invasion of other tissues, ultimately leading to the demise of the insect (Pedrini, 2018).

This chapter aims to provide a comprehensive overview of the contribution of entomopathogenic fungi to pest control, from their biological properties to their practical application in pest management and their broader implications for environmental sustainability.

2. Insect-Pathogen Interactions

An essential consideration in the use of entomopathogenic fungi for the biological control of insect pests is the insect microbiome. Insects, like other animals and plants, interact with a wide variety of microorganisms. These interactions are complex and depend heavily on a variety of factors, with particular emphasis on environmental context (Hajek et al., 2019). They play a crucial role in the survival and development of insects. Microorganisms can inhabit various parts of the insect, including the gut, hemolymph, and specialized cells (e.g., endosymbionts), or they can be present externally on the insect host (ectosymbionts) (Wang et al., 2023). Symbiotic interactions may be either obligatory or facultative. Obligate symbionts generally fulfil crucial functions in the nutrition and physiology of insect hosts, such as amino acid synthesis and lignocellulose degradation. On the other hand, facultative symbionts may provide advantages in particular circumstances, such as protection against adverse environmental conditions and parasites or pathogens (Hajek et al., 2019). Consequently, mutualistic symbionts of insects have a wide range of ecological and evolutionary impacts on their hosts and ecosystems by regulating interactions across multiple trophic levels, particularly those involving insects and their natural enemies (Zytynska and Meyer, 2019).

The diversity and resilience of insect microbiomes may vary across taxonomic groups, feeding habits, and habitats (Pessotti et al., 2021). However, specific insect species host highly specialized microbial communities that are unique to them, such as termite and aphid microbiomes (Douglas, 2015). Research utilizing culture-dependent and metagenomic methods to investigate bacterial communities in different insect orders indicates that insects principally form associations with Proteobacteria (62-66%), Firmicutes (7-20%), Actinobacteria (5-11%), and Bacteroidetes (6-10%) (Douglas, 2015). Actinobacteria, in this context, are mostly responsible for protecting against fungal entomopathogens (Pessotti et al., 2021). Actinobacteria primarily reside in soil, comprising more than 30% of the total soil microbial population. They frequently inhabit the rhizosphere and have been identified as one of the most prevalent classes of endophytic bacteria (Silva et al., 2022a). Therefore, it is not surprising that they have a close association with insects. The obligatory symbiotic relationship between defensive actinobacteria and insects has been observed in fungus-farming ants (Attine), beewolf wasps, and the southern pine beetle *Dendroctonus frontalis* (Kaltenpoth et al., 2014). As well as an interaction with the wood-feeding beetle *Odontotaenius disjunctus* (Pessotti et al., 2021).

The insect-associated actinobacterial genus Streptomyces, which accounts for over 95% of all soil actinobacteria (Silva et al., 2022a), has recently undergone a comprehensive survey. This study assessed the presence of Streptomyces in 2,561 species of insects and other arthropods from 15 taxonomic orders. Through 51,050 individual antimicrobial bioactivity assays using 2003 isolates (Chevrette et al., 2019), researchers identified 10,178 strains from 1,445 insect species spanning 13 different orders. Notable insect orders such as Hymenoptera, Diptera, Lepidoptera, and Coleoptera yielded 2,934, 2,920, 1,326, and 1,139 isolates, respectively, highlighting the broad distribution of Streptomyces-insect interactions. Crucially, this survey revealed that Streptomyces strains associated with insects exhibited enhanced bioactivity against Gram-positive and Gram-negative bacteria and fungi, including prominent entomopathogens such as Metarhizium and Trichoderma, compared to strains from soil and water.

Furthermore, soil-derived Streptomyces have shown significant potential as biocontrol agents against various important agricultural fungal pathogens. These pathogens include *Alternaria alternata, Botryts cinerea, Colletotrichum gloeosporioides, C. musae, Corynespora cassiicola, Fusarium moniliforme, F. oxysporum, F. solani, F. verticillioides, Macrophomina phaseolina, Penicillium digitatum, Pythium aphanidermatum, Phytophthora cinnamomi, Rhizoctonia solani,* and *Sclerotinia sclerotiorum,* (Silva et al., 2022a). Moreover, additional actinobacterial genera, including *Pseudonocardia* (Currie et al., 2003), *Dietzia, Gordonia, Leucobacter,* and *Stenotrophomonas* (Liu et al., 2020), have been documented for their antifungal properties in insect hosts.

In addition to the fungicidal effects of actinobacteria, which are characterized by the synthesis of a diverse range of metabolites such as actinomycins, angucyclinons, nactins, cycloheximidine, mycangimycin, selvamycin, sceliphrolactam, dentogerumycin, cyphomycin, and pierycin, among others (Pessotti et al., 2021), and the secretion of extracellular enzymes including glucanases, cellulases, chitinases, and proteases, it has been observed that bacteria from other phyla, particularly Proteobacteria and Firmicutes, as well as yeasts associated with insects, also have an impact on fungal entomopathogens (Table 1). The mechanisms utilized by these microorganisms to combat fungi may vary from those employed by actinobacteria. In certain cases, these processes may entail the production of antifungal phenolic compounds, colonization resistance, and/or non-specific immunization (Dillon and Charnley, 2002).

The evidence underscores the significance of investigating the interactions between insect microbiomes and natural enemies in various environmental contexts, particularly for species of agricultural importance. Despite their significance, these interactions have received relatively little attention in current research (Shao et al., 2024). However, insights from non-agricultural insect pests strongly suggest the essential role of these interactions in insect control strategies. Understanding these connections is an intricate matter; insects can evolve new resistance mechanisms when subjected to significant selective pressure (Douglas, 2018). Additionally, a trade-off exists between the protective advantages and the fitness costs associated with harboring multiple symbionts in specific insects, such as aphids (Zytynska and Meyer, 2019). Consequently, the results of particular interactions may differ under various environmental conditions and among different insect species (Hajek et al., 2019). For example, a recent study investigated the influence of gut bacterial microbiota on *B. bassiana* infection in the European gypsy moth *Lymantria dispar* (Lepidoptera: Erebidae) and found a notably higher mortality rate in non-axenic *L. dispar* larvae compared to axenic larvae (Bai et al., 2023). Furthermore, the research emphasizes the impact of plant diversity and predator exposure on the protective interactions between insects and symbionts (Miller et al., 2020). Therefore, it is crucial to understand the

Table 1 Non-actinobacterial mutualists engaged in protective interactions with insects (Proteobacteria and Firmicutes were recently renamed Pseudomonadota and Bacillota, respectively).

Insect species	*Microbial mutualists*	*Microbial function*	*Reference*
Acyrthosiphuon pisum (Hemiptera: Aphididae)	*Regiella* sp., *Regiella insecticola, Rickettsiella viridis*, and *Rickettsia* (Proteobacteria) *Spiroplasma* (Mycoplasmatota), and co-infection by *Fukatsuia symbiotica* (Proteobacteria) and *Spiroplasma*	Protection from the fungal entomopathogens *Zoophthora occidentalis* and *Pandora neoaphidis*	Scarborough et al., 2005; Lukasik et al., 2013; Parker et al., 2013
Bactrocera dorsalis (Diptera: Tephritidae)	*Citrobacter* sp.	Insecticides have resistance to trichlorphon, but it is unknown if they can confer protection from entomopathogen infection	Cheng et al., 2017
Blattella germanica (Blattodea: Blattellidae)	*Pseudomonas aeruginosa, P. reactans, Pseudomonas* sp., *Delftia* sp. (Proteobacteria), *Bacillus subtilis* and *Bacillus* sp. (Firmicutes)	Antifungal activity against *B. bassiana* and *M. anisopliae* entomopathogens	Zhang et al., 2018
Dendroctonus brevicomis (Coleoptera: Curculionidae)	*Ogataea pini* (Ascomycota: Saccharamycetales)	Volatiles produced by the yeast inhibited the growth of *B. bassiana*	Davis et al., 2011
Drosophila melanogaster (Diptera: Drosophilidae)	*Wolbachia pipientis* (Proteobacteria)	Protection against the fungal entomopathogen *B. bassiana*	Panteleev et al., 2007
Plutella xylostella (Lepidoptera: Plutellidae)	*Enterococcus* sp. (Firmicutes), *Enterobacter* sp. (Proteobacteria), and *Serratia* sp. (Proteobacteria)	Insecticides have resistance to fenitrothion, but it is unknown if they can confer protection from entomopathogen infection.	Xia et al., 2017, 2018

Contd.

Table 1 *Contd.*

Insect species	*Microbial mutualists*	*Microbial function*	*Reference*
	Pseudomonas sp. Strain PRGB06 (Proteobacteria)	Antagonistic activity towards entomopathogenic fungi, including *B. bassiana, H. thompsonii, M. anisopliae, Paecilomyces* sp., and *P. tenuipes*	Indiragandhi et al., 2007
Riptortus pedestris (Hemiptera: Alydidae)	*Burkholderia* (Proteobacteria)	Insecticides have resistance to chlorpyrifos, but it is unknown if they can confer protection from entomopathogen infection	Kikuchi et al., 2012
Schistocerca gregaria (Orthoptera: Acrididae)	Enterobacteriacea gut bacterial community (Proteobacteria)	Inhibition by the gut microbiota prevents invasion by *Metarhizium robertsii* through the gut	Dillon and Charnley, 2002
Spodoptera exigua (Lepidoptera: Noctuidae)	*Methylobacterium, Marinomonas, Paenochrobactrum, Pseudomonas, Acinetobacter, Delftia* and *Paracoccus* (Proteobacteria)	Probiotics against *B. bassiana* entomopathogen infection	Liu et al., 2020

ecological and evolutionary foundations of multitrophic interactions that encompass insect pests and their microbiomes. Such insights play a crucial role in promoting sustainable and environmentally friendly agricultural practices. Moreover, examining these interactions would facilitate the recognition of potential impacts on non-target organisms that are vital for essential ecosystem services such as pollination, soil decomposition, and natural pest control.

3. Diversity of Entomopathogenic Fungi Used in Agriculture

Entomopathogenic fungi, such as *B. bassiana* and *M. anisopliae*, play a crucial role in the biological control of agricultural pests (Pacheco-Hernández et al., 2019). Molecular evolution has unveiled complexity in the classification of these organisms, resulting in the discovery of multiple cryptic species and genetic diversity. For example, *M. anisopliae* comprises approximately 14 cryptic species that can only be identified through molecular analysis (Luz et al., 2019). Similarly, *B. bassiana* forms a complex of cryptic phylogenetic species (Serna-Domínguez et al., 2019), which makes identification using morphological methods challenging (Imoulan et al., 2017). The existence of hidden lineages presents a challenge to the identification of species, leading to a growing dependence on molecular techniques like targeted genomic sequencing and the analysis of polymorphic markers (Medo et al., 2016).

These fungi offer significant advantages over synthetic insecticides, including their specificity to the target insect, safety for humans and vertebrates, sustained pest control, and disruption of the insect life cycle. In addition, certain species, such as *B. bassiana*, can produce compounds that repel specific pests (Jalinas et al., 2022) and act as endophytes within plants, promoting the synthesis of metabolites that enhance pest resistance and plant growth (Homayoonzadeh et al., 2022). However, they face challenges such as slower efficacy compared to chemical products, requiring more stable and cost-effective formulations, and their limited spectrum of action restricts their use. Furthermore, their combination with botanical insecticides can result in incompatibility (Sain et al., 2019). Specific environmental conditions are required for the application and storage of these products, making them more delicate when compared to chemical products (Guzmán et al., 2019).

3.1 Major Pests Controlled by Entomopathogenic Fungi

Entomopathogenic fungi, such as *B. bassiana* and *M. anisopliae*, are extensively utilized for controlling significant pests because of their insecticidal activity (Table 2). *Beauveria bassiana*, as a generalist, demonstrates the ability to target a wide range of insects, with documented reports of its impact on over 1000 species (Bhattacharyya et al., 2023). *M. anisopliae* naturally targets over 300 insect pest species (Monzón, 2001).

4. Biocontrol of Pests

4.1 Entomopathogenic Fungi against Scyphophorus acupunctatus

Entomopathogenic fungi have historically played a natural role in regulating insect populations. Humans have harnessed their capacity to colonize and induce mortality in hosts, employing them as effective biological control agents against agricultural pests. *Beauveria bassiana* and *M. anisopliae, Aschersonia, Isaria (Paecilomyces), Akanthomyces* (*Lecanicillium*), *Verticillium, Hirsutella citriformis,* and *Entomophthora* are among the most used entomopathogenic fungi in agriculture (Guizar-Guzmán and Sánchez-Peña, 2013; Pérez-González et al., 2016; Liu et al., 2023; Bohatá et al., 2023; Yun et al., 2023).

The agave crop in Mexico holds significant cultural, social, and economic importance as it is used in the production of beverages such as tequila, mezcal, and pulque. Scyphophorus acupunctatus, commonly known as the agave weevil, has caused a 13% decline in agave production in Oaxaca State, Mexico. Synthetic chemical products often prove to be ineffective because the larvae and adult insects shield them inside the stem of the plant. A biologically and environmentally friendly

Table 2 Major entomopathogenic fungi for the control of diverse agricultural pests.

Fungus	*Pest*	*Crop*	*Reference*
B. bassiana	*Hypothenemus hampei* (Coleoptera: Scolytinae)	*Coffea arabica* (Rubiaceae)	Chang et al., 2023
	Lygus lineolaris nymphs (Hemiptera: Miridae)	*Gossypium herbaceum* (Malvaceae)	Portilla et al., 2019
		Fragaria ananassa (Rosaceae)	González-Santarosa et al., 2010
	Sphenarium purpurascens (Orthoptera: Pyrgomorphidae)	*Zea mays* (Poaceae)	Romero-Arenas et al., 2020
	Brachystola magna, B. mexicana (Orthoptera: Romaleidae)	*Phaseolus vulgaris* (Fabaceae)	Gutiérrez and Luna, 2006
	Bemisia tabaci (Hemiptera: Aleyrodidae)	*Solanum lycopersicum* (Solanaceae)	Wei et al., 2018
	Cydia funebrana (Lepidoptera: Tortricidae)	*Prunus domestica* (Rosaceae)	Petrova, 2023
	Phyllophaga obsoleta (Coleoptera: Melolonthinae)	*Z. mays*	Dávila-Orozco et al., 2021
	Anastrepha obliqua (Diptera: Tephritidae)	*Mangifera indica* (Anacardiaceae)	Martínez-Barrera et al., 2020
	Bactericera cockerelli (Hemiptera: Triozidae)	*Capsicum annuum* (Solanaceae)	Liu et al., 2020
	Rhynchophorus ferrugineus (Coleoptera: Curculionidae)	Palms (Arecaceae)	Jalinas et al., 2022
	Carpophilus davidsoni (Coleoptera: Nitidulidae)	*Juglans regia* (Juglandaceae)	Boston et al., 2020
	Leptinotarsa decemlineata (Coleoptera: Chrysomelidae)	*Solanum tuberosum* (Solanaceae)	Baki et al., 2021
	Drosophila suzukii (Diptera: Drosophilidae)	*F. ananassa*	Furuie et al., 2022
	Macrosiphum rosae (Homoptera: Aphididae)	*Rosa* spp. (Rosaceae)	Sayed et al., 2019
	Aphis gossypii (Hemiptera: Aphididae)	*Solanum muricatum* (Solanaceae)	Homayoonzadeh et al., 2022
	Ceraeochrysa valida, Eremochrysa punctinervis (Neuroptera: Chrysopidae)	*Citrus* spp. (Rutaceae)	Gandarilla et al., 2013
	Epilachna varivestis (Coleoptera: Coccinellidae)	*P. vulgaris*	Castrejón-Antonio et al., 2017
	Pachycoris torridus (Hemiptera: Scutelleridae)	*Jatropha curcas* (Euphorbiaceae)	Chávez et al., 2016
	Metamasius spinolae (Coleoptera: Curculionidae)	*Opuntia ficus-indica* (Cactaceae)	Sánchez-Pérez et al., 2016

Contd.

Table 2 *Contd.*

Fungus	*Pest*	*Crop*	*Reference*
M. anisopliae	*Nilaparvata lugens* (Hemiptera: Delphacidae), *Cnaphalocrocis medinalis* (Lepidoptera: Crambidae), *Chilo suppressalis* (Lepidoptera: Pyralidae)	*Oryza sativa* (Poaceae)	Peng et al., 2021
	S. purpurascens, *Melanoplus differentialis* (Orthoptera: Acrididae)	*Z. mays*, *P. vulgaris*	Quesada-Béjar et al., 2023
	Schistocerca piceifrons piceifrons (Orthoptera: Acrididae)	*Z. mays*, *Sorghum bicolor*, *Saccharum officinarum* (Poaceae), *P. vulgaris*, *Glycine max*, *Arachis hypogaea* (Fabaceae), *Gossypium hirsutum* (Malvaceae), *Sesamum indicum* (Pedaliaceae), *Musa acuminata* (Musaceae)	Barrientos et al., 2005
B. bassiana and *M. anisopliae*	*Diaphorina citri* (Hemiptera: Liviidae)	*Citrus* spp.	Corallo et al., 2021
	B. tabaci	More than 500 ornamental plants	García-Ramírez et al., 2013
	Scyphophorus interstitialis (Coleoptera: Curculionidae)	*Agave* spp. (Asparagaceae)	Aquino-Bolaños et al., 2007
	Diatraea magnifactella (Lepidoptera: Crambidae)	*S. officinarum*	Castro et al., 2013
	Phyllophaga crinita (Coleoptera: Melolonthidae)	*Z. mays* (Poaceae), and green areas	Nájera et al., 2005
	Aulacaspis tubercularis (Hemiptera: Diaspididae)	*M. indica*	Pérez-Salgado et al., 2013
	A. obliqua	*M. indica*	Díaz-Ordaz et al., 2010
	B. cockerelli	*S. tuberosum*	Villegas et al., 2017
	Icerya seychellarum (Hemiptera: Monophlebidae), *A. tubercularis* (Hemiptera: Diaspididae)	*M. indica*	Sayed and Dunlap, 2019
	Kuschelorhynchus macadamiae (Coleoptera: Curculionidae)	*Macadamia* spp. (Proteaceae)	Khun et al., 2021
	Callosobruchus maculatus (Coleoptera: Chrysomelidae)	*Vigna unguiculata* (Fabaceae)	Ozdemir et al., 2020
	Demotispa neivai (Coleoptera: Chrysomelidae)	Palms (Arecaceae)	Martínez et al., 2022

Contd.

Table 2 *Contd.*

Fungus	*Pest*	*Crop*	*Reference*
B. bassiana and *Isaria javanica*	*Diaphania indica* (Lepidoptera: Crambidae)	*Citrus sinensis* (Rutaceae)	Augier et al., 2023
Cordyceps fumosorosea	*D. citri*	*Citrus* spp.	Pick et al., 2022
M. anisopliae and entomopa-thogenic nematodes	*Phyllophaga vetula* (Coleoptera: Melolonthidae)	*Z. mays*	Ruiz et al., 2012
B. bassiana, Lecanicillium lecanii and *P. fumosoroseus*	*Toxoptera citricida* (Hemiptera: Aphididae)	*Citrus* spp.	Hernández-Torres et al., 2007
B. bassiana, M. anisopliae, Nomuraea rileyi, P. fumosoroseus and *P. javanicus*	*Spodoptera frugiperda* (Lepidoptera: Noctuidae)	*Z. mays*	Lezama et al., 1996 (a,b)
N. rileyi and *P. fumosoroseus*	*S. frugiperda* (Lepidoptera: Noctuidae)	*Z. mays*	Lezama et al., 1994
Hirsutella citriformis	*D. citri*	*Citrus* spp.	Pérez-González et al., 2013

approach to controlling *Scyphophorus acupunctatus* involves using entomopathogenic fungi that specifically target the adult stage of the insect. These fungi are applied to the stems, leaves, and soil surrounding the plant. As adult insects migrate from one plant to another, they become infected with the fungus and transport it inside the stem, where larvae and pupae are located (Aquino-Bolaños et al., 2007).

Studies have shown that entomopathogenic fungi such as *B. bassiana* and *M. anisopliae* can infect and kill both larvae and adults of *S. acupunctatus* (Aquino-Bolaños et al., 2023). Soil samples from cultivated areas in Oaxaca, Mexico, have revealed native strains of B. bassiana and Metarhizium. These strains have shown pathogenicity against adults of *S. acupunctatus* in laboratory studies. These fungi caused over 70% mortality in *S. acupunctatus* when applied through conidial dipping and spraying at a specific concentration. The lethal effect was observed after 14 days of inoculation. The images depict adult *S. acupunctatus* that have been infected and killed by *B. bassiana* and *Metarhizium* sp. (López-López, 2024) (Fig. 1). In addition, these entomopathogenic fungi secrete extracellular enzymes, such as lipases, proteases, and chitinases, which assist in the penetration of the fungus through the insect cuticle (El-Maraghy et al., 2023).

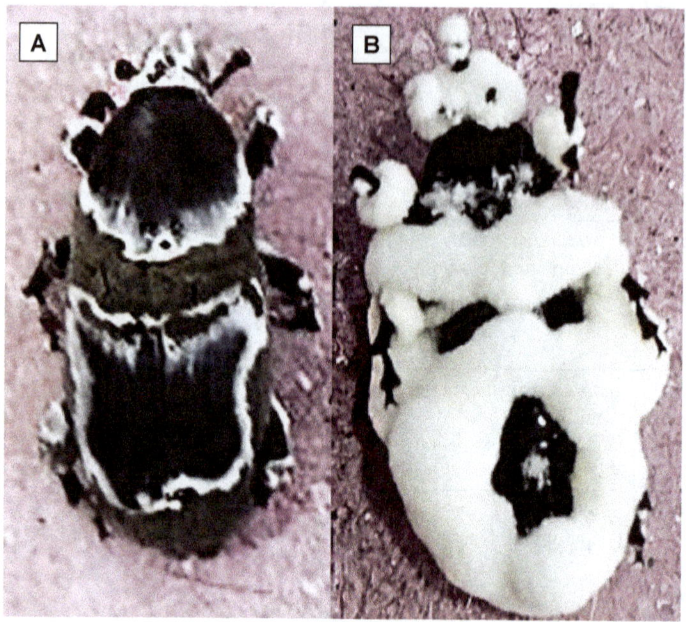

Fig. 1 Adults of *Scyphophorus acupunctatus* collected from *Agave potatorum* plants cultivated in Oaxaca, Mexico, in 2022: (A) Adults inoculated with *B. bassiana*; (B) Adults inoculated with *Metarhizium* spores. Images were taken 14 days after inoculation by spraying conidia at a concentration of 1×10^8 spores/ml. Thesis photograph of Rosa Elvira López-López, graduate student of the Master of Science postgraduate program offered by the IPN CIDIIR Oaxaca, Mexico (López-López, 2024).

Although research on the use of entomopathogenic fungi against *S. acupunctatus* is limited, there is scientific evidence that entomopathogenic fungi can infect and kill insects of the Curculionidae family. In agriculture, these insects pose a serious, devastating threat to many economically important crops. Their populations are controlled with synthetic chemical insecticides. However, it is possible to implement biological control using entomopathogenic fungi in an environmentally friendly way. Previous research has demonstrated the ability of *B. bassiana* and *M. anisopliae* to infect and eliminate insects categorized as "weevils" in several crops, such as *Sitophilus granarius* (Mantzoukas et al., 2019), *R. ferrugineus* (Al-Keridis et al., 2020), *Cosmopolites sordidus* (Membang et al., 2020), *Sitophilus oryzae* (Tarore et al., 2023), *Sternochetus mangiferae* (Silva et al., 2022b), and others. Indigenous strains of entomopathogenic fungi have significant potential to control insect pests in the region where they were originally isolated. Native fungi are adapted to environmental conditions such as temperature, humidity, and light intensity, as well as soil conditions including pH, organic matter content, soil composition, and the presence of local plants and insects.

4.2 Secondary Metabolites of *Lecanicillium* spp.

In the 1960s, the first secondary metabolites with biological activity were discovered: destruxin A and B, which were isolated from *M. anisopliae*. These cyclic peptides inhibit fluid secretion through the Malpighian tubes in arthropods (Skrobek et al., 2008). These metabolites have also been identified in *L. longisporum* (Bult et al., 2009). Oosporein is another metabolite that exhibits entomopathogenic activity. It is produced by species such as *L. aphanocladii* and *L. saksenae* (Higo et al., 2021) and has been tested against hemipterans (Sreeja et al., 2023). Recently, insecticidal compounds have been extracted from *L. lecanii*, including beauvericin and enniatin B. In addition, the production of pyrrocidin A and/or B through solid fermentation in rice was reported for the first time (Blaszczyk et al., 2021). Furthermore, *L. antillanum* was found to produce 11-norbetaenone, which is the first betaenone-like compound for the genus and has anti-angiogenic properties (Li et al., 2017). On the other hand, six new alkaloids have been isolated from *L. kalimantanense* SCSIO41702 named lecanicilliums A-F, along with similar compounds known as emethacin B and versicolor A. Some of these compounds have exhibited cytotoxic and antibacterial activities (Zhong et al., 2023).

The volatile metabolites of *L. saksenae*, including harmine lactone, dipicolinic acid, and octadecanoic acid, demonstrate insecticidal, nematicidal, and antimicrobial properties, indicating promising potential for controlling hemipteran pests (Sreeja and Rani, 2019). Several compounds with antioxidant, insecticidal, antimicrobial, and anticancer properties have been identified from the endophyte *L. aphanocladii,* suggesting numerous potential applications (Asghari et al., 2023). Mitina et al. (2021) found that volatile compounds from six *Lecanicillium* species, including *L. muscarium*, *L. lecanii*, and *L. attenuatum*, demonstrated repellent properties and effectively reduced thrip offspring. The researchers found that the acetic acid in these compounds influenced insect behavior. In addition, extracts of *L. attenuatum,* such as the JEF-145 strain, were found to exhibit insecticidal activity and high levels of juvenile hormone against *Aedes albopictus* and *P. xylostella.* This suggests that they have the potential to be an alternative control method (Woo et al., 2020).

Biopesticide formulations have been developed using the infection and toxicity capacity of *Lecanicillium* to control agricultural pests and promote integrated pest management. Notable commercial products include Vertalec® and Mycotal® from the Kopper Company, which are recommended for the control of *B. tabaci*, as well as Mycotol® and Vertisol® manufactured in Colombia (Mota-Delgado and Murcia-Ordoñez, 2011). Efforts have also been made to explore the potential use of entomopathogenic fungi to develop nanoinsecticides for pest control (Ayilara et al., 2023). Recently, a granulated bioformulation of *L. lecanii* was developed to target *Anopheles culicifacies*, the malaria vector. The formulation showed a median lethal concentration (LC_{50}) of 11,836 mg/μL at third instar mosquito larvae and remained stable at 40 °C for 3 months (Sogan et al., 2023).

Recently, there has been exploration into the capacity of entomopathogenic fungi, such as *Lecanicillium*, to utilize lytic enzymes like esterases, reductases, and cytochrome P450 to metabolize non-toxic compounds from chemical insecticides (Van Eerd et al., 2003). This opens the door to the potential involvement of these fungi in the remediation of pesticide-contaminated soils, offering an alternative approach. The research suggests that *Lecanicillium* species have the potential to be promising candidates not only for agricultural pest control but also for other applications.

4.3 Mechanisms of Action of Lecanicillium spp. Metabolites

The genus *Lecanicillium* (Hypocreales: Cordicipitaceae) was established in the early 21st century through revision of the prostrata section of the genus *Verticillium* (Glomerales: Plectosphaerellaceae) due to morphological inconsistencies, which were supported by molecular analysis (https://www.indexfungorum.org). Taxonomic revisions have been made, sparking controversy over its classification. Kepler et al. (2017) proposed transferring certain species from the genus *Lecanicillium* to *Akanthomyces*, including *L. lecanii*. Wang et al. (2020) proposed the establishment of the genus *Flavocillium* by identifying four species, three of which were previously categorized under *Lecanicillium*. Previous studies have investigated the phylogeny and evolutionary history of *Lecanicillium*, emphasizing the necessity for further research to enhance its classification. Some studies have even suggested the creation of a new genus called *Gamszarea* (Zhang et al., 2021, Zhou et al., 2022). These analyses suggest an intricate evolutionary relationship between *Lecanicillium* and other genera within the Cordicipitaceae family.

Morphologically, the genus *Lecanicillium* is characterized by flat conidiophores that are indistinguishable from the underlying hyphae. The conidiophores are typically short and upright, occurring singly or in terminal or intercalary clusters. The genus has sharp, distinct phialides that may form short aphanophalloids with swollen bases. Conidial dimorphism can be observed, with conidia attached to the hyphal tip in viscid fascicles, chains, or singly. According to Zare and Gams (2001), conidia can take on various shapes, such as ellipsoidal, globose, subglobose, fusiform, falcate, cylindrical, subcylindrical, attenuated, straight, or curved.

The genus *Lecanicillium* is distributed globally, with 881 records documented in the Global Biodiversity Information Facility (GBIF) database (https://www.gbif.org/es/species/2560711). It has been isolated in various regions around the world, including *L. antillanum* in Egypt (Alí and Moharram, 2014), *L. attenuatum* in Korea (Woo et al., 2020), and *L. saksenae* in India (Sreeja et al., 2023). It has also been found to act as a mycoparasite. For example, *L. fungicola* has been observed on edible mushroom crops (Lv et al., 2022), and *L. uredinophilum* has been found on *Pucciniastrum agrimoniae* (Pucciniales: Pucciniastraceae) (Wei et al., 2018).

Lecanicillium is an endophyte that has been found in a wide range of plants, including monocots of the family Gramineae (Poosakkannu et al., 2015), *Laretia acaulis* (Molina-Montenegro et al., 2015), *Taxus baccata* (Behnke-Borowczyk et al., 2019), maize seeds (Blaszczyk et al., 2021), *Solanum mauritianum*, and *Physalis peruviana* (Pelo et al., 2021; Asghari et al., 2023). Studies suggest that specific fungi can function as mutualistic endophytes, safeguarding plants from insects and promoting their growth in a variety of crops, including pumpkin, cotton, tomato, bean, pistachio, grapevine, and millet (Nicoletti and Becchimanzi, 2020; Chaudhary et al., 2023). Other fungi have been discovered on diverse substrates. For example, *L. kalimantanense* was isolated from the exoskeleton of a coleopteran (Sukarno et al., 2009), and found in mangrove sediment (Zhong et al., 2023). Several studies have shown nematicidal activity, including research by Van Nguyen et al. (2007), which demonstrated the effectiveness of *L. antillanum* against *Meloydogyne incognita* (Tylenchida: Heteroderidae) eggs. More recently, Hajji-Hedfi et al. (2023) reported the discovery of *Lecanicillium* sp. The results demonstrated promising potential against *M. javanica.*

Lecanicillium, isolated from various insects, is effective in controlling pests such as aphids, whiteflies, soft scales, thrips, and bed bugs (Zare and Gams, 2001). Several species of *Lecanicillium* have been identified, including *L. lecanii, L. tenuipes, L. aranearum, L. aphanocladii,* and *L. huhutil,* which were isolated from spiders (Zare and Gams, 2001; Chen et al., 2017; Manfrino et al., 2017; Zhou et al., 2022). Additionally, *L. antillanum* has been isolated from *Spodoptera litoralis* (Lepidoptera: Noctuidae), *L. attenuatum* from carpocarp moths, *Aedes albopictus* (Diptera: Culicidae), and *P. xylostella* (Lepidoptera: Plutellidae) (Ali and Moharram, 2014; Woo et al., 2020; Zare and Gams, 2021). Additional research has revealed the presence of *L. aphanocladii* and *L. psalliotae* on *Cameraria ohridella* (Lepidoptera: Gracillariidae) (Nedveckytė et al., 2021), and *L. sabanense* isolated from *Pulvinaria caballeroramosae* (Hemiptera: Coccidae) on ficus branches in Colombia (Chiriví-Salomón et al., 2015).

4.4 Pest Control Mechanisms of Lecanicillium

The mechanism of action of *Lecanicillium* is similar to that of other entomopathogens. The process begins with contact, with conidia adhering to the insect's epicuticle and forming a germ tube. This process is followed by the formation of an appressorium, which is a modified hypha that exerts mechanical force by breaking the epicuticle and initiating penetration. Lipase enzymes produced by the entomopathogenic fungus hydrolyze the ester bonds of fats, lipoproteins, and waxes in the insect's integument. Meanwhile, chitinases and proteinases facilitate the removal of the cuticle (Monzón, 2001; Alí et al., 2009). Subsequently, enzymes such as chitinases, lipases, and proteases degrade the cuticle. The enzyme Pr1 is crucial in this process. Fungal colonization begins with the production of blastospores. Toxins such as beauvericin, bassianolides, dipicolinic acid, and destruxin are produced, leading to the death of the insect. The fungus grows inside the carcass, penetrating all internal tissues and colonizing the arthropod's organs. Under suitable relative humidity, the mycelium emerges through the tegument, grows on the surface, and produces spores approximately 48 to 60 hours after host death (Shinde et al., 2010).

4.5 Mechanisms of Action and Metabolites of Trichoderma

Trichoderma, known for colonizing fungi and oomycetes, has also been documented as an agent that controls eggs and early larvae of nematodes and insects, and can even cause mycosis in humans (Ibrahim et al., 2020; Hatvani et al., 2019). It acts as an entomopathogen through direct colonization, the production of secondary insecticidal metabolites, feeding inhibitory compounds, and repellent metabolites (Kaushik et al., 2020). Like other entomopathogenic fungi, *Trichoderma* can colonize insect bodies and utilize them as a nutrient source to produce new spores. For instance, *T. longibrachiatum* and *T. harzianum* have been demonstrated to parasitize adult *B. tabaci* and *Cimex hemipterus* (Anwar et al., 2016; Zahran et al., 2017), leading to mortality rates of 40% in 5 days and 90% in 14 days.

Trichoderma can quickly colonize plants and stimulate the expression of defense genes, thereby eliciting immune responses against pathogens. This process involves the interaction of molecular patterns associated with microbes or pathogens with plant cell receptors, triggering plant immunity to limit pathogen development. Effectors initiate the plant's resistance response, leading to the localized death of infected cells, a process known as effector-triggered immunity. *Trichoderma*, particularly *T. harzianum* TH12, induces rhizobacteria-mediated induced systemic resistance (ISR), while its cell filtrate induces systemic acquired resistance (SAR) against *Sclerotinia sclerotiorum* (Dodds and Rathjen, 2010; Alkooranee et al., 2015; Bartels and Boller, 2015; Vlot et al., 2021).

Activation of local and systemic immune responses is achieved through the action of jasmonic acid (JA), salicylic acid (SA), and ethylene (ET) (Glazebrook, 2005). *Trichoderma*-induced systemic resistance (ISR) and systemic acquired resistance (SAR) are crucial for preventing and controlling airborne and soilborne pathogens, particularly when both types of diseases coexist (Srivastava et al., 2015). Jasmonic acid and salicylic acid play a critical role in regulating plant defense mechanisms against disease and stimulating systemic resistance through their interaction with the JA and ET pathways (Van Wees et al., 2000). In studies on cucumbers, it was found that treating the roots with *T. longibrachiatum* inhibited *Botrytis cinerea* by activating the JA, SA, and ET pathways (Yuan et al., 2019).

Trichoderma employs various mechanisms to inhibit pathogens, such as producing secondary metabolites with fungicidal properties and degrading cell walls using hydrolytic enzymes like chitinases and cellulases (Saravanakumar et al., 2017). These metabolites include volatile antibiotics, water-soluble compounds, and peptaibols such as gliotoxins, gliovirins, terpenes, polyketides, and pyrones. They are capable of controlling various yeasts, bacteria, and filamentous phytopathogenic fungi (Mukherjee et al., 1995; Monte, 2001; Missbach et al., 2023).

Trichoderma's capacity to control pests is closely associated with antibiotic synthesis. This is demonstrated by strains of *T. virens* that overproduce gliovirin. Gliovirin-deficient mutants were unable to defend cotton seedlings from *Pythium ultimum* (Peronosporales: Pythiaceae) (Chet et al., 1997). The strains of *T. virens* and *T. harzianum* that are most effective against *Gaeumannomyces graminis* (Magnaporthales: Magnaporthaceae) and *R. solani* (Agonomicetales: Agonomicetaceae), produce antibiotics such as pyrones and glyotoxin (Lumsden et al., 1992; Howell, 1998). *Trichoderma asperellum* produces peptaibols, bioactive polypeptides, trichotoxins, and asperellins that exhibit antagonistic potential against several pathogenic fungi (Brito et al., 2014; Degenkolb et al., 2015).

Studies by Howell (1998) and Monte (2001) have shown that the concurrent use of antibiotics and hydrolytic enzymes significantly enhances the effectiveness of *Trichoderma* biocontrol compared to individual use. This combination has demonstrated better control of pathogens such as *Botrytis*, indicating a synergy that enhances the biological control capacity of *Trichoderma*. It is important to note that the interaction between antibiotics and enzymes is not consistent and may vary depending on the specific combination.

5. Production of Conidia by Entomopathogenic Fungi

Plant pathogens, harsh growing conditions, and insect pests cause stress in crop plants. Insects, which consume both above- and below-ground plant parts, are a significant factor in crop losses (Mascarin et al., 2022). Culliney (2014) reported that these factors can affect global crop productivity by up to 26%. Agriculture heavily depends on chemical insecticides and synthetic fertilizers to achieve high crop yields, but their excessive use damages soil pH, nutrients, and non-target organisms. As a result, there is a push to replace these chemicals with environmentally friendly alternatives, such as plant extracts and microbial agents. Research on microbial agents, which encompass various species of bacteria, viruses, protozoa, and fungi, highlights their regulatory role in controlling pest infestations (Sharma and Sharma, 2021). Among these, fungi produce spores that penetrate the insect hosts' surfaces and cause irreparable damage. More than 750 fungal species can infect insects, but the rate of infection varies depending on the specific fungal species and spores.

5.1 Mass Production of Entomopathogenic Fungi

Since the 1980s, several techniques have emerged for the mass production of fungal mycelia, or conidia. These methods include in-vivo production, diphasic fermentation, stirred tank reactors, and submerged and surface cultures. The method chosen depends on the growth requirements of the fungus and the anticipated final product. The most widely used approach for conidial production worldwide is in-vivo systems using rice and bran grains or inert supports as growth substrates.

Mascarin et al. (2010) demonstrated the in-vitro production of *M. anisopliae* and *B. bassiana,* using rice as a substrate, to control spittlebugs and grasshoppers. Diphasic fermentation, an efficient liquid-solid system, increases the yield of industrial-scale production (Santos et al., 2021). Cultures initially develop in solid media before transitioning to a liquid phase, producing various propagules such as blastospores, submerged conidia, mycelia, microsclerotia, and chlamydospores (Mascarin et al., 2015). This approach supports fungal growth in fermenters until the logarithmic phase, which is crucial for producing large numbers of conidia. Both in-vivo and diphasic fermentation methods promote the growth of entomopathogenic fungi, but they are constrained by various technical and economic limitations (Jackson, 1997).

Submerged fermentation allows for precise control of factors such as dissolved oxygen, pH, temperature, osmotic pressure, foaming, and inoculum volume (Zhao et al., 2023). This method creates optimal conditions for the mass production of entomopathogenic fungi (EPF) for pest control. Studies indicate that strains such as *B. bassiana* produce large quantities of conidia, particularly blastospores, through submerged fermentation. These submerged conidia and blastospores have demonstrated increased efficacy against pests such as *S. frugiperda* and *A. grandis* (Iwanicki et al., 2023). In addition, *Clonostachys rosea,* another strain, can be efficiently mass-produced through submerged liquid fermentation. This method offers a cost-effective approach for large-scale production of effective entomopathogenic fungi (Javar et al., 2023).

During submerged liquid fermentation, filamentous fungi display diverse growth patterns that are influenced by various factors, including shear stress. Stirred tank bioreactors optimize mechanical agitation to minimize shear stress, thereby significantly reducing cultivation time while achieving high fungal propagule production (Carvalho et al., 2018). Comparisons between the cultures of *Paecilomyces tenuipes* (now *Isaria japonica*) in stirred tank bioreactors and airlift systems revealed similar conidia production, demonstrating the resilience of some entomopathogenic fungi under hydrodynamic conditions (Xu et al., 2006). Lohse et al. (2014) scaled up the bioprocessing of *B. bassiana* from shake flasks to a 2 L stirred tank reactor. They observed a higher concentration of conidia under high agitation of 600 rpm (revolutions per minute) and 1 vvm oxygen aeration (air flow volume per minute). *Beauveria bassiana* conidia from a bioreactor culture exhibited a 77% mortality rate against *P. xylostella* larvae, demonstrating the advantageous growth and functionality of entomopathogenic fungi in stirred tank bioreactors under hydrodynamic conditions (Xu and Yun, 2003). Furthermore, studies on *Auricularia polytricha* in stirred tank bioreactors have demonstrated the importance of standardizing optimal operating conditions, leading to a tenfold increase in conidia production.

Stirred tank bioreactors offer several advantages over other fermentation processes for the production of environmentally friendly polymers. They create ideal operating conditions by regulating oxygen levels, agitation, pH, carbon, nitrogen, and temperature, leading to substantial reductions in operating costs. Furthermore, this system can adjust to the specific metabolic requirements of various microorganisms. In industrial fermentations, the main goals are to maintain consistent optimal conditions, minimize microbial stress, improve metabolic precision to achieve higher yields, and ensure consistent product quality. Several low-cost bioreactor types are utilized for entomopathogenic fungi biomass production, as illustrated in Table 3.

Table 3 Substrates used for cultivating entomopathogenic fungi for biomass production.

Strain	*Type of bioreactor*	*Substrate*	*Reference*
M. anisopliae ICB 425	Tray bioreactor	Rice	da Cunha et al., 2019
M. brunneum F52	Stirred-tank	Basal medium supplemented with 80 g/L glucose and 9 g/L casein	Jaronski and Jackson, 2012
B. bassiana	Cylindrical packed-column aerobic	Palm oil residues	do Nascimento Silva et al., 2023
B. bassiana CECT 20374	Solid state fermentation (SSF) operating under batch strategy	Rice husk and beer draft	Sala et al., 2023
M. anisopliae CP-OAX	Packed bed bioreactor	Steamed rice	Méndez-González et al., 2018
B. bassiana isolate ATP-02	Stirred tank	TKI medium with: 5% sugar beet molasses as a carbon source	Lohse et al., 2014
B. bassiana PQ2	Biofilm reactor	Czapek-Dox mineral medium	Lara-Juache et al., 2021

Contd.

Table 3 *Contd.*

Strain	*Type of bioreactor*	*Substrate*	*Reference*
M. robertsii (ENCB-MG-81)	spherical-style packed column	Rice	Méndez-González et al., 2022
B. bassiana IBCB 868	Stirred-tank	soy extract	Basso et al., 2023
Clonostachys rosea CMAA1284	Laboratory benchtop (stirred tank)	Rice	Mascarin et al., 2022

Conclusion

Entomopathogenic fungi are becoming increasingly important in biological pest control and have shown outstanding efficacy in protecting agricultural and forestry crops. This alternative offers significant advantages over chemical pesticides. It not only reduces pesticide residues in food but also ensures human safety and preserves ecosystems by promoting biodiversity and preserving natural enemies. The study of the secondary metabolites produced by these fungi is a promising area of research. These compounds can significantly influence the development of new pest control agents as researchers explore their properties and mechanisms of action. A thorough comprehension of these metabolites could lead to the creation of more effective and sustainable strategies for controlling agricultural pests while minimizing adverse environmental impacts. The study of entomopathogenic fungi, their secondary metabolites, and their ability to control pests is a research area with significant potential. Further study of these aspects may not only enhance current agricultural practices but also lead to the discovery of innovative and environmentally friendly solutions to agricultural pest problems.

References

Abdel-Raheem, M. (2020). Isolation, Mass Production and Application of Entomopathogenic Fungi for Insect Pests Control. In: Cottage Industry of Biocontrol Agents and Their Applications: Practical aspects to deal biologically with pests and stresses facing strategic crops, El-Wakeil, N., Saleh, M. and Abu-hashim, M. (Eds) Springer, Cham. pp 231–251.

Alatorre-Rosas, R. (2007). Hongos entomopatógenos. En Teoría y aplicación del control biológico In Ciudad de México, México: Sociedad Mexicana de Control Biológico1, Rodríguez-del-Bosque, L.A. and Arredondo-Bernal, H. (Eds.), 127–143. (In Spanish).

Alí, S., Huang, Z. and Ren, S.X. (2009). Production and extraction of extracellular lipase from the entomopathogenic fungus *Isaria fumosoroseus* (Cordycipitaceae: Hypocreales). Biocontrol. Sci. Technol., 19: 81–89.

Ali, S.S. and Moharram, A.M. (2014). Biodiversity and enzymatic profile of some entomopathogenic fungi. Egypt. Acad. J. Biol. Sci., F Toxicol. Pest Control, 6(1): 73–80.

Al-Keridis, L.A., Gaber, N.M. and Aldawood, A.S. (2020). Pathogenicity of Saudi Arabian fungal isolates against egg and larval stages of *Rhynchophorus ferrugineus* under laboratory conditions. Int. J. Trop. Insect Sci., 40: 845–853.

Alkooranee, J.T., Yin, Y., Aledan, T.R., Jiang, Y., Lu, G., Ju, J. and Li, M. (2015). Systemic resistance to powdery mildew in *Brassica napus* (AACC) and *Raphanus alboglabra* (RRCC) by *Trichoderma harzianum* TH12. PLoS One, 10(11): 0142177.

Anwar, W., Subhani, M.N., Haider, M.S., Shahid, A.A., Mushatq, H., Rehman, M.Z. and Javed, S. (2016). First record of *Trichoderma longibrachiatum* as entomopathogenic fungi against *Bemisia tabaci* in Pakistan. Pakistan J. Phytopathol., 28(2): 287–294

Aquino-Bolaños, T., Iparraguirre-Cruz, M. and Ruiz-Vega, J. (2007). *Scyphophorus acupunctatus* (=interstitialis) Gyllenhal (Coleoptera: Curculionidae). Plaga del agave mezcalero: Pérdidas y daños en Oaxaca, México. Rev. Cient. UDO Agric., 7(1): 175–180. (In Spanish).

Aquino-Bolaños, T., Ortiz-Hernández, Y.D., Bautista-Cruz, A. and Acevedo-Ortiz, M.A. (2023). Viability of entomopathogenic fungi in oil suspensions and their effectiveness against the agave pest *Scyphophorus acupunctatus* under laboratory conditions. Agronomy, 13: 1468.

Asghari, A., Tajick, G.M.A., Bakhshi, M. and Babaeizad, V. (2023). Bioactive potential and GC-MS fingerprinting of extracts from endophytic fungi associated with seeds of some medicinal plants. Mycol. Iran, 10(1): 55–67.

Augier, L., Arredondo-Bernal, H.C., Pérez, D., Martínez, D., Lizondo, M., Mellín-Rosas, M.A. and Gastaminza, G. (2023). Respuesta de *Diaphorina citri* Kuwayama a Cepas de Hongos Entomopatógenos en Argentina. Southwest. Entomol., 48(1): 179–188. (In Spanish).

Ayilara, M.S., Bartholomew, S.A., Saheed, A.A., Chris, A.F., Uswat, T.A., Lanre, A.G., Richard, K.O., Remilekun, M.J., Qudus, O.U. and Olubukola, O.B. (2023). Biopesticides as a promising alternative to synthetic pesticides: A case for microbial pesticides, phytopesticides, and nanobiopesticides. Front. Microbiol., 14:1040901.

Bai, J., Xu, Z., Li, L., Zhang, Y., Diao, J., Cao, J., Xu, L. and Ma, L. (2023). Gut bacterial microbiota of Lymantria dispar asiatica and its involvement in *Beauveria bassiana* infection. J. Invertebr. Pathol., 197:107897.

Baki, D., Tosun, H.S. and Erler, F. (2021). Efficacy of indigenous isolates of *Beauveria bassiana* (Balsamo) Vuillemin (Deuteromycota: Hyphomycetes) against the Colorado potato beetle, *Leptinotarsa decemlineata* (Say) (Coleoptera: Chrysomelidae). Egypt. J. Biol. Pest Control, 31: 1–11.

Barrientos, L., Hunter, D.M., Ávila- Valdéz, J., García- Salazar, P. and Horta-Vega, J.V. (2005) Control biológico de la langosta centroamericana *Schistocerca piceifrons piceifrons* Walker (Orthoptera: Acrididae) en el noreste de México. Vedalia 12(2): 119–128. (In Spanish).

Bartels, S. and Boller, T. (2015). Quo vadis Pep Plant elicitor peptides at the crossroads of inmunity, stress and developtment. J. Exp. Bot., 66(17): 5183–5193.

Basso, V., Fontana, C.R., Montipó, S., Pinheiro, J. and Dillon, A. (2023). High concentration of spores and colony forming units of the biocontrol agent *Beauveria bassiana* via *optimization* of submerged cultivation. Biocatal. Agric. Biotechnol., 47, 102607.

Behnke-Borowczyk, J., Kwaśna, H. and Kulawinek, B. (2019). Fungi associated with Cyclaneusma needle cast in Scots pine in the west of Poland. For. Pathol., 49: e12487

Bhattacharyya, P.N., Sarmah, S.R., Roy, S., Sarma, B., Nath, B.C. and Bhattacharyya, L.H. (2023). Perspectives of *Beauveria bassiana*, an entomopathogenic fungus for the control of insect-pests in tea [*Camellia sinensis* (L.) O. Kuntze]: opportunities and challenges. Int. J. Trop. Insect Sci., 43(1): 1–19.

Bidochka, M.L., Kamp, A.S., Lavender, T.M., Dekoning, J. and De Croos, J.N.A. (2001). Habitat association in two genetic groups of the insectpathogenic fungus *Metarhizium anisopliae*: uncovering cryptic species? Appl. Environ. Microbiol., 67: 1335–1342.

Bidochka, M.L., Menzies, F.V, and Kamp, A.M. (2002). Genetic groups of the insect-pathogenic fungi *Beauveria bassiana* are associated with habitat and thermal growth preferences. Arch. Microbiol., 178: 531–537.

Blaszczyk, L., Waskiewicz, A., Gromadzka, K., Milkolaiczak, K., and Chelkowaski, J. (2021). *Sarocladium* and *Lecanicillium* associated with maize seeds and their potential to form selected secondary metabolites. Biomolecules, 11(1): 98.

Bohatá, A., Folorunso, E.A., Lencová, J., Osborne, L.S. and Mraz, J. (2023). Control of sweet potato whitefly (*Bemisia tabaci*) using entomopathogenic fungi under optimal and suboptimal relative humidity conditions. Pest. Manag. Sci., 80(3): 1065–1075.

Boston, W., Leemon, D. and Cunningham, J.P. (2020). Virulence screen of *Beauveria bassiana* isolates for Australian Carpophilus (Coleoptera: Nitidulidae) beetle biocontrol. Agronomy, 10(8): 1207.

Brito, J.P., Ramada, M.H., de Magalhães, M.T., Silva, L.P. and Ulhoa, C.J. (2014). Peptaibols from *Trichoderma asperellum* TR356 strain isolated from Brazilian soil. Springerplus, 3: 600.

Bult, T.M., El Hadj, N.B., Skrobek, A., Ravensberg, W.J., Wang, C., Lange, C.M., Vey, A., Sha, U.K. and Dudley, E. (2009). Mass spectrometry as a tool for selective profiling of destrucins; their firs identification in *Lecanicillium longisporum*. Rapid Commun. Mass. Spectrom., 23(10): 1426–1434.

Butt, T.M., Coates, C.J., Dubovskiy, I.M. and Ratcliffe, N.A. (2016). Entomopathogenic fungi: new insights into host–pathogen interactions. Adv. Genet., 94: 307–364.

Carvalho, A.L., Castro de Rezende, L., Costa, L.B., Halfeld-Vieira, B.A., Pinto, Z.V., Morandi, M.A.B., Vasconcelos de Medeiros, F.H. and Bettiol, W. (2018). Optimizing the mass production of *Clonostachys rosea* by liquid-state fermentation. Biol. Control, 118: 16–25.

Castrejón-Antonio, J.E., Nuñez-Mejia, G., Iracheta, M.M., Gomez-Flores, R., Tamayo-Mejia, F., Ocampo-Hernandez, J.A. and Tamez-Guerra, P. (2017). *Beauveria bassiana* blastospores produced in selective medium reduce survival time of *Epilachna varivestis* Mulsant larvae. Southwestern Entomol., 42(1): 203–220.

Castro, O., Fonseca, G., Trejo, L., Hernández, V., Rodríguez del B. and Lina, G. (2013). Evaluación de seis cepas de hongos entomopatógenos sobre *Diatraea magnifactella* Dyar (Lepidoptera: Crambidae). Entomol. Mex., 12(1): 430–433. (In Spanish).

Chang, F.M., Lu, H.L. and Nai, Y.S. (2023). Evaluation of potential entomopathogenic fungus, *Beauveria bassiana*, for controlling the coffee berry borer *Hypothenemus hampei* (Ferrari) (Coleoptera: Curculionidae) in Taiwan. J. Asia Pac. Entomol., 26(4): 102118.

Chaudhary, R., Kumar, V., Gupta, S. and Naik, B. (2023). Finger millet (*Eleusine coracana*) Plant–endophyte dynamics: Plant growth, nutrient uptake, and zinc biofortification. Microorganisms, 11(4): 973.

Chávez, B.Y., Rojas, J.C., Barrera, J.F. and Gómez, J. (2016). Evaluación de la patogenicidad de *Beauveria bassiana* sobre *Pachycoris torridus* en laboratorio. Southwestern Entomol., 41(3): 783–790. (In Spanish).

Chen, W., Han, Y., Liang, Z. and Jin, D. (2017) *Lecanicillium araneogenum* sp. Nov., a new araneogenous fungus. Phytotaxa, 305(1): 29–34

Cheng, D., Guo, Z., Riegler, M., Xi, M., Liang, G. and Xu, Y. (2017). Gut symbiont enhances insecticide resistance in a significant pest, the oriental fruit fly *Bactrocera dorsalis* (Hendel). Microbiome, 5: 1–12

Chet, I., Inbar, J. and Hadar, I. (1997). Fungal antagonists and mycoparasites. In: Wicklow DT, Söderström B (eds) The Mycota IV: Environmental and microbial relationships. Springer-Verlag, Berlin, pp 165–184.

Chevrette, M.G., Carlson, C.M., Ortega, H.E. and Thomas, C. (2019) The antimicrobial potential of *Streptomyces* from insect microbiomes. Nat. Commun., 10(1): 516.

Chiriví-Salomón, J.S., Danies, G., Restrepo, S. and Sanjuan, T. (2015). *Lecanicillium sabanense* sp. nov. (Cordycipitaceae) a new fungal entomopathogen of coccids. Phytotaxa, 234 (1): 063–074.

Corallo, A.B., Pechi, E., Bettucci, L. and Tiscornia, S. (2021). Biological control of the Asian citrus psyllid, *Diaphorina citri* Kuwayama (Hemiptera: Liviidae) by entomopathogenic fungi and their side effects on natural enemies. Egypt. J. Biol. Pest Control, 31(1): 1–9.

Corporación Colombiana de Investigación Agropecuaria (CORPOICA). (2011). Evaluación y validación de bioplaguicidas a base de hongos entomopatógenos para el manejo de mosca blanca *Bemisia tabaci* en algodón, tabaco y berenjena en Tolima, Córdoba y Huila (Informe técnico). Bogotá, Colombia: CORPOICA. (In Spanish).

Culliney, T. (2014). Crop Losses to Arthropods. In: Pimentel, D., Peshin, R. (eds) Integrated Pest Management. Springer, Dordrecht. Pp 201–225.

Currie, C.R., Wong, B., Stuart, A.E., Schultz, T.R., Rehner, S.A., Mueller, U.G., Sung, G.H., Spatafora, J.W. and Straus, N.A. (2003). Ancient tripartite coevolution in the attine ant-microbe symbiosis. Science, 299(5605): 386–388.

da Cunha, L.P., Casciatori, F.P., de Cenço Lopes, I. and Thoméo, J.C. (2019). Production of conidia of the entomopathogenic fungus *Metarhizium anisopliae* ICB 425 in a tray bioreactor. Bioprocess Biosyst. Eng., 42: 1757–1768.

Dávila-Orozco, G., Cruz-Salazar, B. and Ruiz-Montoya, L. (2021). How does the application of *Beauveria bassiana* and compost on corn crops affect the survival and genetic diversity of *Phyllophaga obsoleta* (Coleoptera: Melolonthinae)? Environ. Entomol., 50(5): 1227–1240.

Davis, T.S., Hofstetter, R.W., Foster, J.T., Foote, N.E. and Keim, P. (2011). Interactions between the yeast *Ogataea pini* and filamentous fungi associated with the western pine beetle. Microb. Ecol., 61: 626–634.

Degenkolb, T., Fog Nielsen, K., Dieckmann, R., Branco-Rocha, F., Chaverri, P., Samuels, G.J., Thrane, U., von Döhren, H., Vilcinskas, A. and Brückner, H. (2015). Peptaibol, secondary-metabolite, and hydrophobin pattern of commercial biocontrol agents formulated with species of the *Trichoderma harzianum* complex. Chem. Biodivers., 12(4): 662–684.

Díaz-Ordaz, N.H., Pérez, N., and Toledo, J. (2010). Pathogenecity of three strains of entomopathogenic fungus on *Anastrepha obliqua* adults (Macquart) (Diptera: Tephritidae) under laboratory conditions. Acta Zool. Mex., 26(3): 481–494.

Dillon, R. and Charnley, K. (2002). Mutualism between the desert locust Schistocerca gregaria and its gut microbiota, Res. Microbiol., 153(8): 503–509.

do Nascimento Silva, J., Mascarin, G.M., Lopes, R.B. and Freire, D.M.G. (2023). Production of dried *Beauveria bassiana* conidia in packed-column bioreactor using agro-industrial palm oil residues. Biochem. Eng. J., 198: 109022.

Dodds, P.N. and Rathjen, J.P. (2010). Plant immunity: towards an integrated view of plant pathogen interactions. Nat. Rev. Genet., 1: 39–548.

Douglas, A.E. (2015). Multiorganismal insects: diversity and function of resident microorganisms. Annu. Rev. Entomol., 60: 17–34.

Douglas, A.E. (2018). Strategies for enhanced crop resistance to insect pests. Annu. Rev. Plant Biol., 69: 637–660.

El-Maraghy, S.S.M., Abdel-Rahman, M.A.A., Hassan, S.H.M. and Hussein, K.A. (2023). Pathogenicity and other characteristics of the endophytic *Beauveria bassiana* strain (Bals.) (Hypocreales: Cordycipitaceae). Egypt. J. Biol. Pest Control, 33: 79.

Escamilla, B.F., Bautista G.A.Y., Nava G.S.B. and Bibbins M.M. (2022). Los hongos entomopatógenos, aliados de la Agricultura sustentable en el control de plagas. Frontera Biotecnológica México, 10(23): (In Spanish).

Furuie, J.L., da Costa Stuart, A.K., Baja, F., Voidaleski, M.F., Zawadneak, M.A.C. and Pimentel, I.C. (2022). Pathogenicity of *Beauveria bassiana* strains against *Drosophila suzukii* (Diptera: Drosophilidae). Res. Soc. Dev., 11(2).

Gandarilla, P., FL, Quintero, Z.I., Rodríguez, G.R., Elías, S.M., Sandoval, C., CF. and Galán, W.L.J. (2013). Efecto de hongos entomopatógenos sobre *Ceraeochrysa valida y Eremochrysa punctinervis* (Neuroptera: Chrysopidae) depredadores de *Diaphorina citri* Kuwayama (Hemiptera: Liviidae) en México. Entomol. Mex., 12(1): 415–419.

García-Ramírez, E., Pérez-Pacheco, R., León-Enríquez, B.L. and Pliego-Marín, L. (2013) Patogenicidad de *Metarhizium anisopliae* y *Beauveria bassiana* sobre mosca blanca (*Bemisia tabaci*). Rev. Mexicana Cienc. Agric., 4(6): 1129–1138. (In Spanish).

Glazebrook, J. (2005). Contrasting mechanisms of defence against biotrophic and necrotrophic pathogens. Annu. Rev. Phytopathol., 43: 205–227.

Global Biodiversity Information Facility (GBIF). (s.f.). Species. https://www.gbif.org/es/species.

Gómez, H., Zapata, A., Torres, E. and Tenorio, M. (2014). Manual de producción y uso de hongos entomopatógenos. Laboratorio de entomopatógenos SCB, Servicio Nacional de Sanidad Agraria-SENASA, Lima. Perú, 37. (In Spanish).

González-Santarosa, M.G., Salazar-Torres, J.C., Jaimes-Albíter, F., Ramírez-Alarcón, S. and González-Santarosa, R. (2010). Eficacia de *Beauveria bassiana* (Balsamo) Vuillemin en el control de *Lygus lineolaris* (Palisot de Beauvois) en fresa. Rev Chapingo Ser. Hortic., 16(3): 189–193. (In Spanish).

Guizar-Guzmán, L. and Sánchez-Peña, S.R. (2013). Infection by *Entomophthora sensu* stricto (entomophthoromycota: entomophthorales) in *Diaphorina citri* (hemiptera: liviidae) in Veracruz, Mexico. Fla. Entomol., 96(2): 624–627.

Gutiérrez, J.L. and Luna, M.E. (2006). Enemigos naturales y control biológico de *Brachystola magna* (Girard) y B. mexicana (Bruner) (Orthoptera: Acrididae) con Beauveria bassiana en Zacatecas, México. Vedalia, 13(2): 91–96. (In Spanish).

Guzmán, D.R., Almuiña, Y., Hernández, D.G. and los Baños, A. (2019) Los hongos entomopatógenos y su posible uso en el manejo de *Lasioderma serricorne*. Cuba Tabaco, 20(1): 66–71. (In Spanish).

Hajek, A.E. and St Leger, R.J. (1994). Interactions between fungal pathogens and insect hosts. Annu. Rev. Entomol., 39(1): 293–322.

Hajek, A.E., Morris, E.E. and Hendry, T.A. (2019). Context-dependent interactions of insects and defensive symbionts: insights from a novel system in siricid woodwasps. Curr. Opin. Insect. Sci., 33: 77–83.

Hajji-Hedfi, L., Hlaoua, W., Rhouma, A., Al-Judaibi, A.A., Arcos, S.C., Robertson, L., Ciordia, S., Horrigue-Raouani, N., Navas, A. and Abdel-Azeem, A.M. (2023). Biological and proteomic analysis of a new isolate of nematophagous fungus *Lecanicillium* sp. BMC Microbol., 23: 108.

Hatvani, L., Homa, M., Chenthamara, K., Cai, F., Kocsubé, S., Atanasova, L., Mlinaric-Missoni, E., Manikandan, P., Revathi, R., Dóczi, I. and Bogáts, G. (2019). Agricultural systems as potential sources of emerging human mycoses caused by *Trichoderma*: a successful, common phylotype of *Trichoderma longibrachiatum* in the frontline. FEMS Microbiol. Lett., 366 (21).

Hernández-Torres, I., Berlanga Padilla, A., López Arroyo, J.I., Loera Gallardo, J. and Acosta Díaz, E. (2007). Evaluación de hongos entomopatógenos para el control del pulgón café de los cítricos, *Toxoptera citricida* (Kirkaldy), en México. INIFAP. CIRNE. Campo Experimental General Terán. Folleto Científico Núm. 1. General Terán, N L, México. 19 p. (In Spanish).

Hibbett, D.S., Binder, M., Bischoff, J.F., Blackwell, M., Cannon, P.F., Eriksson, O.E., Huhndorf, S., James, T., Kirk, P.M., Lücking, R. and Lumbsch, H.T. (2007). A higher-level phylogenetic classification of the Fungi. Mycol Resh, 111: 509–547.

Higo, M., Kojima, H., Iwatsuki, M., Tanabe, T., Yaguchi, T., Okuda, T., Kasuya, T. and Nonaka, K. (2021). *Lecanicillium aphanocladii* isolated from Tengu-no-Mugimeshi found in Mount Kurohime, Nagano Perfecture, central Japan. Microb. Resour. Syst., 37(1): 1–12.

Homayoonzadeh, M., Esmaeily, M., Talebi, K., Allahyarm, H., Reitz, S. and Michaud, J.P. (2022). Inoculation of cucumber plants with *Beauveria bassiana* enhances resistance to *Aphis gossypii* (Hemiptera: Aphididae) and increases aphid susceptibility to pirimicarb. Eur. J. Entomol., 119: 1–11.

Howell, C.R. (1998). The role of antibiosis in biocontrol. In: *Trichoderma* and *Gliocladium*, vol. 2. Harman GE, Kubicek CP (Eds) Taylor and Francis, Padstow, pp 173–184.

Ibrahim, D.S.S., Elderiny, M.M., Ansari, R.A., Rizvi, R., Sumbul, A. and Mahmood, I. (2020). Role of Trichoderma spp. in the Management of Plant-Parasitic Nematodes Infesting Important Crops. In: Management of Phytonematodes: Recent Advances and Future Challenges, Ansari, R., Rizvi, R. and Mahmood, I. (eds) Springer, Singapore. Pp. 259–278.

Imoulan, A., Hussain, M., Kirk, P.M., El Meziane, A. and Yao, Y.J. (2017). Entomopathogenic fungus *Beauveria*: Host specificity, ecology and significance of morpho-molecular characterization in accurate taxonomic classification. J. Asia-Pac. Entomol., 20(4): 1204–1212.

Index Fungorum. (s.f.). Index Fungorum. https://www.indexfungorum.org.

Indiragandhi, P., Anandham, R., Madhaiyan, M., Poonguzhali, S., Kim, G.H., Saravanan, V.S. and Sa, T. (2007). Cultivable bacteria associated with larval gut of prothiofos-resistant, prothiofos-susceptible and field-caught populations of diamondback moth, *Plutella xylostella* and their potential for, antagonism towards entomopathogenic fungi and host insect nutrition. J. Appl. Microbiol., 103(6): 2664–2675.

Iwanicki, N.S., Lopes E.C.M., de Lira, A.C., Poletto, T., Fonceca, L.Z. and Júnior ID. (2023). Comparative analysis of *Beauveria bassiana* submerged conidia with blastospores: yield, growth kinetics, and virulence. Biol. Control, 185: 1–9.

Jaber, L.R. and Salem, N.M. (2014). Endophytic colonization of squash by the fungal entomopathogen *Beauveria bassiana* (Ascomycota: Hypocreales) for managing Zucchini yellow mosaic virus in curcurbits. Biocontrol Sci. Technol., 24: 1096–1109.

Jackson, M.A., Dunlap, C.A. and Jaronski, S.T. (2010). Ecological considerations in producing and formulating fungal entomopathogens for use in insect biocontrol. BioControl, 55(1): 129–145.

Jackson, M.A. (1997). Optimizing nutritional conditions for the liquid culture production of effective fungal biological control agents. J. Ind. Microbiol. Biotechnol., 19(3): 180–187.

Jackson, M.A. and Jaronski, S.T. (2012). Development of pilot-scale fermentation and stabilisation processes for the production of microsclerotia of the entomopathogenic fungus *Metarhizium brunneum* strain F52. Biocontrol Sci.Technol., 22: 915–930.

Jalinas, J., Lopez-Moya, F., Marhuenda-Egea, F.C. and Lopez-Llorca, L.V. (2022). *Beauveria bassiana* (Hypocreales: Clavicipitaceae) volatile organic compounds (VOCs) repel *Rhynchophorus ferrugineus* (Coleoptera: Dryophthoridae). J. Fungi, 8(8): 843.

Jaronski, S. and Jackson, M. A. (2012). Mass production of entomopathogenic hypocreales. En L. A. Lacey (Ed.), Manual of Techniques in Invertebrate Pathology (pp. 255–284). Nueva York, EE. UU.: Academic Press.

Jaronski, T.S. (2010). Ecological factors in the innundative use of fungal entomopathogens. BioControl, 55: 159–185.

Javar, S., Farrokhi, S., Naeimi, S. and Jooshani, M.K. (2023). Comparison of survival and pathogenicity of *Beauveria bassiana* A1-1 spores produced in solid and liquid state fermentation on whitefly nymph, *Trialeurodes vaporariorum*, J. Plant Prot. Res., 63(3): 308–317.

Kaltenpoth, M., Roeser-Mueller, K., Koehler, S. and Strohm, E. (2014). Partner choice and fidelity stabilize coevolution in a Cretaceous-age defensive symbiosis. Proc. Natl. Acad. Sci. USA, 111(17): 6359–6364.

Kaushik, N., Díaz, C.E., Chhipa, H., Julio, L.F., Andres, M.F. and Gonzalez-Coloma, A. (2020). Chemical composition of an aphid antifeedant extract from an endophytic fungus, *Trichoderma* sp. EFI671. Microorganisms, 8: 420.

Kepler, R.M., Luarngasa-ard, J.J., Hywel-Jones, N.L., Quandt, C.A., Sung, G-H., Rehner, S.A., Aime, M.M., Hankel, T.W., Sanjuan, T., Zare, R., Chen, M., Li, Z., Rossman, A.Y., Spatafora, S.W. and Shrestha, B. (2017). A phylogenetic-based nomenclature for cordicipitaceae (Hypocreales). IMA Fungus, 8(2): 335–353.

Khan, B.A., Nadeem, M.A., Nawaz, H., Amin, M.M., Abbasi, G.H., Nadeem, M. and Ayub, M. A. (2023). Pesticides: impacts on agriculture productivity, environment, and management strategies. In Emerging Contaminants and Plants: Interactions, Adaptations and Remediation Technologies (pp. 109–134). Cham: Springer International Publishing.

Khun, K.K., Ash, G.J., Stevens, M.M., Huwer, R.K. and Wilson, B.A. (2021). Transmission of *Metarhizium anisopliae* and *Beauveria bassiana* to adults of *Kuschelorhynchus macadamiae* (Coleoptera: Curculionidae) from infected adults and conidiated cadavers. Sci. Rep., 11(1): 2188.

Kikuchi, Y., Hayatsu M, Hosokawa T, and Fukatsu T. (2012). Symbiont-mediated insecticide resistance. Proc Natl Acad Sci USA, 109(22): 8618–8622.

Lara-Juache, H.R., Ávila-Hernández, J.G., Rodríguez-Durán, L.V., Michel, M.R., Wong-Paz, J.E. et al., (2021). Characterization of a biofilm bioreactor designed for the single-step production of aerial conidia and oosporein by *Beauveria bassiana* PQ2. J. Fungi, 7: 582.

Lezama, R., Alatorre, R. and Sánchez, F. (1994). Evaluación de cepas de Nomuraea rileyi y *Paecilomyces fumosoroseus* contra *Spodoptera frugiperda* (Lepidoptera: Noctuidae). Vedalia, 1: 19–22. (In Spanish).

Lezama, R., Alatorre, R., Bojalil, J., Molina, O., Arenas, V., González, R. and Rebolledo, D. (1996a). Virulence of five entomopathogenic fungi (Hyphomycetes) against *Spodoptera frugiperda* (Lepidoptera: Noctuidae) eggs and neonate larvae. Vedalia, 3: 35–40.

Lezama, R., Molina O., Contreras O., González R., Trujillo de la L.A. and Rebolledo D.O. (1996b). Susceptibilidad de larvas de *Anastrepha ludens* (Diptera: Tephritidae) a diversos nematodos entomopatógenos (Steinernematidae y Heterorhabditidae). Vedalia, 3: 31–34. (In Spanish).

Li, C.Y., Lo, I.W., Wang, S.W., Hwang, T.L., Chung, Y.M., Cheng, Y.B. and Wu, Y.C. (2017). Novel 11-norbetaenone isolated from an entomopathogenic fungus *Lecanicillium antillanum*. Bioorg. Med. Chem. Lett., 27(9): 1978–1982.

Liu, J.F., Zhang, Z.Q., Beggs, J.R., Paderes, E., Zou, X. and Wei, X.Y. (2020). Lethal and sublethal effects of entomopathogenic fungi on tomato/potato psyllid, *Bactericera cockerelli* (Šulc) (Hemiptera: Triozidae) in capsicum. Crop Prot., 129: 105023.

Liu, Y.C., Chen, T.H., Huang, Y.F., Chen, C.L. and Nai, Y.S. (2023). Investigation of the fall armyworm (*Spodoptera frugiperda*) gut microbiome and entomopathogenic fungus-induced pathobiome. J. Invertebr. Pathol., 200: 107976.

Lohse, R., Jakobs-Schönwandt, D. and Patel, A.V. (2014). Screening of liquid media and fermentation of an endophytic *Beauveria bassiana* strain in a bioreactor. AMB Express, 4(1): 1–11.

López-López, R.E. (2024). Identificación y efectividad biológica de hongos entomopatógenos, nativos del estado de Oaxaca, sobre *Scyphophorus acupunctatus* del Agave spp. Master Degree Thesis. Instituto Politécnico Nacional, CIIDIR-Oaxaca, Mexico. 60p. (In Spanish).

Lukasik, P., Guo, H., van Asch, M., Ferrari, J. and Godfray, H.C. (2013). Protection against a fungal pathogen conferred by the aphid facultative endosymbionts *Rickettsia* and *Spiroplasma* is expressed in multiple host genotypes and species and is not influenced by co-infection with another symbiont. J. Evol. Biol., 26(12): 2654-2661.

Lumsden, R.D., Locke, J.C., Adkins, S.T., Walter, J.F. and Ridout, C.J. (1992). Isolation and localization of the antibiotics gliotoxin produced by *Gliocladium virens* from alginate prill in soil and soilless media. Phytopathol., 82(2).

Luz, C., Rocha, L.F., Montalva, C., Souza, D.A., Botelho, A.B.R., Lopes, R.B. and Júnior, I.D. (2019). *Metarhizium humberi* sp. nov. (Hypocreales: Clavicipitaceae), a new member of the PARB clade in the *Metarhizium anisopliae* complex from Latin America. J. Invertebr. Pathol., 166: 107216.

Lv, B., Yu, S., Chen, Y., Yu, H. and Mo, Q. (2022). First Report of *Lecanicillium aphanocladii* Causing Rot of *Morchella sextelata* in China. Plant Dis., 106(12).

Maina, U.M., Galadima, I.B., Gambo, F.M. and Zakaria, D. (2018). A review on the use of entomopathogenic fungi in the management of insect pests of field crops. J. Entomol. Zool. Stud., 6(1): 27–32.

Maistrou, S., Natsopoulou, M.E., Jensen, A.B. and Meyling, N.V. (2020). Virulence traits within a community of the fungal entomopathogen *Beauveria*: Associations with abundance and distribution. Fungal Ecol., 48: 100992.

Manfrino, R., González, A., Bernache, J., Tornesello Galván, J., Hywell-Jones, N. and López-Lastra, C.C. (2017). Contribution to the knowledge of pathogenic fungo of spiders in Argentina Southemmost record in the world. Rev. Argent. Microbiol., 49(2): 197–200.

Mantzoukas, S., Lagogiannis, I., Mpekiri, M., Pettas, I. and Eliopoulos, P.A. (2019). Insecticidal action of several isolates of entomopathogenic fungi against the granary weevil *Sitophilus granarius*. Agriculture, 9: 222.

Martínez, L.C., Plata-Rueda, A., Ramírez, A. and Serrão, J.E. (2022). Susceptibility of *Demotispa neivai* (Coleoptera: Chrysomelidae) to *Beauveria bassiana* and *Metarhizium anisopliae* entomopathogenic fungal isolates. Pest Manag. Sci., 78(1): 126–133.

Martínez-Barrera, O.Y., Toledo, J., Cancino, J., Liedo, P., Gómez, J., Valle-Mora, J. and Montoya, P. (2020). Interaction Between *Beauveria bassiana* (Hypocreales: Cordycipitaceae) and *Coptera haywardi* (Hymenoptera: Diapriidae) for the Management of *Anastrepha obliqua* (Diptera: Tephritidae). J. Insect Sci., 20(2): 6.

Mascarin, G.M., Alves, S.B. and Lopes, R.B. (2010). Culture media selection for mass production of *Isaria fumosorosea* and *Isaria farinose*. Braz. Arch. Biol. Technol., 53(4): 753–761.

Mascarin, G.M., da Silva A.V.R., da Silva T.P., Kobori N.N., Morandi M.A.B. and Bettiol W. (2022) *Clonostachys rosea*: Production by Submerged Culture and Bioactivity Against *Sclerotinia sclerotiorum* and *Bemisia tabaci*. Front. Microbiol., 13(1).

Mascarin, G.M., Jackson, M.A., Kobori, N.N., Behle, R.W. and Delalibera, J.I. (2015). Liquid culture fermentation for rapid production of desiccation tolerant blastospores of *Beauveria bassiana* and *Isaria fumosorosea* strains. J. Invertebr. Pathol., 127: 11–20.

Medo, J., Michalko, J., Medová, J. and Cagáň, Ľ. (2016). Phylogenetic structure and habitat associations of Beauveria species isolated from soils in Slovakia. J. Invertebr. Pathol., 140: 46–50.

Membang, G., Ambang, Z., Mahot, H.C., Kuate, A.F., Fiaboe, K.K.M. and Hanna, R. (2020). *Cosmopolites sordidus* (Germar) susceptibility to indigenous Cameroonian *Beauveria bassiana* (Bals.) Vuill. and *Metarhizium anisopliae* (Metsch.) isolates. J. Appl. Entomol., 144: 468–480.

Méndez-González, F., Figueroa-Montero, A., Saucedo-Castañeda, G., Loera, O. and Favela-Torres, E. (2022). Addition of spherical-style packing improves the production of conidia by *Metarhizium robertsii* in packed column bioreactors. J. Chem. Technol. Biotechnol., 97: 1517–1525.

Méndez-González, F., Loera, O. and Favela-Torres, E. (2018). Conidia Production of *Metarhizium anisopliae* in Bags and Packed Column Bioreactors. Curr. Biotechnol., 7: 65–69.

Meyling, N.V., Lübeck, M., Buckley, E.P., Eilenberg, J. and Rehner, S.A. (2009). Community composition, host range and genetic structure of the fungal entomopathogen *Beauveria* in adjoining agricultural and seminatural habitats. Mol Ecol., 18: 1282–1293.

Meyling, N.V., Pell, J.K. and Eilenberg, J. (2006). Dispersal of *Beauveria bassiana* by the activity of nettle insects. J. Invertebr. Pathol., 93: 121–126.

Miller, T., Crossley M.S., Fuc Z., Meier AR., Crowder DW. and Snyder WE. (2020). Exposure to predators, but not intraspecific competitors, heightens herbivore susceptibility to entomopathogens. Biol. Control, 151: 104403.

Missbach, K., Flatschacher, D., Bueschl, C., Samson, J. M., Leibetseder, S., Marchetti-Deschmann, M., and Schuhmacher, R. (2023). Light-Induced changes in secondary metabolite production of *Trichoderma atroviride*. J. Fungi, 9(8): 785.

Mitina, G.V., Krasavina, L.P. and Trapeznikova, O.V. (2021). Compatibility of the fungus Lecaicillium muscarium and the predatory mite *Amblyseius swirskii* for their comind application against the greenhouse whitefly *Trialeurodes vaporariorum*. Plant Prot. News, 104(3): 163–170.

Molina-Montenegro, M.A. Oses, R. Torres-Díaz, C. Atala, C. Núñez, M.A. and Armas, C. (2015). Fungal endophytes associated with roots of nurse cushion species have positive effects on native and invasive beneficiary plants in an alpine ecosystem. Perspect. Plant Ecol. Evol. Syst., 17: 218–226.

Monte, E. (2001). Understanding *Trichoderma*: between biotechnology and microbial ecology. Int. Microbiol., 4: 1–4.

Monzón, A. (2001). Producción, uso y control de calidad de hongos entomopatógenos en Nicaragua, Manejo Integrado de Plagas. (CATIE, Costa Rica), 63: 95–103. (In Spanish).

Mota-Delgado, P.A. and Murcia-Ordoñez, B. (2011). Hongos entomopatógenos para el control de plagas, Rev. Ambient. Água, 6(2): 77–90. (In Spanish).

Mukherjee, P.K., Mukhopadhyay, A.N., Sarmah, D.K. and Shrestha, S.M. (1995). Comparative antagonistic properties of *Gliocladium virens* and *Trichoderma harzianum* on *Sclerotium rolfsii* and *Rhizoctonia solani* its relevance to understanding the mechanisms of biocontrol. J. Phytopathol., 143: 275–279.

Nájera R., M.B., M. García M., R.L. Crocker, V. Hernández V. and L.A. Rodríguez del B. (2005). Virulencia de *Beauveria bassiana* y *Metarhizium anisopliae*, nativos del occidente de México, contra larvas de tercer estadio de *Phyllophaga crinita* (Coleoptera: Melolonthidae) bajo condiciones de laboratorio. Fitosanidad, 9(1): 33–36. (In Spanish).

Nedveckytė, I., Pečiulytė, D. and Būda, V. (2021). Fungi associated with horse-chestnut leaf miner moth *Cameraria ohridella* Mortality, Forests, 12(1): 58.

Nicoletti, R. and Becchimanzi, A. (2020). Endophytism of *Lecanicillium* and *Akanthomyces*. Agriculture, 10: 205.

Ozdemir, I.O., Tuncer, C., Erper, I. and Kushiyev, R. (2020). Efficacy of the entomopathogenic fungi; *Beauveria bassiana* and *Metarhizium anisopliae* against the cowpea weevil, *Callosobruchus maculatus* F. (Coleoptera: Chrysomelidae: Bruchinae). Egypt. J. Biol. Pest Control, 30(1): 1–5.

Pacheco-Hernández, M., Reséndiz Martínez, J. and Arriola Padilla, V.J. (2019). Organismos entomopatógenos como control biológico en los sectores agropecuario y forestal de México: una revisión. Rev. Mex. Cienc. Forestales, 10(56): 4–32. (In Spanish).

Páez, D.J. and Fleming-Davies, A.E. (2020). Understanding the evolutionary ecology of host–pathogen interactions provides insights into the outcomes of insect pest biocontrol. Viruses, 12(2): 141.

Panteleev, D.Y., Goryacheva I.I., Andrianov B.V., Reznik N.L., Lazebny O.E. and Kulikov A.M. (2007). The endosymbiotic bacterium Wolbachia enhances the nonspecific resistance to insect pathogens and alters behavior of *Drosophila melanogaster*. Russ. J. Genet., 43: 1066-1069.

Parker, B.J., Spragg, C.J., Altincicek, B. and Gerardo, N.M. (2013). Symbiont-mediated protection against fungal pathogens in pea aphids: a role for pathogen specificity? Appl. Environ. Microbiol., 79(7): 2455–2458.

Pedrini, N. (2018). Molecular interactions between entomopathogenic fungi (Hypocreales) and their insect host: Perspectives from stressful cuticle and hemolymph battlefields and the potential of dual RNA sequencing for future studies. Fungal Biol., 122(6): 538–545.

Pelo, S.P., Adebo, O.A. and Green, E. (2021). Chemotaxonomic profiling of fungal endophytes of *Solanum mauritianum* (alien weed) using gas chromatography high resolution time-of-flight mass spectrometry (CG-HRTOFMS). Metabolomics, 17(5): 43.

Peng, G., Xie, J., Guo, R., Keyhani, N.O., Zeng, D., Yang, P. and Xia, Y. (2021). Long-term field evaluation and large-scale application of a *Metarhizium anisopliae* strain for controlling major rice pests. J. Pest Sci., 94: 969–980.

Pérez-González, O., Maldonado-Blanco, M.G., Torres-Acosta, R.I., Rodríguez-Guerra, R., Elías-Santos, M. and López-Arroyo, J.I. (2013). Evaluación de aislados mexicanos de *Hirsutella citriformis* Speare contra *Diaphorina citri* Kuwayama (Hemiptera: Liividae) en Laboratorio. Entomol. Mex., 12(1): 381–386. (In Spanish).

Pérez-González, O., Rodríguez-Guerra, R., López-Arroyo, J.I., Sandoval-Coronado, C.F. and Maldonado-Blanco, M.G. (2016). Effect of Mexican *Hirsutella citriformis* (Hypocreales: Ophiocordycipitaceae) strains on *Diaphorina citri* (Hemiptera: Liviidae) and the predators *Chrysoperla rufilabris* (Neuroptera: Chrysopidae) and *Hippodamia convergens* (Coleoptera: Coccinellidae). Fla. Entomol., 99 (3): 509–515.

Pérez-Salgado, J., Ángel-Ríos, M.D., Arteaga-Deloya, A., Hernández-Castro, E., Damián-Nava, A., Carretera, A.D.C.N.U. and Rancho, C.P. (2013). Hongos entomopatógenos y extractos vegetales contra escama blanca (*Aulacaspis tubercularis Newstead*) en cultivo de mango en San Luis La Loma, Municipio de Tecpan de Galeana, Gro., México. Entomol. Mex., 12(1): 452- 455. (In Spanish).

Pessotti, R.D.C., Hansen, B.L., Reaso, J.N., Ceja-Navarro, J.A., El-Hifnawi, L., Brodie, E.L. and Traxler, M.F. (2021). Multiple lineages of *Streptomyces* produce antimicrobials within passalid beetle galleries across eastern North America. Elife, 10: e65091.

Petrova, V. (2023). Control of plum fruit moth *Cydia (Grapholita) funebrana* in organic plum production. Agric. Sci. Technol., 15(3): 57–60.

Pick, D.A., Avery, P.B., Qureshi, J.A., Arthurs, S.P. and Powell, C.A. (2022). Field persistence and pathogenicity of *Cordyceps fumosorosea* for management of *Diaphorina citri*. Biocontrol Sci. Technol. 32(2): 151–162.

Poosakkannu, A., Nissinen, R. and Kytöviita, M.M. (2015). Culturable endophytic microbial communities in the circumpolar grass, *Deschampsia flexuosa* in a sub-Arctic inland primary succession are habitat and growth stage specific. Environ. Microbiol. Rep., 7: 111–122.

Portilla, M., Abbas, H. K., Accinelli, C. and Luttrell, R. (2019). Laboratory and field investigations on compatibility of *Beauveria bassiana* (Hypocreales: Clavicipitaceae) spores with a sprayable bioplastic formulation for application in the biocontrol of tarnished plant bug in cotton. J. Econ. Entomol., 112(2): 549–557.

Quesada-Béjar, V., Rincón, M.B.N., Reyes-Novelo, E., Real-Santillán, R.O., Wies, G. and González-Esquivel, C.E. (2023). Molting Patterns and Mortality of *Sphenarium purpurascens purpurascens* (Orthoptera: Pyrgomorphidae) Inoculated with *Metarhizium anisopliae*. Southwestern Entomol., 48(1): 161–168.

Ramanujam, B., Rangeshwaran, R., Sivakmar, G., Mohan, M. and Yandigeri, M.S. (2014). Management of Insect Pests by Microorganisms. Proc. Indian Natl. Sci. Acad., 80(2): 455–471.

Rodríguez, S. Marta, Gerding P. Marcos and France, A. (2006). Selección de Aislamientos de Hongos Entomopatógenos para el Control de Huevos de la Polilla del Tomate, *Tuta absoluta* Meyrick (Lepidoptera: gelechiidae). Agr. Tec., 66(2): 151–158. (In Spanish).

Romero-Arenas, O., Amaro-Leal, L.J., Rivera, A., Parraguirre-Lezama, C., Sánchez-Morales, P. and Villa-Ruano, N. (2020). Formulations of *Beauveria bassiana* MABb1 and mesoporous materials for the biological control of *Sphenarium purpurascens* in maize crops from Puebla, Mexico. J. Asia-Pac. Entomol., 23(3): 653–659.

Ruiz, V.J., Aquino, B.T., Rivera, M.E.S. and Girón Pablo, S. (2012). Control integrado de la gallina ciega *Phyllophaga vetula* Horn (Coleoptera: Melolonthidae) con agentes entomopatógenos en Oaxaca, México. Rev. Cient. UDO Agric., 12(3): 609–616. (In Spanish).

Sain, S.K., Monga, D., Kumar, R., Nagrale, D.T., Hiremani, N.S. and Kranthi, S. (2019). Compatibility of entomopathogenic fungi with insecticides and their efficacy for IPM of *Bemisia tabaci* in cotton. J. Pestic. Sci., 44(2): 97–105.

Sala, A., Barrena, R., Meyling, N.V. and Artola, A. (2023). Conidia production of the entomopathogenic fungus *Beauveria bassiana* using packed-bed bioreactor: Effect of substrate biodegradability on conidia virulence. J. Environ. Manage., 341: 118059.

Sánchez-Pérez, L., Rodríguez-Navarro, S., Marín-Cruz, V.H., Ramos-López, M.Á., Ramos, A.P. and Barranco-Florido, J.E. (2016). Assessment of *Beauveria bassiana* and their enzymatic extracts against *Metamasius spinolae* and *Cyclocephala lunulata* in laboratory. Adv. Enzyme Res., 4: 98–112.

Santos, P. de S., Abati, K., Mendoza, N.V.R., Mascarin, G.M. and Júnior, I.D. (2021). Nutritional impact of low-cost substrates on biphasic fermentation for conidia production of the fungal biopesticide *Metarhizium anisopliae*. Bioresour. Technol. Rep., 13: 100619.

Saravanakumar, K., Li, Y., Yu, C., Wang, Q.Q., Wang, M., Sun, J., Gao, J.X. and Chen, J. (2017). Effect of *Trichoderma harzianum* on maize rhizosphere microbiome and biocontrol of *Fusarium* Stalk rot. Sci. Rep., 7: 1–13.

Saroop, S. and Tamchos, S. (2024). Impact of pesticide application: Positive and negative side. Pesticides in a Changing Environment, 155–178.

Sayed, A.M. and Dunlap, C.A. (2019). Virulence of some entomopathogenic fungi isolates of *Beauveria bassiana* (Hypocreales: Cordycipitaceae) and *Metarhizium anisopliae* (Hypocreales: Clavicipitaceae) to *Aulacaspis tubercularis* (Hemiptera: Diaspididae) and *Icerya seychellarum* (Hemiptera: Monophlebidae) on mango crop. J. Econ. Entomol., 112(6): 2584–2596.

Sayed, S. M., Ali, E. F. and Al-Otaibi, S. S. (2019). Efficacy of indigenous entomopathogenic fungus, *Beauveria bassiana* (Balsamo) Vuillemin, isolates against the rose aphid, *Macrosiphum rosae* L. (Hemiptera: Aphididae) in rose production. Egypt. J. Biol. Pest Control, 29: 1–7.

Scarborough, C. L., Ferrari, J. and Godfray, H. C. J. (2005). Aphid protected from pathogen by endosymbiont. Science, 310(5755): 1781–1781.

Serna-Domínguez, M.G., Andrade-Michel, G.Y., Rosas-Valdez, R., Castro-Félix, P., Arredondo-Bernal, H.C. and Gallou, A. (2019). High genetic diversity of the entomopathogenic fungus *Beauveria bassiana* in Colima, Mexico. J. Invertebr. Pathol., 163, 67–74.

Shahid, A. A., Rao, Q. A., Bakhsh, A. and Husnain, T. (2012). Entomopathogenic fungi as biological controllers: new insights into their virulence and pathogenicity. Arch. Biol. Sci., 64(1): 21-42.

Shamim, M., Kumar, D., Kumar, M., Srivastava, D., Kumar, S., Ranjan, T. and Jha, V.B. (2024). Role of Fungi in Biocontrol of Diseases in Cereal Crops. In Applied Mycology for Agriculture and Foods (pp. 79–96). Apple Academic Press.

Shao, Y., Mason, C.J. and Felton, G.W. (2024). Toward an integrated understanding of the lepidoptera microbiome. Annu. Rev. Entomol., 69:117–137.

Shapiro-Ilan D.I., Fuxa J.R., Lacey J.A., Onstad D.W. and Kaya Y.H.K. (2005). Definitions of pathogenicity and virulence in invertebrate pathology. J. Invertebr. Pathol., 88(1): 1–7.

Sharma, R. and Sharma, P. (2021). Fungal entomopathogens: a systematic review. Egypt. J. Biol. Pest Control, 31: 1–13.

Shinde, S.V., Purohit, M.S., Sabalpara, A.N. and Patel, M.B. (2010). First report of enthomopathogenic fungus *Lecanicillium ecanii* (Zimm.) Zare and Gams on sugarcane whitefly *Aleurolobus barodensis* (Maskell) from Gujart. Trends Biosci., 3(1): 76–79.

Silva, A.C., Ricalde, M.P., Scalzer, R.R.C., Zilli, J.E. and Lopes, R.B. (2022b). Natural occurrence of *Beauveria bassiana* on adults of the invasive mango seed weevil *Sternochetus mangiferae* (Coleoptera: Curculionidae) in Brazil. J. Plant Dis. Prot., 129: 79–84.

Silva, G.D.C., Kitano I.T., Ribeiro, I.A.F. and Lacava, P.T. (2022a). The potential use of actinomycetes as microbial inoculants and biopesticides in agriculture. Frontiers in Soil Science, Front. Soil Sci., 2: 833181.

Skrobek, A., Shah, F.A. and Butt, T.M. (2008). Destruxin production by entompgenous fungi *Metarhizium anisopliae* in insects and factors influencing their degradation. BioControl, 53: 361–373.

Sogan, N., Kala, S., Kapoor, N., Negpal, BN., Ramlal, A. and Nautiyal, A. (2023). Novel development of *Lecanicillium lecanii* based granules as a plataform against malaria vector *Anopheles culicifacies*. World J. Microbiol. Biotechnol., 39: 142.

Sreeja, P. and Rani, R.O.P. (2019). Volatil metabolites of *Lecanicillium saksenae* (Kushwaha) Kurihara and Sukarno and their toxicity to brinjal mealybug *Coccidohystrix insolita*(G). Entomon., 44(3): 183–190.

Sreeja, P., Rani, R.O.P. and Chellappan, M. (2023). Characterization of insecticidal metabolite oosporein in *Lecanicillium saksenae* (Kushwaha) Kurihara and Sukarno, a geographically distinct isolate from Kerala, India. Toxicon., 230: 107176.

Srivastava, M., Mohammad, S., Pandey, S., Vipul, K., Singh, A., Shubha, T., Srivastava, Y. and Shivram, S. (2015). *Trichoderma*: A scientific approach against soil borne pathogens. Afr. J. Microbiol. Res., 9: 2377–2384.

Sukarno, N., Kurihara, Y., Park, JY., Inaba, S., Ando, K., Harayama, S., Ilyas, M., Mangunwardoyo, W., Sjamsuridzal, W., Yuniarti, E., Saraswati. R. and Widyastuti, Y. (2009). *Lecanicillium* and

Verticillium species from Indonesia and Japan including three new species. Mycoscience, 50(5): 369-379.

Tapfuma, K.I., Nyambo, K., Baatjies, L., Keyster, M., Mekuto, L., Smith, L. and Mavumengwana, V. (2022). Compuestos derivados de hongos y nanopartículas micogénicas con actividad antimicobacteriana: una revisión. SN Ciencias Aplicadas, 4(5): 134. (In Spanish).

Tarore, D., Manueke, J., Montong, V.B. and Semuel, M.Y. (2023). Control of rice powder pest *Sitophilus oryzae* L. with entomopathogenic fungus *Metarhizium anisopliae* (Metch.). Pak. J. Biol. Sci., 26(2): 56–62.

Thomas, M.B. (1997). Fungal Ecology and its application to the practical use of mycoinsecticides. BCPC Symposium Microbial Proceedings N° 68.

Valero-Jiménez, C.A., Wiegers, H., Zwaan, B.J., Koenraadt, C.J. and van Kan, J.A. (2016). Genes involved in virulence of the entomopathogenic fungus *Beauveria bassiana*. J. Invertebr. Pathol., 133: 41–49.

Van Eerd, L.L., Hoagland, R.E., Zablotowicz, R.M. and Hall, J.C. (2003). Pesticide metabolism in plants and microorganisms. Weed Sci., 51(4): 472–495.

Van Nguyen, N., Kim, Y.J., Oh, K.T., Jung, W.J. and Park, R.D. (2007). The role of chitinase from *Lecanicillium antillanum* B-3 in parasitism to root-knot nematode (*Meloidogyne incognita*) eggs. Biocontrol Sci. Technol., 17: 1047–1058.

Van Wees, S.C.M., De Swart, E.A.M., Van Pelt, J.A., Van Loon, L.C. and Pieterse, C.M.J. (2000). Enhancement of induced disease resistance by simultaneous activation of salicylate- and jasmonate-dependent defense pathways in *Arabidopsis thaliana*. Proc. Natl. Acad. Sci. USA, 97(15): 8711–8716.

Vidhate, R.P., Dawkar, V.V., Punekar, S.A. and Giri, A.P. (2023). Genomic determinants of entomopathogenic fungi and their involvement in pathogenesis. Microbial Ecol., 1–12.

Villegas R.F., Díaz G., Casas F., Monreal V., Tamayo, M. and Aguilar, M. (2017). Actividad de dos hongos entomopatógenos, identificados molecularmente, sobre *Bactericera cockerelli*. Rev. Colomb. Entomol. 43(1): 27–33. (In Spanish).

Vlot, A.C., Sales, J.H., Lenk, M., Bauer, K., Brambilla, A., Sommer, A., Chen, Y. Wenig, M. and Nayem, S. (2021). Systemic propagation of immunity in plants. New Phytol., 229(3): 1234–1250.

Wang, C. and Wang, S. (2017). Insect pathogenic fungi: genomics, molecular interactions, and genetic improvements. Annu. Rev. Entomol., 62, 73–90.

Wang, Y.B., Wang, Y., Fan, Q., Duan, D.E., Zhang, G.D., Dai, R.Q., Zeng, W. et al. (2020). Multigene phylogeny of the family Cordycipitaceae (Hypocreales): new taxa and the new systematic position of the Chinese cordycipitoid fungus *Paecilomyces hepialid*. Fungal Divers.,103: 1–46.

Wang, Z., Yong, H., Zhang, S., Liu, Z. and Zhao, Y. (2023). Colonization resistance of symbionts in their insect hosts. Insects, 14(7), 594.

Wei, D.P., Wanasinghe, D.N., Chaiwat, T.A. and Hyde, K.D. (2018). *Lecanicillium uredinophilum* known from rusts, also occurs on animal hosts with chitinous bodies. Asian J Mycol., 1(1): 63–73.

Woo, R.M., Park, M.G., Choi, J.Y., Park, D.H., Kim, J.Y., Wang, M., Kim, H.J., Woo, S.D., Kim, J.S. and Je, Y.H. (2020). Insecticidal and insect growth regulatory activities of secondary metabolites for entomopathogenic fungi, *Lecanicillium attenuatum*. J. Appl. Entomol., 144(7): 655–663.

Xia, X., Gurr, G.M., Vasseur, L., Zheng, D., Zhong, H. and Qin, B. (2017). Metagenomic sequencing of diamondback moth gut microbiome unveils key holobiont adaptations for herbivory. Front. Microbiol., 8: 663.

Xia, X., Sun, B., Gurr, G.M., Vasseur, L., Xue, M. and You, M. (2018). Gut microbiota mediate insecticide resistance in the diamondback moth, *Plutella xylostella* (L.). Front. Microbiol., 9:25.

Xu, C.P. and Yun, J.-W. (2003). Optimization of submerged-culture conditions for mycelial growth and exo-biopolymer production by *Auricularia polytricha* (wood ears fungus) using the methods of uniform design and regression analysis. Biotechnol. Appl. Biochem., 38(2): 193.

Xu, C.P., Kim, S.W., Hwang, H.J. and Yun, J.W. (2006). Production of exopolysaccharides by submerged culture of an enthomopathogenic fungus, *Paecilomyces tenuipes* C240 in stirred-tank and airlift reactors. Bioresour. Technol., 97(5): 770–777.

Yuan, M., Huang, Y., Ge, W., Jia, Z., Song, S., Zhang, L. and Huang, Y. (2019). Involvement of jasmonic acid, ethylene and salicylic acid signaling pathways behind the systemic resistance induced by *Trichoderma longibrachiatum* H9 in cucumber. BMC Genomics, 1–13.

Yun, K.S., Hyon, J.H., Kim, H.S., and Jang, S.H. (2023). Biological control of the small leafhopper, *Empoasca flavescens* F. (Homoptera: Cicadellidae) using the entomopathogenic fungus, *Verticillium lecanii*. Egypt. J. Biol. Pest Control, 33(1): 1–5.

Zahran, Z., Nor, N.M.I.M., Dieng, H., Satho, T. and Ab Majid, A.H. (2017). Laboratory efficacy of mycoparasitic fungi (*Aspergillus tubingensis* and *Trichoderma harzianum*) against tropical bed bugs (*Cimex hemipterus*) (Hemiptera: Cimicidae). Asian Pac. J. Trop. Biomed., 7: 288–293.

Zare, R. and Gams, W. (2001). A revision of *Verticillium* section Prostata. IV: genera *Lecanicillium* and *Simplicillium* gen. nov. Nova Edwigia, 73(1-2): 1–50.

Zhang, F., Sun, X.X., Zhang, X.C., Zhang, S., Lu, J., Xia, Y.M., Huang, Y.H. and Wang, X.J. (2018). The interactions between gut microbiota and entomopathogenic fungi: a potential approach for biological control of *Blattella germanica* (L.). Pest Manag. Sci., 74(2): 438–447.

Zhang, Z.F., Zhou, S.Y., Eurwilaichitr, L., Ingsriswang, S., Raza, M., Chen, Q. and Cai, L. (2021). Culturable mycobiota from Karst caves in China II, with descriptions of 33 new species. Fungal Divers., 106: 29–136.

Zhao, X., Chai, J., Wang, F. and Jia, Y. (2023). Optimization of submerged culture parameters of the aphid pathogenic fungus *Fusarium equiseti* based on sporulation and mycelial biomass. Microorganisms, 11:190.

Zhong, L.F. Juan, L. Luo, L.X. Yang, C.N. and Xiao, L. (2023). Lecanicilliums A–F, thiodiketopiperazine-class alkaloids from a mangrove sediment-derived fungus *Lecanicillium kalimantanense*. Mar. Drugs, 21: 575.

Zhou, Y.M., Zhi, J.R., Qu, J.J. and Zou, X. (2022). Estimate divergence times of *Lecanicillium* in the family Cordycipitaceae provide insights into the attribution of *Lecanicillium*. Front. Microbiol. 13:859886.

Zimmermann, G. (2007). Review on safety of the entomopathogenic fungi Beauveria bassiana and *Beauveria brongniartii*. Biocontrol Sci. Technol., 17: 553–596.

Zytynska, S.E. and Meyer, S.T. (2019). Effects of biodiversity in agricultural landscapes on the protective microbiome of insects–a review. Entomol. Exp. Appl., 167: 2–13.

14 Mass Production of *Beauveria bassiana*

Parthasarathy Seethapathy[1*]

1. Introduction

Fungal biopesticides that target insect pests and mites primarily comprise soil-dwelling entomopathogenic fungi from the Hypocreales order, in Ascomycota. Most of these fungi generate anamorphic air-borne spores that are highly resistant to environmental conditions. These spores are then dispersed in the atmosphere and can infect arthropod communities (do Nascimento Silva et al., 2023). *Beauveria bassiana* (Balsamo) Vuillemin (Hypocreales: Clavicipitaceae) is a versatile soil saprophytic fungus that can effectively control pest populations in soil and build symbiotic relations with flora as a beneficial endophyte, making it an ideal choice for integrated pest management (Solano-González et al., 2023). In 1835, Italian scientist Agostino Bassi conducted the initial infection studies and identified the fungus responsible for causing muscardine disease. He observed that the fungus was able to kill and mummify silk moth larvae. This discovery subsequently established that the disease could be transmitted through airborne carriers.

The term "*Beauveria*" was coined by French physician Jean Paul Vuillemin (1861-1932) as a tribute to the French physicist Jean Beauverie (1874-1938). The species epithet is named after Bassi, the individual who made the discovery. Vuillemin designated it the type species for his newly established genus (Qiu et al., 2021). The term *B. bassiana* has traditionally been employed to denote a group of isolates that are morphologically and genetically identical, forming a species complex. *B. bassiana* is composed of numerous separate heredities that should be acknowledged as distinctive phylogenetic types (Rehner et al., 2011). In 2011, the genus *Beauveria* was redefined, and a suggested type for *B. bassiana* was established. *B. bassiana* is the asexual form (anamorph) of *Cordyceps bassiana* (Bustamante et al., 2019; Rehner et al., 2011). The latter teleomorph refers to the polyphasic taxonomy form, with two widely distributed complexes of *Beauveria*: *B. asiatica* and *B. bassiana* (Kobmoo et al., 2021).

Beauveria spp., an entomopathogenic fungus species, plays a crucial role in naturally controlling insect populations and is environmentally beneficial (Li et al., 2023). *Beauveria* is a predominant source of insect pathology, impacting a diverse array of more than 700 insect species. It can thrive in various ecological habitats, such as soil, organic composts, plants, and insects, while adapting to a multifunctional existence (Imoulan et al., 2017). *Beauveria* spp. widely persists in the

[1] Department of Plant Pathology, Amrita School of Agricultural Sciences, Amrita Vishwa Vidyapeetham, Coimbatore 642109, India.

* Corresponding author: spsarathyagri@gmail.com

soil as saprotrophic mycelia or inactive propagules until it attaches to a suitable host in the nearby micro-environment or forms associations with plants as endophytes. Furthermore, it can generate dormant cells known as asexual conidia, enabling it to disperse and endure different environments. *Beauveria* is a precious fungal resource due to the tremendous commercial and ecological value found in specific species (Wang et al., 2022). Global biopesticide ventures have assessed *B. bassiana* as a potential biocontrol agent against plant pests and diseases. *B. bassiana* and *B. brongniartii* are well-known ecologically beneficial alternatives to pesticides for controlling crop pests. *B. pseudobassiana*, another species, has shown considerable efficacy in managing various insect pests (Imoulan et al., 2017; Wang et al., 2022). These species can either exist as endophytes or epiphytes which reside in the soil as root inhabitants. *Beauveria* species can synthesize diverse chemicals that inhibit the growth of certain insect pests and fungal plant diseases and induce systemic resistance in plants.

B. bassiana is a prominent entomopathogenic fungus used in integrated pest management approaches due to its effectiveness. The broad spectrum of insect pests that it can infect and manipulate and its ecologically beneficial characteristics have established it as a vital asset for promoting sustainable agriculture. This chapter briefly summarizes the methods and factors involved in the large-scale production and culture of *B. bassiana*. The efficient large-scale production of *B. bassiana* depends on several criteria, such as the careful selection of strains, appropriate growth medium, optimal environmental conditions, and the design of bioreactors. The chapter covers the approaches and strategies used to enhance the spread and sporulation of the fungus to ensure a top-notch product for pest management purposes. Additionally, the text examines the production of *B. bassiana* using liquid fermentation, solid-state fermentation, and submerged fermentation methods, highlighting their benefits and constraints. These growth methods necessitate variation in substrates, food, and aeration factors, emphasizing the need for precise knowledge of appropriate methodologies for specific pest management.

2. Biological Potential of *Beauveria*

Many studies have been conducted on *B. bassiana* to understand its ecological interactions, molecular mechanisms, and genomic diversity (Proietti et al., 2023). *B. bassiana* is widely used against pests because it produces insecticidal compounds such as enzymes, secreted effectors, and toxins, including Beauvericin ($C_{45}H_{57}N_3O_9$) and other secondary metabolites, which cause disease in the host insect. Beyond its role in agriculture, *B. bassiana* has garnered interest for various biotechnological applications in medicine and bioenergy. *Beauveria* strains secrete metabolites comprising carboxylic acids, amino acids, phenolics, and siderophore substances. These metabolites possess the capacity to liberate protons and form complexes with metals. They contribute to the breakdown and synthesis of minerals, as well as the conversion and purification of harmful metals like cadmium, copper, lead, zinc, and uranium (Kozłowska et al., 2018). The fungus produces enzymes with industrial significance, including chitinases and proteases, which have applications in the degradation of organic materials and bioconversion processes (Elawati et al., 2018).

In 1995, the initial commercial strains of *B. bassiana* were made using the ATCC74040 and GHA strains. These strains are cultivated in large quantities through fermentation and marketed as products like BotaniGard® 22WP, Mycotrol® WPO, and Naturalis® L. A similar strain, ANT-03, was discovered in 2000 and commercially released in 2013 as BioCeres® WP (Rodriguez-Saona et al., 2022). Various commercially accessible strains are available to target a range of insect pests. This fungus is a promising alternative to chemical pesticides, and its versatility and effectiveness make it an essential tool for sustainable agriculture (Qiu et al., 2021). Understanding the molecular mechanisms underlying the entomopathogenicity of *B. bassiana* is crucial for optimizing its efficacy as a biopesticide. Recent studies have delved into the expression profiles of essential genes involved in host recognition, penetration, and immune evasion (Li et al., 2023). As the demand for sustainable pest management practices grows, assessing the environmental impact and safety of *B. bassiana* becomes paramount. Research has focused on evaluating its ecological interactions, persistence in the environment, and potential non-target effects. Its genome has also been sequenced, which is helping researchers better understand its biocontrol abilities (Solano-González et al., 2023). This information is crucial for developing guidelines to ensure the safe and responsible use of this biopesticide. Commercial production methods and formulations for aerial conidia and immersed spores of pathogenic fungi have been developed (Awan et al., 2021; Kassa et al., 2008). Studies have investigated formulation strategies to enhance the stability and shelf life of the fungus (Oliveira et al., 2018). Field trials and case studies have provided insights into the practical efficacy and challenges of applying *B. bassiana* in agroecosystems (Bamisile et al., 2021; Hamzah et al., 2021).

3. Entomopathogenic Action of *Beauveria*

B. bassiana is a global species that naturally resides in soil. It can cause mycotoxicosis, which can induce tissue death in various arthropod hosts, including insects from several orders, ticks, and mites. Research has shown that the pathogenesis and virulence of *Beauveria* isolates are associated with the production of *in vivo* toxicogenic metabolites, cuticle-degrading and anti-oxidant enzymes, and active vegetative development within the host, which leads to physiological starvation (Pedrini, 2022). The spores can disseminate through the air or adhere to insects as they traverse their environment. The spores exhibit hydrophobic characteristics, enabling them to comply readily with the insect exoskeleton. *B. bassiana* spores can germinate within a few hours of a carbon source being present. To maintain growth, a nitrogen source is essential. The infection process of *B. bassiana* spores begins upon contact with the insect epidermis.

Although the typical mode of infection is through the integument, evidence suggests that *B. bassiana* can also infect insects through oral ingestion, particularly those with chewing mouthparts. Upon contact, the fungal conidia produce various enzymes, such as chitinases, lipases, and proteases. These enzymes act as catalysts for breaking down the layers of the insect cuticle (Huarte-Bonnet et al., 2018). The cuticle is mainly comprised of chitin and serves as a carbon and nitrogen source. Proteases are first secreted, suggesting the presence of a protein layer on the chitin fibrils, followed by the release of chitinases. Chitin is associated with cuticle proteins, which define the mechanical properties of the cuticle. Following this, the fungus spores germinate, forming germ tubes and appressoria. These structures attach the spores to the insect, apply mechanical pressure to the cuticle, and invade the hemocoel. Upon reaching the insect hemolymph, the germ tubes, now called hyphae, initiate the development of specialized thin-walled, budding single-celled spores known as yeast-like blastospores. Subsequently, the mature haustorium liberates blastospores and releases toxic metabolites, initiating the germination process and causing the destruction of internal tissues within the host (Swathy et al., 2024).

Upon entering the insect body, *B. bassiana* generates a diverse array of toxins. These toxins, classified as secondary metabolites, include bassianin, bassianolide, beauvericin, beauverolides, oosporein, oxalic acid, and tenellin. The toxins play a crucial role in facilitating the parasitization and subsequent death of the host (Wang et al., 2021). The blastospores exploit the nutrient-rich insect blood, establish themselves in the tissues, and produce toxic compounds that lead to the insect's demise. Blastospores are occasionally described as the yeast-like stage of the fungus. After the insect cuticle has been penetrated, blastospores emerge from the hypha. They are capable of reproductive budding. Colonization is accompanied by the production of various toxic metabolites (antimicrobial peptides) that contribute to host immune suppression, destruction of host internal tissues, nutrient depletion, and ultimately host mortality. Subsequently, spores are produced by the fungus, which colonizes the desiccated remnants of the deceased insect (Huang et al., 2021) (Fig. 1).

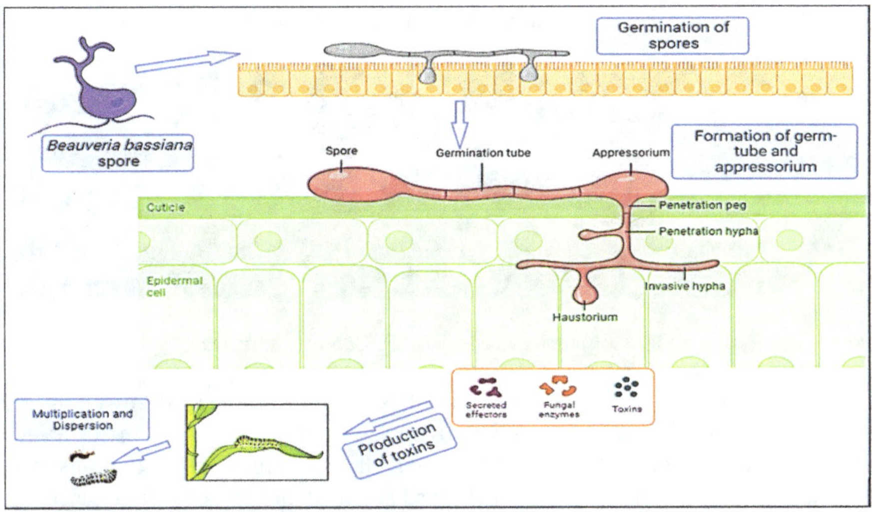

Fig. 1 Entomopathogenic cycle of *Beauveria bassiana.* (Created with BioRender.com)

4. Biopesticide Potential of *Beauveria*

B. bassiana, a globally distributed fungus, demonstrates parasitic activity towards many arthropod species, leading to white muscardine disease. Consequently, it can be classified as an entomopathogenic fungus. It is used as a biological formulation to control a range of pests, such as aphids, beetles, termites, whiteflies, and various thrips. The effectiveness of this substance in managing bedbugs and malaria-carrying mosquitoes has been well documented. In recent years, many research publications, including thorough reviews, have been published on the diversity and virulence properties of *B. bassiana* (Wang et al., 2022). Most of this research primarily investigates cytotoxicity and insecticidal efficacy (Wang et al., 2021).

The insecticidal efficacy of *B. bassiana* is commonly attributed to the synergistic action of several toxic substances. The insecticidal mechanisms consist of diverse processes that enhance their efficiency. These processes include the secretion of factors to sustain constant virulence, hindering the initiation of the host immune response, disturbing the transmission of nerve impulses, damaging the insect outer layer to facilitate penetration of hyphae, blocking the respiratory openings of the insect host, and absorbing water and nutrients from the host organism, among others (Pedrini, 2022). According to recent research findings, it has been observed that *B. bassiana* conidia exhibit a higher level of pathogenicity towards *Popillia japonica*, *Halyomorpha halys*, and *Tenebrio molitor* species after a 10-day treatment period. Notably, *H. halys* and *T. molitor* experienced complete mortality within this 10-day time frame. The concentration (LC50) at which mortality occurred was found to be significantly lower in *H. halys* (9.5×10^3 conidia/mL), *T. molitor* (2.6×10^3 conidia/mL), and *P. japonica* (8.3×10^4 conidia/mL). Furthermore, action with *B. bassiana* resulted in a notable reduction in insect lifespan, with *H. halys* experiencing an average lifespan of 6 days, *T. molitor* 5 days, and *P. japonica* 7 days. According to the findings, the efficacy of *B. bassiana* conidia was observed to be higher when targeting these three prominent coleopteran and hemipteran pests (Swathy et al., 2024) (Fig. 2).

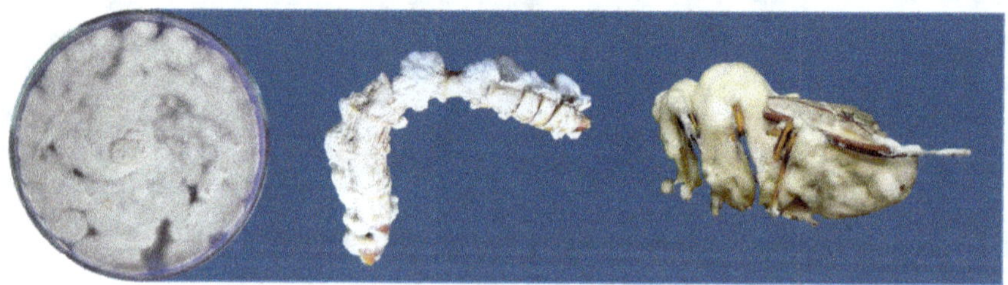

Fig. 2 Entomopathogenicity of *Beauveria bassiana*.

B. bassiana targets a broad spectrum of hosts and can infect soil-dwelling and aerial insect pests (Table 1). Due to its soil persistence and capacity to infect several insect species, there may be competitive interactions between invasive strains of *B. bassiana* and indigenous microorganisms and potential unintended consequences for beneficial insects. Nonetheless, there can be significant variations in the biological host range (host range that can be infected in controlled laboratory conditions) and the environmental host range (range of insects affected under native environmental or introduced agricultural circumstances) among different fungal strains (Celar and Kos, 2016; Wang et al., 2022). The multifaceted cycles in *Beauveria* are regarded as a virulence characteristic, as blastospores have developed the ability to avoid immunological defences and efficiently utilize host nutrients. The lack of hydrophobin elements or hydrophobic rodlet layers in aerial conidia indicates that the electrostatic charges on the surface of blastospores could substantially affect the interaction between the host and the infectious agent. The blastospores have demonstrated comparable or even greater virulence against various insect pests than airborne conidia, or submerged conidia (Mascarin and Jaronski, 2016).

Table 1 Targeted host diversity of *Beauveria* spp.

Major Arthropod Host	*Beauveria* spp.	*Reference*
Orthoptera: Acrididae	*Beauveria acridophila*	Sanjuan et al., 2014
Hymenoptera: Formicidae	*Beauveria amorpha*	Rehner et al., 2011
Araneae	*Beauveria araneola*	Chen et al., 2017
Coleoptera: Cerambycidae	*Beauveria asiatica*	Rehner et al., 2011; Wang et al., 2022
Coleoptera: Chrysomelidae	*Beauveria baoshanensis*	Chen et al., 2017
Lepidoptera: Arctiidae; Hymenoptera: Pamphiliidae; Coleoptera: Scarabaeidae	*Beauveria bassiana*	Rehner et al., 2011; Wang et al., 2022
Blattodea: Blattidae	*Beauveria blattidicola*	Kepler et al., 2017
Coleoptera: Scarabaeidae	*Beauveria brongniartii*	Rehner et al., 2011; Wang et al., 2022
Coleoptera: Cerambycidae	*Beauveria caledonica*	Wang et al., 2022
Phasmatodea: Diapheromeridae	*Beauveria diapheromeriphila*	Sanjuan et al., 2014
Coleoptera: Melolonthidae	*Beauveria hoplocheli*	Robène-Soustrade et al., 2015
Homoptera: Delphacidae	*Beauveria kipukae*	Kepler et al., 2017
Coleoptera: Coccinellidae	*Beauveria lii*	Zhang et al., 2012
Orthoptera: Romaleidae	*Beauveria locustiphila*	Sanjuan et al., 2014
Coleoptera: Cerambycidae; Coleoptera: Scarabaeidae	*Beauveria majiangensis*	Rehner et al., 2011; Wang et al., 2022
Coleoptera	*Beauveria medogensis*	Wang et al., 2022
Coleoptera: Curculionidae	*Beauveria peruviensis*	Bustamante et al., 2019
Hymenoptera: Formicidae	*Beauveria polyrhachicola*	Wang et al., 2022
Lepidoptera: Tortricidae; Coleoptera: Scarabaeidae	*Beauveria pseudobassiana*	Rehner et al., 2011; Wang et al., 2022
Lepidoptera: Bombycidae	*Beauveria rudraprayagi*	Agrawal, 2014
Coleoptera: Scarabaeidae	*Beauveria scarabaeidicola*	Rehner et al., 2011; Wang et al., 2022

Contd.

Table 1 *Contd.*

Major Arthropod Host	*Beauveria* spp.	*Reference*
Orthoptera: Grylloidea; Lepidoptera: Geometridae	*Beauveria sinensis*	Chen et al., 2017
Coleoptera: Scarabaeidae	*Beauveria songmingensis*	Wang et al., 2022
Coleoptera: Staphylinidae; Coleoptera: Cerambycidae	*Beauveria staphylinidicola*	Wang et al., 2022
Coleoptera: Scarabaeidae	*Beauveria subscarabaeidicola*	Wang et al., 2022
Coleoptera: Curculionidae	*Beauveria varroae*	Rehner et al., 2011
Lepidoptera; Coleoptera: Scarabaeidae	*Beauveria yunnanensis*	Wang et al., 2022

An endophytic form of *B. bassiana* can colonise various crops in a systemic way, thereby improving crop resistance. A study investigated the efficacy of isolates of *B. bassiana* in managing damping-off pathogens, such as *Rhizoctonia solani* and *Pythium myriotylum,* in tomato seedlings. The study revealed the activation of induced systemic resistance through peroxidases (POX), phenylalanine ammonia-lyase (PAL), and phenolic compounds. (Azadi et al., 2016). The species *B. bassiana* and *B. pseudobassiana*, along with the mycotoxin beauvericin, were found to have a significant impact on the suppression of *Bursaphelenchus xylophilus*, also known as the pine wood nematode. These substances effectively reduced the nematode survival rate, resulting in mortality rates of up to 100% (Sánchez-Gómez et al., 2023). They also emphasised the application of *Beauveria* species and the mycotoxin beauvericin in an integrated pest management approach for effectively controlling nematodes and pests. These cases, and the extensive literature, prove that *Beauveria* is an excellent biocontrol agent for combating insect pests, fungal pathogens, and nematodes. Additionally, it has been shown to induce systemic resistance when it interacts with the host as an epiphyte, endophyte, or through rhizospheric colonization.

In a 2017 study, a commercial preparation of *B. bassiana* resulted in a mortality rate of over 94% for bed bugs that were resistant to pyrethroids (Barbarin et al., 2017). Aprehend® is an innovative fungal biopesticide specifically designed to effectively manage bed bug infestations. The formulation comprises fungal spores of *B. bassiana,* which are suspended in a specialised oil formulation. Chemical insecticides are frequently used to treat bed bugs, and it may be necessary to apply them numerous times to completely eliminate the problem (Shikano et al., 2021). The inverted emulsion formulation of *B. bassiana* JN5R1W1 fungal conidia successfully controlled adult mosquitoes of *Aedes albopictus* and *Culex pipiens* by demonstrating high virulence, perseverance, and efficiency (Yong Lee et al., 2023). It is apparent that *Beauveria* is an excellent biocontrol agent for globally threatening household nuisance insects such as bedbugs, cockroaches, and mosquitoes.

5. Culturing and Mass production of *Beauveria*

The efficacy of microbial biopesticides such as *B. bassiana* depends on identifying, characterizing, and assessing their infectivity and on the streamlined production of the

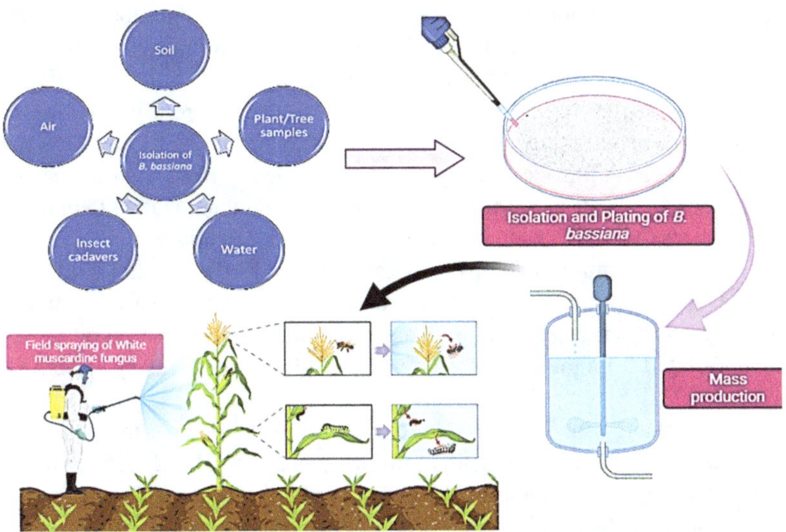

Fig. 3 Mass production and application of *Beauveria.* (Created with BioRender.com)

microbial agents in a controlled laboratory environment (Kassa et al., 2008) (Fig. 3). The highest possible growth of *B. bassiana* was achieved using the SDA medium, with a recorded mycelial development of 85.3 mm. The production of *B. bassiana* formulations can be approached in two different manners. One approach involves generating a limited amount of inoculum for lab and field trials through biopesticidal formulation. The other approach focuses on establishing a fundamental manufacturing process for large-scale production, utilizing labor-intensive and economically viable methods suitable for smaller markets. A reliable and efficient production system can be achieved using basic multiplication techniques. These techniques include submerged liquid fermentation for producing short-lived blastospores and hydrophilic or solid-state fermentation for producing aerial conidia. However, the most practical strategies for large-scale production involve utilizing a two-phase method. This consists of generating fungal inoculums in a liquid culture and then inoculating the solid substrate(s) to produce conidia (do Nascimento Silva et al., 2023).

The development of an appropriate formulation is an essential aspect of commercializing entomopathogenic fungi as biological pesticides. Suitable mixtures can be determined by comprehensively comprehending the entire system, encompassing the organism, diverse production processes, and the application environment (Oliveira et al., 2018). Specific formulation technologies can be used to solve critical variables. The primary focus of formulation research is typically on biological factors, such as viability and efficacy, as well as the physical characteristics of natural pesticide solutions, including form, mixing, and usage.

5.1 Isolation of B. bassiana

Obtaining pure cultures of *B. bassiana* from environmental samples often requires specialized techniques. Selective media are frequently employed to isolate *B. bassiana*. Various media formulations, including Sabouraud dextrose agar (SDA) or other specialized nutritional media, can be augmented with antibiotics or additional ingredients to prevent the growth of unwanted contaminants and facilitate the development of *Beauveria* (Kassa et al., 2008).

According to one study, newly formulated media more effectively separated *B. bassiana* than commonly used selective media such as Veen and Ferron media (Posadas et al., 2012). Insect baiting entails strategically placing insect cadavers or other appropriate substrates in the field, aiming to attract and isolate *Beauveria* from infected insects (Masoudi et al., 2020). This method proves advantageous for isolating strains that have a natural relationship with arthropod hosts. The soil dilution plating technique entails diluting soil samples and subsequent plating onto selective media. *B. bassiana* has been detected in both external and internal plant parts. *B. bassiana* has been isolated from the bark of walnut trees and nearby soil using selective media (Gürlek et al., 2018).

Scientists have found *B. bassiana* in its natural habitat on the phylloplane of different hedgerow plants using leaf imprinting and a specialized medium to study its presence (Meyling and Eilenberg, 2006). Additionally, the occurrence of *B. bassiana* as an endophyte in various plant species has been observed using the 'Galleria-bait' method and various selective media (Fergani and Yehia, 2020).

5.2 Characterization of B. bassiana

Beauveria strains can be characterized and confirmed using molecular techniques, such as polymerase chain reaction (PCR) and DNA sequencing. PCR can be applied using primers specifically designed to target conserved areas of the *Beauveria* genome. Through the process of sequencing the amplified DNA, a conclusive identification can be acquired. A meticulous analysis of *B. bassiana* growth under varying pH conditions revealed that the fungus demonstrated robust viability even at pH values surpassing 10. The fungus exhibited resilience to decreased sugar concentration and $CuCl_2$ (Mishra and Malik, 2013). Further, a selective medium enriched with $CuCl_2$, reduced sugar, and brilliant green dye was developed to isolate *B. bassiana* from soil samples (Posadas et al., 2012).

Multiple researchers have investigated the analysis of the ITS gene, uncovering a correlation between global climate zones and the genetic groupings of *B. bassiana* (Kobmoo et al., 2021). Furthermore, examining nuclear ITS sequences has demonstrated the genetic unity of *Beauveria* and the presence of at least two distinct lineages within *B. bassiana* (Awan et al., 2021). They studied the genomic sequences of *B. bassiana* isolates to understand their inherent variations in virulence as biocontrol agents. The investigation involved sequencing the ITS1-5.8S-ITS2 region of ten *B. bassiana* strains, yielding significant insights into the genetic variation and allowing for their categorization into three primary groups based on similarities (de Costa Eula Maria et al., 2011).

In another study (Sayed et al., 2018), a total of 94 soil samples were collected from several places in Taif. Among the samples, the presence of the *B. bassiana* fungus was detected in only 11 of them, resulting in a ratio of 11.7%. Further, the examination of the partial *COI* and *ITS* gene sequences revealed that the four isolates (Sp1 to Sp4) exhibited a variety of genetic makeups. The genetic distances among the four isolates under investigation ranged from 0.002 to 0.008, suggesting a significant genetic correlation.

5.3 Culturing of B. bassiana

The majority of mycopesticides currently on the global market are manufactured using solid-state fermentation (SSF) or liquid-state fermentation (LSF). The type of myco-propagules produced depends on the characteristics and circumstances of the fermentation process (Oliveira et al., 2018). *B. bassiana* has the ability to thrive in several types of synthetic and defined media under solid, liquid, and semisolid submerged cultivation conditions. *B. bassiana* grows well on liquid and solid media, including potato dextrose and sabouraud dextrose broth and agar (Feng et al., 1994).

Aerial conidia formed on solid surfaces were similar in morphology and efficiency to those produced on the surface of insect cadavers. Numerous natural solid substrates have been evaluated for production of the entomopathogenic *Beauveria*, including sugar by-products (bagasse, press mud, and molasses), barley, beetroot, broken rice, carrot tubers, coconut cake, cottonseed cake, finger millet, groundnut cake, maize, maize bran, neem cake, sesame cake, pearl millet, potato tubers, prawn waste, press mud, rice, rice bran, rice hulls, rice husk, tapioca rind and tubers, wheat, and wheat bran (Jaronski, 2014).

In submerged cultures, six developmental stages were observed. Blastospores, resembling an insect haemocoel, were often produced by schizolytic dissociation at the septa or mechanical hyphae fragmentation caused by shearing pressures in the liquid medium (Huarte-Bonnet et al., 2018). Furthermore, blastospores could be produced by yeast-like budding from parent single cells (Pedrini, 2022). Numerous studies on blastospore growth in submerged cultures have been undertaken (Huarte-Bonnet et al., 2018). Because the growth method is functionally comparable to short hyphal cells, *B. bassiana* blastospores are correctly referred to as "hyphal bodies" (Lohse et al., 2014). *B. bassiana* has been successfully sub-cultured at an optimal temperature of 25°C on an SDA agar medium supplemented with 1% casein peptone, 2% glucose, and 1.5% agar at a pH of 5.5 (Lohse et al., 2014). A study provides a procedure for culturing *B. bassiana* from the Hatay yellow strain cadaver on artificial Sabouraud CAF agar media containing chloramphenicol (40 µg/ml) (Gençer et al. 2023). The cultures are placed in a controlled environment with a temperature range of 26-28°C for a period of 1-2 weeks to facilitate the growth of mycelium and the formation of blastospores. Cultures are cultivated on CAF agar medium and preserved at –20°C in glycerol for extended periods (Gençer et al., 2023). *B. bassiana* blastospores had thin walls and were unstable when applied in the field and after fermentation. Producing large quantities of aerial conidia through diphasic fermentation, which involves producing vegetative mycelia in liquid batch culture and then allowing the mycelia to surface conidia on a nutrient or inert carrier, was considered labor-intensive and incompatible with the traditional fermentation methods.

5.4 Diphasic Fermentation Methods

Earlier studies pointed out that LUBILOSA diphasic fermentation combined the advantages of both solid and liquid substrates. The *B. bassiana* was cultured in fermenters until the end of the log phase to maximize mycelial biomass output. It was then transferred to nutrient- or inert-rich substrates to develop aerial conidia in the form of a naturally occurring inoculum. Despite its simplicity, the diphasic technique was considered the most expensive and labour-intensive (Posada-Flórez, 2008). The study utilized a biphasic approach (Roswanjaya et al., 2022), where *B. bassiana* was first cultured in a submerged environment and prompted to produce conidia in a SSF using three distinct cultivation media such as potato dextrose broth (PDB), malt extract broth (MEB), and yeast and malt extract broth (YMEB). Also, use five distinct solid compositions, such as 100% rice, 100% maize, a blend containing 75% rice and 25% maize, a blend containing 50% rice and 50% maize, and a blend containing 25% rice and 75% maize. It was determined that potato dextrose broth demonstrated the highest efficacy as a culture medium for the synthesis of blastophore and beauvericin. The synthesis of beauvericin, a bioactive molecule with insecticidal effects, was effectively accomplished in this medium. The selection test revealed that utilizing 100% rice as the solid medium yielded the highest conidia output. During a rice storage test, the quantity of *B. bassiana* conidia was constant for a duration of 3.5 months when maintained at ambient temperature. Today, the dual cyclone technology for collecting conidia is becoming increasingly popular in commercial production systems due to its ability to remove large particles, effectively separate high-quality spores, ensure operator safety, provide cost-effective processing, facilitate storage, and achieve high conidia productivity, particularly for *B. bassiana* (Velozo et al., 2023).

5.5 Solid Phase Fermentation

Biopesticides derived from fungi can be manufactured using solid-state fermentation (SSF). *Beauveria* and other hypocrealean entomopathogenic fungi utilize aerial conidia to spread infection (Velozo et al., 2023). The production of this hydrophobic asexual spore can be accomplished with minimal effort and expense. A prevalent method for producing aerial conidia entails the utilization of moist, sterilized cereal grains (Roswanjaya et al., 2022). SSF does not require significantly more time and effort and offers an ideal substitute for low-tech artisanal manufacturing. Conventional large-scale production facilities employed the spawn bag or tray technique to cultivate *Beauveria* blastospores. However, the yield of blastospores varied significantly based on factors such as substrate, oxygen concentration, starting moisture content, and isolation. In Brazil, steamed rice has traditionally been used as the main starch source for SSF to produce blastospores of *Beauveria* spp. (do Nascimento Silva et al., 2023). However, production of aerial conidia can be accomplished through two-stage fermentation or one stage fermentation in the industrial process (Oliveira et al., 2018). Instantaneously, upon commencing the manufacturing procedure, conidia cultivated in the solid culture are introduced onto the substrate. Submerged liquid fermentation utilized during the second stage of production to generate an inoculum, typically composed of blastospores. This inoculum is then incorporated into the solid-phase bulk manufacturing procedure. SSF remains the sole technique capable of economically producing airborne blastospores. In addition, SSF provides the advantage of efficiently harnessing agro-industrial wastes as substrates, thereby promoting fungal growth at a reduced cost (Mattedi et al., 2023). The described fermentation procedure, consisting of two stages and utilizing non-absorbent inert assistance, shows promise for producing blastospores of *B. bassiana* (Oliveira et al., 2018). A study examined a range of substrates with different levels of biodegradability, such as rice husk and beer draff, through rigorous testing. The experimental procedure encompassed the progression from 1.5 L to 22 L bioreactors, yielding positive results, and utilizing beer draff as a substrate led to a notable rise in conidia production (2.5×10^9 and 6.0×10^8 conidia/g dry matter) in both reactors. The significance of air-free porosity as a scaling-up criterion is highlighted. However, variations in quality were observed between conidia cultivated on agar plates and fermented yields. The observed disparities between plate and fermented samples suggest that the processes of fermentation, extraction, and preservation could possibly cause a decline in quality. It is crucial to optimize these steps to achieve the highest possible preservation of virulence (Sala et al., 2023).

Further research has explored the use of affordable palm oil meals as the primary material in a circular packed-column aerobic fermenter model specifically designed for the cultivation of *B. bassiana*. They assessed the effects of temperature, airflow, substrate moisture percentage, and the proportion of palm kernel cake to fiber, on conidial harvest and dehydration tolerance through fermentation treatments. The results demonstrated a substantial increase in production (2×10^{10} conidia/g) and a decrease in fermentation period (from 168 h to 120 h) using this approach, compared to the standard tray fermenter. The ideal substrate environment comprised 60% opening moisture, 30% palm fiber and 70% palm kernel cake. The fermentation was carried out at a temperature of 26°C with an aeration rate of at least 0.2 L/min. Asexual conidia that were dried in the column resulted in a germination rate of 95%. The study has demonstrated the viability of utilizing a packed-column bioreactor to efficiently produce large quantities of *B. bassiana* conidia using inexpensive agricultural leftovers. This approach contributes to developing a sustainable technique for making this biopesticide (do Nascimento Silva et al., 2023).

5.6 Liquid Fermentation under Submerged Conditions

LSF typically produces mycelia and blastospores. However, immersed conidia and microsclerotia can also be generated under specific conditions (Fig. 4). LSF offers enhanced automation of procedures and scalability to create many cells efficiently within regulated nutrition and environmental conditions. However, the capacity to withstand challenging industrial and environmental circumstances, such as drying out after harvesting, being stable without refrigeration, or persisting in the field, is lower than that seen with aerial blastospores. Prior efforts to produce significant amounts of blastospores using LSF yielded fewer than 1×10^9 blastospores/mL, required fermentation times of over five days, exhibited insufficient durability during drying, and required greater integrity during preservation (Lohse et al., 2014). *B. bassiana* has been cultivated in shake flask cultures to produce greater quantities of spores, including both blastospores and submerged conidiospores (SCS) (Lohse et al., 2014). A TKI broth with 5% sugar beet molasses, containing 50% sucrose, yielded the highest concentration of 1.33×10^9 spores/ml and the most significant yield of

5.32×10^{10} spores/g. The scale-up process for increasing the reactor volume to a 2 L BIOSTAT® Bplus stirred tank fermenter was used under the following conditions: a temperature of 25°C, agitation speeds ranging from 200 to 600 rpm, an aeration rate of 1 vvm, and a maintained pH at 5.5. After 216 hours, a total of 5.2×10^{10} spores per g of sucrose were collected, yielding an SCS yield of 0.2×10^{10} SCS/per g of sucrose. This study provides the initial documentation of the fermentation process of *B. bassiana*, demonstrating the production of very high quantities of total spores.

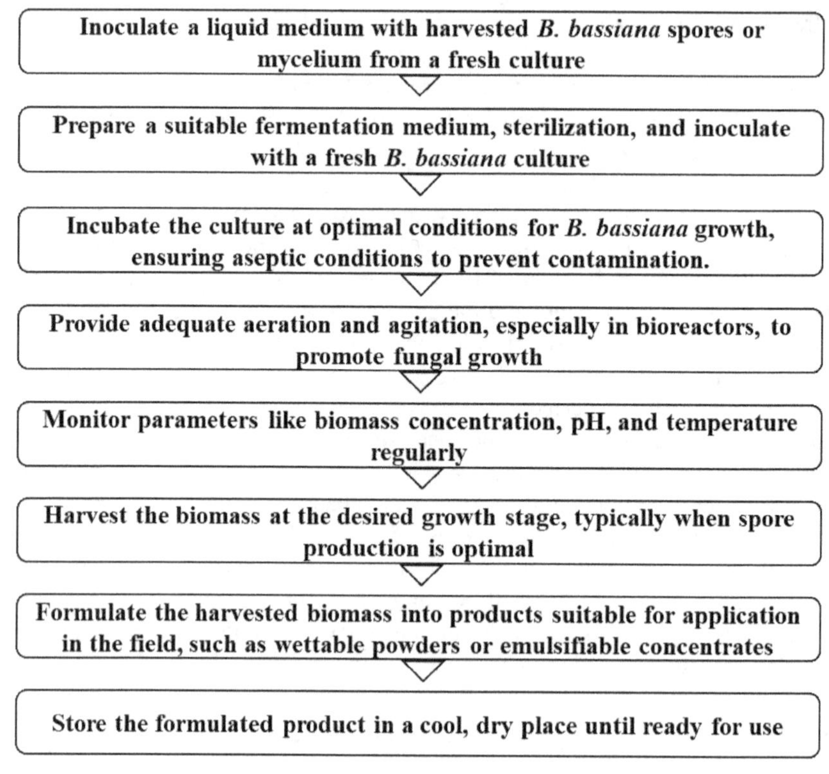

Fig. 4 Flow chart steps; liquid fermentation of *B. bassiana.*

A study identified critical environmental and nutritional factors necessary for rapidly producing large quantities of blastospores of *Beauveria* and comparable species in optimal dietary conditions (Mascarin et al., 2015). After two to three days of fermentation, blastospore yields increased dramatically when oxygenation rates were raised to better oxygenate the mixture by adjusting the volume-to-surface ratio and agitation speed. Blastospores grown in liquid media supplemented with 10% glucose exhibited enhanced resistance to desiccation. Subsequently, a blastospore concentration of 2×10^9/ml was achieved, with over 60% of these cells retaining viability following air or spray drying. Blastospores can withstand the 0.7-1.2 MPa hyperosmotic environment of insect hemolymphs.

One notable difference between the diminished production of aerial conidia on solid culture media subjected to osmotic stress conditions and the enhanced formation of blastospores in the liquid medium appears to be the consequence of heightened osmotic pressure (Mascarin et al., 2015). Blastospore production was further enhanced by increasing the osmotic pressure of LSF with nonionic or ionic osmolytes, such as sugars and ions. The capacity of blastospores to amass endogenous osmoprotectant mixtures, including glyceraldehyde, to preserve osmotic equilibrium has been associated with the high osmolarity glycerol pathway, wherein osmosensing proteins mediate alteration to the bug hemocoel (Mascarin et al., 2021).

6. Formulation and Quality Control

Commercial biopesticide formulations are characterized by their intricate composition and are offered in solid and liquid states. The initial phase consists of fermentation, formulation with fillers, adjuvants that synergistically function, and other necessary elements for formulating a stable treatment. A schematic overview of fermentation and its downstream processes for formulations is given here (Fig. 5). The requirement for organic (biodegradable) and cost-effective formulation components poses challenges to both environmental issues and the overall cost of the product. In recent times, a significant number of mycoinsecticides have been produced that allow the use of a wide range of entomopathogenic fungus species, with at least 12 out of the vast pool of over 800 species being utilized.

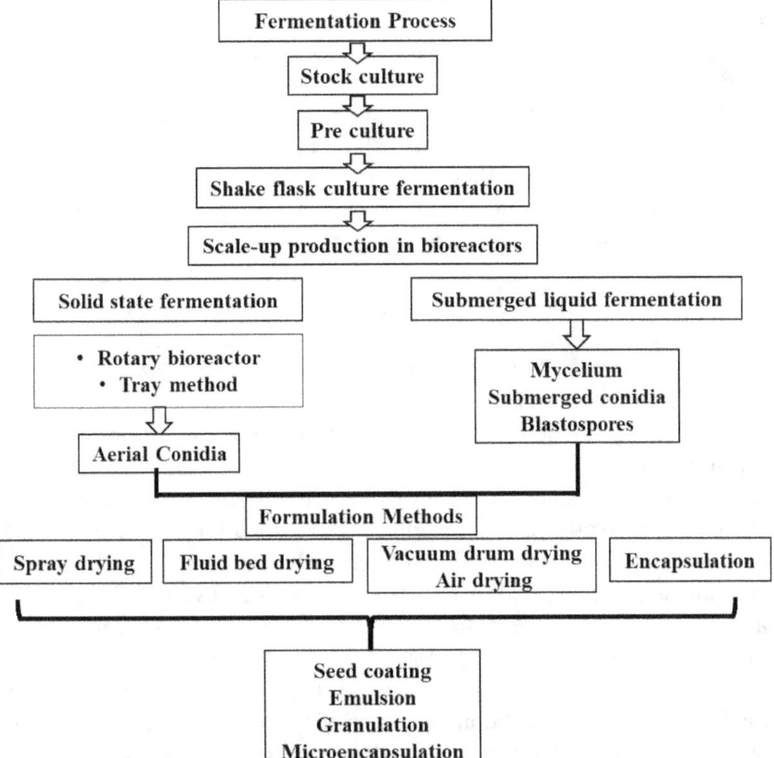

Fig. 5 Downstream processing of microbial fermentation.

The hypocrealean fungal entomopathogens *B. bassiana, Isaria fumosorosea,* and *B. brongniartii* are extensively used worldwide for pest control. One of the most used genera is *Beauveria,* which can be found in various formulations, including fungi-colonized substrates, wettable powders, and oil dispersions (Velozo et al., 2023). *B. bassiana* alone accounts for 34% of the biopesticide market, followed by *I. fumosorosea* at 5.8% and *B. brongniartii* at 4.1%. Strains from the genus *Beauveria* and *Metarhizium* combined dominate almost 70% of the global market for mycopesticides (do Nascimento Silva et al., 2018).

B. bassiana is typically incorporated into three primary formulations: bait/solid (commonly composed of tea residue), encapsulation, and emulsion (Masoudi et al., 2020). The primary component in the bait formulation is *B. bassiana* conidia, which is mixed with nourishment or other enticing elements. This method is financially viable characterized by a simple preparation process. Under controlled laboratory conditions, various wettable powder formulations of *B. bassiana* have demonstrated a slightly higher conidial mortality rate compared to a similar formulation in the form of an emulsifiable suspension. Enclosed formulations of *B. bassiana* often enhance the effectiveness and durability of the fungus while safeguarding its spores from unfavourable environmental conditions (Yong Lee et al., 2023). Additives such as skimmed milk powder, polyvinyl pyrrolidone K-90, and glucose improve formulation control and aid in the even dispersion of conidia in *B. bassiana.* Nevertheless, it is essential to acknowledge that the encapsulating procedure negatively affects conidial survivability.

Using vegetable oil in the emulsion formulation of entomopathogenic fungi is a highly feasible solution. Emulsions improve the ability of fungal conidia to stick to the outer layer of insect pests and protect them from harmful UV radiation (Oliveira et al., 2021; Swathy et al., 2024). This enhances the ability of the conidia to cause disease and effectively control these pests. Moreover, they are easy to implement. Emulsion formulations of entomopathogenic fungi are often prepared with several oils, such as almond oil, coconut oil, castor oil, eucalyptus oil, gingelly oil, and mustard oil, along with commonly used linseed oil, olive oil, soybean oil, sunflower oil, and tile oils. The efficacy of most of these vegetable oils (and synthetic oils) in terms of their compatibility with conidia from *B. bassiana* has been assessed. This assessment considered criteria such as germination proportion, asexual development, and conidial production (Oliveira et al., 2021; Velozo et al., 2023).

Unsurprisingly, about 90% of the entomopathogenic agents supplied worldwide consist of aerial conidia grown on solid substrates. On the other hand, vegetative propagules, which include blastospores, mycelium, and microsclerotia, make up less than 10% of available formulations. Talc and kaolin are the most frequently used carriers in wettable particle formulations, whereas oil is the preferred carrier in liquid formulations (Feng et al., 1994). By nature of their hydrophobic and lipophilic characteristics, *B. bassiana* conidia disperse readily in oils (de Oliveira et al., 2023). The biological and physical characteristics of the formulation must demonstrate long-term stability, preferably exceeding 18 months, before commercialization takes place (Bamisile et al., 2021). In addition to desiccated mycelium, the entomopathogen can also be preserved as blastospores or conidia. Preserving conidial specimens is more desirable owing to their enhanced stability and virulence (Swathy et al., 2024). The efficiency of formulations containing *B. bassiana* is influenced by various elements, such as the strength of UV light, humidity levels, temperature, the life stage of the insects, and the virulence of the fungal strain (Meyling and Eilenberg, 2006). Both humidity and temperature significantly influence the survival and germination of fungi.

Generally, higher temperatures and increased humidity promote greater survival and germination rates (Shikano et al., 2021). The efficacy of *B. bassiana* is significantly influenced by exposure to UV radiation. Moreover, the targeted insect species can influence the rate at which insect mortality and fungal sporulation occur (Gençer et al., 2023; Posadas et al., 2012). A comprehensive quality assurance component is essential to producing a high-quality mycoinsecticide. Although the fermentation variables appear under strict control, the production process involving solid or liquid substrates is only moderately regulated, mainly when operating on a large scale (Jaronski, 2014). A recent study details the creation of tablet formulations incorporating dried conidia of *B. bassiana*. The tablet preparation, consisting of 30% conidia and 70% cornflour, exhibited favourable outcomes regarding its mechanical strength and ability to disperse in water. The tablet demonstrated a hardness of 173.94 N and a friability of 0.34%. The tablet rapidly disintegrated in water within 2 minutes, resulting in a dispersion of 10^9 conidia mL^{-1}. When stored in a polyethylene container containing polymerized silica, the tablet maintained its viability of at least 80% for a storage period of 180 days at temperatures ranging from 25 to 28°C (Almeida et al., 2023).

7. Commercialization

In 1995, the initial commercial products of *B. bassiana* were successfully developed. The fungus is commercially produced through fermentation and marketed under the trade names BotaniGard® 22WP, Mycotrol® WPO, and Naturalis® L. Since then, over 250 commercial products have been marketed all over the world; the leading products on the global market are listed in Table 2. Among them, Boveril® WP ESALQ-PL63 is produced by solid-state fermentation with moist rice by the Koppert company in Brazil. The Koppert company manufactures pure, dehydrated conidia for spray applications, primarily targeting coffee and eucalyptus. *B. bassiana* was shown to control coffee berry borers, whiteflies, eucalyptus snout beetles, and spider mites. The market for *Beauveria* is projected to reach approximately 40 million hectares because of the presence of whiteflies on soybean crops in Brazil. The *Beauveria* strain GHA is extensively employed for the management of various pests such as aphids, leafhoppers, mealybugs, psyllids, plant bugs, thrips, whiteflies, and a range of Coleopteran and Orthopteran pests (Lacey et al., 2015).

Only a limited number of *Beauveria* products, such as BiolisaMadar® for beetles and Bb Moscas® for fruit flies, are designed for automated dispersal towards specified targets. The combined use of *Beauveria* and other botanical pesticides derived from natural sources has been demonstrated to be beneficial in many circumstances. Some companies, including LAM International, USA, are incorporating specific botanical chemicals, like azadirachtin, into their treatments based on *Beauveria*. This is to create synergy and ultimately enhance the efficacy of pest management in cropped field conditions.

Table 2 Commercial products that are licensed to contain *B. bassiana* propagules for biological control of different insect pests.

Trade Name/ Strain Name	*Target Pests*
ABG 6178	White grub
Abn Bb102 WP PHC® BEA TRON®	Coffee berry borer, and Mexican rice borer
BioPower®, Naturalis®, Biosoft®, Ostrinil®	Coleoptera, Hemiptera, Diptera, and Lepidoptera
AgroNova®	Coleoptera, and Lepidoptera
balEnce™	Flies and beetles in animals and birds
Ballveria	Whitefly biotype B
Bassianil	Coffee berry borer, leaf hoppers, Mexican cotton boll weevil, two-spotted spider mite, and whitefly biotype B
Bazam	Chrysomelidae, Curculionidae (Coleoptera), Aleyrodidae, Aphididae (Hemiptera), Noctuidae, Plutellidae (Lepidoptera), and Tetranychidae (Acari)
Bauveril	Coleoptera, and Lepidoptera
Bb-Protec	Herbivorous mites, wire worms, and whiteflies.
Beauveria JCO	Banana root borer, corn leafhopper, red spider mite, and whitefly biotype B
Beaublast	Aphid, psylla, and whitefly biotype B
Beauvedieca	Coleoptera (Curculionidae)
Beaugenic	Aphid, thrips, and whitefly biotype B
Bea-Sin	Whitefly biotype B
Becan	Coleoptera, Diptera, Hemiptera, Lepidoptera, and Orthoptera
Bouveriz WP Biocontrol	Banana root borer, corn leafhopper, red spider mite, and whitefly biotype B
Bovebio®	Banana root borer, corn leafhopper, red spider mite, and whitefly biotype B
Boverin®	Colorado potato beetle, Hemiptera, Thysanoptera, and Acari
Bovemax EC	Asian citrus psyllid, coffee berry borer, green mate borer
Boveril WP PL63	Coffee berry borer, citrus root weevil, and two-spotted spider mite
Biolrol	European com borer
Bio-Fung	Orthoptera
Bio-Power	Mite and Coffee green bug
BioGuard Rich	Coleoptera, Hemiptera, Lepidoptera, and Thysanoptera
Biostop F	Beet flea leaf aphid, cabbage aphids, codling moth, cabbage moth, potato beetle, gipsy moth, rosa moth, and tobacco thrips
Boverol®	Colorado beetle and other Chrysomelidae
Boverosil	Curculionidae and stored pests

Contd.

Table 2 *Contd.*

Trade Name/ Strain Name	*Target Pests*
BotaniGard ES; Botani-Gard 22WP	Aphids, European corn borer, lepidoptera, leafhoppers, leaf-feeding beetles, grasshoppers, locusts, mole cricket, mealybugs, mormon cricket, plant hoppers, psyllids, plant bugs, stem-boring lepidoptera, scarab beetles, thrips, weevils, and whiteflies Ticks include the brown dog tick, the castor bean tick, the lone star tick, the ornate cow tick, and the winter tick.
BroadBrand	Diamondback moth, false codling moth, California red scale, two-spotted spider mite, potato tuber moth, stinkbug, thrips, and whiteflies
Brocaril 50 WP	Coleoptera (Curculionidae)
Conidia	Coleoptera (Curculionidae)
Ostrinil®; CornGuard®	European corn borer
IPL Daman WP	Caterpillars, Colorado potato beetles, locusts, root borers, and sucking pests
Entostat Bb38	Grain storage insects
Granada	Banana root borer, corn leafhopper, red spider mite, and whitefly biotype B
FBB	Pine caterpillar
MicosPlag®	Coffee berry borer and plant parasitic nematodes
Mirabiol®	Coleoptera (Curculionidae)
Mycotrol ES; Mycotrol-O	Aphids, European corn borer, lepidoptera, leafhoppers, leaf-feeding beetles, grasshoppers, locusts, mole cricket, mealybugs, mormon cricket, plant hoppers, psyllids, plant bugs, stem-boring lepidoptera, scarab beetles, thrips, weevils, and whiteflies
Naturalis®	Ants, Colorado potato beetle, European chafer, European crane fly larvae, Northern masked chafer, Tetranychid mite, aphids, budworm, bollworm, chinch bug, citrus blackfly, corn borer, armyworm, mole cricket, cutworm, Elateridae, fleahopper, grasshopper, green June beetle, Japanese beetle, leaf-feeding caterpillar, fungus gnat, leafhopper, looper, Lygus bug, millipede, sod webworm, shore fly, mealybug, mite, psyllid, pear psylla, root weevil, sowbug, spittle bug, tarnished plant bug, thrips, weevil, tomato fruit worm, and whitefly.
Naturalis-L; Bb-Protec	Cotton pests, including bollworms
Nagestra	Caterpillars, locusts, potato beetles, root borers, and sucking pests
Ostrinil®	European corn borer, and Asiatic corn borer
Racer BB	*Helicoverpa*, *Spodoptera*, fruit borers, rice leaf folders, loopers, coffee berry borers, cotton boll worms, root grubs, leaf-eating caterpillars, mealy bugs, surface living larvae, and nymphs.
Proecol	Coleoptera
Seremoni WP	Greenhouse whitefly, and two spotted spider mites
Trichobass-L; Trichobass-P	Curculionidae, Scarabaeidae, Lepidoptera, Aleyrodidae, Thysanoptera, and Tetranychidae

Conclusion

B. bassiana is a highly adaptable entomopathogenic fungus known for its effectiveness in biological pest management and its production of insecticidal substances, making it an excellent alternative to chemical pesticides. Recent discoveries into the life cycles of entomopathogenic Ascomycetes, specifically *Beauveria*, have revealed that they can also function as endophytes and rhizobiomes. This entirely novel comprehension opens up opportunities for further study and utilization of these organisms as versatile bioagents capable of inducing systemic resistance, stimulating plant growth, and acting as biopesticides. Future research needs to emphasize biotechnological applications, molecular mechanisms of pathogenicity and synergism, environmental effects, safety considerations, and the development of marketable bioconsortium formulations.

References

Agrawal, Y. (2014). Multi-gene genealogies reveal cryptic species *Beauveria rudraprayagi* sp. nov. from India. Mycosphere, 5(6): 719–736. https://doi.org/10.5943/mycosphere/5/6/3

Almeida, M.G.B., Varize, C.S., Santos, T.S., Rezende, C.S., López, J.A. et al. (2023). Technology of a novel conidia-tablet formulation and packaging type to increase *Beauveria bassiana* (Hypocreales: Ophiocordycipitaceae) shelf life at room temperature. Biotechnol. Agron. Soc. Environ., 27(2): 109–118. https://doi.org/10.25518/1780-4507.20372

Awan, U.A., Xia, S., Meng, L., Raza, M.F., Zhang, Z. et al. (2021). Isolation, characterization, culturing, and formulation of a new *Beauveria bassiana* fungus against *Diaphorina citri*. Biol. Control, 158: 104586. https://doi.org/10.1016/j.biocontrol.2021.104586

Azadi, N., Shirzad, A. and Mohammadi, H. (2016). A study of some biocontrol mechanisms of *Beauveria bassiana* against *Rhizoctonia* disease on tomato. Acta Biol. Szeged., 60(2): 119–127.

Bamisile, B.S., Siddiqui, J.A., Akutse, K.S., Aguila, L.C.R. and Xu, Y. (2021). General limitations to endophytic entomopathogenic fungi use as plant growth promoters, pests and pathogens biocontrol agents. Plants, 10(10): 1–23. https://doi.org/10.3390/plants10102119

Barbarin, A.M., Bellicanta, G.S., Osborne, J.A., Schal, C. and Jenkins, N.E. (2017). Susceptibility of insecticide-resistant bed bugs (*Cimex lectularius*) to infection by fungal biopesticide. Pest Man. Sci., 73(8): 1568–1573. https://doi.org/10.1002/ps.4576

Bustamante, D.E., Oliva, M., Leiva, S., Mendoza, J.E., Bobadilla, L., et al. (2019). Phylogeny and species delimitations in the entomopathogenic genus *Beauveria* (Hypocreales, Ascomycota), including the description of *B. peruviensis* sp. Nov. MycoKeys, 58: 47–68. https://doi.org/10.3897/mycokeys.58.35764

Celar, F.A. and Kos, K. (2016). Effects of selected herbicides and fungicides on growth, sporulation and conidial germination of entomopathogenic fungus *Beauveria bassiana*. Pest Manag. Sci., 72(11): 2110–2117. https://doi.org/10.1002/ps.4240

Chen, W.H., Han, Y.F., Liang, Z.Q. and Jin, D.C. (2017). A new araneogenous fungus in the genus *Beauveria* from guizhou, China. Phytotaxa, 302(1): 57–64. https://doi.org/10.11646/phytotaxa.302.1.5

de Costa Eula Maria, M.B., Pimenta, C.F., Luz, C., de Oliveira, V., Oliveira, M., et al. (2011). *Beauveria bassiana*: Quercetinase production and genetic diversity. Braz. J. Microbiol., 42(1): 12–21. https://doi.org/10.1590/s1517-83822011000100002

de Oliveira, J.L., de Oliveira, T.N., Savassa, S.M., Della Vechia, J.F., de Matos, S.T.S., et al. (2023). Encapsulation of *Beauveria bassiana* in polysaccharide-based microparticles: a promising carrier system for biological control applications. ACS Agric. Sci. Technol., 3(9): 785–794. https://doi.org/10.1021/acsagscitech.3c00135

do Nascimento Silva, J., Mascarin, G.M., dos Santos Gomes, I.C., Tinôco, R.S., Quintela, E.D., et al. (2018). New cost-effective bioconversion process of palm kernel cake into bioinsecticides based on *Beauveria bassiana* and *Isaria javanica*. Appl. Microbiol. Biotechnol., 102(6): 2595–2606. https://doi.org/10.1007/s00253-018-8805-z

do Nascimento Silva, J., Mascarin, G.M., Lopes, R.B. and Freire, D.M.G. (2023). Production of dried *Beauveria bassiana* conidia in packed-column bioreactor using agro-industrial palm oil residues. Biochem. Eng. J., 198: 1–8. https://doi.org/10.1016/j.bej.2023.109022

Elawati, N.E., Pujiyanto, S. and Kusdiyantini, E. (2018). Production of extracellular chitinase *Beauveria bassiana* under submerged fermentation conditions. J. Phys.: Conf. Ser., 1025(1): 1–6. https://doi.org/10.1088/1742-6596/1025/1/012074

Feng, M.G., Khachatourians, G.G. and Poprawski, T.J. (1994). Production, formulation and application of the entomopathogenic fungus *Beauveria bassiana* for insect control: current status. Biocontrol Sci. Technol., 4(1): 3–34. https://doi.org/10.1080/09583159409355309

Fergani, Y.A. and Yehia, R.S. (2020). Isolation, molecular characterization of indigenous *Beauveria bassiana* isolate, using ITS-5.8 s rDNA region and its efficacy against the greatest wax moth, *Galleria mellonella* L.(Lepidoptera: Pyralidae) as a model insect. Egyptian J. Biol. Pest Con., 30: 1–7. https://doi.org/10.1186/s41938-020-00298-x

Gençer, D., Ulaşli, B., Can, F. and Demır, İ. (2023). Isolation and identification of a fungal pathogen, *Beauveria bassiana* (Bals.-Criv.) Vuill. (Ascomycota: Hypocreales) from the Hatay yellow strain of silkworm, *Bombyx mori* L., 1758 (Lepidoptera: Bombycidae) in Türkiye. Turkiye Entomoloji Dergisi, 47(2): 189–197. https://doi.org/10.16970/entoted.1218790

Gürlek, S., Sevim, A., Sezgin, F.M. and Sevim, E. (2018). Isolation and characterization of beauveria and metarhizium spp. From walnut fields and their pathogenicity against the codling moth, *Cydia pomonella* (L.) (Lepidoptera: Tortricidae). Egyptian J. Biol. Pest Con., 28(1): 1–6. https://doi.org/10.1186/s41938-018-0055-y

Hamzah, A.M., Mohsin, A., Naeem, M. and Khan, M.A. (2021). Efficacy of *Beauveria bassiana* and *Metarhizium anisopliae* (Ascomycota: Hypocreales) against *Bactrocera cucurbitae* (Coquillett) (Diptera: Tephritidae) under controlled and open-field conditions on bitter gourd. Egyptian J. Biol. Pest Con., 31(1): 0–7. https://doi.org/10.1186/s41938-021-00490-7

Huang, W., Tang, R., Li, S., Zhang, Y., Chen, R., et al. (2021). Involvement of epidermis cell proliferation in defense against *Beauveria bassiana* infection. Front. Immunol., 12: 1–12. https://doi.org/10.3389/fimmu.2021.741797

Huarte-Bonnet, C., Paixão, F.R.S., Ponce, J.C., Santana, M., Prieto, E.D. et al. (2018). Alkane-grown *Beauveria bassiana* produce mycelial pellets displaying peroxisome proliferation, oxidative stress, and cell surface alterations. Fungal Biol., 122(6): 457–464. https://doi.org/10.1016/j.funbio.2017.09.003

Imoulan, A., Hussain, M., Kirk, P.M., El Meziane, A. and Yao, Y.J. (2017). Entomopathogenic fungus *Beauveria*: Host specificity, ecology and significance of morpho-molecular characterization in accurate taxonomic classification. J. Asia-Pac. Entomol., 20(4): 1204–1212. https://doi.org/10.1016/j.aspen.2017.08.015

Jaronski, S.T. (2014). Mass production of entomopathogenic fungi: state of the art. In *Mass production of beneficial organisms* (pp. 357–413). Elsevier. https://doi.org/10.1016/B978-0-12-391453-8.00011-X

Kassa, A., Brownbridge, M., Parker, B.L., Skinner, M., Gouli, V., et al. (2008). Whey for mass production of *Beauveria bassiana* and *Metarhizium anisopliae*. Mycol. Res., 112(5): 583–591. https://doi.org/10.1016/j.mycres.2007.12.004

Kepler, R.M., Luangsa-Ard, J.J., Hywel-Jones, N.L., Quandt, C.A., Sung, G.H., et al. (2017). A phylogenetically-based nomenclature for Cordycipitaceae (Hypocreales). IMA Fungus, 8(2): 335–353. https://doi.org/10.5598/imafungus.2017.08.02.08

Kobmoo, N., Arnamnart, N., Pootakham, W., Sonthirod, C., Khonsanit, A., et al. (2021). The integrative taxonomy of *Beauveria asiatica* and *B. bassiana* species complexes with whole-genome sequencing, morphometric and chemical analyses. Pers.: Mol. Phylogeny Evol., 47: 136–150. https://doi.org/10.3767/persoonia.2021.47.04

Kozłowska, E., Urbaniak, M., Hoc, N., Grzeszczuk, J., Dymarska, M., et al. (2018). Cascade biotransformation of dehydroepiandrosterone (DHEA) by *Beauveria* species. Sci. Rep., 8(1): 1–11. https://doi.org/10.1038/s41598-018-31665-2

Li, J.X., Fernandez, K.X., Ritland, C., Jancsik, S., Engelhardt, D.B., et al. (2023). Genomic virulence features of *Beauveria bassiana* as a biocontrol agent for the mountain pine beetle population. BMC Genomics, 24(1): 1–17. https://doi.org/10.1186/s12864-023-09473-4

Lohse, R., Jakobs-Schönwandt, D. and Patel, A.V. (2014). Screening of liquid media and fermentation of an endophytic *Beauveria bassiana* strain in a bioreactor. AMB Exp., 4(1): 1–11. https://doi.org/10.1186/s13568-014-0047-6

Mascarin, G.M., Iwanicki, N.S., Ramirez, J.L., Delalibera, Í. and Dunlap, C.A. (2021). Transcriptional Responses of *Beauveria bassiana* Blastospores Cultured Under Varying Glucose Concentrations. Front. Cell. Infect. Microbiol., 11: 1–18. https://doi.org/10.3389/fcimb.2021.644372

Mascarin, G.M., Jackson, M.A., Kobori, N.N., Behle, R.W. and Delalibera Júnior, Í. (2015). Liquid culture fermentation for rapid production of desiccation tolerant blastospores of *Beauveria bassiana* and *Isaria fumosorosea* strains. J. Invertebr. Pathol., 127: 11–20. https://doi.org/10.1016/j.jip.2014.12.001

Mascarin, G.M. and Jaronski, S.T. (2016). The production and uses of *Beauveria bassiana* as a microbial insecticide. World J. Microbiol. Biotechnol., 32(11): 1–26. https://doi.org/10.1007/s11274-016-2131-3

Masoudi, A., Wang, M., Zhang, X., Wang, C., Qiu, Z., et al. (2020). Meta-Analysis and evaluation by insect-mediated baiting reveal different patterns of hypocrealean entomopathogenic fungi in the soils from two regions of China. Front. Microbiol., 11: 1–23. https://doi.org/10.3389/fmicb.2020.01133

Mattedi, A., Sabbi, E., Farda, B., Djebaili, R., Mitra, D., et al. (2023). Solid-State Fermentation: Applications and Future Perspectives for Biostimulant and Biopesticides Production. Microorganisms, 11(6). https://doi.org/10.3390/microorganisms11061408

Meyling, N.V. and Eilenberg, J. (2006). Isolation and characterisation of *Beauveria bassiana* isolates from phylloplanes of hedgerow vegetation. Mycol. Res., 110(2): 188–195. https://doi.org/10.1016/j.mycres.2005.09.008

Mishra, S. and Malik, A. (2013). Nutritional optimization of a native *Beauveria bassiana* isolate (HQ917687) pathogenic to housefly, *Musca domestica* L. J. Parasitic Dis., 37(2): 199–207. https://doi.org/10.1007/s12639-012-0165-5

Oliveira, B.B., Veigas, B. and Baptista, P.V. (2021). Isothermal amplification of nucleic acids: the race for the next "gold standard." Front. Sens., 2: 1–22. https://doi.org/10.3389/fsens.2021.752600

Oliveira, D.G.P. de, Lopes, R.B., Rezende, J.M. and Delalibera, I. (2018). Increased tolerance of *Beauveria bassiana* and *Metarhizium anisopliae* conidia to high temperature provided by oil-based formulations. J. Invertebr. Pathol., 151: 151–157. https://doi.org/10.1016/j.jip.2017.11.012

Pedrini, N. (2022). The Entomopathogenic fungus *Beauveria bassiana* shows its toxic side within insects: expression of genes encoding secondary metabolites during pathogenesis. J. Fungi, 8(5). https://doi.org/10.3390/jof8050488

Posada-Flórez, F.J. (2008). Production of *Beauveria bassiana* fungal spores on rice to control the coffee berry borer, *Hypothenemus hampei*, in Colombia. J. Insect Sci., 8(41): 1–13. https://doi.org/10.1673/031.008.4101

Posadas, J.B., Comerio, R.M., Mini, J.I., Nussenbaum, A.L. and Lecuona, R.E. (2012). A novel dodine-free selective medium based on the use of cetyl trimethyl ammonium bromide (CTAB) to isolate *Beauveria bassiana, Metarhizium anisopliae* sensu lato and *Paecilomyces lilacinus* from soil. Mycologia, 104(4): 974–980. https://doi.org/10.3852/11-234

Proietti, S., Falconieri, G.S., Bertini, L., Pascale, A., Bizzarri, E., et al. (2023). *Beauveria bassiana* rewires molecular mechanisms related to growth and defense in tomato. J. Exp. Bot., 74(14): 4225–4243. https://doi.org/10.1093/jxb/erad148

Qiu, L., Nie, S.X., Hu, S.J., Wang, S.J., Wang, J.J. and Guo, K. (2021). Screening of *Beauveria bassiana* with high biocontrol potential based on ARTP mutagenesis and high-throughput FACS. Pe Pestic. Biochem. Physiol., 171: 104732. https://doi.org/10.1016/j.pestbp.2020.104732

Rehner, S.A., Minnis, A.M., Sung, G.H., Luangsa-ard, J.J., Devotto, L. et al. (2011). Phylogeny and systematics of the anamorphic, entomopathogenic genus *Beauveria*. Mycologia, 103(5): 1055–1073. https://doi.org/10.3852/10-302

Robène-Soustrade, I., Jouen, E., Pastou, D., Payet-Hoarau, M., Goble, T., et al. (2015). Description and phylogenetic placement of *Beauveria hoplocheli* sp. nov. used in the biological control of the sugarcane white grub, *Hoplochelus marginalis*, on Reunion Island. Mycologia, 107(6): 1221–1232. https://doi.org/10.3852/14-344

Rodriguez-Saona, C., Holdcraft, R. and Kyryczenko-Roth, V. (2022). Control of blunt-nosed leafhopper with biological insecticides in cranberries. Arthropod Manag. Tests, 47(1): 1–2. https://doi.org/10.1093/amt/tsac012

Roswanjaya, Y.P., Saryanah, N.A. and Devy, L. (2022). Conidia production of *Beauveria bassiana* in solid substrate fermentation using a biphasic system. KnE Life Sciences, 2022: 648–663. https://doi.org/10.18502/kls.v7i3.11169

Sala, A., Barrena, R., Meyling, N.V. and Artola, A. (2023). Conidia production of the entomopathogenic fungus *Beauveria bassiana* using packed-bed bioreactor: Effect of substrate biodegradability on conidia virulence. J. Environ. Manage., https://doi.org/10.1016/j.jenvman.2023.118059

Sánchez-Gómez, T., Harte, S.J., Zamora, P., Bareyre, M., Díez, J.J., et al. (2023). Nematicidal effect of *Beauveria* species and the mycotoxin beauvericin against pinewood nematode *Bursaphelenchus xylophilus*. Front. For. Glob. Change., 6: 1–10. https://doi.org/10.3389/ffgc.2023.1229456

Sanjuan, T., Tabima, J., Restrepo, S., Læssøe, T., Spatafora, J.W. et al. (2014). Entomopathogens of amazonian stick insects and locusts are members of the *Beauveria* species complex (*Cordyceps* sensu stricto). Mycologia, 106(2): 260–275. https://doi.org/10.3852/13-020

Sayed, S.M., Ali, E.F., El-Arnaouty, S.A., Mahmoud, S.F. and Amer, S.A. (2018). Isolation, identification, and molecular diversity of indigenous isolates of *Beauveria bassiana* from Taif region, Saudi Arabia. Egypt. J. Biol. Pest Control, 28(1): 1–6. https://doi.org/10.1186/s41938-018-0054-z

Shikano, I., Bellicanta, G.S., Principato, S. and Jenkins, N.E. (2021). Effects of chemical insecticide residues and household surface type on a *Beauveria bassiana*-based biopesticide (Aprehend®) for bed bug management. Insects, 12(3): 1–17. https://doi.org/10.3390/insects12030214

Solano-González, S., Castro-Vásquez, R. and Molina-Bravo, R. (2023). Genomic Characterization and Functional Description of *Beauveria bassiana* Isolates from Latin America. J. Fungi, 9(7). https://doi.org/10.3390/jof9070711

Swathy, K., Parmar, M.K. and Vivekanandhan, P. (2024). Biocontrol efficacy of entomopathogenic fungi *Beauveria bassiana* conidia against agricultural insect pests. Environ. Qual. Manag., 1–11. https://doi.org/10.1002/tqem.22174

Velozo, S.G.M., Velozo, M.R., Domingues, M.M., Becchi, L.K., de Carvalho, V.R., et al. (2023). From the dual cyclone harvest performance of single conidium powder to the effect of *Metarhizium anisopliae* on the management of *Thaumastocoris peregrinus* (Hemiptera: Thaumastocoridae). PLoS One, 18(3): 1–13. https://doi.org/10.1371/journal.pone.0283543

Wang, H., Peng, H., Li, W., Cheng, P. and Gong, M. (2021). The Toxins of *Beauveria bassiana* and the Strategies to Improve Their Virulence to Insects. Front. Microbiol., 12: 1–11. https://doi.org/10.3389/fmicb.2021.705343

Wang, Y., Fan, Q., Wang, D., Zou, W.Q., Tang, D.X., et al. (2022). Species Diversity and Virulence Potential of the *Beauveria bassiana* Complex and *Beauveria scarabaeidicola* Complex. Front. Microbiol., 13: 1–15. https://doi.org/10.3389/fmicb.2022.841604

Yong Lee, J., Mi Woo, R. and Dong Woo, S. (2023). Formulation of the entomopathogenic fungus *Beauveria bassiana* JN5R1W1 for the control of mosquito adults and evaluation of its novel applicability. J. Asia-Pacific Ento., 26(2): 102056. https://doi.org/10.1016/j.aspen.2023.102056

Zhang, S., He, L., Chen, X. and Huang, B. (2012). *Beauveria lii*. Mycotaxon, 121(2011): 199–206.

15 Solid-State Culture Spore Production of Entomopathogenic Fungi

Fernando Méndez-González,[1] José Juan Buenrostro-Figueroa[1] and Ernesto Favela-Torres[2*]

1. Introduction

Entomopathogenic fungi (EF) are important biological control agents, with their effectiveness driving the development of diverse global commercial products (Castillo-Minjarez, 2022; Méndez-González et al., 2022a; Deans and Krischik, 2023). Most formulations of bioinsecticides derived from EF incorporate spores as the active ingredient. EF spores are produced by liquid (blastospores) and solid (conidia) culture processes (Méndez-González et al., 2018b). Conidia generated via solid-state culture (SSC) demonstrate high resistance to environmental stress, making them highly favored in the industry (Méndez-González et al., 2022a). Various processes using different substrates, bioreactors, and operating conditions have been developed to produce EF conidia. These processes encompass two levels of technology: (i) simple low-tech bioreactors (traditional processes); and (ii) high-tech bioreactors, equipped with instruments for monitoring and controlling the production process. Among traditional bioreactors are small containers (fabricated in metal, wood, or plastic) (Goettel, 1984; Roberts and St. Leger, 2004; Grzywacz et al., 2014) and plastic bags (PBB) (Jaronski, 2023). These processes are popular as they can be implemented irrespective of technological capability and entail low equipment costs (Méndez-González et al., 2022a). On the other hand, the high-tech processes encompass trays (TB), packed bed columns (PBCB), and agitated bioreactors (AB). These bioreactors facilitate a controlled process with high production and quality yields (Méndez-González et al., 2020), but entail higher costs for acquisition, installation, maintenance, and operation. That said, TB, PBCB, and AB can be scaled up and automated, providing a considerable industrial advantage.

The diverse processes designed to produce conidia via SSC have varying levels of technology, making them suitable for implementation in different socioeconomic regions worldwide (Grzywacz et al., 2014; Jaronski, 2023). Bioreactor suitability and required operating conditions depend on the microorganism being produced. This chapter begins by addressing the fundamental requirements to produce conidia via solid-state culture. Later, we analyze the characteristics and operating conditions of

[1] Laboratorio de Biotecnología y Bioingeniería, Centro de Investigación en Alimentación y Desarrollo, Delicias, Chihuahua, Mexico, 33088.

[2] Departamento de Biotecnología, Universidad Autónoma Metropolitana, Iztapalapa, Mexico City, Mexico, 09340.

* Corresponding author: favela@xanum.uam.mx

traditional and high-tech bioreactors. This will provide an overview to help select the optimal conidia production system.

2. Production of EF Conidia by Solid-State Culture

Solid-state culture (SSC) is an aerobic or anaerobic process in which the microorganism grows on a solid matrix (substrate or support) (Viniegra-González, 1997). The solid matrix contains the carbon source, nutrients, and water required for the microorganism's culture (Manan and Webb, 2017). The most commonly used substrates are rice, wheat, sorghum, and millet grains. However, the use of agro-industrial by-products (bagasse, bran, husk, fibers, and fruit peels) and inert supports enriched with a culture medium is possible. The water content in the solid matrix is essential for microbial growth in SSC, since it serves as the vehicle for nutrient transport (Oriol et al., 1988; Viniegra-González, 1997). Substrates with a water activity of 0.98 favor the growth of important EF from the genera *Metarhizium*, *Beauveria*, and *Paecilomyces* (Hallsworth and Magan, 1999). Besides containing the compounds necessary for microbial growth, the bed must have interparticle spaces that allow growth of the mycelium, production of conidia, removal of metabolic heat, and gas exchange (Méndez-González et al., 2022).

The temperature and composition of the gaseous atmosphere affect the production and quality of conidia. Most EF distributed worldwide grow in a temperature range of 20 to 32 °C (Goettel, 1984; Ouedraogo et al., 1997; Ekesi et al., 1999; Hallsworth and Magan, 1999). Generally, temperatures outside this range significantly affect microbial growth (Ouedraogo et al., 1997; Ekesi et al., 1999), production (Méndez-González et al., 2020), and quality of conidia (Ekesi et al., 1999; Hallsworth and Magan, 1999). Furthermore, the content of O_2, CO_2, and volatile compounds in the gaseous atmosphere plays an essential role in the production and quality of conidia. Aerobic microorganisms (such as EF) require O_2 to obtain energy from the substrate; therefore, its low availability limits microbial growth (Finger et al., 1976). On the other hand, atmospheres with high levels of CO_2 (~5% v/v) negatively affect conidia production (observed in *Beauveria bassiana*) (Garza-López et al., 2011), while volatile compounds such as 1-Octen-32-ol diminish conidia quality (Muñiz-Paredes et al., 2017). It follows that, the design of SSC systems must consider mechanisms for temperature maintenance and gas exchange.

2.1 *Solid-State Culture Bioreactors*

The bioreactor is the central part of the SSC system. Bioreactors provide the environment for microbial growth (Pandey, 1991). SSC bioreactors present two configurations: static bed and mixed bed (Méndez-González et al., 2018b). These bioreactors can be classified by their capacity: laboratory (a few milliliters to < 2 L capacity), bench (2 to <5 L capacity), pilot (5 to <50 L capacity), and industrial scale (>50 L capacity) (Méndez-González et al., 2022a). Due to the degrees of difference in monitoring and control instrumentation, they may also be classified as traditional or high-tech. SSC bioreactors have different designs and operating conditions that can promote growth and conidia production from specific microorganisms (Méndez-

González et al., 2022a). The following sections describe the main characteristics of SSC bioreactors used for entomopathogenic fungi conidia production.

2.2 Traditional Bioreactors

Traditional processes were developed in the middle of the last century, primarily involving two types of bioreactors: small rigid containers (crafted in wood, metal, glass, or plastic) and plastic bags (Table 1). Low-capacity covered containers (CC), such as covered pal bioreactors (Goettel, 1984) and wooden baskets (Roberts and St. Leger, 2004), are used to produce conidia from *Tolypocladium cylindrosporum*, *Metarhizium anisopliae*, *Beauveria bassiana*, *Lecanicillium lecanii*, and *Culicinomyces clavispurum*. In this type of bioreactor, a thin layer of inoculated substrate (4-5 cm in height) is spread inside the container and covered with a porous or non-porous material (Lonsane et al., 1985). When the covering material is porous, gas exchange occurs through diffusion across the membrane. In contrast, with non-porous materials, it is necessary to not cover the bed surface completely to allow for aeration (Goettel, 1984).

Plastic bag bioreactors (PBB), employed for EF conidia production, are generally made of high-density polypropylene (Méndez-González et al., 2020) with a capacity for between 200-1000 g of substrate (Méndez-González et al., 2018b; Jaronski, 2023). The manufacturing material contains micropores that allow gas transfer by diffusion (Durand, 2003). However, the low velocity of diffusive transport through the membrane can lead to CO_2 accumulation (Méndez-González et al., 2020). In some cases, to improve gas exchange, bags are left partially open or plugs with greater porosity than the bag material are used (usually made of cotton). Some PBBs made of non-porous plastics (such as polyethylene) require the use of these porous plugs.

Due to the limited capacity of CC and PBB ($\leq$1 kg of wet mass (kg_{wm})), processing a large number of units per batch is necessary, requiring intense labor and large incubation areas (Lonsane et al., 1985; Méndez-González et al., 2022a). Additionally, due to the characteristics of the bioreactor, the use of monitoring and control instruments is challenging (Méndez-González et al., 2018b). Lack of monitoring and control in the process can lead to significant variations in the production and quality of conidia (Luz and Fargues, 1998; Jenkins and Grzywacz, 2000), in turn affecting production costs and marketing prices (Cherry et al., 1999). However, the culture processes in CC and PBB allow the production of propagules of different EFs (Table 1) with low capital and technology requirements (Swanson, 1997). Therefore, these bioreactors are ideal for establishing small production units with a capacity of 300 to 350 kg of conidia per year (sufficient for treating 7000 ha) (Grzywacz et al., 2014). That said, to meet the growing demand for fungal bioinsecticides, higher levels of industrialization are necessary.

2.3 High-Tech Bioreactors

High-tech bioreactors can be equipped with instruments for monitoring and controlling culture conditions, addressing critical variables including moisture content, temperature, and gaseous concentrations of O_2 and CO_2. Regulation of these properties ensures a process that is reproducible, has high production yields, and

Table 1 Conidia production of biological control agents in traditional solid-state culture bioreactors (subscripts 'dm' and 'wm' indicate 'dry mass' and 'wet mass', respectively).

Fungus	*Bioreactor and culture conditions*	*Conidia production (conidia × $10^9/g_{dm}$)*	*Reference*
Akanthomyces lecanii	Polypropylene bag Packed mass: 100 g_{wm} Substrate: Wheat bran and grains (70:30) Incubation time: 15 d (in dark) Incubation temperature: 25 ± 0.5°C	1.45	Machado et al., 2010
	Polypropylene bag (31 × 21 cm) Packed mass: 300 g_{wm} Substrate: Pearl millet, sucrose, and urea Moisture content: 50% Incubation time: 9 d Incubation temperature: 25 ± 2°C Incubation RH: 45 ± 2%	1.0	Chaparro et al., 2021
Beauveria bassiana	Plastic bag Packed mass: 200 g_{wm} Substrate: Cooked rice Moisture content: 50–60% Incubation time: 15 d	0.27	Posada-Flórez, 2008
	Polypropylene bag (25 × 35 cm) with cotton plugs Packed mass: 92 g_{wm} Substrate: Rice grains, wheat bran, and yeast extract Moisture content: 35% Incubation time: 15 d Incubation temperature: 28°C	29	Dhar and Kaur, 2011
	Polypropylene bag (25 × 35 cm) with cotton plugs Packed mass: 92 g_{wm} Substrate: Crushed sorghum and yeast extract Moisture content: 20% Incubation time: 15 d Incubation temperature: 28°C	9.63	Dhar and Kaur, 2011
	Polyethyne bag Packed mass: 500 g_{wm} Substrate: Rice grains and nutrient solution Incubation time: 7 d Incubation temperature: 25°C	3.5	Adane et al., 1996
	Covered pal fermenter (36 × 25 × 4) Packed mass: 770 g_{wm} Substrate: Wheat bran Moisture content: 90% Incubation time: 14 d Incubation temperature: 20°C Photoperiod: 17 h	15 ± 3.8	Goettel, 1984

Contd.

Table 1 *Contd.*

Fungus	*Bioreactor and culture conditions*	*Conidia production (conidia × $10^9/g_{dm}$)*	*Reference*
Coniothyrium minitans	Spawn bags (22.5 × 56 × 15 cm) Substrate: Oat grains Moisture content: 400 mL tap water per L of oat grains Incubation time: 28 d Incubation temperature: 18 – 30°C	0.75	McQuilken and Whipps, 1995
	Spawn bags (22.5 × 56 × 15 cm) Substrate: Maizemeal-perlite Incubation time: 28 d Incubation temperature: 18 – 30°C	0.93	McQuilken and Whipps, 1995
	Spawn bags (22.5 × 56 × 15 cm) Substrate: Rice grains Moisture content: 400 mL tap water per 600 mL of rice grain Incubation time: 28 d Incubation temperature: 18 – 30°C	0.54	McQuilken and Whipps, 1995
Cordyceps fumosoreosea	Polyethyne bag Packed mass: 500 g_{wm} Substrate: Parboiled rice Incubation time: 10 d Incubation temperature: 26 ± 1°C Photoperiod: 12 h	3.51	Rojas et al., 2023
Culicinomyces clavisporus	Covered pal fermenter (36 × 25 × 4) Packed mass: 770 g_{wm} Substrate: Wheat bran Moisture content: 90% Incubation time: 21 d Incubation temperature: 25°C Photoperiod: 17 h	0.76 ± 0.14	Goettel, 1984
Lecanicillium lecanii	Covered pal fermenter (36 × 25 × 4) Packed mass: 770 g_{wm} Substrate: Wheat bran Moisture content: 90% Incubation time: 21 d Incubation temperature: 25°C Photoperiod: 17 h	28.0 ± 3.0	Goettel, 1984
Metarhizium anisopliae	Polypropylene bag (50 × 39 cm) Packed mass: 1 kg_{wm} Substrate: Parboiled rice Moisture content: 40% Incubation time: 8 – 13 d Incubation temperature: 25 ± 2°C Incubation RH: 67 – 81%	1.51 – 1.61	Cherry et al., 1999

Contd.

Table 1 *Contd.*

Fungus	*Bioreactor and culture conditions*	*Conidia production (conidia × $10^9/g_{dm}$)*	*Reference*
	Polypropylene bag (18 × 25 cm) Packed mass: 0.5 kg_{wm} Substrate: Rice grains Moisture content: 40% Incubation time: 15 d Incubation temperature: 28°C	0.8 – 1.0	Alcantara-Vargas et al., 2020
	Polypropylene bag (32.5 × 43.5 cm) Packed mass: 0.5 kg_{wm} Substrate: Parboiled rice Moisture content: 40% Incubation time: 20 d Incubation temperature: 25 ± 1°C	1.29 – 2.56	Barra-Bucarei et al., 2016
	Polypropylene bag (20 × 30 cm) Packed mass: 250 g_{wm} Substrate: Rice grains Moisture content: 30% Incubation time: 9 d Incubation temperature: 27°C	0.88	Méndez-González et al., 2018a
	Covered pal fermenter (36 × 25 × 4) Packed mass: 770 g_{wm} Substrate: Wheat bran Moisture content: 90% Incubation time: 14 d Incubation temperature: 20°C Photoperiod: 17 h	6.1 ± 0.9	Goettel, 1984
Metarhizium robertsii	Polypropylene bag (25 × 35 × 0.6 cm) Packed mass: 100 g_{wm} Substrate: Parboiled rice Moisture content: 40% Incubation time: 20 d Incubation temperature: 27°C Photoperiod: 12 h	2.1	Ferreira et al., 2021
	Polypropylene bag (30 × 40 cm) Packed mass: 1 kg_{wm} Substrate: Rice grains Moisture content: 30% Incubation time: 10.29 d Incubation temperature: 30°C	0.9 – 1.3	Méndez-González et al., 2020
	Plastic bags with cotton plugs Packed mass: 1.3 kg_{wm} Substrate: Parboiled rice Moisture content: 23% Incubation time: 28 d Incubation temperature: 25°C	0.03 – 0.04	Mathulwe et al., 2022

Contd.

Table 1 *Contd.*

Fungus	*Bioreactor and culture conditions*	*Conidia production (conidia × $10^9/g_{dm}$)*	*Reference*
Tolypocladium cylindrosporum	Covered pal fermenter (36 × 25 × 4) Packed mass: 770 g_{wm} Substrate: Wheat bran Moisture content: 90% Incubation time: 14 d Incubation temperature: 20°C Photoperiod: 17 h	34.0 ± 3.7	Goettel, 1984

delivers high-quality products. Bioreactors in this category include static bed systems such as TB and PBCB, as well as those with mechanical mixing like Rotating Drums (RRB) and Internal Stirred Bioreactors (ISB). The following sections describe the characteristics and operating conditions of these bioreactors.

2.4 Tray Bioreactors

Among the high-tech SSC bioreactors, tray bioreactors (TB) are widely used in the industry (Pandey, 1991). TBs feature a simple design consisting of a rectangular surface, a tray, where a thin layer of substrate is spread (Lonsane et al., 1985; Méndez-González et al., 2018b). Trays can be crafted from various materials, including wood, plastic, and metal (Pandey, 1991). Depending on heat removal and gas exchange requirements, trays may or may not have perforations (Méndez-González et al., 2022a). In the classic form (non-perforated trays), the transport of CO_2 and O_2 between the bed and headspace occurs by diffusion and natural convection (Raghava Rao et al., 1993; Rajagopalan and Modak, 1994; Méndez-González et al., 2020). Heat removal occurs by conduction and evaporation (Figueroa-Montero et al., 2011; Diniz-da Silva et al., 2022).

On the other hand, perforated trays allow forced aeration through the bed, improving O_2 supply and the removal of heat, CO_2, and volatile compounds (Lonsane et al., 1985). Heat and mass transfer in TBs are hindered by an increasing bed height (Ghildyal et al., 1994). Studies of *Beauveria bassiana* (Xie et al., 2012) and *Metarhizium anisopliae* (da Cunha et al., 2019) demonstrate that increasing bed height from 2 to 8 cm leads to heat accumulation, resulting in decreased conidia production (between 63 and 76%). Therefore, maintaining an appropriate bed height of 2 to 4 cm is crucial to prevent overheating and anaerobic zones (Ye et al., 2006; Xie et al., 2012; Méndez-González et al., 2020). Mathematical estimation of the critical bed height is possible; Da Cunha et al. (2019) used the model developed by Rajagopalan and Modak (1994) to estimate the kinetic temperature profile in the bed based on increased height. Although the temperature profile was slightly underestimated, the model proved helpful in determining the bed height in a TB. TBs are incubated in temperature-regulated rooms (similar to traditional bioreactors) (Lonsane et al., 1985) or closed chambers (Méndez-González et al., 2020), where trays are arranged one above the other with a few centimeters of separation. Closed chambers facilitate the use of instruments for monitoring and controlling process conditions (temperature, humidity, and composition of the gaseous atmosphere, among others) (Méndez-González et al., 2022a).

In these systems, forced aeration improves the removal of metabolic heat, aiding temperature control (Ye et al., 2006; Figueroa-Montero et al., 2011). However, high aeration rates cause dehydration in the bed, even with a saturated air supply (Mitchell et al., 1999). Spraying sterile water inside the incubation chamber can prevent dehydration (Ye et al., 2006; Dallastra et al., 2023). Another helpful strategy is increasing air turbulence in the headspace using fans inside the chamber, allowing temperature control with minimal water losses (Figueroa-Montero et al., 2011). Therefore, TBs enable the development of controlled processes to produce conidia, achieving yields similar to those obtained in traditional bioreactors (Table 2). It was previously mentioned that there are limitations to increasing the height of the beds in TBs. Thus, to increase the batch size, the number of trays is increased by decreasing the distance between them (Brand et al., 2010). TBs have some drawbacks, such as high labor requirements (Pandey, 1991) and a need for a large process area (Lonsane et al., 1985). However, Dallastra et al. (2023) presented an ingenious system to produce conidia in TB, where most of the process unit operations are carried out in the incubation chamber. The system's design facilitates automation, thereby reducing operator and process area requirements.

Table 2 Conidia production of biological control agents in high-tech solid state culture bioreactors (subscripts 'dm' and 'wm' indicate 'dry mass' and 'wet mass', respectively).

Fungus	*Bioreactor and culture conditions*	*Production of conidia (conidia × 10^9/gdm)*	*Reference*
Beauveria bassiana	Perforated tray bioreactor (60 × 40 cm) Bed height: 2 cm Packed mass: 1 kg_{wm} Substrate: Rice grains Moisture content: 40% Incubation time: 8 d Incubation temperature: 25°C Incubation RH: 80 – 100%	2.5 – 2.9	Xie et al., 2012
	Perforated tray bioreactor (58 × 58 × 4 cm) Bed height: 2.5 cm Packed mass: 3 kg_{wm} Substrate: Rice grains Moisture content: 38% Incubation time: 7 d Incubation temperature: 25°C Incubation RH: 100%	1.8 – 2.8	Ye et al., 2006
	Tray bioreactor (28 × 39 × 10 cm) Substrate: Rice grains Incubation time: 10 d Incubation temperature: 27 ± 1°C Photoperiod: 16 h	3.72	Alves and Pereira, 1989

Contd.

Table 2 *Contd.*

Fungus	*Bioreactor and culture conditions*	*Production of conidia (conidia × 10^9/gdm)*	*Reference*
	Glass packed bed column Bioreactor height: 20 cm Bed height: 12 cm Ø: 4 cm Substrate: Potato residues and sugarcane bagasse (60:40 w/w) Moisture content: 65% Incubation time: 10 d Incubation temperature: 26°C Aeration: 60 mL/min	9.8	Dalla Santa et al., 2004
	PVC packed bed column bioreactor Bioreactor height: 13 cm Ø: 4 cm Packed mass: 100 g_{wm} Working volume: 0.45 L Substrate: Potato Peel Moisture content: 89% Incubation time: 7 – 8 d Incubation temperature: 25°C Aeration rate: 0.2 L/ kg_{wm} min Bed porosity: 0.59	1.3	Sala et al., 2021
	PVC packed bed column bioreactor Bioreactor height: 13 cm Ø: 4 cm Packed mass: 100 g_{wm} Working volume: 0.45 L Substrate: Beer draff and rice husk Moisture content: 80% Incubation time: 8 d Incubation temperature: 25°C Aeration rate: 0.2 L/ kg_{wm} min Bed porosity: 66	2.5	Sala et al., 2023
	Packed bed column Ø: 40 cm Bed height: 40 cm Working volume: 50 L Substrate: Clay granules and nutrient solution Moisture content: 40% Incubation time: 3 d Incubation temperature: 25°C Aeration rate: 10 L/ kg_{dm} min	2	Desgranges et al., 1993

Contd.

Table 2 *Contd.*

Fungus	*Bioreactor and culture conditions*	*Production of conidia (conidia × 10^9/gdm)*	*Reference*
Coniothyrium minitans	Stirred packed bed Large: 2 m Bioreactor height: 0.8 m Bed height: 40 cm Working volume: 1600 L Substrate: Clay granules and nutrient solution Moisture content: 40% Incubation time: 2 d Incubation temperature: 25°C Aeration rate: 10 L/kg_{dm} min	0.01	Desgranges et al., 1993
	Stirred packed bed Total volume: 15 L Substrate: Hemp, glucose, and yeast extract Incubation time: 18 d	12.8	De Vrije et al., 2001
	Polypropylene packed bed column Bed height: 50 cm Ø: 20 cm Substrate: Oat grains Incubation time: 22.7 d Incubation temperature: 27.5°C Aeration rate: 1.6 kg/m^3 min (saturated air at 18°C)	2.6 – 5.1	Jones et al., 2004
	Polypropylene packed bed column Bed height: 50 cm Ø: 20 cm Substrate: Hemp, glucose, and yeast extract Incubation time: 14 d Incubation temperature: 27.5°C Aeration rate: 1.6 kg/m^3 min (saturated air at 18°C)	3.7 – 6.4	Jones et al., 2004
	Glass scraped-drum L: 20 cm Ø: 5 cm Packed mass: 138 g_{dm} Substrate: Oat grains Moisture content: 54% Incubation time: 19 d Incubation temperature: 20°C Aeration: 2.17 L/kg_{dm} min Agitation: 0.2 rpm (continuously mixed)	5	Oostra et al., 2000

Contd.

Table 2 *Contd.*

Fungus	*Bioreactor and culture conditions*	*Production of conidia (conidia × 10^9/gdm)*	*Reference*
Metarhizium anisopliae	Aluminum tray bioreactor (20 × 30 × 3 cm) Bed height: 2.5 cm Packed mass: 500 g_{dm} Substrate: Rice grains Moisture content: 30% Incubation time: 14 d Incubation temperature: 30°C Aeration rate: 1 L/kg_{wm} min (saturated air)	0.88	Méndez-González et al., 2018b
	Aluminum tray bioreactor (40 × 29 × 12 cm) Bed height: 2 cm Packed mass: 1 kg_{dm} Substrate: Rice grains Incubation time: 10 d Incubation temperature: 28°C Aeration: 0.8 L/min (saturated air)	3.6	da Cunha et al., 2019
	Perforated stainless-steel tray (24 × 47.5 × 3 cm) Bed height: 2 cm Packed mass: 2 kg_{wm} Substrate: Rice grains Moisture content: 30% Incubation time: 7 d Incubation temperature: 28°C Aeration: 1.5 L/kg_{wm} min Relative humidity: 91%	1.6	Dallastra et al., 2023
	Aluminum modular packed bed column Height: 33 cm Ø: 10 cm Packed mass: 400 g_{dm} Substrate: Rice grains Incubation time: 10 d Incubation temperature: 28°C Aeration rate: 2.5 L/ kg_{dm} min (saturated air)	1.7 – 2.2	da Cunha et al., 2020
	Aluminum modular packed bed column Bioreactor height: 33 cm Ø: 10 cm Packed mass: 400 gdm Substrate: Rice grains and sugarcane bagasse (9:1 w/w) Moisture content: 48% Incubation time: 10 d Incubation temperature: 28°C Aeration rate: 2.5 L/kg_{dm} min (saturated air)	3.2 – 3.7	da Cunha et al., 2020

Contd.

Table 2 *Contd.*

Fungus	*Bioreactor and culture conditions*	*Production of conidia (conidia × 10^9/gdm)*	*Reference*
	Aluminum modular packed bed column Bioreactor height: 69 cm Ø: 10 cm Packed mass: 4.5 kg_{dm} Substrate: Rice grains and sugarcane bagasse (9:1 w/w) Moisture content: 48% Incubation time: 10 d Incubation temperature: 28°C Aeration rate: 2.6 L/ kg_{dm} min (saturated air)	3.4 – 3.8	da Cunha et al., 2020
	Glass-packed bed column Volume: 250 mL Packed mass: 30 g_{dm} Substrate: Rice grains and maize oil Moisture content: 37.5% Incubation time: 7 d Incubation temperature: 25°C Aeration rate: 2 L/kg_{dm} min (saturated air)	6	Arzumanov et al., 2005
	Glass-packed bed column Bed height: 32 cm Ø: 15 cm Packed mass: ~1 kg_{dm} Working volume: 5.85 L Substrate: Rice bran and husk (1:1 w/w) Moisture content: 47% Incubation time: 9 d Incubation temperature: Packing density: 0.27 g/mL Aeration rate: 5.6 L/kg_{dm} min (saturated air)	1.6	Dorta and Arcas, 1998
	Glass-packed bed column Bed height: 15 cm Ø: 5 cm Packed mass: ~130 gdm Substrate: Rice grains Moisture content: 30% Incubation time: 7 d Incubation temperature: 27°C Aeration rate: 0.11 L/kg_{dm} min (saturated air)	1.5	Méndez-González et al., 2018a
	Glass-packed bed column Bed height: 8 cm Ø: 2 cm Working volume: 25 mL Substrate: Hemp, glucose, and yeast extract (8:1:1) Moisture content: 66% Incubation time: 12.5 d Incubation temperature: 25°C Aeration: 50 mL/min (saturated air)	8	Van Breukelen et al., 2011

Contd.

Table 2 *Contd.*

Fungus	*Bioreactor and culture conditions*	*Production of conidia (conidia × 10^9/gdm)*	*Reference*
	Glass-packed bed column Bed height: 8 cm Ø: 2 cm Working volume: 25 mL Substrate: Wheat bran Moisture content: 66% Incubation time: 12.5 d Incubation temperature: 25°C Aeration: 50 mL/min (saturated air)	5.6	Van Breukelen et al., 2011
	Glass packed bed column Bed height: 8 cm Ø: 2 cm Working volume: 25 mL Substrate: Rice bran Moisture content: 66% Incubation time: 12.5 d Incubation temperature: 25°C Aeration: 50 mL/min (saturated air)	3.4	Van Breukelen et al., 2011
	Glass-packed bed column Bed height: 8 cm Ø: 2 cm Working volume: 25 mL Substrate: Wheat grains Moisture content: 50% Incubation time: 12.5 d Incubation temperature: 25°C Aeration: 50 mL/min (saturated air)	0.9	Van Breukelen et al., 2011
Metarhizium robertsii	Aluminum tray bioreactor (20 × 30 × 3 cm) Bed height: 2.5 cm Packed mass: 500 g_{dm} Substrate: Rice grains Moisture content: 30% Incubation temperature: 14 d Incubation temperature: 30°C Aeration rate: 1 L/kg_{wm} min (saturated air)	0.7 – 1.0	Méndez-González et al., 2018b
	Aluminum tray bioreactor (5 × 10 × 3 cm) Bed height: 3 cm Packed mass: 56 g_{dm} Substrate: Rice grains Moisture content:30% Incubation time: 12.49 d Incubation temperature: 30°C Aeration rate: 0.1 L/kg_{wm} min (saturated air)	0.7 – 1.1	Méndez-González et al., 2020

Contd.

Table 2 *Contd.*

Fungus	*Bioreactor and culture conditions*	*Production of conidia (conidia × 10^9/gdm)*	*Reference*
	Glass-packed bed column Bed height: 20 cm Ø: 2 cm Packed mass: 21 g_{dm} Substrate: Rice grains Moisture content: 30% Incubation time: 8 d Incubation temperature: 30°C Aeration rate: 0.66 L/kg_{wm} min	1.3 – 1.8	Méndez-González et al., 2020
	PVC packed bed column Bed height: 30 cm Ø: 10.1 cm Packed mass: 1.4 kg_{dm} Substrate: Rice grains and spherical style packings Moisture content: 30% Incubation time: 8.3 d Incubation temperature: 30°C Packing density: 0.75 g/mL Aeration rate 0.66 L/kg_{wm} min (RH 86%)	1.0 – 1.5	Méndez-González et al., 2022

2.5 Packed Bed Column Bioreactors

Packed bed column bioreactors (PBCB) are cylindrical systems supplied with airflow that passes through the bed (Mitchell et al., 1993). Forced aeration promotes heat removal and gas exchange by convective mechanisms (Lopes-Perez et al., 2019; Méndez-González et al., 2020) and can be supplied from either the top or bottom of the bioreactor (Lonsane et al., 1985). The design of PBCBs facilitates the monitoring and controlling of process conditions (da Cunha et al., 2020; Méndez-González et al., 2022b; Henrique et al., 2022). Therefore, these systems allow the controlled production of high concentrations of conidia. In some cases, the number of conidia produced in the PBCB is higher than that obtained in other bioreactor types (Table 2). However, PBCB scale-up is complicated, limiting its use in the industry.

One of the main problems with the PBCB scale-up is metabolic heat accumulation. In laboratory-scale PBCBs, heat removal in radial and axial directions allows for temperature control (Saucedo-Castañeda et al., 1990). However, in bioreactors with higher capacities, heat transfer through the wall is negligible (Finkler et al., 2021). Thus, heat removal is mainly carried out by mechanisms in the axial direction (conduction, evaporation, and convection) (Mitchell et al., 1999). The effectiveness of these mechanisms in maintaining the bed temperature is set by the maximum dimensions of the bioreactor.

There are only two possibilities to increase the capacity of a PBCB: increase the height or increase the diameter. In PBCBs aerated from the bottom, bed temperature rises steadily as a function of height (Mitchell et al., 2023). Since microorganisms grow in a specific temperature range, there is a maximum or critical height in PBCBs related to the maximum temperature allowing for microbial growth (the critical temperature). This height depends on the microorganism and the operating conditions of the bioreactor. Mitchell et al. (1999) generated an accurate tool to estimate critical height by modifying the expression of the Damköhler number (Da_M). Additionally, Da_M also allows the evaluation of the effect of the velocity and moisture content of the supplied air on the temperature distribution in the bed, which is useful for selecting operating conditions. However, it is necessary to consider that increasing the air velocity or decreasing its moisture content causes drying in the bed (Lonsane et al., 1985), which could affect microbial growth. Since the monotonic increase in bed temperature limits the operating height, increasing the diameter of the bioreactor is a better strategy for scaling up (Finkler et al., 2021; Mitchell et al., 2023).

Increasing the diameter of a PBCB causes an increase in temperature from the wall to the center of the bed. Saucedo-Castañeda et al. (1990) developed a useful model to estimate radial temperature profiles using the dimensionless numbers of Peclet (Pe), Biot (Bi), and Damköhler (Da_{III}). This increase in temperature in the radial direction limits the critical diameter of the bioreactor. One of the causes of temperature gradients in the radial direction is poor air distribution (Mitchell et al., 2023). In a packed bed, the increase in diameter generates a gradual decrease in porosity (from the wall to the center of the bioreactor) (Benenati and Brosilow, 1962; Ridgway and Tarbuck, 1968). This heterogeneous structure of the solid matrix causes poor air distribution and insufficient heat removal (Méndez-González et al., 2022). Furthermore, these conditions avoid operating the bioreactor up to a critical height estimated in lab-scale bioreactors. Therefore, improving the solid matrix structure is the key to PBCB scale-up. To do this, two main strategies have been evaluated: (i) the use of low-density substrates (or supports) and (ii) the addition of texturizers.

The first strategy allows for a significant increase in bed porosity, facilitating the operation of higher-capacity bioreactors to produce EF conidia (De Vrije et al., 2001; Van Breukelen et al., 2011). However, using these materials affects the volumetric production of the bioreactor (Jones et al., 2004; Méndez-González et al., 2022a). On the other hand, adding texturizers promotes the opening of empty spaces in the bed, improving air distribution and heat removal (Méndez-González et al., 2022). Da Cunha et al. (2020) added sugarcane bagasse to the rice bed, thereby increasing the production of conidia of *M. anisopliae* by up to 40%. Furthermore, this strategy allowed for increasing the capacity of the PBCB from 400 to 4500 g of dry mass (g_{dm}) without decreasing conidia production. Similarly, the addition of plastic spherical-style packings evaluated by Méndez-González et al. (2020) allowed for reducing the packing density in rice beds by ~8%. This allowed for increasing the mass capacity of the PBCB from 18 to 1400 g_{dm}, while maintaining conidia production. Both studies showed that it is possible to increase the capacity of the bioreactor with the addition of texturizers without affecting volumetric production. However, the temperature distribution observed by Da Cunha et al. (2020) and Méndez-González et al. (2020) suggests that adding texturizers does not completely solve the channeling problem in the bed.

Additionally, the loss of void space in the culture bed hinders the O_2 supply and CO_2 removal (and other volatile compounds). As previously mentioned, an effective gas exchange allows for aerobic conditions favoring EF growth. In aerobic SSC, microorganisms consume O_2 and produce CO_2 (Martínez-Ramírez et al., 2021). Therefore, in PBCBs aerated from the bottom, the concentration of O_2 in the gas phase decreases due to the increase in bed height, and an opposite profile occurs for CO_2 (Saucedo-Castañeda et al., 1992). *Metarhizium robertsii* cultures in 30 cm beds reach a maximum CO_2 concentration of less than 2% while the temperature increases by ~3°C (Méndez-González et al., 2022). This suggests that heat removal is a bigger problem than CO_2 removal. However, the substrate's shrinkage and agglomeration can cause anaerobic zones in the culture bed. Finkler et al. (2021) suggest that PBCBs should be designed to allow intermittent stirring to avoid problems associated with a solid matrix structure. The following (AB) section describes the effects of agitation on the EF culture.

2.6 Agitated Bioreactors

Agitated bioreactors (AB) present two configurations: rotating drum (RDB) and internal stirred (ISB) (De Vrije et al., 2001). Continuous or intermittent mixing in these bioreactors contributes to gas exchange and heat removal (Brand et al., 2010). Like PBCB, forced aeration can be supplied to AB (Lopez-Ramirez et al., 2018; Martínez-Ramírez et al., 2021), improving heat and mass transfer in the bioreactor. ABs allow controlled processes and have been used on an industrial scale (up to 25 t) to produce enzymes (Durand and Chereau, 1988) and livestock feed (Xue et al., 1992). RDBs are horizontally rotating cylinders (Mitchell et al., 1992). The smooth rotation of the drum causes moderate damage to mycelium but is ineffective in breaking up substrate agglomerations (Brand et al., 2010). ISBs are cylindrical (horizontal and vertical) and have a system of blades for mixing the substrate (Durand, 2003). Stirring causes mechanical damage to the mycelium and substrate but contributes to bed uniformity (Lonsane et al., 1992; Durand, 2003).

The mechanical damage caused to the mycelium has limited the use of agitated bioreactors to produce EF conidia (Méndez-González et al., 2022a). Some important genera, such as *Metarhizium* (Van Breukelen et al., 2011) and *Beauveria* (Desgranges et al., 1993), are highly sensitive to damage caused by agitation. On the other hand, *Coniothyrium. minitans* and *Paecilomyces lilacinus* present resistance to agitation (even continuous) and have been successfully cultivated in RDB (Oostra et al., 2000) and ISB (De Vrije et al., 2001). In fact, in an ISB with a 15 L capacity and hemp-glucose-yeast extract as the substrate, the production of *C. minitans* conidia is higher than that achieved in a PBCB with the same capacity (Table 2). Considering that the traditional process in PBB includes a period of manual agitation (Méndez-González et al., 2018a), an appropriate adjustment of the speed and period of agitation may be the key to producing conidia of other EFs using ABs.

Conclusion

Since the middle of the last century, the demand for EF conidia has mainly been met through simple processes using low-capacity bioreactors. However, the increasing demand for fungal bioinsecticides necessitates deploying more sophisticated processes with a higher degree of industrialization. Controlled processes can be implemented in trays, packed columns, and agitated bioreactors. These bioreactors operate on a large scale, ensuring standardized production with high-quality products. In many cases, they demonstrate higher production and productivity than traditional bioreactors. Among these, tray bioreactors can be easily integrated into the industry. However, it is also possible to achieve higher production yields with lower labor and process area requirements by utilizing packed bed columns and agitated bioreactors.

The scale-up of packed bed column bioreactors poses challenges, limiting their large-scale application. The primary difficulties are associated with porosity gradients in the culture bed. Nonetheless, strategies such as adding texturizers have enabled the expansion of bioreactor dimensions without compromising conidia production. Agitation is another strategy that allows for solving the problems presented by the packed bed column. Agitated bioreactors can operate on a large scale, but their use is restricted to fungal species resistant to mechanical stress. Adjusting the speed and duration of agitation could be an available solution for cultivating most entomopathogenic fungi. Although conidia production in high-tech bioreactors has made significant advances, further improvements are necessary for successful integration into the industry. Developing these processes will facilitate the intensification of sustainable agricultural practices for producing chemical-free foods.

References

Adane, K., Moore, D. and Archer, S.A. (1996). Preliminary studies on the use of *Beauveria bassiana* to control *Sitophilus zeamais* (Coleoptera: Curculionidae) in the laboratory. J. Stored Prod. Res., 32: 105–113. 10.1016/0022-474X(96)00009-4.

Alcantara-Vargas, E., Espitia-López, J., Garza-López, P.M. and Angel-Cuapio, A. (2020). Producción y calidad de conidios de cepas de entomopatógenos del género *Metarhizium anisopliae*, aislados en zonas agrícolas del Estado de México. Rev. Mex. Biodivers., 91:912912. 10.22201/ib.20078706e.2020.91.2912.

Alves, S. and Pereira, R. (1989). Production of *Metarhizium anisoplae* (Metsch.) Sorok. and *Beauveria bassiana* (Bals.) Vuill. in plastic trays. Ecossistema, 14: 188–192.

Arzumanov, T., Jenkins, N. and Roussos, S. (2005). Effect of aeration and substrate moisture content on sporulation of *Metarhizium anisopliae* var. acridum. Process Biochem., 40:1037–1042. 10.1016/j.procbio.2004.03.013.

Barra-Bucarei, L., Vergara, P. and Cortes, A. (2016). Conditions to optimize mass production of *Metarhizium anisopliae* (Metschn.) Sorokin 1883 in different substrates. Chil. J. Agric. Res., 76: 448–454. 10.4067/S0718-58392016000400008.

Benenati, R.F. and Brosilow, C.B. (1962). Void fraction distribution in beds of spheres. AIChE Journal, 8: 359–361. 10.1002/aic.690080319.

Brand, D., Soccol, C.R., Sabu, A. and Roussos, S. (2010). Production of fungal biological control agents through solid state fermentation: a case study on *Paecelomyces lilacinus* againts root-knot nematodes. Micol. Apli. Int., 22: 31–48.

Castillo-Minjarez, J.M. (2022). Contexto de los bioplaguicidas comerciales de base fúngica en México. Mex. J. Tech. Eng., 1: 15–27. 10.61767/mjte.001.1.1527.

Chaparro, M.L., Sanabria, P.J., Jiménez A.M. et al. (2021). A circular economy approach for producing a fungal-based biopesticide employing pearl millet as a substrate and its economic evaluation. Bioresour. Technol. Rep., 16:100869. 10.1016/j.biteb.2021.100869.

Cherry, A.J., Jenkins, N.E., Heviefo, G. et al. (1999). Operational and economic analysis of a west African pilot-scale production plant for aerial conidia of *Metarhizium* spp. for use as a mycoinsecticide against locusts and grasshoppers. Biocontrol Sci. Technol., 9: 35–51. 10.1080/09583159929893.

da Cunha, L.P., Casciatori, F.P., de Cenço-Lopes, I. and Thoméo, J.C. (2019). Production of conidia of the entomopathogenic fungus *Metarhizium anisopliae* ICB 425 in a tray bioreactor. Bioprocess Biosyst. Eng., 42: 1757–1768. 10.1007/s00449-019-02172-z.

da Cunha, L.P., Casciatori, F.P., Vicente, I.V. et al. (2020). *Metarhizium anisopliae* conidia production in packed-bed bioreactor using rice as substrate in successive cultivations. Process Biochem., 97: 104–111. 10.1016/j.procbio.2020.07.002.

Dalla-Santa, H.S., Sousa, N.J., Brand, D. et al. (2004). Conidia production of *Beauveria* sp. by solid-state fermentation for biocontrol of *Ilex paraguariensis* Caterpillars. Folia Microbiol. (Praha), 49: 418–422. 10.1007/BF02931603.

Dallastra, E.D.G., Dias, A.C.P., de Morais, P.B. et al. (2023). Development of a novel pilot-scale tray bioreactor for solid-state fermentation aiming at process intensification. Chem. Eng. and Process. - Process Intensification, 193:109526. 10.1016/j.cep.2023.109526.

De Vrije, T., Antoine, N., Buitelaar, R.M. et al. (2001). The fungal biocontrol agent *Coniothyrium minitans*: Production by solid-state fermentation, application and marketing. Appl. Microbiol. Biotechnol., 56: 58–68. 10.1007/s002530100678.

Deans, C. and Krischik, V. (2023). The current state and future potential of microbial control of scarab pests. Appl. Sci., 13: 766. 10.3390/app13020766.

Desgranges, C., Vergoignan, C., Léréec, A. et al. (1993). Use of solid state fermentation to produce *Beauveria bassiana* for the biological control of European corn borer. Biotechnol. Adv., 11: 577–587. 10.1016/0734-9750(93)90026-J.

Dhar, P. and Kaur, G. (2011). Response surface methodology for optimizing process parameters for the mass production of *Beauveria bassiana* conidiospores. Afr. J. Microbiol. Res., 5(17): 2399–1406. 10.5897/AJMR10.088.

Diniz-da Silva, M.P., Dutra, R.S., Casciatori, F.P. and Grajales, L.M. (2022). A two-phase model for simulation of water transfer during lipase production by solid-state cultivation in a tray bioreactor using babassu residues as substrate. Chem. Eng. Process. - Process Intens., 177: 108981. 10.1016/j.cep.2022.108981.

Dorta, B. and Arcas, J. (1998). Sporulation of *Metarhizium anisopliae* in solid-state fermentation with forced aeration. Enzyme Microb. Technol., 23:501–505. 10.1016/S0141-0229(98)00079-9.

Durand, A. (2003). Bioreactor designs for solid state fermentation. Biochem. Eng., J 13:113–125.

Durand, A. and Chereau, D. (1988) A new pilot reactor for solid-state fermentation: Application to the protein enrichment of sugar beet pulp. Biotechnol. Bioeng., 31: 476–486. 10.1002/bit.260310513.

Ekesi, S., Maniania, N.K. and Ampong-Nyarko, K. (1999). Effect of temperature on germination, radial growth and virulence of *Metarhizium anisopliae* and *Beauveria bassiana* on *Megalurothrips sjostedti*. Biocontrol Sci. Technol., 9: 177–185. 10.1080/09583159929767.

Ferreira, J.M., Pinto, S.M.N. and Soares, F.E.F. (2021). *Metarhizium robertsii* protease and conidia production, response to heat stress and virulence against *Aedes aegypti* larvae. AMB Express, 11: 166. 10.1186/s13568-021-01326-1.

Figueroa-Montero, A., Esparza-Isunza, T., Saucedo-Castañeda, G. et al. (2011). Improvement of heat removal in solid-state fermentation tray bioreactors by forced air convection. J. Chem. Technol. Biotechnol., 86: 1321–1331. 10.1002/jctb.2637.

Finger, S.M., Hatch, R.T. and Regan, T.M. (1976). Aerobic microbial growth in semisolid matrices: Heat and mass transfer limitation. Biotechnol. Bioeng., 18: 1193–1218. 10.1002/bit.260180904.

Finkler ATJ, de Lima Luz LF, Krieger N, et al (2021) A model-based strategy for scaling-up traditional packed-bed bioreactors for solid-state fermentation based on measurement of O2 uptake rates. Biochem. Eng. J., 166: 107854. 10.1016/j.bej.2020.107854.

Garza-López, P.M., Konigsberg, M., Saucedo-Castañeda, G. and Loera, O. (2011). Differential profiles of *Beuveria bassiana* (Bals.-Criv.) Vuill. as a response to CO2: Production of conidia and amylases. Agrociencia, 45: 761–770.

Ghildyal, N.P., Gowthaman, M.K., Raghava-Rao, K.S.M.S. and Karanth, N.G. (1994). Interaction of transport resistances with biochemical reaction in packed-bed solid-state fermentors: Effect of temperature gradients. Enzyme Microb. Technol., 16: 253–257. 10.1016/0141-0229(94)90051-5.

Goettel, M.S. (1984). A simple method for mass culturing entomopathogenic hyphomycete fungi. J. Microbiol. Methods, 3: 15–20. 10.1016/0167-7012(84)90040-X.

Grzywacz, D., Moore, D. and Rabindra, R.J. (2014). Mass production of entomopathogens in less industrialized countries. pp. 519–561. In: Morales-Ramos, J., Rojas, G. and Shapiro-Ilan, D. (eds.). Mass Production of Beneficial Organisms. Elsevier, Amsterdam.

Hallsworth, J.E. and Magan, N. (1999). Water and temperature relations of growth of the entomogenous fungi *Beauveria bassiana*, *Metarhizium anisopliae*, and *Paecilomyces farinosus*. J. Invertebr. Pathol., 74: 261–266. 10.1006/jipa.1999.4883.

Henrique, J.P., Casciatori, F.P. and Thoméo, J.C. (2022) Automatic system for monitoring gaseous concentration in a packed-bed solid-state cultivation bioreactor. Chem. Eng. Sci., 259: 117793. 10.1016/j.ces.2022.117793.

Jaronski, S.T. (2023). Mass production of entomopathogenic fungi—state of the art. pp. 317–357. In: Morales-Ramos, J., Rojas, G. and Shapiro-Ilan. (eds.). Mass Production of Beneficial Organisms. Elsevier, Amsterdam.

Jenkins, N.E. and Grzywacz, D. (2000). Quality control of fungal and viral biocontrol agents - Assurance of product performance. Biocontrol. Sci. Technol., 10: 753–777. 10.1080/09583150020011717.

Jones, E.E., Weber, F.J., Oostra, J. et al. (2004). Conidial quality of the biocontrol agent *Coniothyrium minitans* produced by solid-state cultivation in a packed-bed reactor. Enzyme Microb. Technol., 34: 196–207. 10.1016/j.enzmictec.2003.10.002.

Lonsane, B.K., Ghildyal, N.P., Budiatman, S. and Ramakrishna, S.V. (1985). Engineering aspects of solid state fermentation. Enzyme Microb. Technol., 7: 258–265. 10.1016/0141-0229(85)90083-3.

Lonsane, B.K., Saucedo-Castañeda, G., Raimbault, M. et al. (1992). Scale-up strategies for solid state fermentation systems. Process Biochem., 27: 259–273.

Lopes-Perez, C., Casciatori, F.P. and Thoméo, J.C. (2019). Strategies for scaling-up packed-bed bioreactors for solid-state fermentation: The case of cellulolytic enzymes production by a thermophilic fungus. Chem. Eng. J., 361: 1142–1151. 10.1016/j.cej.2018.12.169.

Lopez-Ramirez, N., Volke-Sepulveda, T., Gaime-Perraud, I. et al. (2018). Effect of stirring on growth and cellulolytic enzymes production by *Trichoderma harzianum* in a novel bench-scale solid-state fermentation bioreactor. Bioresour. Technol., 265: 291–298. 10.1016/j.biortech.2018.06.015.

Luz, C. and Fargues, J. (1998). Factors affecting conidial production of *Beauveria bassiana* from fungus-killed cadavers of *Rhodnius prolixus*. J. Invertebr. Pathol., 72: 97–103. 10.1006/jipa.1998.4774.

Machado, A.C.R., Monteiro, A.C., Almeida, A.M.B. and de Martins, M.I.E.G. (2010). Production technology for entomopathogenic fungus using a biphasic culture system. Pesqui. Agropecu. Bras., 45: 1157–1163. 10.1590/S0100-204X2010001000015.

Manan, M.A. and Webb, C. (2017). Design aspects of solid state fermentation as applied to microbial bioprocessing. J. Appl. Biotechnol. Bioeng., 4: 511–532. 10.15406/jabb.2017.04.00094.

Martínez-Ramírez, C., Esquivel-Cote, R., Ferrera-Cerrato, R. et al. (2021). Solid-state culture of *Azospirillum brasilense*: a reliable technology for biofertilizer production from laboratory to pilot scale. Bioprocess Biosyst. Eng., 44: 1525–1538. 10.1007/s00449-021-02537-3.

Mathulwe, L.L., Malan, A.P. and Stokwe, N.F. (2022). Mass production of entomopathogenic fungi, *Metarhizium robertsii* and *Metarhizium pinghaens*, for commercial application against insect pests. J. Vis. Exp., 181: e63246. 10.3791/63246.

McQuilken, M.P. and Whipps, J.M. (1995). Production, survival and evaluation of solid-substrate inocula of *Coniothyrium minitans* against *Sclerotinia sclerotiorum*. Eur. J. Plant Pathol., 101: 101–110. 10.1007/BF01876098.

Méndez-González, F., Castillo-Minjarez, J.M., Loera, O. and Favela-Torres, E. (2022a). Current developments in the resistance, quality, and production of entomopathogenic fungi. World J. Microbiol. Biotechnol., 38(7): 115. 10.1007/s11274-022-03301-9.

Méndez-González, F., Chávez-Escalante, G., Loera, O. and Favela-Torres, E. (2022b). What does respirometric analysis tell us about *Metarhizium robertsii*. J. Appl. Biotechnol. Bioeng., 9: 94–96. 10.15406/jabb.2022.09.00292.

Méndez-González, F., Figueroa-Montero, A., Saucedo-Castañeda, G. et al. (2022). Addition of spherical-style packing improves the production of conidia by *Metarhizium robertsii* in packed column bioreactors. J. Chem. Technol. Biotechnol., 97: 1517–1525. 10.1002/jctb.6993.

Méndez-González, F., Loera, O. and Favela-Torres, E. (2018a). Conidia production of *Metarhizium anisopliae* in bags and packed column bioreactors. Curr. Biotechnol., 7: 65–69. 10.2174/2211550105666160926123350.

Méndez-González, F., Loera, O., Saucedo-Castañeda, G. and Favela-Torres, E. (2020). Forced aeration promotes high production and productivity of infective conidia from *Metarhizium robertsii* in solid-state fermentation. Biochem. Eng. J., 156: 107492. 10.1016/j.bej.2020.107492.

Méndez-González, F., Loera-Corral, O., Saucedo-Castañeda, G. and Favela-Torres, E. (2018b). Bioreactors for the production of biological control agents produced by solid-state fermentation. pp. 109–121. In: Pandey, A., Larroche, C. and Soccol, C.R. (eds.). Current Developments in Biotechnology and Bioengineering. Elsevier, Amsterdam. 10.1016/B978-0-444-63990-5.00007-4.

Mitchell, D.A., Lonsane, B.K., Reanud, R. et al. (1993). General principles of bioreactor design and operation for SSC. pp. 115–136. In: Doelle, H., Mitchell, D.A. and Rolz, C.E. (eds.). Solid State Cultivation. Elsevier Applied Science, London.

Mitchell, D.A., Pandey, A., Sangsurasak, P. and Krieger, N. (1999). Scale-up strategies for packed-bed bioreactors for solid-state fermentation. Process Biochem., 35: 167–178. 10.1016/S0032-9592(99)00048-5.

Mitchell, D.A., Ruiz, H.A. and Krieger, N. (2023). A critical evaluation of recent studies on packed-bed bioreactors for solid-state fermentation. Processes, 11: 872–903. 10.3390/pr11030872.

Muñiz-Paredes, F., Miranda-Hernández, F. and Loera, O. (2017). Production of conidia by entomopathogenic fungi: from inoculants to final quality tests. World J. Microbiol. Biotechnol., 33: 57–66. 10.1007/s11274-017-2229-2.

Oostra, J., Tramper, J. and Rinzema, A. (2000). Model-based bioreactor selection for large-scale solid-state cultivation of *Coniothyrium minitans* spores on oats. Enzyme Microb. Technol., 27: 652–663. 10.1016/S0141-0229(00)00261-1.

Oriol, E., Raimbault, M., Roussos, S. and Viniegra-Gonzalez, G. (1988). Water and water activity in the solid state fermentation of cassava starch by *Aspergillus niger*. Appl. Microbiol. Biotechnol., 27: 498–503. 10.1007/BF00451620.

Ouedraogo, A., Fargues, J., Goettel, M.S. and Lomer, C.J. (1997). Effect of temperature on vegetative growth among isolates of *Metarhizium anisopliae* and *M. flavoviride*. Mycopathologia, 137: 37–43. 10.1023/A:1006882621776.

Pandey, A. (1991). Aspects of fermenter design for solid-state fermentations. Process Biochem., 26: 355–361. 10.1016/0032-9592(91)85026-K.

Posada-Flórez, F.J. (2008). Production of *Beauveria bassiana* fungal spores on rice to control the coffee berry borer, Hypothenemus hampei, in Colombia. J. Insect Sci., 8: 1–13. 10.1673/031.008.4101.

Raghava-Rao, K.S.M.S., Gowthaman, M.K., Ghildyal, N.P. and Karanth, N.G. (1993). A mathematical model for solid state fermentation in tray bioreactors. Bioprocess Eng., 8: 255–262. 10.1007/BF00369838.

Rajagopalan, S. and Modak, J.M. (1994). Heat and mass transfer simulation studies for solid-state fermentation processes. Chem. Eng. Sci., 49: 2187–2193. 10.1016/0009-2509(94)E0012-F.

Ridgway, K. and Tarbuck, K.J. (1968). Voidage fluctuations in randomly-packed beds of spheres adjacent to a containing wall. Chem. Eng. Sci., 23: 1147–1155. 10.1016/0009-2509(68)87099-X.

Roberts, D.W. and St. Leger, R.J. (2004) *Metarhizium* spp., cosmopolitan insect-pathogenic fungi: mycological aspects. pp. 1–70. In: Laskin, A., Bennett, J. and Gadd, G. (eds.). Advances in Applied Microbiology. Academic Press, London.

Rojas, V.M.A., Iwanicki, N.S., D'Alessandro, C.P. et al. (2023). Characterization of brazilian *Cordyceps fumosorosea* isolates: Conidial production, tolerance to ultraviolet-B radiation, and elevated temperature. J. Invertebr. Pathol., 197: 107888. 10.1016/j.jip.2023.107888.

Sala, A., Barrena, R., Meyling, N.V. and Artola, A. (2023). Conidia production of the entomopathogenic fungus *Beauveria bassiana* using packed-bed bioreactor: Effect of substrate biodegradability on conidia virulence. J. Environ. Manage., 341: 118059. 10.1016/j.jenvman.2023.118059.

Sala A, Vittone S, Barrena R, et al (2021). Scanning agro-industrial wastes as substrates for fungal biopesticide production: Use of *Beauveria bassiana* and *Trichoderma harzianum* in solid-state fermentation. J. Environ. Manage., 295: 113113. 10.1016/j.jenvman.2021.113113.

Saucedo-Castañeda, G., Gutiérrez-Rojas, M., Bacquet, G. et al. (1990). Heat transfer simulation in solid substrate fermentation. Biotechnol. Bioeng., 35: 802–808. 10.1002/bit.260350808.

Saucedo-Castañeda, G., Lonsane, B.K., Navarro, J.M. and Roussos, S. (1992). Control of carbon dioxide in exhaust air as a method for equal biomass yields at different bed heights in a column fermentor. Appl. Microbiol. Biotechnol., 37: 580–582. 10.1007/BF00240729.

Swanson, D. (1997). Economic feasibility of two technologies for production of mycopesticide in Magadascar. Memoirs oEntomol. Soc. Can., 129: 101–113. 10.4039/entm129171101-1.

Van Breukelen, F.R., Haemers, S., Wijffels, R.H. and Rinzema, A. (2011). Bioreactor and substrate selection for solid-state cultivation of the malaria mosquito control agent *Metarhizium anisopliae*. Process Biochem., 46: 751–757. 10.1016/j.procbio.2010.11.023.

Viniegra-González, G. (1997). Solid state fermentation: Definition, characteristics, limitations and monitoring. pp. 5–22. In: Roussos, S., Lonsane, B.K., Raimbault, M., Viniegra-González, G. (eds.). Advances in Solid State Fermentation. Springer Netherlands, Dordrecht.

Xie, L., Chen, H.M. and Yang, J.B. (2012). Conidia production by *Beauveria bassiana* on rice in solid-state fermentation using tray bioreactor. Adv. Mat. Res., 610–613: 3478–3482. 10.4028/www.scientific.net/AMR.610-613.3478.

Xue, M., Liu, D., Zhang, H. et al. (1992). A pilot process of solid state fermentation from sugar beet pulp for the production of microbial protein. J. Ferment. Bioeng., 73: 203–205. 10.1016/0922-338X(92)90161-M.

Ye, S.D., Ying, S.H., Chen, C. and Feng, M.G. (2006). New solid-state fermentation chamber for bulk production of aerial conidia of fungal biocontrol agents on rice. Biotechnol. Lett., 28: 799–804. 10.1007/s10529-006-9004-z.

Index

For Product Safety Concerns and Information please contact our EU representative GPSR@taylorandfrancis.com
Taylor & Francis Verlag GmbH, Kaufingerstraße 24, 80331 München, Germany

www.ingramcontent.com/pod-product-compliance
Lightning Source LLC
LaVergne TN
LVHW020610110826
845149LV00002B/433

* 9 7 8 1 0 3 2 8 2 5 7 6 2 *